Recent Landform Evolution

Springer Geography

Volume

For further volumes:
http://www.springer.com/series/10180

Dénes Lóczy • Miloš Stankoviansky
Adam Kotarba
Editors

Recent Landform Evolution

The Carpatho-Balkan-Dinaric Region

Editors
Dénes Lóczy
Department of Environmental
Geography and Landscape Conservation
Institute of Environmental Sciences
University of Pécs
Ifjúság utja 6
7624 Pécs
Hungary
loczyd@ttk.jpte.hu

Miloš Stankoviansky
Department of Physical Geography
and Geoecology
Comenius University in Bratislava
Mlynská dolina
842 15 Bratislava 4
Slovakia
milos.Stankoviansky@test.fns.uniba.sk

Adam Kotarba
Polish Academy of Sciences
Institute of Geography and Spatial
Organization
PL-31018 Kraków, ul. Sw. Jana 22
Poland
kotarba@zg.pan.krakow.pl

ISBN 978-94-007-2447-1   e-ISBN 978-94-007-2448-8
DOI 10.1007/978-94-007-2448-8
Springer Dordrecht Heidelberg London New York

Library of Congress Control Number: 2011942321

© Springer Science+Business Media B.V. 2012
No part of this work may be reproduced, stored in a retrieval system, or transmitted in any form or by any means, electronic, mechanical, photocopying, microfilming, recording or otherwise, without written permission from the Publisher, with the exception of any material supplied specifically for the purpose of being entered and executed on a computer system, for exclusive use by the purchaser of the work.

Printed on acid-free paper

Springer is part of Springer Science+Business Media (www.springer.com)

Foreword

Let me share some personal observations as I introduce this important compilation of collaborative work on the Carpatho–Balkan–Dinaric Region. As a geomorphology student in the English-speaking world in the 1960s, I had little knowledge and contact with the research that emanated from Eastern Europe. From basic geography I knew a little of those vast mountain lands that wound their way like some large serpent from Czechoslovakia, through Poland, Hungary, Romania (where it took an extraordinary turn), and down through Bulgaria and into former Yugoslavia, but its people and its geomorphology remained a mystery. Fortunately, my professors knew some eminent geomorphologists from that region (largely as a result of contact made through the International Geographic Union) and thus, I and my fellow students were introduced to the very few papers that were available in English. For example, to understand geomorphological mapping we were directed to Polish examples, which appeared much more advanced than anything I had seen elsewhere. Hungary was also clearly advanced in this aspect of geomorphology where geomorphological mapping by Márton Pécsi was used not only to highlight the distribution of periglacial and other landforms but also in terms of applied geomorphology – the latter approach being well ahead of its time. We were also made aware of the advances being made in unraveling the continental glaciations as well as the pioneering work of Dylik on periglacial processes and forms. Of course, we knew that the former Yugoslavia was the home of karst geomorphology but, although aware of the work of Cvijić, much of the research was not available in English until the publications of Gams in the 1970s. More specifically, my own interests in slope stability and erosion soon had me seeking out the work of Professor Starkel from Poland while work in Czechoslovakia became known to me through the remarkable textbook of Záruba and Mencl. But in general, our contact with the "East" was very limited. We knew that some great geomorphology was being carried out in the region, but tantalizingly we saw very few papers in English.

In the 1970s and 1980s, as we began to meet a trickle of geomorphologists from the region at international conferences, we realized that we were seeing only a fraction of the work of a long-standing geomorphological research tradition firmly embedded in the region. We were also aware that only a few scientists were able to engage with the "West"and that they did so under considerable financial and sometimes political restraints. In recent years I have had the privilege of meeting some of these colleagues and their students in easier times.

Again from a personal perspective, in the last few years I have had the good fortune of being introduced to the region by some very knowledgeable local geomorphologists. My first experience in the region was associated with the first European Conference on Landslides held in Prague in 2002. As a result of field trips associated with this meeting, I was able to view many of the slope stability issues associated with the region, from the Czech Republic through into the Tatra Mountain areas of Slovakia. Until then I had not realized the extent of research into landslides that existed in this region and the advances that had been made to our understanding of slope stability.

So with my curiosity heightened, it was not surprising that I accepted Professor Dan Balteanu's invitation to attend the regional IAG meeting in the Transylvanian mountains. Again I came away from that visit thinking that there is much innovative research being carried out that has not been completely shared with the international community.

More recently, I was invited by my Polish colleagues to visit and provide a number of lectures. My principal hosts were Professor Piotr Migoń (Wroclaw), and Professor Zbigniew Zwoliński (Poznań). On a short visit to Kraków I was welcomed by Professor Zofia Rączkowska, Head of the Department of Geomorphology and Hydrology, Institute of Geography, Polish Academy of Sciences and a number of her colleagues including Professor Kotarba, and Professor Starkel. In Poznań, I was welcomed by Professor Kostrzewski, Head of the Institute of Geoecology and Geoinformation, Adam Mickiewicz University. I saw much of the region including the Polish Carpathians. Again my eyes were opened, not only to the strength of the geomorphological community but the vast array of research that has and is going on and is increasingly being directed to issues of social concern. You may therefore understand why I consider it a great personal privilege to write the foreword to this multinational contribution to geomorphology.

As the President of the International Association of Geomorphologists (IAG) I am delighted that one of our long-standing working groups has been instrumental in producing this monograph. Indeed, the Carpatho–Balkan–Dinaric Regional Working Group of IAG has up until recently has been the only working group with a regional rather than systematic focus. Some have argued that as an international association we should not favor specific regions but rather focus on process and forms from a generic perspective. However, the long history of successful collaborative research, the multicultural and multinational nature of this region, and indeed the publication of this monograph has demonstrated the unique value of this group.

The publication of this monograph represents a significant milestone for international geomorphology. It is a work that reflects the common dedication of many earth scientists to their discipline and their willingness to pass on and share their knowledge with the wider scientific community. But his has not come easily because the Carpatho–Balkan–Dinaric region is one that has been characterized by cultural, linguistic, and political differences. The twentieth century, when many of the contributors to this volume received their academic training and embarked on their research careers, has been one of hardship, deprivation, and international isolation for many of those involved. This publication is a testament to the perseverance of authors and their institutions and the belief that the curiosity of mankind has universality that overrides political considerations and should be pursued without prejudice.

I congratulate the editors and all the contributors for their research efforts and their willingness to contribute to this volume, which I am sure will represent a major benchmark in our understanding of the geomorphology of this diverse and fascinating region.

Michael Crozier
President of the International Association
of Geomorphologists
Wellington, New Zealand

Contents

Part I General

1 Geological and Tectonic Setting 3
   János Haas

2 Climate 19
   Tadeusz Niedźwiedź

3 Rivers 31
   Pavol Miklánek

4 Land Cover and Land Use 39
   Ján Feranec and Tomáš Soukup

Part II Recent Geomorphic Evolution: National Chapters

5 Recent Landform Evolution in the Polish Carpathians 47
   Zofia Rączkowska, Adam Łajczak, Włodzimierz Margielewski,
   and Jolanta Święchowicz

6 Recent Landform Evolution in the Moravian–Silesian
   Carpathians (Czech Republic) 103
   Jaromir Demek, Jan Hradecký, Karel Kirchner, Tomáš Pánek,
   Aleš Létal, and Irena Smolová

7 Recent Landform Evolution in Slovakia 141
   Miloš Stankoviansky, Ivan Barka, Pavel Bella, Martin Boltižiar,
   Anna Grešková (†), Jozef Hók, Pavol Ištok, Milan Lehotský,
   Monika Michalková, Jozef Minár, Martin Ondrášik, Rudolf Ondrášik,
   Jozef Pecho, Peter Pišút, Milan Trizna, and Ján Urbánek

8 Recent Landform Evolution in the Ukrainian Carpathians 177
   Ivan Kovalchuk, Yaroslav Kravchuk, Andriy Mykhnovych,
   and Olha Pylypovych

9	**Recent Landform Evolution in Hungary**	205
	Dénes Lóczy, Ádám Kertész, József Lóki, Tímea Kiss, Péter Rózsa, György Sipos, László Sütő, József Szabó, and Márton Veress	
10	**Recent Landform Evolution in the Romanian Carpathians and Pericarpathian Regions**	249
	Dan Bălteanu, Marta Jurchescu, Virgil Surdeanu, Ion Ionita, Cristian Goran, Petre Urdea, Maria Rădoane, Nicolae Rădoane, and Mihaela Sima	
11	**Recent Landform Evolution in Slovenia**	287
	Blaž Komac, Mauro Hrvatin, Drago Perko, Karel Natek, Andrej Mihevc, Mitja Prelovšek, Matija Zorn, and Uroš Stepišnik	
12	**Recent Landform Evolution in the Dinaric and Pannonian Regions of Croatia**	313
	Andrija Bognar, Sanja Faivre, Nenad Buzjak, Mladen Pahernik, and Neven Bočić	
13	**Recent Landform Evolution in Serbia**	345
	Radmila Pavlović, Jelena Ćalić, Predrag Djurović, Branislav Trivić, and Igor Jemcov	
14	**Recent Landform Evolution in Bulgaria**	377
	Krasimir Stoyanov and Emil Gachev	
15	**Recent Landform Evolution in Macedonia**	413
	Dragan Kolčakovski and Ivica Milevski	

Part III   Conclusions

16	**Conclusions**	445
	Dénes Lóczy, Miloš Stankoviansky, and Adam Kotarba	

Index 453

Contributors

Dan Bălteanu Institute of Geography, Romanian Academy, D. Racovița Str. 12, 023993 Bucharest, Romania, igar@geoinst.ro

Ivan Barka Department of Ecology and Biodiversity of Forest Ecosystems, Forest Research Institute, National Forest Centre, T.G. Masaryka 22, 960 92 Zvolen, Slovakia, barka@nlcsk.org

Pavel Bella State Nature Conservancy of the Slovak Republic, Slovak Caves Administration, Hodžova 11, 031 01 Liptovský Mikuláš, Slovakia

Department of Geography, Pedagogical Faculty, Catholic University, Hrabovská cesta 1, 034 01 Ružomberok, Slovakia, bella@ssj.sk; bella@ku.sk

Neven Bočić University of Zagreb, Faculty of Science, Department of Physical Geography, Marulićev trg 19/II, 10 000 Zagreb, Croatia, nbocic@geog.pmf.hr

Andrija Bognar University of Zagreb, Faculty of Science, Department of Physical Geography, Marulićev trg 19/II, 10 000 Zagreb, Croatia, andrija.bognar@zg.htnet.hr

Martin Boltižiar Department of Geography and Regional Development, Faculty of Natural Sciences, Constantine the Philosopher University in Nitra, Trieda A. Hlinku 1, 949 74 Nitra, Slovakia, mboltiziar@ukf.sk

Nenad Buzjak University of Zagreb, Faculty of Science, Department of Physical Geography, Marulićev trg 19/II, 10 000 Zagreb, Croatia, nbuzjak@geog.pmf.hr

Jelena Ćalić Geographical Institute "Jovan Cvijić", Serbian Academy of Sciences and Arts, Djure Jakšića 9, 11000 Belgrade, Serbia, j.calic@gi.sanu.ac.rs

Jaromir Demek The Silva Tarouca Research Institute for Landscape and Ornamental Gardening, Průhonice, Czech Republic

Department of Landscape Ecology, Lidická 25/27, 602 00 Brno, Czech Republic, DemekJ@seznam.cz

Predrag Djurović Faculty of Geography, University of Belgrade, Studentski trg 3, 11000 Belgrade, Serbia, geodjura@eunet.rs

Sanja Faivre University of Zagreb, Faculty of Science, Department of Physical Geography, Marulićev trg 19/II, 10 000 Zagreb, Croatia, sfaivre@geog.pmf.hr

Ján Feranec Institute of Geography, Slovak Academy of Sciences, Štefánikova 49, 814 73 Bratislava, Slovakia, feranec@savba.sk

Emil Gachev Southwestern University "Neofit Rilski", Faculty of Mathematics and Natural Sciences, Department of Geography, Ecology and Environmental Protection, 66 Ivan Mihailov str., 2700, Blagoevgrad, Bulgaria

Institute of Geography, Bulgarian Academy of Sciences, Sofia, Bulgaria, e_gachev@yahoo.co.uk

Cristian Goran "Emil Racoviță" Institute of Speleology, Romanian Academy, Calea 13 Septembrie, Str. 13, 050711 Bucharest, Romania, cristian.goran@gmail.com

Anna Grešková (†) Institute of Geography, Slovak Academy of Sciences, Štefánikova 49, 814 73 Bratislava, Slovakia

János Haas Department of Geology, Faculty of Sciences, Eötvös Loránd University, Pázmány Péter sétány 1/c, H-1117 Budapest, Hungary, haas@ludens.elte.hu

Jozef Hók Department of Geology and Paleontology, Faculty of Natural Sciences, Comenius University in Bratislava, Mlynská dolina, 842 15 Bratislava 4, Slovakia, hok@fns.uniba.sk

Jan Hradecký Department of Physical Geography and Geoecology, Faculty of Science, University of Ostrava, Chittussiho 10, 710 00 Ostrava – Slezská Ostrava, Czech Republic, jan.hradecky@osu.cz

Mauro Hrvatin Anton Melik Geographical Institute, Scientific Research Centre of the Slovenian Academy of Sciences and Arts, Novi trg 2, SI–1000 Ljubljana, Slovenia, mauro@zrc-sazu.si

Ion Ionita Department of Geography, "Alexandru Ioan Cuza" University of Iasi, Carol I Blvd., 20 A, 700505 Iasi, Romania, ion.ionita72@yahoo.com

Pavol Ištok D. Štúra 758/3, 926 01 Sereď, Slovakia, pavol.istok@gmail.com

Igor Jemcov Faculty of Mining and Geology, University of Belgrade, Djušina 7, 11000 Belgrade, Serbia, jemcov@gmail.com

Marta Jurchescu Institute of Geography, Romanian Academy, D. Racovița str., 12, 023993 Bucharest, Romania, marta_jurchescu@yahoo.com

Ádám Kertész Geographical Research Institute, Hungarian Academy of Sciences, Budaörsi út 45, H-1012 Budapest, Hungary, kertesza@helka.iif.hu

Karel Kirchner Institute of Geonics, Academy of Sciences of the Czech Republic, Branch Brno, Drobného 28, 602 00 Brno, Czech Republic, kirchner@geonika.cz

Tímea Kiss Department of Physical Geography and Geoinformatics, University of Szeged, Egyetem utca 2-6, H-6722 Szeged, Hungary, kisstimi@gmail.com

Dragan Kolčakovski Institute of Geography, University "Ss. Cyril and Methodius", Skopje, Gazi Baba bb., 1000 Skopje, Macedonia, kolcak@iunona.pmf.ukim.edu.mk

Blaž Komac Anton Melik Geographical Institute, Scientific Research Centre of the Slovenian Academy of Sciences and Arts, Novi trg 2, SI–1000 Ljubljana, Slovenia, blaz.komac@zrc-sazu.si

Adam Kotarba Department of Geomorphology and Hydrology of Mountains and Uplands, Institute of Geography and Spatial Organisation, Polish Academy of Sciences, ul. Św. Jana 22, 31-018 Kraków, Poland, kotarba@zg.pan.krakow.pl

Ivan Kovalchuk Department of Geodesy and Cartography, Faculty of Land Use, National University of Nature Use and Life Sciences of Ukraine, Vasylkivs'ka Str. 17/114, 03040 Kyiv, Ukraine, kovalchukip@ukr.net

Yaroslav Kravchuk Department of Geomorphology, Faculty of Geography, Ivan Franko National University of Lviv, Doroshenko Str. 41, 79000 Lviv, Ukraine, 2andira@ukr.net

Adam Łajczak Institute of Geography, Jan Kochanowski University, Świętokrzyska Str. 15, 25-406 Kielce, Poland, alajczak@o2.pl

Milan Lehotský Institute of Geography, Slovak Academy of Sciences, Štefánikova 49, 814 73 Bratislava, Slovakia, geogleho@savba.sk

Aleš Létal Department of Geography, Faculty of Science, Palacký University in Olomouc, 17. listopadu 12, 771 46 Olomouc, Czech Republic, ales.letal@upol.cz

Dénes Lóczy Institute of Environmental Sciences, University of Pécs, Ifjúság útja 6, H-7624 Pécs, Hungary, loczyd@gamma.ttk.pte.hu

József Lóki Department of Physical Geography and Geoinformatics, University of Debrecen, Egyetem tér 1., H-4010 Debrecen, Hungary, loki.jozsef@science.unideb.hu

Włodzimierz Margielewski Institute of Nature Conservation, Polish Academy of Sciences, A. Mickiewicza Av. 33, 31-120 Kraków, Poland, margielewski@iop.krakow.pl

Monika Michalková Department of Physical Geography and Geoecology, Faculty of Natural Sciences, Comenius University in Bratislava, Mlynská dolina, 842 15 Bratislava 4, Slovakia, michalkovam@fns.uniba.sk

Andrej Mihevc Karst Research Institute, Scientific Research Centre of the Slovenian Academy of Sciences and Arts, Titov trg 2, SI–6230 Postojna, Slovenia, mihevc@zrc-sazu.si

Pavol Miklánek Institute of Hydrology, Slovak Academy of Sciences, Račianska 75, 831 02 Bratislava 3, Slovakia, miklanek@uh.savba.sk

Ivica Milevski Institute of Geography, University "Ss. Cyril and Methodius", Skopje, Gazi Baba bb., 1000 Skopje, Macedonia, ivica@iunona.pmf.ukim.edu.mk

Jozef Minár Department of Physical Geography and Geoecology, Faculty of Natural Sciences, Comenius University in Bratislava, Mlynská dolina, 842 15 Bratislava 4, Slovakia

Department of Physical Geography and Geoecology, Faculty of Science, University of Ostrava, Chittussiho 10, 710 00 Ostrava, Czech Republic, minar@fns.uniba.sk, josef.minar@osu.cz

Andriy Mykhnovych Department of Applied Geography and Cartography, Faculty of Geography, Ivan Franko National University of Lviv, Doroshenko Str. 41, 79000 Lviv, Ukraine, 2andira@ukr.net

Karel Natek Department of Geography, Faculty of Arts, University of Ljubljana, Aškerčeva cesta 2, SI–1000 Ljubljana, Slovenia, karel.natek@guest.arnes.si

Tadeusz Niedźwiedź Department of Climatology, Faculty of Earth Sciences, University of Silesia, Bedzinska 60, 41-200 Sosnowiec, Poland, tadeusz.niedzwiedz@us.edu.pl

Martin Ondrášik Department of Geotechnics, Slovak University of Technology in Bratislava, Radlinského, 11 813 68 Bratislava, Slovakia, martin.ondrasik@stuba.sk

Rudolf Ondrášik Department of Engineering Geology, Faculty of Natural Sciences, Comenius University in Bratislava, Mlynská dolina, 842 15 Bratislava 4, Slovakia, ondrasik@fns.uniba.sk

Mladen Pahernik Croatian Defence Academy, Ilica 256 b, 10 000 Zagreb, Croatia, mladen.pahernik@zg.t-com.hr

Tomáš Pánek Department of Physical Geography and Geoecology, Faculty of Science, University of Ostrava, Chittussiho 10, 710 00 Ostrava - Slezská Ostrava, Czech Republic, tomas.panek@osu.cz

Radmila Pavlović Faculty of Mining and Geology, University of Belgrade, Djušina 7, 11000 Belgrade, Serbia, prada@rgf.bg.ac.rs

Jozef Pecho Institute of Atmospheric Physics, Academy of Sciences of the Czech Republic, Boční II 1401, 141 31 Praha 4, Czech Republic, pecho@ufa.cas.cz

Drago Perko Anton Melik Geographical Institute, Scientific Research Centre of the Slovenian Academy of Sciences and Arts, Novi trg 2, SI–1000 Ljubljana, Slovenia, drago@zrc-sazu.si

Peter Pišút Department of Physical Geography and Geoecology, Faculty of Natural Sciences, Comenius University in Bratislava, Mlynská dolina, 842 15 Bratislava 4, Slovakia, pisut@fns.uniba.sk

Mitja Prelovšek Karst Research Institute, Scientific Research Centre of the Slovenian Academy of Sciences and Arts, Titov trg 2, SI–6230 Postojna, Slovenia, mitja.prelovsek@zrc-sazu.si

Olha Pylypovych Department of Applied Geography and Cartography, Faculty of Geography, Ivan Franko National University of Lviv, Doroshenko Str. 41, 79000 Lviv, Ukraine, olha@tdb.com.ua

Zofia Rączkowska Department of Geoenvironmental Research, Institute of Geography and Spatial Organisation, Polish Academy of Sciences, Św. Jana Str. 22, 31-018 Kraków, Poland, raczk@zg.pan.krakow.pl

Maria Rădoane Faculty of History and Geography, "Ştefan cel Mare" University of Suceava, Universitatii Str., 13, 722029 Suceava, Romania, radoane@usv.ro

Nicolae Rădoane Faculty of History and Geography, "Ştefan cel Mare" University of Suceava, Universitatii Str., 13, 722029 Suceava, Romania, nicolrad@yahoo.com

Péter Rózsa Department of Mineralogy and Geology, University of Debrecen, Egyetem tér 1., H-4010 Debrecen, Hungary, rozsa.peter@science.unideb.hu

Mihaela Sima Institute of Geography, Romanian Academy, D. Racoviţa Str. 12, 023993 Bucharest, Romania, simamik@yahoo.com

György Sipos Department of Physical Geography and Geoinformatics, University of Szeged, Egyetem utca 2-6, H-6722 Szeged, Hungary, gysipos@geo.u-szeged.hu

Irena Smolová Department of Geography, Faculty of Science, Palacký University in Olomouc, 17. listopadu 12, 771 46 Olomouc, Czech Republic, irena.smolova@upol.cz

Tomáš Soukup GISAT s.r.o., Milady Horákové 57, 170 00 Praha 7, Czech Republic, tomas.soukup@gisat.cz

Miloš Stankoviansky Department of Physical Geography and Geoecology, Faculty of Natural Sciences, Comenius University in Bratislava, Mlynská dolina, 842 15 Bratislava 4, Slovakia, milos.Stankoviansky@test.fns.uniba.sk

Uroš Stepišnik Department of Geography, Faculty of Arts, University of Ljubljana, Aškerčeva cesta 2, SI–1000 Ljubljana, Slovenia, uros.stepisnik@gmail.com

Krasimir Stoyanov Southwestern University "Neofit Rilski", Faculty of Mathematics and Natural Sciences, Department of Geography, Ecology and Environmental Protection, 66 Ivan Mihailov Str., 2700 Blagoevgrad, Bulgaria

Institute of Geography, Bulgarian Academy of Sciences, Sofia, Bulgaria, krasi_sto@yahoo.com

Virgil Surdeanu Faculty of Geography, Babeş-Bolyai University of Cluj-Napoca, Clinicilor Str. 5-7, 400006 Cluj-Napoca, Romania, surdeanu_v@yahoo.com

László Sütő Department of Geography and Tourism, College of Nyíregyháza, Sóstói út 31/B, H-4400 Nyíregyháza, Hungary, sutolaci@zeus.nyf.hu

Jolanta Święchowicz Department of Geomorphology, Institute of Geography and Spatial Management, Jagiellonian University, Gronostajowa Str. 7, 30-387 Kraków, Poland, j.swiechowicz@geo.uj.edu.pl

József Szabó Department of Physical Geography and Geoinformatics, University of Debrecen, Egyetem tér 1, H-4010 Debrecen, Hungary, szabo.jozsef@science.unideb.hu

Branislav Trivić Faculty of Mining and Geology, University of Belgrade, Djušina 7, 11000 Belgrade, Serbia, btrivic@rgf.bg.ac.rs

Milan Trizna Department of Physical Geography and Geoecology, Faculty of Natural Sciences, Comenius University in Bratislava, Mlynská dolina, 842 15 Bratislava 4, Slovakia, trizna@fns.uniba.sk

Ján Urbánek Institute of Geography, Slovak Academy of Sciences, Štefánikova 49, 814 73 Bratislava, Slovakia, jan.urbanek@zoznam.sk

Petru Urdea Department of Geography, Faculty of Chemistry, Biology and Geography, West University of Timişoara, Pestalozzi Blvd. 16, 300115 Timişoara, Romania, urdea@cbg.uvt.ro

Márton Veress Department of Physical Geography, University of West-Hungary, Károlyi Gáspár tér 4, H-9701 Szombathely, Hungary, vmarton@ttmk.nyme.hu

Matija Zorn Anton Melik Geographical Institute, Scientific Research Centre of the Slovenian Academy of Sciences and Arts, Novi trg 2, SI–1000 Ljubljana, Slovenia, matija.zorn@zrc-sazu.si

Introduction

Monographs are necessary to summarize the stage reached by research in a particular field of knowledge. The present globalization also implies that this knowledge cannot be restricted to certain countries. The accession of Eastern European countries in the European Union raised attention to this region, politically compartmentalized in the twentieth century.

The region of the southeastern European mountain ranges and the enclosed basins, however, has always been recognized as a physical geographical unit. This is proved by the fact that earlier contacts between geomorphologists were established under the aegis of the Carpatho-Balkanic Geomorphological Commission (CBGC). It all started in September 1963, when an international symposium on the geomorphology of the Carpathians were organized by the Department of Geomorphology and Hydrology of Mountains and Uplands of the Institute of Geography, Polish Academy of Sciences, and the Institute of Geography of the Slovak Academy of Sciences. The program started in Krakow and continued with excursions from Poland to Slovakia (then part of Czechoslovakia) and finished in Bratislava. Its participants decided to found the CBGC according to the model of in that time already existing Carpathian–Balkan Geological Association. The CBGC was an unofficial grouping, initiated by the directors of national geographical institutes of the national academies. Later senior university lecturers also involved. The main objective of the Commission from its beginning was to provide a platform not only for the exchange of knowledge and experience concerning the Carpatho-Balkanic mountain system and adjacent depressions, but also for research cooperation. Its membership gradually extended to Bulgaria, Czechoslovakia, Hungary, Poland, Romania, the Soviet Union, and Yugoslavia, and leading geomorphologists from universities also became involved in its work. An important motivation was that at that time the founding socialist countries were isolated to various degrees from academic life in Western Europe, contacts with institutions outside the "Soviet block" were not encouraged.

The presidents of the CBGC have been prominent geomorphologists, who served terms of various length: Professors Mieczyslaw Klimaszewski, Poland (1963–1978); Emil Mazúr, Czechoslovakia (1978–1987); Zoltán Pinczés, Hungary (1987–1998); Dan Balteanu, Romania (1998–2003); Miloš Stankoviansky, Slovakia (2003–2007), and Dénes Lóczy, Hungary (2007–). (The editors have learned with deep sorrow that one of the past presidents, Zoltán Pinczés, Professor Emeritus of the University of Debrecen, died just weeks before his 85th birthday during the editing of this book, in June 2011.) Successful and memorable conferences followed the first (of 1963) in Sofia (1966), Bucharest (1970), Budapest (1975), Prešov (1982), Debrecen (1987), Baile Herculane–Oršova (1989), Bratislava (2003), Pécs (2007), and Ostravice (2011).

The annually issued official journal of the CBGC, *Studia Geomorphologica Carpatho–Balcanica,* edited by Adam Kotarba (Kraków) since 1991 and now by Zofia Rączkowska, published a multitude of research papers on the geomorphology of the Carpathian and Balkan Mountains – and recently of other high mountain environments. (A special volume of Studia Geomorphologica Carpatho-Balcanica, Volume 40, was particularly concerned with the topic of the present monograph with the subtitle *Recent geomorphological hazards in the Carpatho–Balkan–Dinaric region.*) Another predecessor of the monograph was the two-volume *Geomorphological problems of the Carpathians,* the first issued in Bratislava in 1965, and as a special issue of Geographia Polonica in 1966.

Commission meetings have been held regularly and friendly personal contacts built but few joint projects have been implemented. One major exception was the geomorphological map of the region with uniform legend at 1:1,000,000 scale, a collective work of authors from Hungary, Bulgaria, Czechoslovakia, Poland, and Yugoslavia, edited by Professor Márton Pécsi, was included in the *Atlas der Donauländer,* published in Vienna.

The changes that happened in the political and economic systems around 1990 also affected scientific life and academic cooperation. The attention of researchers turned towards the West, bilateral contacts between the former socialist countries became looser. As a consequence, the 1990s are not counted among the most successful periods of the Commission. In the meantime, however, international cooperation in geomorphology became global and covered most countries in the world (partly in the framework of the International Association of Geomorphologists, IAG/AIG), irrespective of their political systems. This fortunate development had an indirect positive impact on the activities of the CBGC, too. In 2001, at its Tokyo Conference, IAG/AIG set up a Working Group focusing on Human Impact on the Landscape (abbreviated as HILS). It had been a Hungarian idea, later joined by geomorphologists from most of the countries within the Carpatho–Balkan–Dinaric region. At the next conference, in Zaragoza in 2005, at the initiation of the Slovak national group, the first Regional Working Group of the same international organization was formed for the study of the geomorphology of the CBD region. As an end product of this Carpatho–Balkan–Dinaric Regional Working Group (CBDRWG), a comprehensive monograph on recent landform evolution was envisaged. The first term of the Working Group served for the preparatory work and now, after six years, the volume is ready. More than 70 researchers under national coordination from 11 countries have been working on it. A great achievement of our authors is that they managed to collect the most important research findings on recent evolution of landforms that had been only available in the national languages of the CBD region.

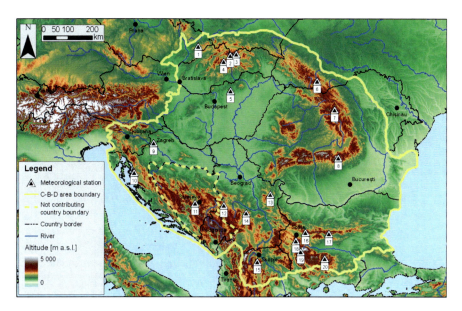

Fig. 1 Physical geographical delineation of the Carpatho-Balkan-Dinaric macroregion (yellow line) on a digital elevation model. Dashed yellow line marks the boundaries of the areas and countries which are not included among the national chapters

As far as the geographical delineation of the Carpatho–Balkan–Dinaric macroregion is concerned, the editors had to agree that – while the northern and eastern boundaries are more or less marked along the forehills of the Carpathians – to establish the western and particularly the southern boundaries seems to be an impossible task. For instance, the World Wide Fund for Nature (WWF) delimits the Carpathians in the framework of the Carpathians Ecoregion Initiative (CERI) on the basis of complex criteria – mainly geomorphological (elevation, slope, exposition, geology) and ecological. Accordingly, the Ecoregion stretches along 210,000 km^2 from Vienna to the Danube Gorges (the Iron Gate Dam reservoir between Romania and Serbia), including the Transylvanian Depression and adjacent hill regions. In order to find an elegant solution, in our case the physical boundaries of morphostructures have been adjusted to political borders: thus, Austria (the Hainburg Hills) is "excluded" from the region but Slovenia and Croatia are fully included down to the Adriatic coast. In the southwest and south the macroregion is delimited – rather arbitrarily – along the southern borders of Serbia, Macedonia, and Bulgaria. The region was delimited specially for the purpose of this publication, taking into consideration, first of all, the countries participating in the project, and thus it is not fully in conformity with natural boundaries of megaunits under study, namely the Carpathians, Balkanides, Dinarides, and the Pannonian Basin. The boundaries represent a compromise between the natural borders of the megaunits concerned and national frontiers. Thus, two boundaries are shown on the delineation map (see Fig. 1): one marks more or less precisely the physical geographical limits of the macroregion, while the other presents the area of the countries included among the national chapters and from where the description of recent geomorphic processes was available. From a practical point of view and for the sake of integrity, the region incorporates the territory of Bosnia and Herzegovina, Montenegro, and northernmost Albania, though these countries could not be involved in the work. On the contrary, the relevant part of Austria was not included.

The monograph consists of three two parts. Part I provides background information to the geomorphology of the Carpatho–Balkan–Dinaric region. In four chapters the geological-tectonic, climatic conditions, hydrography, and land cover are briefly summarized. The core of the monograph is Part II, i.e., 11 national chapters of variable size and coverage. Some of them only deal with parts of their territories that belong to some of the assessed megaunits, as, for example, in the case of Czechia, Poland, and Ukraine (different sections of the Carpathians). The Romanian chapter presents geomorphic processes in the Carpathians and the Pericarpathian Regions. For the remaining countries the coverage comprises their whole areas. It is possible to refer the countries into two groups: some fully lie in the assessed megaunits (like Slovakia, Hungary, Croatia, and Serbia), while others have regions outside the particular megaunit or their position is doubted (Slovenia, Macedonia, and Bulgaria).

Chapters are arranged in a uniform structure: the history of geomorphological research in each country is followed by brief introductions into the geological-tectonic setting, climatic and hydrological conditions. Then the geomorphological districts delimited and described. Geomorphic processes and landforms are treated in the order of their relative significance emphasizing the consequences of human interventions into the natural environment. In some chapters special attention is also paid to the predictable impact of climate change on exogenous geomorphic processes in the particular country.

However, such an ambitious project has a series of difficulties – some of which are perhaps not at all possible to overcome. It is only fair to mention some of them. As research continues to be performed in national frameworks, it is not easy to imagine how all the information contained could be integrated at the present level of knowledge. However, the editors tried to draw some comparisons as conclusions in the final chapter of the monograph.

The interpretation of recent landform evolution also varies with countries. Some of the summaries focus on millennia, while others draw the boundary at the first major human intervention into the operation of natural processes (settlement, deforestation) or even at a later point in time. The date of such events does not only varies with countries but also with altitudinal zones. In some swampy lowlands human settlement only began to influence natural processes fundamentally in the last two centuries. It is not only the time scales but the spatial dimensions that vary from country to country. The geomorphological subdivisions of the mountain systems cannot be unified either.

Political divisions (particularly on the territory of the former Yugoslavia) appear on the maps as conceived by the individual countries, not modified by editors. In spite of all efforts from editors, unfortunately, a full coverage of the Dinaric mountain system could not be accomplished: Bosnia-Herzegovina and Montenegro (Crna Gora) could not be included among the national chapters – although they are naturally treated in the chapters of the general part of the book. The Republic of Macedonia appears as Macedonia, the complicated circumscriptional denotation FYROM (Former Yugoslav Republic of Macedonia) is not applied. At the time of writing, the political unit of uncertain status, which is called Kosovo (Kosova), still unambiguously belonged to Serbia, and, therefore, its geomorphology is described in the Serbian chapter. The editors apologize if this kind of solution hurt anyone's national feeling.

The editors hope that the book will be equally useful for researchers in geomorphology and generally in geosciences and university lecturers, particularly at postgraduate level, and that all readers, both in CBD countries and outside the region, will be satisfied with the overview provided by this monograph, and encouraged to search for more information on recent geomorphic processes in the region.

Dénes Lóczy, Pécs
Miloš Stankoviansky, Bratislava
Adam Kotarba, Kraków

Part I
General

Chapter 1
Geological and Tectonic Setting

János Haas

Abstract The present-day geological structure of the Carpatho–Balkan–Dinaric region is controlled by the evolution of the Tethys and Atlantic Oceans and the related movements of the European and the African Plates. The Carpathians, a continuation of the Eastern Alps, formed along the margin of the European Plate to effect the closure of the Neotethys and the Atlantic related Penninic Oceans. The Balkanides are the continuation of the Southern Carpathians of nappe structure. The Dinarid mountain system is also characterized by a multiple nappe structure developed from the Middle Jurassic to the Tertiary. The basement of the Pannonian Basin, enclosed by the mountain systems, is composed of the Tisza and ALCAPA (Alpine–Carpathian–Pannon) Mega-units with markedly different geological structure and evolution history. The Late Neogene basin system was formed as a result of indentation of the Adria Microplate and the subduction of the European Plate margin.

Keywords Geological evolution • Carpathian foredeep • Carpathians • Balkanides • Dinarides • Pannonian Basin

1.1 Introduction

The Carpatho–Balkan–Dinaric region is part of the *Mediterranean Mountain System*. It was formed during the last plate-tectonic cycle, which commenced in the latest Paleozoic times. In Europe it is an about 300–800 km wide belt (Neo-Europe) accreted to the previously consolidated parts of Europe (Variscan Europe or Meso-Europe) as

---

J. Haas (✉)
Department of Geology, Faculty of Sciences, Eötvös Loránd University,
Pázmány Péter sétány 1/c, H-1117 Budapest, Hungary
e-mail: haas@ludens.elte.hu

a result of the *Alpine orogeny* caused by the convergence of the European (Eurasian) and African Plates. The present-day geological structure of the region is mostly determined by the evolution of the Tethys and Atlantic Ocean systems. It involved the dismembering of the European and African continental plate margins during the early evolutionary stages and their tectonic deformation and uplifting as consequences of *plate and microplate collisions*. Plate-tectonic processes led to the formation of the large Pannonian Basin system in the Late Tertiary.

The main stages of the evolution of the geological structure can be summarized as follows:

1. *Pre-Alpine*, mostly Variscan, evolution that determined the geological structure of the plate margins at the beginning of the Alpine plate-tectonic cycle. Large fragments of the Variscan Belt dismembered from the margins and incorporated into the Alpine orogenic system.
2. The *early* stage of the *Alpine* plate-tectonic cycle is characterized by the *opening of oceanic basins*: the western Neotethys Ocean basin from east to west from the Middle Triassic to the Early Jurassic and the Penninic branch of the Atlantic Ocean from west to east from the Middle Jurassic to the Early Cretaceous.
3. The stage of *mountain building* processes: closure of the Neotethys basin from the Middle Jurassic to the Late Cretaceous–earliest Tertiary and of the Penninic branch from the early Late Cretaceous to the Early Miocene.
4. *Development of molasse basins* in the foreland of the Alpine nappe stacks and in back-arc setting related to the subduction of the European Plate in the Late Tertiary.

From the above-discussed plate-tectonic processes the following structural units can be deduced for the European segment of the Mediterranean Mountain System:

1. The external Alpine–Carpathian foredeep molasse basins;
2. The Alpine Mountain System, developed along the margin of the previously consolidated European craton, including the ranges of the Alps, Carpathians, and Balkanides;
3. The Dinaric Mountain System, developed along the margin of the Adriatic microplate dismembered from the African continent in the early stage of the Alpine cycle, including the Appenines, Southern Alps, Dinarides, and Hellenides;
4. The internal (back-arc) molasse basins.

This classification will be followed in the description of the main features of the geological structure of the region discussed. The major structural units are presented in Fig. 1.1.

1.2 Carpathian Foredeep – External Molasse Zone

The Carpathian Molasse Zone is the direct continuation of the Eastern Alpine Molasse Zone (Fig. 1.1). The molasse basins were formed by *flexular downbending* of the European Plate (Bohemian Massif, East European Platform, Moesian

1 Geological and Tectonic Setting

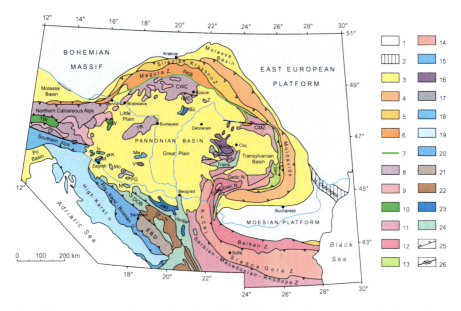

Fig. 1.1 Major structural units of the Carpathic–Balkan–Dinaric region (compiled by J. Haas using the maps of Schmid et al. 2008; Zagorchev 1994; Dimitrijević 1997). *1* Precambrian–Paleozoic platforms, *2* North Dobrogea Unit, *3* molasse basins, *4–6* Carpathian Flysch Zone, *4* Moldavides, *5* Silesian–Krosno Zone, Outer Dacides (OD), *6* Magura Zone, *7* Pieniny Klippen Belt (PKB), *8* Upper Austroalpine Unit, Transdanubian Range Unit, Fatric, Hronic and Silicic Units, *9* Lower Austroalpine Unit, Tatric, Veporic and Gemeric Units, *10* Penninic Unit, *11* Crystalline-Mesozoic Zone (CMZ), Serbian–Macedonian–Rhodope Zone, Biharia Unit (Bih), *12* Danubian Nappes, Balkan Zone, *13* Severin Nappe (Sev), *14* Getic Nappes, Kučaj–Sredna Gora Zone, *15* Mecsek Zone (Me), *16* Villány (V)–Bihor Zone, *17* Papuk (P)–Codru (Cod) Zone, *18* Southern Alpine Units, *19* High Karst Unit, *20* Pre-Karst–Bosnian Unit, *21* East Bosnian–Durmitor Unit (EBD), *22* Drina–Ivanjuca Unit (Dr-Iv), *23* Jadar Unit (Jad), Bükk Unit (Bü), *24* Vardar Zone, Transylvanian Nappes (Trans), Dinaric Ophiolite Belt (DOB), *25* overthrust, *26* strike-slip fault. Further abbreviations: *Tw* Tauern window, *Rw* Rechnitz window, *Dr* Drau Range, *K* Kalnik, *Iv* Ivanscica, *Mo* Moslavačka Gora, *PG* Požekša Gora, *Z* Zemplén Mountains

Platform) under the load of the progressing nappe stacks (Kováč and Plašienka 2002). Precambrian, Caledonian, and Variscan rocks and slightly deformed and undeformed Mesozoic and older Cainozoic sedimentary sequences constitute the basement of the molasse deposits. The oldest molasse formations are uppermost Oligocene to Lower Miocene terrestrial sediments. A marked transgression in the Early Miocene led to the establishment of brackish-water and normal marine conditions and deposition of siliciclastic successions. The subsequent transgression-regression cycles were mostly controlled by the nappe tectonics. The last transgression took place in the Middle Miocene (Badenian) when argillaceous sediments and shallow-marine limestones deposited, followed by the formation of evaporites in some restricted sub-basins. The uppermost Miocene and Pliocene are represented by lacustrine and fluvial deposits.

1.3 The Carpathians

The Carpathians were developed along the margin of the European Plate as a result of the *closure of the Neotethys and the Penninic* ("Alpine Tethys") *Oceans*. The Western Carpathians can be considered as the continuation of the Eastern Alps and the South Carpathians continue in the Balkanides. The external belt of the mountain system is characterized by the predominance of flysch facies of Cretaceous to Paleogene age. In contrast, the internal belts are more varied: there are significant differences in the geological setting and geodynamic evolution along the strike of the mountain range. Accordingly, the Carpathians are usually subdivided into *three main* structural *units*: the Western, Eastern, and Southern Carpathians (Fig. 1.1).

1.3.1 Western Carpathians

The mountain ranges of the Western Carpathians extend from the eastern end of the Eastern Alps to the northwestern termination of the Eastern Carpathians on the territory of the Slovak Republic, Poland, and Ukraine.

The *Flysch Zone* is built of a system of rootless nappes (Rakús et al. 1998). The inner zone consists predominantly of Upper Jurassic to Lower Miocene flysch complexes, several 1,000 m in thickness. In the external units Upper Jurassic to Lower Cretaceous carbonates and volcanites are also common (Kováč and Plašienka 2002). The nappes are detached from their subducted basement and stacked in an accretionary wedge of 70–80 km width and reach down to 15–20 km depth (Tomek 1993).

The *Silesian–Krossno Belt* is the outermost unit of the Flysch Zone that can be followed from the eastern termination of the Alps (Waschberg Zone) to the units of the Moldavids in the East Carpathians. It is overthrust onto the foreland continental plate (Bohemian Massif, East European Platform). The nappe stack was accreted in the period from the latest Oligocene to the earliest Miocene. Its stratigraphic record encompasses the Jurassic to Upper Oligocene interval (Kováč and Plašienka 2002). The belt is built of several nappes among which the Silesian Unit, consisting of an extremely thick Upper Jurassic to Oligocene succession, is most prevalent. The thickness of the Upper Jurassic to Lower Cretaceous flyschoid succession with alkaline basalts may reach 6,000 m. Large blocks of Upper Jurassic reef limestone occur at the base of this unit. The Upper Cretaceous and Paleogene are mostly represented by flysch in a thickness up to 5,000 m.

The *Magura Belt* is overthrust onto the Silesian–Krossno Belt as a continuation of the Alpine Rheno-Danubian Flysch Zone. It is predominantly built of Cretaceous to Paleogene flysch formations, detached from their subducted, probably oceanic substrate. Based on stratigraphical and lithological characteristics, several units are distinguished within the Magura Belt. In the most extended Rača Unit of Upper Jurassic to Lower Cretaceous carbonate turbidites represents the oldest part of the succession, overlain by terrigenous siliciclastic turbidite formations of mid-Cretaceous to Early Oligocene age. Miocene trachyandesite and basalt sills crosscut the flysch nappes in many parts of the Magura Belt.

1.3.1.1 Central Western Carpathians

The very complex geodynamic evolution of the Central Western Carpathians during the Middle to Late Cretaceous resulted in a complicated structure. There are major crustal units that are made up of rocks of the *pre-Alpine basement* together with their Upper Paleozoic to Mesozoic cover (Tatric, Veporic, and Gemeric Units) and there are relatively thin detached cover *nappe systems* consisting of Upper Paleozoic and Mesozoic formations (Fatric, Hronic, and Silicic Units) (Plašienka et al. 1997; Kováč and Plašienka 2002).

The *Pieniny Klippen Belt* is a narrow (only a few kilometers wide), intensely deformed zone running along the boundary of the Outer and Central West Carpathians. It can be followed from the Vienna Basin to Poiana Botizi, northern Romania. The belt is built of blocks ("klippen") of various Mesozoic rocks, mostly limestones of rarely Triassic, predominantly Jurassic and Early Cretaceous age, Middle to Upper Cretaceous marls, and Upper Cretaceous to Paleogene flysch (Birkenmajer 1977). The tectonically disintegrated competent limestone blocks are enveloped by the incompetent marls and flysch. During the Paleocene to Early Miocene the Pieniny Klippen Belt was subject to transpressional tectonic movements, followed by transtension-caused development of small pull-apart basins (e.g., the Orava Basin).

The *Tatra–Fatra Belt* comprises the pre-Alpine crystalline rocks of the Tatric Unit and its sedimentary cover and the nappe system of the Fatric and Hronic Units with their Paleogene, Neogene, and Quaternary cover. The pre-Alpine (mostly Variscan) crystalline complex contains medium- to high-grade metamorphic formations that are intruded by several granitoid plutons. The Alpine nappe structure is well documented in the Malé Karpaty Mountains, where the Bratislava Nappe contains a thick Lower Paleozoic volcano-sedimentary succession affected by low-grade metamorphism (Plašienka et al. 1991). The sedimentary cover of the Tatric Unit is made up of Germanic-type terrestrial and shallow-marine Triassic formations; the Jurassic is represented by hemipelagic marls and deep-marine limestones, and the Cretaceous consist mostly of shallow-marine carbonates. The Fatric Unit includes a pre-Alpine basement (phyllites, micaschists, amphibolites, and granites) and its Permian to Lower Triassic sedimentary cover. The detached complex (Križna Nappe) contains Middle Triassic to Cretaceous sedimentary rocks. The Hronic Unit is the highest nappe system in the Tatra–Fatra Belt. It corresponds with parts of the Upper Austroalpine Nappe System. Upper Paleozoic–Lower Triassic series form the basal part of the Hronic Unit. The Middle Triassic is represented by shallow-marine carbonates of 2–3 km thickness. They are overlain by Lower and Middle Jurassic limestones of swell facies and Upper Jurassic–Lower Cretaceous deep-sea carbonates and marls.

The *Vepor Belt* is thrust onto the Tatric Unit and it is overthrust by the Gemer Unit. The pre-Alpine basement of this unit is similar to that of the Tatric Unit. Its Upper Paleozoic to Mesozoic cover is preserved only locally. There are large plutons of complex lithology (Vepor Pluton) and several smaller granitoid plutons (Vozárová et al. 2009). The basement of the South Veporic Unit was affected by paleo-Alpine tectonometamorphic processes, i.e., low- to medium-grade metamorphism. The sedimentary cover is similar to that of the paleogeographically related Fatric Unit.

The *Gemer Belt* is in the highest structural position among all the basement nappe systems. It is made up of several north-vergent thrust imbrications and partial nappes. The largest unit is the Volovec Unit, which consists of low-grade metamorphic rocks of thousands of meters in thickness and Lower Paleozoic volcano-sedimentary complexes intruded by Permian granitoids (Vozárová et al. 2009).

The highest unmetamorphosed nappes in the Vepor and Gemer Belts are referred to as the Silicic Unit. They contain Upper Permian evaporites at their base, Lower Triassic shallow-marine carbonates and siliciclastics, thick Middle Triassic platform carbonates, and Jurassic shallow- (swell) and deep-marine formations.

1.3.1.2 Internal Western Carpathians

The Rožňava Line ("Meliatic Suture") separates the Internal Western Carpathians from the Central Western Carpathians.

The *Meliatic Unit* includes the deepest structural elements of the Slovak Karst and consists of remnants of the Neotethys Ocean. It is built of mélange, i.e., olistostromes in shale and radiolarite matrix (Mock et al. 1998). The mélange complex is exposed in tectonic windows from beneath the Turnaic or Silicic Nappes.

The *Turnaic Unit* is a rootless nappe system between the Meliatic and Silicic Nappes containing Middle Carboniferous to Upper Triassic formations in the area of the Slovak Karst and Rudabánya Hills. The formations were affected by low-grade metamorphism.

1.3.2 Eastern Carpathians

The Western Carpathians continue southwards into the Eastern Carpathians. The continuation of the Outer Carpathian units in the Flysch Zone of the Eastern and Southern Carpathians is straightforward. The Pieniny Klippen Belt that separates the Outer and the Central Western Carpathians can be followed to Poiana Botizi, northern Romania, and the geological boundary between the Western and the Eastern Carpathians can be drawn here at the North Transylvanian Fault (Fig. 1.1).

The *Flysch Zone* is made up of nappes of predominantly sedimentary rocks detached from their original basement and overthrust eastward onto the foreland. The innermost nappes (Outer Dacides: Black Flysch, Ceahlau, etc.) consist only of uppermost Jurassic to Cretaceous formations; the post-tectonogenic cover is of Paleogene age. The Moldavids (Convolute Flysch, Macla, etc.) contain Cretaceous, Paleogene, and Lower to Middle Miocene sedimentary rocks. The outermost overthrust unit is the Subcarpathian Nappe, which is made up mostly of Miocene formations. This structure is the result of a multi-stage development started in the mid-Cretaceous. The second overthusting stage is dated to the period from the latest Cretaceous and to the Miocene.

The *Crystalline-Mesozoic Zone* is built of basement nappes, which contain very thick metamorphic complexes. They are covered by Permo-Mesozoic or Mesozoic sedimentary sequences. These nappe-stacks were formed during the mid-Cretaceous orogenic event. The Infrabucovinian Nappe is the deepest one; it is overlain by the Subbucovinian and the Bucovinian Nappes (Săndulescu and Dimitrescu 2004). The metamorphic complexes consist of superposed metamorphic series (Kräutner 1997). The prevalent part of them is mesometamorphic and Precambrian in age, while the youngest series is epimetamorphic and of Lower Paleozoic age. The Mesozoic successions of the nappes show significant differences from the Middle Triassic. The Middle Triassic of the Bucovinian and Subbucovinian nappes are characterized by shallow-marine carbonates (mostly dolomites) whereas deep-marine limestones occur in the Infrabucovinian Nappes. Shallow-marine formations prevail in the Lower Jurassic and deeper-water siliciclastic sediments in the Middle Jurassic. Then in the latest Jurassic and Early Cretaceous a flysch trough developed in the external belt of the Bucovinian domain whereas in the central part of the Bucovinian and in the Infrabucovinian domains calcareous pelagic sediments deposited. The Aptian-Albian wildflysch with large olistoliths completes the Bucovinian sedimentary sequence.

The *Transylvanian nappes* are overthrust onto the wildflysch of the Bucovinian nappe. These nappes (Olt, Perşani, Hăghimaş) formed by obduction of the basement of the Transylvanian branch of the Neotethys ocean, and it can be considered as the eastward continuation of the Vardar Zone (Săndulescu and Dimitrescu 2004). The lithological features and the ages of the formations in the three nappes show significant differences. In the Olt Nappe Triassic ophiolites are overlain by Middle to Upper Triassic deep-marine limestones. The succession of the Perşani Nappe begins with shallow-marine Lower Triassic and is made up mostly of Middle to Upper Triassic limestones of deep-sea and slope facies. In the Hăghimaş Nappe Jurassic ophiolites occur that are covered by Upper Jurassic pelagic limestones and uppermost Jurassic to Lower Cretaceous shallow-marine limestones.

1.3.3 Southern Carpathians

The Southern Carpathians extend from the Prahova Valley in the east to the Timok Valley in the south (Fig. 1.1). It comprises Alpine nappes and pre-Alpine structural units that are incorporated into the Alpine nappes. All of the Alpine units – with the exception of the Severin Nappe – have a *metamorphic basement* of Precambrian and Lower Paleozoic age and an Upper Paleozoic to Mesozoic sedimentary series

(Săndulescu and Dimitrescu 2004). Upper Cretaceous magmatites, mostly *calc-alkaline plutons* (banatites) occur in many parts of the Southern Carpathians; andesitic *volcano-sedimentary successions* cover the Upper Cretaceous basins in the Getic and Supragetic Units (Berza et al. 1998).

The *Lower Danubian Unit* is composed of several pre-Alpine nappes with Upper Paleozoic to Mesozoic post-orogenic cover. The most important pre-Alpine unit (Retezat-Parâng Nappe) consists of polymetamorphic rocks (predominantly amphibolites) intruded by granitoids and covered by slightly metamorphosed Lower Paleozoic sequences. The post-Variscan series is characterized by Permian continental red-beds, Lower Jurassic shallow-marine conglomerates and sandstones, Middle Jurassic marine sandstones, Upper Jurassic to Lower Cretaceous shallow-marine limestones, Middle Cretaceous shales, and Upper Cretaceous flysch and volcaniclastic sandstones.

In the *Upper Danubian Unit* the pre-Alpine basement consists of a polymetamorphic complex with amphibolites and mica-gneisses, and a predominantly quartzite series. Both metamorphic series are intruded by Variscan granites. The crystalline basement is overlain by very low-grade Lower and Middle Paleozoic formations. The post-Variscan succession begins with Upper Carboniferous coal-bearing sandstones, followed by Permian continental red-beds and rhyolites. Coal-bearing sandstones occur in the basal part of the Lower Jurassic, overlain by Middle to Upper Jurassic volcano-sedimentary formations or marine carbonates and shales. The Upper Cretaceous is represented by flysch.

The *Severin Nappe* of moderate thickness occurs between the Danubian and the Getic units. It is built of obducted slices of Upper Jurassic ophiolitic mélange, overlain by Upper Jurassic to Lower Cretaceous flysch (Săndulescu and Dimitrescu 2004).

The basement of the *Getic Nappes* consists of polymetamorphic rocks, mostly micaschists and gneisses and slabs of metamorphosed ophiolites. High-temperature metamorphosis and ductile deformation took place during the Late Variscan orogenic stage (Dallmeyer et al. 1994). The crystalline complex is overlain by Lower Paleozoic low-grade metamorphic formations (mostly metamorphosed terrigenous sedimentary rocks) and granites. Upper Carboniferous coal-bearing conglomerates and sandstones and Permian terrestrial sandstones and black shales occur in the basal part of the post-Variscan series. Transgression in the Early Jurassic led to the deposition of coal-bearing sandstones, and in the Middle Jurassic marine sandstones and marls were formed. The Upper Jurassic and Lower Cretaceous are represented by shallow marine limestones and marls. The Upper Cretaceous succession consists of red marls, sandstones, conglomerates, and volcano-sedimentary rocks.

In the *Supragetic Nappes* Western and Eastern nappes are distinguished (Săndulescu and Dimitrescu 2004). The former occur in the Banat area. Their metamorphic basement consists mainly of gneisses and amphibolites. The Paleozoic sequence contains low-grade metamorphosed Devonian to Lower Carboniferous volcano-sediments and carbonates, overlain by Upper Jurassic–Lower Cretaceous limestones. The Eastern Supragetic nappes are exposed in the Făgăraş Mountains. Their crystalline basement consists of gneisses and micaschists, crystalline carbonates, and graphite schists.

1.4 The Balkanides

1.4.1 Balkan Zone

The Danubian Units continue southwards and southeastwards into the Balkan Zone, namely in the Stara Planina, Miroč, and West Poreč Units (Krstić and Karamata 1992; Kräutner and Krstić 2002).

1.4.2 Kučaj–Sredna Gora Zone

The continuation of the Getic Nappes was recognized on the territory of Serbia (Kučaj and Timok) and Bulgaria (Sredna Gora). Upper Proterozoic *crystalline schists* are the oldest formations of the Kučaj Zone. They are covered by Lower Paleozoic low-grade metamorphic series. Variscan granites and Permian red-beds are locally also present. The older formations are discordantly overlain by an Upper Jurassic carbonate succession, which is followed by Cretaceous *sedimentary* rocks and banatites in the western part of the Kučaj Zone. In the Timok Zone the oldest exposed rocks are Lower Cretaceous shallow-marine limestones, overlain by a flysch succession. In the Late Cretaceous a very thick andesitic *volcano-sedimentary complex* was formed. The Getic Nappes and the superimposed volcano-sedimentary succession can be followed in eastern direction to the Black Sea in the Sredna Gora Zone (Fig. 1.1).

1.4.3 Serbian-Macedonian-Rhodope Zone

The Serbian-Macedonian massif is usually considered as the continuation of the Southern Carpathian Supragetic Unit (Săndulescu 1984; Zagorchev 1994; Dimitrijević 1997) – although they are separated from that by a major shear zone (Kräutner and Krstić 2002). It can be followed in southeastern direction to the Aegean Sea. The structural unit is built of *two crystalline complexes*: the older (Upper Proterozoic in age) consisting of gneisses, micaschists, quartzites, migmatites, and marble and the younger composed of Lower Paleozoic sandstones, argillites, and basic magmatites subject to low-grade metamorphosis. Locally higher-grade Variscan metamorphic complexes also occur. In some places outliers of the Upper Paleozoic and Mesozoic formations cover the crystalline massifs. Erosional remnants of Cretaceous coal-bearing beds and the overlying Eocene deposits occur along the Morava Valley. Neogene formations locally fill the tectonically controlled intramountain basins, with dacitic and andesitic volcanic suites.

1.5 The Dinarides

The mountain system of the Dinarides is characterized by *nappes* developed by a *multiple* structure evolution from the Middle Jurassic to the Tertiary (Fig. 1.1). The strikes of the structure zones are roughly northwest to southeast and the nappe systems are usually southwest vergent (Dimitrijević 1997). The closure of the sub-basins of the Neotethys Ocean induced the first nappe movement. Then nappe formation and tectonic deformations continued during the Cretaceous and the Tertiary till the Early Miocene.

1.5.1 Budva Zone

It is a narrow and highly deformed unit of relatively short extension (about 100 km). The oldest exposed rocks are Lower Triassic terrestrial clastic sediments that are followed by a Middle Triassic marine succession with volcanites and volcaniclastics, then deep-water limestones and radiolarites. Deep-sea conditions prevailed during the Late Triassic to Early Paleogene period, with flysch deposition during the Paleocene to Early Eocene (Dimitrijević 1997).

1.5.2 High Karst Unit

It mostly includes Upper Carboniferous and Permian marine formations (Vozárová et al. 2009), shallow-marine siliciclastics and carbonates developed in Early Triassic and shallow and deep marine formations in the Middle Triassic. Subsequently, in the Late Triassic an *immense carbonate platform* developed (Adriatic Carbonate Platform) (Kovács et al. 2010). The carbonate platform conditions prolonged for a very long time from the Late Triassic to the end of Cretaceous leading to the accumulation of shallow-marine limestones in a thickness of several kilometers (Dragičević and Velić 2002). Flysch-type sediments deposited in the Eocene.

1.5.3 Pre-Karst and Bosnian Unit

From a paleogeographical aspect, the Pre-Karst Unit represents the foreslope of the Adriatic Carbonate Platform (High Karst Unit) whereas the Bosnian Unit was a flysch trough during the Late Jurassic to Cretaceous period (Schmid et al. 2008). The Paleozoic series exposed in the Bosnian Schist Mountains exhibits no or only low-grade metamorphism (Hrvatovic and Pamić 2005). The appearance of deep-sea facies in the Middle Triassic indicates the onset of differentiation of this zone from the Adriatic Carbonate Platform already at this time. Carbonate slope facies are typical in the Jurassic (Tilšjar et al. 2002). The *Bosnian Flysch* began to deposit in the latest Jurassic and its accumulation continued during the Cretaceous and even in the Paleogene.

1.5.4 East Bosnian–Durmitor Unit

It is a large *composite thrust sheet* that is made up of Paleozoic formations and overlying Triassic to Lower Jurassic platform carbonates that are covered by Middle Jurassic deep marine radiolarites (Dimitrijević 1997). It is overlain by an Upper Jurassic ophiolitic mélange complex formed during the obduction of the overlying West Vardar Unit (Schmid et al. 2008). In the southwest the East Bosnian–Durmitor Unit is thrust over Upper Cretaceous siliciclastic turbidites of the Bosnian Unit.

1.5.5 Dinarid Ophiolite Belt

This unit consists of detached blocks (olistoliths) and dismembered thrust sheets of various lithology and ages set in a sheared sandstone/siltstone matrix (mélange) and ultramafic bodies of various size (Robertson et al. 2009). In the mélange there are blocks originated from the Paleozoic basement of a continental domain, from a carbonate platform, and also from the oceanic basement. Accretion of the mélange complex was related to the *subduction zone of the Neotethys,* which was active from the Middle to the Late Jurassic.

1.5.6 Drina–Ivanjica Unit

It is a *thrust sheet* that was probably emplaced onto the East Bosnian–Durmitor Unit in the Early Cretaceous (Schmid et al. 2008). It consists of metamorphic and intensively folded Lower Paleozoic formations, overlain by a slightly metamorphosed Carboniferous deep-marine pelitic and turbiditic succession and covered by terrestrial Lower Triassic deposits. The upper part of the Lower Triassic is represented by shallow-marine marls and carbonates.

1.5.7 Jadar Unit

The Jadar Unit is an *enclave* within the Vardar Zone. It consists of Devonian to Upper Carboniferous siliciclastic and shallow-marine carbonate sediments that are overlain by Middle Permian terrestrial siliciclastic rocks. Evaporites and dolomites and above them shallow-marine limestones characterize the Upper Permian. The marine carbonate sedimentation continued in the Early Triassic (Dimitrijević 1997).

1.5.8 Vardar Zone

The Vardar Zone is built of discontinuous bodies of *ophiolites* and *deep-sea sediments*: a mélange with basaltic rocks, gabbros, ultramafic rocks, and olistostromes containing fragments of various sedimentary and magmatic rocks. The mélange complex was formed in the Middle Jurassic as a result of subduction of the Neotethys Ocean and obduction of its oceanic basement onto the Adriatic margin. Two sub-units are distinguished (Karamata 2006): the *Western Vardar Belt* (largely equivalent to the Sava Zone defined by Pamić 2002 and Schmid et al. 2008) and the Main Vardar Zone. The Western Belt extends from Belgrade to Zagreb. Its mélange commonly contains Triassic limestone, greenschist, and metabasalt blocks. Blueschist blocks of mid-Cretaceous age also occur and constrained Cretaceous subduction at least of parts of the Western Belt (Schmid et al. 2008). The *Main Vardar Zone* is a north-northwest–south-southeast-trending narrow unit containing a typical ophiolite mélange of Middle to Upper Jurassic age. There are blocks of Jurassic volcanic rocks and granitoids and fragments of a Late Paleozoic volcanic arc (Robertson et al. 2009).

1.6 The Pannonian Basin

The Pannonian (Carpathian) Basin *system* consists *of several basins*, separated by isolated ranges (inselbergs) made up of Paleozoic, Mesozoic, and Paleogene sedimentary sequences and Tertiary igneous rocks. The pre-Neogene structure of the Pannonian Basin exhibits a complex mosaic-like collage structure (Kovács et al. 2000). The *basement* is divided into two large structural units by the Mid-Hungarian Shear Zone trending east-northeast to west-southwest (Csontos et al. 1992; Fodor et al. 1999; Haas et al. 2001, 2010; Schmid et al. 2008). These large units, namely the Tisza and ALCAPA (Alpine-Carpathian-Pannon) Mega-units, show markedly different geological features and evolution history (Fig. 1.2).

The *Tisza Mega-unit* consists of blocks accreted during the Late Paleozoic Variscan orogenic phases, when it formed part of the European Variscan Belt (Ebner et al. 2009). It separated from this belt in the Middle Jurassic and since the late Early Cretaceous it has moved as a separate entity, i.e., a *microcontinent*. The Variscan crystalline complexes are covered by Upper Paleozoic continental siliciclastic series, and continental, and shallow-marine Triassic formations (Bleahu et al. 1994). The Jurassic is typified by facies diversification with coal-bearing and then neritic and deep-marine siliciclastic formations in the Lower and lower Middle Jurassic

1 Geological and Tectonic Setting

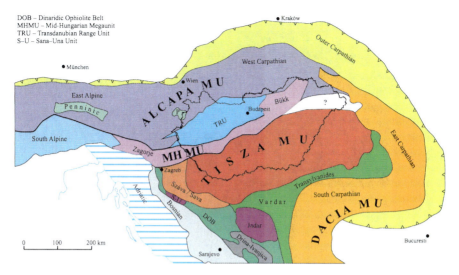

Fig 1.2 Basement structural units of the Carpatho–Balkan–Dinaric region (After Haas et al. 2010)

and deep-sea cherty limestones in the upper Middle and Upper Jurassic in the Mecsek Zone and condensed swell facies in the Villány–Bihor Zone. The Lower Cretaceous is characterized by basic volcanites and volcano-sedimentary rocks in Mecsek Zone and shallow-marine limestones in the Villány–Bihor Zone (Haas and Péró 2004). The Paleozoic–Mesozoic series of the Tisza Mega-unit are exposed in the Papuk Mountains in Croatia, in the Mecsek and Villány Mountains in South Hungary, and in the Apuseni Mountains in Romania. They were also recognized in a great number of boreholes in the basement of the Pannonian Basin (Fig. 1.2).

Parts of the *ALCAPA Mega-unit* constitute the basement of the northwestern segment of the Pannonian Basin as a *continuation of the Austroalpine units* exposed in the Eastern Alps (Fig. 1.2). The Transdanubian Range Unit is considered as an Upper Austroalpine-type nappe. It is built of a Lower Paleozoic low-grade metamorphic complex, Permian fluvial sandstones, and Triassic shallow marine sedimentary formations, mostly shallow-marine carbonates. The mostly deep-marine Jurassic–Lower Cretaceous formations are overlain by continental to deep-marine sediments formed by tectonically controlled transgression-regression cycles during times between the Late Cretaceous and the Paleogene.

The *Mid-Hungarian Zone* (Fig. 1.2) contains strongly sheared, displaced elements of the South Alpine and Dinaridic origin. The Bükk Unit, exposed in the Bükk Mountains, northeast Hungary is considered as a part of the Mid-Hungarian Zone. It is composed of low-grade metamorphosed Upper Paleozoic to Jurassic sedimentary and volcanic rocks, overthrust by mélange with fragments of the Neotethys accretionary complex (Haas and Kovács 2001).

Large-scale *strike-slip movements* and coeval *opposed rotation* of the mega-units of the basement led to their juxtaposition during the Early Tertiary (Csontos et al. 1992;

Fodor et al. 1999; Márton and Fodor 2003). These motions were controlled by the *indentation of the Adria Microplate* and *rollback of the subducting slabs* along the European Plate margin, which led to the formation of a young basin system through crustal thinning beneath the area (Horváth et al. 2006). The phase that formed the Pannonian Basin was initiated by *attenuation of the crust*, leading to intense *volcanism* and significant but uneven *subsidence* during the Miocene. An andesite-dacite stratovolcanic chain, sub-parallel to the Carpathian arc was formed 17 to 12 million years ago. It was followed by intense subsidence and sedimentary upfill of the basins between 11.5 and 5 million years ago. Coevally, due to uplift of the Carpathian arc, the connection with the Black Sea had ceased to exist and a large lake came into being, the *Pannonian Lake* (Magyar et al. 1999). Parallel to the intense subsidence, basalt volcanism started in some parts of the Pannonian Basin about 7.5 million years ago. Sediments derived from the uplifting Alps and Carpathians gradually *filled up* the lake *through advancing deltas* (Juhász et al. 2007). By the Pliocene, the former lake a fluvial-lacustrine system had been replaced by extensive swamps and wetlands. Five million years ago an intense uplift of Transdanubia, the western part of the Danube-Tisza Interfluve, and of the present day mountains began, whereas the subsidence of the deep basins continued, giving rise to the deposition of thick fluvial sediments during the Pleistocene.

References

Berza T, Constantinescu E, Vlad S (1998) Upper Cretaceous magmatic series and associated mineralization in the Carpathian-Balkan fold belt. J Res Geol 48:291–306

Birkenmajer K (1977) Jurassic and Cretaceous lithostratigraphic units of the Pieniny Klippen Belt, Carpathians, Poland. Stud Geol Pol 45:1–159

Bleahu M, Mantea Gh, Bordea S, Panin Şt, Ştefănescu M, Šikić K, Haas J, Kovács S, Péró Cs, Bérczi-Makk A, Konrád Gy, Nagy E, Rálisch-Felgenhauer E, Török Á (1994) Triassic facies types, evolution and paleogeographic relations of the Tisza Megaunit. Acta Geol Hung 37:187–234

Csontos L, Nagymarosy A, Horváth F, Kovács M (1992) Tertiary evolution of the Intra-Carpathian area: a model. Tectonophysics 208:221–241

Dallmeyer RD, Neubauer F, Pană D, Fritz H (1994) Variscan vs. Alpine tectonothermal evolution within the Apuseni Mountains, Romania: evidence from $^{40}Ar/^{39}Ar$ mineral ages. Rom J Tech Reg Geol 75(2):65–76

Dimitrijević MD (1997) Geology of Yugoslavia. Geological Institute Gemini, Special Publication, Belgrade, p 187

Dragičević I, Velić I (2002) The northeastern margin of the Adriatic Carbonate Platform. Geol Croat 55:185–232

Ebner F, Vozárová A, Kovács S, Kräutner H-G, Kristić B, Szederkényi T, Jamičić D, Balen D, Belak M, Trajanova M (2009) Devonian-Carboniferous pre-flysch and flysch environments in the Circum Pannonian Region. Geol Carp 59:159–195

Fodor L, Csontos L, Bada G, Györfi I, Benkovics L (1999) Tertiary tectonic evolution of the Pannonian Basin system and neighbouring orogens: a new synthesis of paleostress data. In: Durand B, Jolivet L, Horváth F, Séranne M (eds) The Mediterranean Basins: Tertiary Extension within the Alpine orogen, Geological Society of London, Special Publication 156. The Geological Society, London, pp 295–334

Haas J, Kovács S (2001) The Dinaridic–Alpine connection – as seen from Hungary. Acta Geol Hung 44(2–3):345–362

Haas J, Péró Cs (2004) Mesozoic evolution of the Tisza Mega-unit. Int J Earth Sci (Geol Rundsch) 93:297–313

Haas J (ed), Hámor G, Jámbor Á, Kovács S, Nagymarosy A, Szederkényi T (2001) Geology of Hungary. Eötvös University Press, Budapest, 317 p

Haas J, Budai T, Csontos L, Fodor L, Konrád Gy (2010) Pre-Cenozoic geological map of Hungary, 1:500 000. Geological Institute of Hungary, Budapest

Horváth F, Bada G, Szafián P, Tari G, Ádám A, Cloeting S (2006) Formation and deformation of the Pannonian Basin: constraints from observational data. In: Gee DG, Stephenson R (eds) European lithosphere dynamics, Geological Society London, Memoir 32. The Geological Society, London, pp 191–206

Hrvatovic H, Pamić J (2005) Principle thrust-nappe structures of the Dinarides. Acta Geol Hung 48(2):133–151

Juhász Gy, Pogácsás Gy, Magyar I, Vakarcs G (2007) Tectonic versus climatic control on the evolution of fluvio-deltaic systems in a lake basin, Eastern Pannonian Basin. Sediment Geol 202:72–95

Karamata S (2006) The geological development of the Balkan Peninsula related to the approach, collision and compression of Gondwanan and Eurasian units. In: Robertson AHF, Mountrakis D (eds) Tectonic development of the Eastern Mediterranean region, Geological Society London, Special Publications 260. The Geological Society, London, pp 155–178

Kováč M, Plašienka D (2002) Geological structure of the Alpine-Carpathian-Pannonian junction. Comenius University, Bratislava, pp 1–84

Kovács S, Szederkényi T, Haas J, Buda Gy, Császár G, Nagymarosy A (2000) Tectonostratigraphic terranes in the pre-Neogene basement of the Hungarian part of the Pannonian area. Acta Geol Hung 43:225–328

Kovács S, Sudar M, Karamata S, Haas J, Péró Cs, Grădinaru E, Gawlick H-J, Gaetani M, Mello J, Polák M, Aljinicic D, Ogorelec B, Kollar-Jurkovsek T, Jurkovsek B, Buser S (2010) Triassic environments in the Circum-Pannonian region related to the initial Neotethyan rifting stage. In: Vozár J, Ebner F, Vozárová A, Haas J, Kovács S, Sudar M, Bielik M, Péró Cs (eds) Variscan and Alpine terranes of the Circum-Pannonian region. Slovak Academy of Sciences, Geological Institute, Bratislava, pp 87–156

Kräutner H-G (1997) Alpine and pre-Alpine terranes in the Romanian Carpathians and Apuseni Mts. In: Papanikolau D, Sassi FP (eds) IGCP Project 276: terrane maps and terrane descriptions. Ann Géol Pays Helléniques, 1e ser. 37:271–330

Kräutner H, Krstić B (2002) Alpine and pre-Alpine structural units within the Southern Carpathians and the Eastern Balkanids. In: 27th CBGA Congress, Bratislava, pp 1–8

Krstić B, Karamata S (1992) Terranes of the Eastern Serbian Carpatho-Balkanids. C.R.D.S. Soc Serbe Géol livre jubil (1891–1991), Beograd, pp 57–74

Magyar I, Geary DH, Müller P (1999) Paleogeographic evolution of the Late Miocene Lake Pannon in Central Europe. Palaeogeogr Palaeoclimatol Palaeoecol 147:151–167

Márton E, Fodor L (2003) Tertiary paleomagnetic results and structural analysis from the Transdanubian Range (Hungary): sign for rotational disintegration of the Alcapa unit. Tectonophysics 363:201–224

Mock R, Sýkora M, Aubrecht R, Ožvoldová L, Kronome B, Reichwalder P, Jablonský J (1998) Petrology and stratigraphy of the Meliaticum near the Meliata and Jaklovce villages, Slovakia. Slovak Geol Mag 4:223–260

Pamić J (2002) The Sava-Vardar Zone of the Dinarides and Hellenides versus the Vardar Ocean. Eclogae Geol Helv 95:99–113

Plašienka D, Michalik J, Kováč M, Gross P, Putiš M (1991) Paleotectonic evolution of the Malé Karpaty Mts. Geol Carp 42:195–208

Plašienka D, Grecula P, Putiš M, Hovorka D, Kováč M (1997) Evolution and structure of the Western Carpathians: an overview. In: Grecula P, Hovorka D, Putiš M (eds) Geological evolution of the Western Carpathians. Mineralia Slovaca Corp.-Geocomplex, a.s, Bratislava, pp 1–24

Rakús M, Potfaj M, Vozárová A (1998) Basic paleogeographic and paleotectonic units of the Western Carpathians. In: Rakús M (ed) Geodynamic development of the Western Carpathians. Geological Survey Slovak Republic, Bratislava, pp 15–26

Robertson A, Karamata S, Šarić K (2009) Overview of ophiolites and related units in the Late Paleozoic-Early Cenozoic magmatic and tectonic development of Tethys in the northern part of the Balkan region. Lithos 108:1–36

Săndulescu M (1984) Geotectonica Romaniei. Editura Technica, Bucuresti, p 336

Săndulescu M, Dimitrescu R (2004) Geological structure of the Romanian Carpathians. In: Field Trip Guide Book B12, 32nd International Geologial Congress, Florence, 20–28 Aug 2004, pp 1–48

Schmid SM, Bernoulli D, Fügenschuh B, Matenco L, Schuster R, Schefer S, Tischler M, Ustaszewski K (2008) The Alpine–Carpathian–Dinaridic orogenic system: correlation and evolution of tectonic units. Swiss J Geosci 101:139–183

Tilšjar J, Vlahović I, Velić I, Sokač B (2002) Carbonate platform megafacies of the Jurassic and Cretaceous deposits of the Karst Dinarides. Geol Croat 55:139–170

Tomek Č (1993) Deep crustal structure beneath the central and inner West Carpathians. Tectonophysics 226:417–431

Vozárová A, Ebner F, Kovács S, Krautner H-G, Szederkényi T, Kristic B, Sremac J, Aljinovic D, Novak M, Skaberne D (2009) Late Variscan (Carboniferous to Permian) environments in the Circum Pannonian Region. Geol Carp 60:71–104

Zagorchev I (1994) Structure and tectonic evolution of the Pirin–Pangaion Structural Zone (Rhodope Massif, South-Bulgaria and North-Greece). Geol J 29:241–268

Chapter 2
Climate

Tadeusz Niedźwiedź

Abstract This chapter provides a concise description of climatic conditions of the Carpatho–Balkan–Dinaric region. This mountain system covers an area of ca 1,000×1,100 km in Central and Southeastern Europe. Although the predominant part of the region belongs to the zone of temperate warm climates, the southwestern and southern parts have a subtropical (Mediterranean) climate. About 20 meteorological stations deliver data on the higher elevations of the mountains. According to M. Hess' classification, the mountain area is divided into seven vertical climatic zones, distinguished on the basis of mean annual temperature, which varies from 16°C in the southwest, along the Adriatic Sea coast, to −4°C on highest mountain summits. Data on the most important climatic parameters such as absolute maxima and minima of temperature, sunshine duration, precipitation, and snow cover duration are presented in tabulated form.

Keywords Mountain climate • Sunshine hours • Temperature • Precipitation • Snow cover • Vertical zonation

2.1 Introduction

Due to its large meridional extent of about 1,000 km from the latitude of 41°N near the southern boundaries of Macedonia and Bulgaria, to 50°N at the northern edge of the Carpathians in southern Poland, the climate of the Carpatho–Balkan–Dinaric

---

T. Niedźwiedź (✉)
Department of Climatology, Faculty of Earth Sciences, University of Silesia,
Bedzinska 60, 41-200 Sosnowiec, Poland
e-mail: tadeusz.niedzwiedz@us.edu.pl

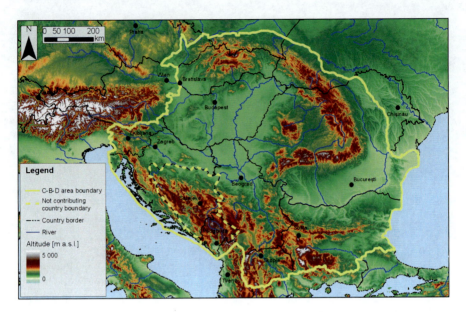

Fig. 2.1 Location of mountain meteorological stations in the Carpatho–Balkan–Dinaric region (Station numbers as listed in Table 2.1)

region, as conceived in the present volume, is very diverse. The longitudinal extent of the region is equally large (ca 1,100 km): from about 13.5°E on the western boundary of Slovenia (Kras Mountains) to 27°E on the margin of the Eastern Carpathians in Romania or even to the Dniester valley in Moldova (29°E). Consequently, solar irradiation and sunshine hours vary significantly within the region. Air temperature, precipitation, and snow cover duration are also modified by orographic factors such as elevation, mountain slope angle, and exposure (aspect). The greatest altitudinal difference occurs in the Balkan region (Musala, 2,925 m). In the Dinaric area the highest peak, Korabi, in the Macedonian Shar Planina reaches 2,764 m. In the Carpathian region the highest elevations occur in the High Tatras (Gerlachovský štít, 2,655 m) and in the Bucegi range in the Southern Carpathians (Moldoveanu, 2,543 m). According to C. Troll's (1973) criterion for the lower limit of a high mountain belt in Central Europe (above 1,600–1,700 m), the mountains in question are classified as high mountains (Barry 2008).

The climatic conditions in the highest parts of the region can be investigated due to the suitably dense coverage of meteorological stations, most of which are included in the synoptic network of the World Meteorological Organization. The geographical locations of the 20 mountain meteorological stations are presented in Fig. 2.1 and Table 2.1.

2 Climate

Table 2.1 List of mountain meteorological stations in the Carpatho–Balkan–Dinaric area

No	WMO No	Station	Latitude N	Longitude E	Elevation, m	Country	Mountain range
1	11787	Lysá Hora	49° 33'	18° 27'	1,324	Czech Republic	Silesian Beskidy, Western Carpathians
2	12650	Kasprowy Wierch	49° 14'	19° 59'	1,991	Poland	Tatra Mountains
3	11930	Lomnický štít	49° 12'	20° 13'	2,633	Slovakia	Tatra Mountains
4	11916	Chopok	48° 59'	19° 36'	2,008	Slovakia	Low Tatra Mountains
5	12851	Kékestető	47° 52'	20° 01'	1,014	Hungary	Matra Mountains
6	-----	Pozyzevskaja Polonina	48° 09'	24° 31'	1,429	Ukraine	Eastern Carpathians
7	15108	Cehlau Toaca	46° 56'	25° 55'	1,897	Romania	Eastern Carpathians
8	15280	Virful Omu	45° 27'	25° 27'	2,504	Romania	Bucegi Mountains, Southern Carpathians
9	14235	Puntijarka	45° 55'	15° 58'	988	Croatia	Medvednica
10	14324	Zavizan	44° 49'	14° 59'	1,598	Croatia	Velebit Mountains
11	14652	Bjelasnica	43° 43'	18° 16'	2,067	Bosnia and Herzegovina	Dinaric Mountains
12	13367	Zlatibor	43° 44'	19° 43'	1,029	Serbia	Dinaric Mountains
13	13289	Crni Vrh	44° 07'	21° 57'	1,037	Serbia	Homolskje Mountains
14	13378	Kopaonik	43° 17'	20° 48'	1,711	Serbia	Kopaonik Mountains
15	14584	Solunska Glava	41° 44'	21° 31'	2,540	Macedonia	Golesnica
16	15600	Mourgash	42° 50'	23° 40'	1,687	Bulgaria	Balkan Mountains (Stara Planina)
17	15627	Botev	42° 40'	24° 50'	2,384	Bulgaria	Balkan Mountains (Stara Planina)
18	15613	Cherni Vrah	42° 35'	23° 16'	2,286	Bulgaria	Vitosha Mountains
19	15615	Musala	42° 11'	23° 35'	2,925	Bulgaria	Rila Mountains Chernatica
20	14726	Rojen	41° 53'	24° 44'	1,754	Bulgaria	Mountains, Western Rhodopes

2.2 The Carpatho–Balkan–Dinaric Region in Climatic Regionalization

In the well-known Köppen–Geiger climate typology, regions are divided between two types of climate: *Df* – cold temperate without a dry season and *Cs* – warm temperate climate with a dry summer, often called Mediterranean climate (Martyn 1992).

According to the more detailed climate classification of W. Okołowicz with additions by D. Martyn (1992), the predominant part of the Carpatho–Balkan–Dinaric region belongs to the *IV-A* zone of *temperate warm climates*. In the Western Carpathians and its northern surroundings, subtype 15, *transitional* between the *maritime and continental* climates is typical, with mean January temperatures between 0°C and −10°C and summer (June to August, JJA) temperatures from 15°C to 20°C. Adjacent to the Eastern Carpathians, on the territories of Ukraine, Moldova, and Romania, subtype 17, a *continental* climate with average summer temperatures ≥20°C and annual temperature range >25 K is characteristic. The Romanian (Walachian or Lower Danube Plain) between the Southern Carpathians and the Balkan Mountains represents subtype 19, i.e., *dry continental* climate. The internal part of the Carpathian arc, the Transylvanian and Pannonian Basins belongs to subtype 16, *warm intermediate climate*.

The southwestern and southern part of the Carpatho–Balkan–Dinaric region belongs to *zone III* of subtropical climate, with the mean temperature of the coolest month ≥10°C. Part of the mountains stretching along the Adriatic Sea coast represent subtype 9, *intermediate subtropical* climate. Most of southern Bulgaria belongs to subtype 10, *continental subtropical* climate.

2.3 Sunshine Hours and Solar Radiation

The annual duration of sunshine varies from less than 1,600 h on the northern side of the Western Carpathians (Kasprowy Wierch: 1,453 h, Kraków: 1,522 h) to ca 2,600 h in the southwestern part of the region (Dubrovnik 2,598 h, Split 2,658 h) (for details see Tables 2.3, 2.4, 2.5 and 2.6). In the Western Carpathians the annual totals of global solar radiation range from about 3,600 MJ m^{-2} at the foot of the mountains to 4,100 MJ m^{-2} at the top of the Tatras ridge (Olecki 1989). Annual solar irradiation reaches 4,400 MJ m^{-2} on the southern side of the Tatras but only 3,500 MJ m^{-2} in the north (Tomlain 1979). In the Eastern Carpathian annual global radiation decreases with altitude from 4,200 MJ m^{-2} in the foreland to ca 3,940 MJ m^{-2} at the mountain ridges (Konstantinov and Gojsa 1966). Relatively high solar radiation income is observed in the Transylvanian Basin and in Moldova (4,600–4,800 MJ m^{-2}). In Bucharest, however, annual radiation totals reach 5,270 MJ m^{-2} (Stoenescu and Tistea 1962). In northern Voivodina the annual total of solar irradiation is ca 4,606 MJ m^{-2} (Martyn 1992), but it increases to more than 5,400 MJ m^{-2} in the southernmost and southeastern parts of Bulgaria.

2.4 Air Temperature and Vertical Climate Zonation

In mountainous regions air temperature is the most important climatic element, which strongly depends on elevation (Hess 1965; Niedźwiedź 1992; Barry 2008). For example, on the northern slopes of the Western Carpathians mean annual temperature ranges from ca 8°C at 200 m to −4°C on the highest peaks of the Tatra ridge (Table 2.3). In Europe a similar temperature range is observed horizontally between Kraków (latitude 50°N) and the islands of Svalbard in the Arctic (77°N). In the Eastern Carpathians (Buchinskij et al. 1971; Niedźwiedź 1987) mean annual temperature varies from 9.6°C in Uzhhorod (at 115 m elevation) on the southwestern side and 7.0°C in Lviv (325 m) on the northeastern side to about 0°C in the Charnohora Massif (Hoverla, 2,061 m).

On the south-exposed slopes of the Southern Carpathians (Table 2.4) mean annual temperature ranges from 11.1°C in Bucharest (at 82 m) to −2.5°C on Virful Omu (2,504 m). In the Balkan lowlands (Table 2.5) temperature reaches ca 12°C and higher (Plovdiv: 12.1°C), whereas on the mountain ridges above 2,000 m it falls below zero and in the Rila Mountains (on Musala, 2,925 m) down to −2.7°C. The situation is more complicated in the Dinaric area (Table 2.6). On the Adriatic Sea shore mean annual temperature varies from 14.4°C in Rijeka to 16.2°C in Dubrovnik (Furlan 1977), while in the basins and lowlands inside the more continental part of the region it is ca 1.2°C. On the peak of Bjelasnica (2,067 m) annual temperature drops to −1.2°C and below 0°C above 1,940 m elevation.

Based on the significant relationship between altitude and mean annual temperature, M. Hess (1965, 1971) distinguished seven vertical climatic zones in the Carpathians (Table 2.2). He also found a significant connection between the climatic zones and vegetation belts. The most evident is the conformity of the upper tree line (timberline) with the +2°C annual isotherm (10°C July isotherm) and the climatic snow line with the −2°C annual isotherm. W. Dobiński (2005) found that climatic conditions sufficient for permafrost development occur at 1,930 m elevation in the Tatra Mountains, above 2,000 m in the Southern Carpathians, and above 2,300 m in the Balkans. Compared to the Western Carpathians, in the Southern Carpathians the upper limit of the cool zone (isotherm 2°C) is about 200 m higher (Hess 1971). In the Balkan (Stara Planina) and Dinaric Mountains the isotherm runs at ca 500 m and in the Rhodopes even 600 m higher than in the Western Carpathians (Table 2.2).

The vertical gradients of annual temperature vary from −0.50 K 100 m^{-1} in the Western Carpathians to −0.62 K 100 m^{-1} in Dinaric Mountains. However, there are seasonal differences in temperature lapse rate from −0.4 K 100 m^{-1} in winter to −0.7 K 100 m^{-1} in summer in the northern region to −0.8 K 100 m^{-1} in the southern part of the Balkan and Dinaric area (Furlan 1977). A significant decrease with altitude is observed in the absolute maximum of temperature and annual temperature range (Tables 2.3, 2.4, 2.5 and 2.6). Absolute minima reach the lowest values in concave relief forms (basins), independent of altitude (Sarajevo: −23.4°C, Belgrade: 25.5°C, Sofia: −27.5°C). In the northeastern basins and lowlands minimum temperature may drop even below −30°C (Sibiu: −34.2°C, Hurbanovo: −35.0°C), while on the mountain slopes it is higher as a result of temperature inversion.

Table 2.2 Vertical climatic zones in the Carpatho–Balkan–Dinaric area (After Hess 1965)

Vertical climatic zones	Mean annual temperature	Elevation above sea level, m					
		Northern Carpathians	Southern Carpathians	Eastern Carpathians	Balkan Mountains	Rhodopes	Dinaric Mountains
Cold	<−2°C	>2,250	>2,430	–	–	>2,800	–
Moderately cold	−2°C (climatic snow line)–0°C	2,250–1,850	2,430–2,050	>1,850	>2,250	2,800–2,450	>2,250
Very cool	0–2°C	1,850–1,450	2,050–1,670	1,850–1,550	2,250–1,930	2,450–2,050	2,250–1,940
Cool	2°C (upper tree line)–4°C	1,450–1,050	1,670–1,300	1,550–1,200	1,930–1,600	2,050–1,700	1,940–1,600
Moderately cool	4–6°C	1,050–650	1,300–920	1,200–850	1,600–1,250	1,700–1,350	1,600–1,270
Moderately warm	6–8°C	<650	920–540	850–500	1,250–930	1,350–980	1,270–950
Warm	>8°C	–	<540	<500	<930	<980	<950

Data for the Northern and Southern Carpathians after M. Hess (1965, 1971)

2 Climate

Table 2.3 Values of selected climatic parameters in the Western and Eastern Carpathians

Station	Elevation, m	T_a	T_c	T_w	ATR	T_{mxa}	T_{mna}	SD h	PRA mm	$d_{1.0}$ days	SCD days
Lomnicky štít	2,633	−3.7	−11.6	4.2	15.8	17.8	−30.5	2,071	1,645	174	236
Kasprowy Wierch	1,991	−0.8	−8.9	7.3	16.2	23.0	−30.2	1,453	1,721	188	223
Pozyzevskaja Polonina	1,429	2.7	–	–	–	–	–	–	1,501	–	–
Stary Smokovec	1,018	4.7	−5.6	14.4	20.0	33.1	−27.0	1,817	946	133	130
Kékestető	1,010	5.4	−5.4	15.4	20.8	30.5	−22.6	1,992	891	107	113
Zakopane	844	4.9	−5.8	14.9	20.7	32.3	−34.1	1,528	1,124	140	117
Ocna Şugatag	498	7.9	−3.8	18.1	21.9	38.5	−30.5	–	761	112	61
Oravsky Podzamok	493	6.5	−4.7	16.4	21.1	36.0	−34.0	1,463	818	125	90
Lviv	325	7.0	−5.3	18.2	23.5	37.0	−34.0	1,792	655	119	80
Kraków	213	8.6	−2.9	19.3	22.2	35.7	−26.6	1,522	645	107	62
Košice	206	8.7	−3.6	19.6	23.2	37.4	−29.8	2,088	636	95	51
Bratislava	133	9.6	−1.6	20.1	21.7	38.6	−31.8	2,168	657	97	39
Debrecen	123	10.3	−2.7	21.8	24.5	39.2	−30.2	2,030	584	96	33
Budapest	120	11.2	−1.1	22.2	23.3	39.5	−23.4	1,988	630	90	40
Hurbanovo	115	9.9	−2.1	20.5	22.6	39.0	−35.0	2,126	574	88	38
Uzhhorod	115	9.6	−2.8	19.9	22.7	–	–	1,952	740	108	–

Sources of data: Konček (1974), Okołowicz (1977), Baucus and Kalb (1982), Niedźwiedź (1992), and WMO (1996)

Explanations of abbreviations: T_a mean annual temperature, °C; T_c mean temperature of the coldest month, °C; T_w mean temperature of the warmest month, °C; ATR annual temperature range ($T_w - T_c$), K; T_{mxa} absolute maximum of temperature, °C; T_{mna} absolute minimum temperature, °C; SD annual sunshine duration, h; PRA annual precipitation, mm; $d_{1.0}$ annual number of days with precipitation ≥1.0 mm; SCD snow cover duration, days

Table 2.4 Values of selected climatic parameters in the Southern Carpathians

Station	Elevation, m	T_a	T_c	T_w	ATR	T_{mxa}	T_{mna}	SD h	PRA	$d_{1.0}$	SCD
Virful Omu	2,504	−2.5	−10.5	5.3	15.8	8.2	−25.6	1,600	1,053	141	–
Predeal	1,093	4.9	−5.4	14.6	20.0	31.7	−33.8	1,648	938	125	126
Gheorghieni	815	5.5	−6.9	16.5	23.4	35.0	−38.0	1,890	626	96	76
Petroşeni	607	7.5	−3.6	17.3	20.9	35.8	−29.9	1,692	739	108	49
Sibiu	416	8.9	−4.0	19.8	23.8	37.4	−34.2	1,950	651	95	45
Tigru Mare	309	9.1	−4.7	19.5	24.2	39.0	−32.8	1,981	601	93	54
Iaşi	101	9.4	−4.1	21.6	25.7	40.0	−30.6	2,015	498	75	55
Bucharest	82	11.1	−2.7	23.3	26.0	41.1	−30.0	2,258	578	76	50

Sources of data: Furlan (1977), Baucus and Kalb (1982), and WMO (1996)
For abbreviations see Table 2.3

Table 2.5 Values of selected climatic parameters in the Balkanic area

Station	Elevation, m	T_a	T_c	T_w	ATR	T_{mxa}	T_{mna}	SD h	PRA	$d_{1.0}$	SCD
Musala	2,925	−2.7	−11.0	5.8	16.8	–	–	–	–	–	200
Botev	2,384	−0.7	−8.9	7.9	16.8	–	–	–	–	–	–
Sofia	550	10.4	−1.7	21.3	23.0	37.5	−27.5	2,082	622	–	47
Sliven	235	12.2	1.0	23.3	22.3	40.8	−18.4	2,324	565	76	10
Plovdiv	161	12.1	−0.4	23.4	23.8	41.9	−31.5	2,247	539	74	21
Pleven	109	11.6	−1.8	23.5	25.3	43.4	−28.3	2,114	556	79	35
Varna	35	12.1	1.2	22.9	24.1	41.4	−24.3	2,105	474	–	8

Sources of data: Furlan (1977), Baucus and Kalb (1982), and WMO 1996
For abbreviations see Table 2.3

Table 2.6 Values of selected climatic parameters in the Dinaric area

Station	Elevation, m	T_a	T_c	T_w	ATR	T_{mxa}	T_{mna}	SD h	PRA	$d_{1.0}$	SCD
Kredarica (Julian Alps)	2,514	−1.7	−9.2	6.0	15.2	18.8	−28.0	1,732	2,143	154	256
Bjelasnica	2,067	1.2	−7.4	10.2	17.6	25.3	−29.6	1,618	1,245	135	174
Bitola	586	11.6	−0.1	22.8	22.9	40.5	29.4	2,378	640	82	31
Sarajevo	537	9.8	−1.4	19.7	21.1	38.1	−23.4	1,818	925	113	60
Ljubljana	299	9.6	−1.6	19.8	21.4	38.8	−25.6	1,687	1,387	116	48
Skopje	240	12.4	−0.1	24.0	24.1	41.2	−23.9	2,140	500	74	22
Niš	202	11.8	−0.5	22.7	23.2	42.2	−23.7	2,000	555	87	31
Zagreb	157	11.6	0.2	22.0	21.8	40.3	−21.7	1,808	864	103	34
Belgrade	132	11.8	−0.2	22.6	22.8	39.4	−25.5	2,049	701	96	37
Split	128	16.1	7.8	25.6	17.8	38.6	−9.0	2,658	816	86	–
Mostar	99	15.1	4.8	25.7	20.9	43.0	−11.1	2,324	1,387	102	5
Podgorica	52	15.4	5.6	26.4	20.8	41.2	−9.7	2,467	1,632	108	6
Dubrovnik	49	16.2	8.6	24.6	16.0	37.2	−7.0	2,598	1,301	98	–

Sources of data: Furlan (1977), Hänle and Kalb (1981), and WMO (1996)
For abbreviations see Table 2.3

2.5 Precipitation and Snow Cover

Precipitation conditions in the Carpatho–Balkan–Dinaric region are characterized by strong variability. Higher precipitation in the mountain chains than in the surrounding lowlands is connected with more frequent and thicker cloud cover (Barry 2008) and strong orographic convection. A good example of thermal convection above the Carpathian ranges is presented in Fig. 2.2. The slopes of the southwestern Dinaric Mountains in Montenegro near the Kotor Bay (Crkvice station, 1,050 m) are exceptional due to the mean annual precipitation, which amounts to more than

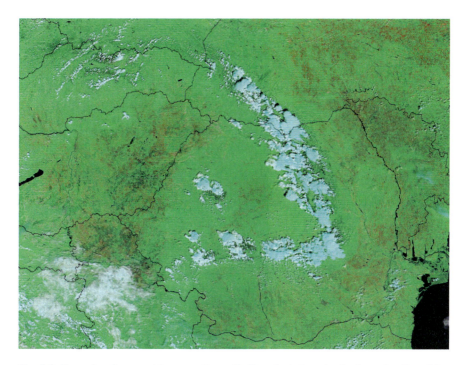

Fig. 2.2 Example of orographic convection with *Cumulonimbus* clouds above the Carpathian mountain chain. Satellite Terra picture from 24 May 2011 09:35 UTC (Source: Modis Rapid Response System, NASA. Picture available online at http://rapidfire.sci.gsfc.nasa.gov/subsets/?subset=AERONET_CLUJ_UBB.2011144)

4,600 mm (Furlan 1977). This station records the highest annual precipitation in Europe. Generally the southwestern slopes of the Dinaric Mountains receive more than 2,000 mm of precipitation annually. Lying in rain-shadow, mountain basins receive much less precipitation, for example, in the Niš Basin the annual total is ca 600 mm (Furlan 1977) or even lower (Niš: 555 mm; Table 2.5). On the top of Bjelasnica (2,067 m) annual precipitation exceeds 1,245 mm. In the highest parts of the Rhodopes, Balkans, and Southern Carpathians annual precipitation is ca 1,500 mm (Furlan 1977). On the northern side of the Tatras range it is 1,700–1,800 mm (Niedźwiedź 1992). The Pannonian Basin (Hurbanovo: 574 mm, Debrecen: 584 mm) and Transylvania (Cluj: 538 mm) are relatively dry.

The number of days with precipitation ≥1.0 mm changes with elevation from about 60–80 days in the lowlands between the Southern Carpathians and the Balkans to 170–180 days on the summits of the Tatras. On the northern slopes of the mountain chain the heaviest summer rainfall is associated with northern or northeastern cyclonic situations. At Hala Gąsienicowa station (1,520 m) on 30 June 1973, the 24-hour rainfall total reached 300 mm, the highest value ever recorded in the Western Carpathians (Niedźwiedź 1992). In the Southern Carpathians (Ciupercenii Vechii in eastern Romania), 24-hour precipitation was as high as 349 mm (Furlan 1977). An extreme daily rainfall of 480 mm was recorded on the mountain slopes at Crkvice, near the Dalmatian coast of Montenegro, on 21 November 1927 (Martyn 1992). D. Furlan (1977) reported 550 mm for the same station.

Prolonged winter snow cover is a characteristic feature of the mountain regions of Central and Southeastern Europe. In the south the number of days with snow cover increases from about 5–6 days near the Dalmatian coast to 30–50 days in the interior. In the Dinaric Mountains it increases to more than 100 days above the altitude of 1,300 m and attains 174 days on the top of Bjelasnica. On Musala in Bulgaria snow cover lasts about 200 days, and longer than 220 days in the Tatra Mountains (Table 2.3). There are localities in the Pannonian Basin where the number of days with snow cover is lower than 30 days, while in Wallachia and Bukovina it varies between 40 and 55 days (Furlan 1977).

References

Barry RG (2008) Mountain weather and climate. Cambridge University Press, Cambridge, 506 p
Baucus M, Kalb M (eds) (1982) Klimadaten von Europa, Teil III: Südost- und Osteuropa. Deutscher Wetterdienst, Offenbach am Main, A93 p
Buchinskij IO, Volevacha MM, Korzov VO (1971) Klimat Ukrainskich Karpat (Climate of the Ukrainian Carpathians). Naukova Dumka, Kijiv (in Russian)
Dobiński W (2005) Permafrost of the Carpathian and Balkan Mountains, eastern and southern Europe. Permafr Periglac Process 16(4):395–398
Furlan D (1977) The climate of Southeast Europe. In: Wallen CC (ed) Climates of central and southern Europe. Elsevier, Amsterdam/Oxford/New York, pp 185–235 (World Survey of Climatology 6)
Hänle B, Kalb M (eds) (1981) Klimadaten von Europa II: Südwesteuropa und Mittelmeerländer. Deutscher Wetterdienst, Offenbach am Main, A176 p
Hess M (1965) Piętra klimatyczne w polskich Karpatach Zachodnich (Vertical climatic zones in the Polish Western Carpathians). Zeszyty Naukowe UJ, Prace Geograficzne 11:1–267 (in Polish)
Hess M (1971) Piętra klimatyczne w Kawrpatach Północnych i Południowych i ich charakterystyka termiczna (Vertical climatic zones in north and south Carpathians and their thermic characteristics). Folia Geogr Ser Geogr Phy Kraków 5:15–23 (in Polish)
Konček M (ed) (1974) Klima Tatier (Climate of the Tatras). Veda, Bratislava, 856 p (in Slovak)
Konstantinov AR, Gojsa NJ (1966) Atlas sostavljajuschich teplovogo i vodnogo balansa Ukrainy (Atlas of heat and water balance components in Ukraine). Gidrometeorologicheskoje Izdatelstvo, Leningrad (in Russian)
Martyn D (1992) Climates of the world. PWN/Elsevier, Warszawa/Amsterdam/Oxford/New York/Tokyo, 435 p
Niedźwiedź T (1987) Main climate forming factors in the Carpathians. In: Proceedings, 13th international conference on Carpathian Meteorology, Busteni, Romania, pp 307–316
Niedźwiedź T (1992) Climate of the Tatra mountains. Mt Res Dev 12(2):131–146
Okołowicz W (1977) The climate of Poland, Czechoslovakia and Hungary. In: Wallen CC (ed) Climates of central and southern Europe. Elsevier, Amsterdam/Oxford/New York, pp 75–124 (World Survey of Climatology 6)
Olecki Z (1989) Bilans promieniowania słonecznego w dorzeczu górnej Wisły (Solar radiation balance in the upper Vistula basin). Rozprawy Habilitacyjne UJ, Kraków 157:1–126 (in Polish)

Stoenescu SM, Tistea D (eds) (1962) Clima Republici populare Romine (Climate of the People's Republic of Romania) I–II. Institutul Meteorologic, București (in Romanian)

Tomlain J (1979) K raspredelenii summarnoj radiacii v gornych rajonach CSSR (On the distribution of global solar radiation in the mountainous regions of Czechoslovakia). In: 7th Mezhdunarodnaja Konferencija po Meteorologii Karpat, Bratislava (in Russian)

Troll C (1973) High mountain belts between the polar caps and the equator: their definition and lower limits. Arct Alp Res 5(3,2):A19–A27

WMO (1996) Climatological normals (CLINO) for the period 1961–1990, WMO publications 847. World Meteorological Organization, Geneva, 768 p

Chapter 3
Rivers

Pavol Miklánek

Abstract The main hydrological axis of the Carpatho–Balkan–Dinaric region is the Danube River. The region consists of extensive drainage basins along the whole length of the river, except for the Upper Danube. The Danube enters the Carpathian (Pannonian) Basin at the Devín Gate, where its middle section begins, and leaves it at the Iron Gate gorge, where the Lower Danube starts. The river carries considerable quantities of solid particles (mostly quartz grains). The damming of the Danube and its tributaries have significantly changed sediment transport. Some peripheral areas of the region are drained by rivers flowing to the Baltic, Black (through the Dniester River), Aegean, and Adriatic Seas.

Keywords Danube Basin • Tributaries • Water discharge • Sediment transport • Channel conditions • Other river catchments

3.1 Introduction

The Carpatho–Balkan–Dinaric (CBD) region is composed of a series of drainage basins. The *main axis* of the region is the Danube River, and the overwhelming part of the region belongs to the system of the Danube. Some peripheral parts of the CBD region, along its boundaries, are not part of the Danube river basin (the Ostrava region in the Czech Republic, Southern Poland, the eastern part of the CBD region in Ukraine, Southern Bulgaria, Macedonia, Istria, and Dalmacia).

P. Miklánek (✉)
Institute of Hydrology, Slovak Academy of Sciences, Račianska 75,
831 02, Bratislava 3, Slovakia
e-mail: miklanek@uh.savba.sk

3.2 The Danube Drainage System

The Danube drainage system is the second largest in Europe and the world's most international river basin, as it extends over the territories of 19 countries. According to the ICPDR (2011) the Danube is 2,857 km long, with a total drainage area of 801,463 km². The channel is locally up to 1.5 km wide and 10 m deep. Other authors mention that the total area of the basin is 817,000 km² (Stančík and Jovanovič 1988 and many others). Of the total river channel length only about 1,800 km belong to the CBD region.

3.2.1 The Danube Sections, Tributaries, and Channel Environments

Based on the geology and geomorphology of the catchment, the river is generally divided into Upper, Middle, and Lower Danube (Sommerhäuser et al. 2003).

The *Upper Danube* basin covers the area from the source tributaries in the Black Forest down to the Devín Gate near Bratislava. With the exception of the catchment of a left-bank tributary, the Morava River, the Upper Danube is not part of the CBD region. The Morava confluence is at river km 1,880 (measured from the Black Sea).

The *Middle Danube* Basin comprises the largest portion of the drainage area. It includes the Danube section between the Devín Gate (Austria/Slovakia) and the Iron Gate barrages (Serbia/Romania). The basin is bordered by the Carpathians in the north and east and parts of the Dinaric mountain range in the west and south. The major tributaries in this region are (from the left) the Váh, Hron, Ipeľ/Ipoly, and Tisa/Tisza, and from the right the Rába, Sió, Drava, Sava, and Velika Morava (Table 3.1). First in the Little Hungarian/Slovak basin, then in the Great Plain, below the Danube Bend gorge, the river built immense composite Pleistocene alluvial fans and gradually filled the depressions of the Pannonian (Carpathian) Basin (Lóczy 2007). In the terraced valley east of Győr the Danube flows in a braided channel with numerous shoals built of the Váh and Hron deposits. South of Budapest it receives no major tributary for a more than 250-km long section. On the right bank the high undercut loess bluffs of the Mezőföld contrast the flat alluvial left bank.

The next and most important gorge is the 117-km long Iron Gate, the lower end of this river section. The Iron Gate (Portile de Fier) Gorges (Orghidan 1966) are separated by an embayment into a small and a large gorge with 250–300 m high walls of Jurassic limestone on both sides. The mean water level after damming is at 77 m altitude, but the deepest points of large potholes in the rocky riverbed are just 7–8 m above sea level. The ongoing uplift of the Carpathian Mountains during the Pleistocene and Holocene resulted in intense erosion of the valley, which is less than 200 m wide here.

Table 3.1 Major tributaries of the Danube river in the CBD region and their basic parameters (mean values for the period 1931–1970 (After Stančík and Jovanovič 1988)

River	Catchment area (km²)	Mean discharge (m³ s⁻¹)	Specific yield (L s⁻¹ km⁻²)	Runoff depth (mm)
Sava	94,778	1,545	16.3	514
Tisa	158,182	888	5.6	177
Drava	41,810	577	13.8	435
Velika Morava	38,233	262	6.8	216
Siret	45,420	226	5.0	157
Olt	24,810	184	7.4	234
Váh	9,714	128	13.2	415
Morava	27,633	124	4.5	141
Jiu	10,713	94	8.8	278
Prut	28,945	88	3.0	96
Argeş	11,814	71	6.0	190
Rába	14,702	61	4.1	130
Iskar	7,811	57	7.3	230
Ialomiţa	10,305	48	4.6	146
Sió	15,129	39	2.6	81
Whole Danube basin	817,000	6,857	8.4	264
Middle and Lower Danube basin[a]	710,734	4,930	6.9	219

[a]Middle and Lower Danube basin – area of the Danube River basin from Devín Gate to delta including the Morava river basin and excluding the Danube Delta

The *Lower Danube* basin is formed by the Romanian-Bulgarian lowland and its plateaus and mountains. It is bordered by the Southern Carpathians (north), the Bessarabian upland plateau (east), and by the Dobrogea Massif and the Balkan Mountains (south). The important *tributaries* here are the Timok, Iskar, and Jantra Rivers from the south, and the Jiu, Olt, Arges, Ialomiţa, Siret, and Prut Rivers from the north (Table 3.1). In the Lower Danube, the river course is more stable. The tributaries from the mountains deposited their bedload material directly in alluvial fans at the mountain footslopes and only small amounts of sediments, mainly suspended load, reached the terraced Danube valley. During the Holocene the Lower Danube valley has been affected by tectonic activities and aeolian action. The thickness of the aeolian sands and the heights of dunes decrease progressively eastwards (Ghenea and Mihailescu 1991). The Lower Danube flows along the southern margin of the Romanian Plain in an asymmetric valley as a braided sand channel 300–1,600 m wide at low water stages. Navigation is hindered by several fords, the shallowest being 0.9 m deep. There are 97 islands of various size along this accumulating section. The marshy Holocene floodplain is 8–12 km wide along the western section and broadens to 20 km at Câlâraşi, dotted by numerous oxbow lakes.

Most significant *changes in course* affected the Olt River, which originally had been a tributary to the Mureș/Maros River, but at present it is a tributary of the Tisza. It was captured by a minor direct Danube tributary that incised deeply into the range of the Southern Carpathians and diverted the Olt towards the Danube (Sommerwerk et al. 2009). Thus, the Transylvanian Basin became divided between the Tisza and Danube basins.

The Danube first entered the sea south of the Dobrogea region, but due to a tectonic uplift of this area in the second half of the Pleistocene, it was forced to follow the northern margins of the Dobrogea (Sommerwerk et al. 2009). The shifts of the *river mouth* have been adjusted to the water level of the Black Sea, which fluctuated by 70–80 m.

Three stages in the evolution of the *Danube Delta* have been identified: a gulf, a lagoon, and a delta phase. Optical luminescence analyses suggest that this evolution happened within a relatively short time interval (over 5,200 years according to Giosan et al. 2006) – at least compared to the previous estimations of 8,500–9,000 years.

Divisions of the Danube channel based on the *channel slope* were identified by Lászlóffy (1965) and resulted in six sections:

1. The *mountain section* (river km 2,780–2,497) from the confluence of the source rivers, Brigach and Breg, down to the confluence of Lech River (average slope is 101 cm km^{-1}). This section is not included in the CBD region.
2. The *Upper Danube* (river km 2,497–1,794) from the confluence of the Lech River to the rejoining of the Moson Danube at Gönyű (average slope is 40 cm km^{-1}). This section is partially included in the CBD region starting from the river km 1,880 at the confluence of the Danube and Morava Rivers at Devín.
3. The *Middle Danube* (river km 1,794–1,048, the longest section in the CBD region) from the confluence of river Rába to the cataract at the Iron Gate (average slope is 6 cm km^{-1}).
4. The *cataract reach* (river km 1,040–941) on the Serbian–Romanian border, on a 100 km stretch there is an altitude difference of 28 m (average slope is 28 cm km^{-1}). This section is also fully included in the CBD region.
5. The *Lower Danube* (river km 941–80) from the Romanian/Walachian Lowland to the Danube delta (average slope is 3.9 cm km^{-1}) is also part of the CBD region as interpreted here.
6. The *Danube Delta* (below river km 80), with a minimum average slope (a few millimeters per kilometer), lies right on the margin of the CBD region.

3.2.2 Water and Sediment Discharge

With the spatial and seasonal distribution of precipitation, the surface runoff and discharge regime varies for the various parts of the drainage basins of Danube and its main tributaries (Sommerwerk et al. 2009). The territories of Austria and Romania contribute most to the water discharge of the Danube (22% and 18% of total flow, respectively), reflecting the abundant precipitation in the Alps and Carpathians. The average annual specific discharge decreases downstream from 25 to 35 L s^{-1} km^{-2} in the Alpine headwaters to 19 L s^{-1} km^{-2} for the Sava, 6.3 L s^{-1} km^{-2} for the Tisza, and to 2.8 L s^{-1} km^{-2} for the rivers draining the eastern slopes of the Carpathians (Belz et al. 2004). There is only a moderate growth of discharge in the northern sector of the Carpathian Basin (at Bratislava: 2,080 m^3 s^{-1}; at Budapest: 2,340 m^3 s^{-1}), while more to the south, with the expansion of the drainage area, the Drava, the Tisza, and particularly the Sava substantially increase discharge (at Orşova: 5,600 m^3 s^{-1}). Water transport along the Romanian–Bulgarian border only shows a minimum growth (at Galaţi: 6,430 m^3 s^{-1}) (Lóczy 2007).

The difference between annual minimum and maximum runoff volume is increasing towards the mouth. At Bratislava the mean annual runoff volume (65 km^3) fluctuates between 48 km^3 in 1972 and 92 km^3 in 1965 (Pekárová et al. 2008). At the upstream end of the Danube Delta the mean annual runoff volume (203.7 km^3) fluctuates between the minimum value of 134 km^3 in 1990 and maximum of 297.1 km^3 in 1941 (Sommerwerk et al. 2009).

The river carries considerable amounts of solid *load*, nearly all of which are quartz grains. The constant shift of deposits in different parts of the riverbed forms shoals along most channel sections. Before barrage constructions, more coarse bedload came from the Western Carpathians through the major left-bank tributaries, the Váh and the Hron, while a major right-bank tributary, the Rába, only transports suspended load. The three largest tributaries downstream are quite different as far as water and sediment discharges are concerned. Along its upper reaches the Drava carries large amounts of bedload, but gravels and coarse sands are deposited in the lowlands and only some sand and suspended silt and clay reach the Danube. The longest tributary, the ("blonde", i.e., light brown colored) Tisza transports large amounts of suspended fine silt from the Eastern Carpathians, while the Sava brings walnut-sized limestone and ophiolite boulders from the Dinaric Mountains. An even greater contrast is found between the sediment loads of the next left-bank tributary the Timiş and a right-bank one, the Velika Morava since the former transports very fine sands and silts, while the latter large (>100 mm diameter) boulders of volcanic rocks (Bendeffy 1979).

Along the Lower Danube downstream the Iron Gate large-scale *deposition* and an accumulating river mechanism is observed. The tributaries in the Romanian Plain show meandering and braided patterns. The damming of the river has changed the proportions of sediment transport and deposition. Large *water management* schemes along the tributaries (the Olt, Argeş, Siret, and Prut Rivers) reduced the suspended sediment load of the Danube from 40 million tons year^{-1} before dam constructions to 7.3 million tons year^{-1} of today. The water stored in reservoirs generally loses its silt load, and the relatively silt-free outflow erodes bed and banks farther downstream.

3.3 Other Catchments in the Carpatho–Balkan–Dinaric Region

The areas of the CBD region that are not part of the Danube basin belong to the following drainage basins:

- The upper Odra River in the Czech Republic and Poland;
- Upper Vistula/Wisła River in Poland;
- Upper Dniester river in Ukraine;
- The Marica and Struma Rivers in Bulgaria and the Vardar River in Macedonia.

Information on the major parameters of major rivers in the CBD region (except the Danube) is summarized in Table 3.2. The data allow basic comparisons between the hydrological characters of the different peripheral parts of the CBD region. The best parameter to quantify water available for feeding river discharge in the basins is specific yield (runoff from unit area of the river basin), which also indicates the impact of hydrological processes on the intensity of erosion and exogenous processes.

The *Odra/Oder River* rises in the Czech Republic. Its upper section in the Czech Republic, above the gauging station at Bohumín, belongs to the CBD region. The CBD region also includes the border river Olše/Olza. It drains the Beskids on the right side of the basin and the Oderské vrchy highland on the left.

The *Vistula* is the longest and the most important river in Poland. Of its catchment area of 194,424 km², ca 40,000 km² lies within the CBD region. The Vistula has its source in the Silesian Beskids (Western Carpathians in southern Poland). It crosses the vast Polish plains towards the Baltic Sea, leaving the CBD region at the city of Kraków. The main right-bank tributaries within the CBD region collect water from the Beskid Maly and Beskid Wyspowy Mountains (Sola, Skawa, Raba, Dunajec), the High Tatras, the Beskid Sadecki (Dunajec, Poprad), and the Beskid Niski Mountains (Wisłoka, San).

The *Dniester River* flows on the boundary of the CBD region in Ukraine. It drains the Eastern Carpathians with the main right-hand tributaries Stryj, Lomnytsya, and Bystrytsya.

The *Marica River* springs in the Rila Mountains in Western Bulgaria, flows southeast between the Balkan and Rhodope Mountains, and then turns to the south to form its delta in the Aegean Sea. The two main tributaries, the Tundzha and Arda, join the Marica River downstream the gauging station Svilengrad. The Tundzha River collects water from the southern slopes of the Stara planina mountains, while the Arda mainly drains the Rhodope Mountains.

The *Struma River* has its source in the Vitosha Mountain in Bulgaria, runs first westward, then southward, and flows into the Aegean Sea in Greece. The tributaries include small streams like the Rila River, the Blagoevgradska Bystrica, the Konska River, and the Sandanska Bystritsa flowing from Pirin Mountains on the left and Osogovo Mountains on the borders with Macedonia.

The *Vardar River* rises in the Šar planina Mountains in Macedonia and flows in a north-northeastern direction. Near Skopje it turns sharply to the southeast into

Table 3.2 Major rivers in the CBD region (other than the Danube) and their main parameters (Compiled from various databases)

River	Station	Country	Area (km^2)	Mean discharge (m^3 s^{-1})	Specific yield (L s^{-1} km^{-2})	Runoff depth (mm)
Odra	Bohumín	Czech Republic	4,665	48	10.3	325
Wisla	Szczucin	Poland	23,901	248	10.4	327
Dniester	Mohyliv-Podilskyi	Ukraine	43,000	278	6.5	204
Marica	Svilengrad	Bulgaria	20,857	103	4.9	156
Struma	Marino Pole	Bulgaria	10,243	70	6.8	216
Vardar	Taor	Macedonia	5,499	70	12.8	403
Cetina	Mouth	Croatia	12,000	105	8.8	276

Greece, where it completes its course in the Aegean Sea. Its major tributaries are the Treska, Pcinja, Bregalnica, and Crna.

In the *Dalmatian coastal zone* and Istria there are only three rivers with lengths over 25 km (but abundant flow), the Zrmanja, Krka, and Cetina, which drain the karstic Dinaric Mountains and flow into the Adriatic Sea.

References

Belz JU, Goda L, Buzás Z, Domokos M, Engel H, Weber J (2004) Das Abflussregime der Donau und ihres Einzugsgebietes, Regionale Zusammenarbeit der Donauländer im Rahmen des IHP der UNESCO. Bundesamt für Gewässerkunde, Koblenz/Baja, 152 p

Bendeffy L (1979) A Duna medrében görgetett hordalék eredete és kőzetminősége (Origin and lithology of Danubian bedload). Földrajzi Értesítö 28(1–2):73–89 (in Hungarian)

Ghenea C, Mihailescu N (1991) Palaeogeography of the Lower Danube Valley and the Danube delta during the last 15 000 years. In: Starkel L, Gregory KJ, Thornes JB (eds) Temperate palaeohydrology. Wiley, Chichester, pp 343–364

Giosan L, Donnelly JP, Constantinescu S, Filip F, Ovejanu I, Vespremeanu-Stroe A, Vespremeanu EI, Duller GAT (2006) Young Danube delta documents stable black sea level since the middle Holocene: morphodynamic, paleogeographic and archaeological implications. Geology 34(9):75–76. doi:10.1130/G22587.1

ICPDR (2011) The Danube River basin. International Commission for Protection of the Danube River, Vienna. 112 p. http://www.icpdr.org/icpdr-pages/river_basin.htm

Lászlóffy W (1965) Die Hydrographie der Donau. Der Fluss als Lebensraum. In: Liepolt R (ed) Limnologie der Donau – Eine monographische Darstellung, Kapitel II. Schweizerbartsche Verlag, Stuttgart, pp 16–57

Lóczy D (2007) Chapter 12 The Danube: morphology, evolution and environmental issues. In: Gupta A (ed) Large rivers: geomorphology and management. Wiley, Chichester, pp 235–260

Orghidan N (1966) Die Donau und das Eiserne Tor. Revue Roumaine Ser Géogr 10(1):29–37

Pekárová P, Onderka M, Pekár J, Miklánek P, Halmová D, Škoda P, Bačová Mitková V (2008) Hydrologic scenarios for the Danube river at Bratislava. Key Publishing, Ostrava, 158 p. http://147.213.145.2/DanubeFlood/PDF/ Pekaovamonograph.pdf

Sommerhäuser M, Robert S, Birk S, Hering D, Moog O, Stubauer I, Ofenböck T (2003) UNDP/ GEF Danube Regional Project "Strengthening the implementation capacities for nutrient reduction and transboundary cooperation in the Danube river basin". Final report, 97 p. http://www.undp-drp.org/drp/en/activities_1-1_eu_wfd_implementation_fr_phase1.html

Sommerwerk N, Hein T, Schneider-Jacoby M, Baumgartner Ch, Ostojić A, Siber R, Bloesch J, Paunović M, Tockner K (2009) Chapter 3 The Danube river basin. In: Tockner K, Uehlinger U, Robinson ChT (eds) Rivers of Europe. Elsevier, London, pp 59–112

Stančík A, Jovanovič S (1988) Hydrology of the river Danube. Publishing House Príroda, Bratislava, 271 p

Chapter 4
Land Cover and Land Use

Ján Feranec and Tomáš Soukup

Abstract Land use and land cover integrated with georelief characteristics, which can be derived from topographic maps or digital elevation models, provide the background for the study of the intensity of landscape processes, changes in the landscape, and identification of objects. The knowledge of land cover is imperative and indispensable in analyzing causes and effects in geomorphic evolution or in estimating human impact on the landscape. From a broader perspective, the maintenance of ecological stability is a primary aspect of decision making in the course of environmental planning.

Keywords Land cover • CORINE • PELCOM • Satellite imagery • Urban areas • Agricultural land • Grassland • Forest

4.1 Introduction

Land cover (LC) portrays the physical status of the present landscape, which is constituted of objects both natural and created or modified by humans. The intensity of landform evolution depends to a large extent on the utilization of land surfaces by humans reflected in LC. LC classes range from urbanized and technically developed areas, through agricultural, forest, and semi-natural areas to classes that are close to

---

J. Feranec (✉)
Institute of Geography, Slovak Academy of Sciences, Štefánikova 49,
814 73 Bratislava, Slovakia
e-mail: feranec@savba.sk

T. Soukup
GISAT s.r.o, Milady Horákové 57, 170 00 Praha 7, Czech Republic
e-mail: tomas.soukup@gisat.cz

the natural areas (like alpine meadows, which are classified as *Grasslands*, while *Wetlands* are included into the class *Other* LC). This order of LC classes also indicates the degree of human impact on the landscape (Feranec and Otahel 2001).

4.2 Data Sources for Land Cover Mapping

Today remote sensing technology allows the synchronous recording of LC even over such extensive areas like the Carpatho–Balkan–Dinaric (CBD) region. The schematic map of LC of this area was compiled from the *CORINE* (Coordination of Information on the Environment) Land *Cover (CLC)* (Heymann et al. 1994; EEA 2011) and *PELCOM (Pan-European Land Cover Monitoring)* databases (PELCOM 2001). Through LC mapping in the framework of the CORINE Land Cover Project with its nomenclature (Heymann et al. 1994) at a scale of 1:100,000 and by using satellite images the basic picture of the landscape can be produced. The PELCOM data (in classes of Urban areas, Arable land, Grasslands, Coniferous forest, Deciduous forest, and Mixed forest) were only used for a minor section of southwestern Ukraine, as no CLC data are available for this country. Both data sets are compatible for the small-scale map (Fig. 4.1).

The CLC database was derived by the interpretation of SPOT and IRS satellite images obtained from 2006 (or a year before or after), while the PELCOM database

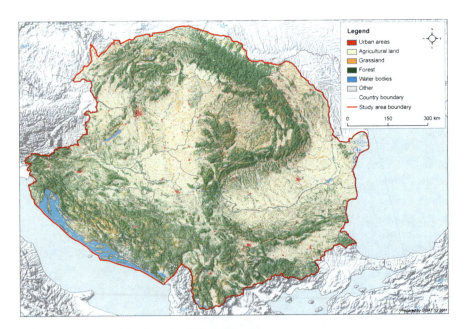

Fig. 4.1 Land cover of the Carpatho–Balkan–Dinaric area

is from NOAA-AVHRR data with 1 × 1 km resolution. The specificity of the CLC database corresponds to the processing scale that is 1:100,000; the minimum size of identifiable areas is 25 ha. It is to be noted that because of this low resolution, the visual interpretation of the CLC can lead to an overestimation of the dominant LC (over most of the CBD area arable land and forest) and an underestimation of the other classes (ICPDR 2004).

The CLC nomenclature includes 44 LC classes in a three-level hierarchy. The five level-one categories are (Heymann et al. 1994; Bossard et al. 2000; Büttner et al. 2004; Feranec et al. 2007):

1. Artificial surfaces;
2. Agricultural areas;
3. Forests and semi-natural areas;
4. Wetlands;
5. Water bodies.

Six classes had been distinguished for the purpose of the schematic representation of LC for the CBD area:

- *Urban areas* (areas of "Urban pattern," "Industrial, commercial and transport units," "Mine dump and construction sites," and "Artificial non-agricultural vegetated areas are aggregated in this class");
- *Agricultural land* (this class covers "Arable land," "Permanent crops," and "Heterogeneous agricultural areas");
- *Grasslands* (this class includes "Pastures" and "Natural grasslands");
- *Forests* (composed of "Broad-leaved forests," "Coniferous forests," "Mixed forest," and "Transitional woodland/shrubs");
- *Water bodies* (this class integrates "Inland water bodies" and "Sea and ocean"); and
- *Other LC* (including, for instance, "Inland wetlands" and "Coastal wetlands," "Sparsely vegetated areas," etc.)

4.3 Land Cover in the Carpatho–Balkan–Dinaric Region

Using the boundaries specially defined for the purposes of the present monograph, the schematic LC map of the CBD region has been prepared (Fig. 4.1) and demonstrates the spatial distribution of the LC classes described above (Table 4.1).

The statistics in Table 4.1 of the LC classes represent a contribution to the knowledge of the present-day landscape structure in the CBD area. *Agricultural land* is the class occupying the largest area (359,879 km^2 or 43.3% of the investigated territory) mostly in lowlands along the Danube River and its tributaries, the Tisa, Drava, Sava, and Marica Rivers, and on loess-mantled hilly terrains.

The second largest area is represented by *Forests* (318,387 km^2, 38.3%), prevalent in the mountain ranges of the Carpathians, Stara Planina, and some members of

Table 4.1 Distribution of LC classes in the CBD region, km^2

Class	CLC	%	PELCOM	%	Total (CLC+ PELCOM)	%
Urban areas	37,263	4.7	157	0.4	37,420	4.5
Agricultural land	352,978	44.3	6,901	19.7	359,879	43.3
Grasslands	74,268	9.3	1,567	4.5	75,835	9.1
Forests	292,499	36.7	25,888	73.8	318,387	38.3
Water bodies	24,908	3.1	15	0.0	24,923	3.0
Other	14,747	1.9	572	1.6	15,319	1.8
Total	796,663	100.0	35,100	100.0	831,763	100.0

the Dinaric mountain range but some riparian forests along the Middle Danube and *Robinia* stands applied for the fixation of blown sand areas are also well visible.

Forests are followed by the class of *Grasslands* (75,835 km^2, 9.1%), where agriculturally used meadows are usually found in lowland landscapes, while natural grasslands unexploited or only sparsely used by humans mostly occur in mountain ranges above the upper timber line and in areas under nature conservation.

LC classes that have smaller extensions include *Urban areas* (37,420 km^2, 4.5%). Even at this overview scale all the capital cities in the CBD area (Bratislava, Budapest, Ljubljana, Zagreb, Bucharest, Belgrade, Sarajevo, Sofia, Podgorica, and Skopje) along with some other major conurbations (like Kosice, Szeged, Timişoara, Novi Sad, Plovdiv, and many more) are clearly identifiable (Fig. 4.1).

In the case of *Water bodies* (24,923 km^2, 3.0%), only the lakes of the Danube delta in Romania, Lake Balaton and the Tisza reservoir in Hungary, Lake Shkodra on the Montenegro–Albanian border, and the coastal waters of the Adriatic Sea are discernible. The *Other* LC class covers, among others, swamps, rocks, and other terrains of particular reflectance. Its area amounts to only 15,319 km^2 (1.8% of the total CBD area).

4.4 Land Cover Changes

Changing LC usually remarkably influences exogenic geomorphic processes. In general, the analyses performed by the European Environment Agency based on CLC data show that in Europe the annual rate of these changes slowed in the period 2000–2006, compared to the period 1990–2000 (Table 4.2 – EEA 2010). However, landscape processes (urbanization, the intensification of agricultural and, in parallel, land abandonment coupled with the natural spreading of bushes and woodlands) are still a very strong trend and expected to continue in the future. Particularly three northern countries of the CBD area, the Czech Republic, Hungary, and Slovakia, showed remarkable changes in LC over the last two decades (Table 4.2). Converting non-forest land to forests has clear environmental benefits: reducing runoff, protecting the soil. In all mountain ranges reforestation is needed to compensate wide-

Table 4.2 LC change on the total territories of the Carpatho–Balkan–Dinaric countries, 1990–2000 and 2000–2006, pointing out trends for 2000–2006 (EEA 2010)

Country	LC change, % of total area 1990–2000	LC change, % of total area 2000–2006	Trends in 2000–2006 Artificial	Agricultural	Forest and nature
Albania	n.a.	0.18	Very high rate of residential sprawl	Loss of agricultural land	Forest gains from agriculture, loss to urbanization
Bosnia and Herzegovina	n.a.	0.12	Diffuse residential sprawl	Loss of pasture, vineyards and orchards	Semi-natural land transitions, fires
Bulgaria	0.11	0.09	Accelerating urban sprawl	Loss of pasture, vineyards and orchards but overall stabilization	Forest expansion replaced by management
Croatia	0.19	0.17	Highway construction	Arable and complex cultivation replacing grassland	Forest management, loss of open spaces, re-growth of burnt areas
Czech Republic	0.81	0.33	Accelerating urban sprawl	Conversion of arable to pasture continued at slower rate	Stabilization of natural landscapes, some loss of natural grasslands
Hungary	0.56	0.48	Expansion of construction and mineral extraction	Withdrawal of farming, some conversion of pasture to arable land	Afforestation of former farmland and grasslands
Macedonia	n.a.	0.14	Residential sprawl, growing mineral extraction	Various transitions, loss of vineyards and orchards	Forest management, new water bodies, loss of natural grasslands
Montenegro	0.2	0.4	Extension of construction sites and residential areas	Loss of pastures to artificial surfaces	Forest transitions, loss of natural areas to economic sites, fires
Poland	0.10	0.10	Increased sprawl of economic sites, highway construction	Loss of agricultural (mostly arable)	Farmland to woodland transition, new water bodies
Romania	0.16	0.05	Accelerating residential sprawl around main cities	Decelerating agricultural transitions, loss of pastures	Recent felling and land transition, some loss of natural open areas
Slovakia	0.51	0.25	Decelerating residential sprawl	Slow-down of changes, loss of agricultural land	Afforestation on abandoned farmland
Slovenia	0.2	0.3	New constructions	Limited changes, loss of agricultural land	Limited changes, forest felling, new water bodies
Ukraine	n.a.	0.2[a]	Residential sprawl	Loss of agricultural land	Afforestation

[a] Estimate from various sources

spread deforestation in earlier centuries. Urban sprawl and the reduction in the area of natural grasslands, recorded for most of the countries, are unfavorable tendencies in terms of geomorphological hazards.

4.5 Conclusion

The availability of CLC data layers for 1990, 2000, and 2006 is a fundamental tool for landscape assessment and identification of landscape changes in the CBD area. The LC map is a basic source of information on the way that land can be used and land use classes spatially organized. Integrated with the georelief characteristics, it renders the opportunity to monitor the intensity of processes and changes observable in landscape.

References

Bossard M, Feranec J, Otahel J (2000) CORINE land cover technical guide – Addendum 2000. Technical report 40, European Environment Agency, Copenhagen. http://www.eea.europa.eu/publications/tech40add

Büttner G, Feranec J, Jaffrain G, Mari L, Maucha G, Soukup T (2004) The CORINE land cover 2000 project. In: Reuter R (ed) EARSeL eProceedings, 3(3). EARSeL, Paris, pp 331–346

EEA (2010) The European environment state and outlook 2010. Land use. European Environment Agency, Copenhagen, p 52. http://www.eea.europa.eu/soer/europe/land-use

EEA (2011) Downloadable data about Europe's environment. European Environment Agency, Copenhagen. http://www.eea.europa.eu/data-and-maps/data

Feranec J, Otahel J (2001) Land cover of Slovakia. VEDA, Bratislava, 123 p

Feranec J, Hazeu G, Christensen S, Jaffrain G (2007) Corine land cover change detection in Europe (case studies of the Netherlands and Slovakia). Land Use Policy 24(1):234–247

Heymann Y, Steenmans Ch, Croissille G, Bossard M (1994) CORINE land cover. Technical guide. Office for Official Publications European Communities, Luxembourg

ICPDR (2004) A brief description of the Danube River Basin. International commission for the protection of the Danube River, Vienna. http://www.icpdr.org/pub/dpcu/brief.html

PELCOM (2001) Pan-European land cover monitoring. Alterra, Centre for Geo-Information, Wageningen. http://www.geo-informatie.nl/projects/pelcom/public/index.htm

Part II
Recent Geomorphic Evolution: National Chapters

Chapter 5
Recent Landform Evolution in the Polish Carpathians

Zofia Rączkowska, Adam Łajczak, Włodzimierz Margielewski, and Jolanta Święchowicz

Abstract The key drivers of relief transformation in the Polish Carpathians (the best studied region of the Carpathians) have been local downpours, continuous rainfalls, and rapid snowmelt. Threshold values are quickly exceeded during such events and powerful morphological processes are initiated. However, due to human activities, geomorphic processes are often accelerated and intensified with serious consequences. Thus, increased human impact is a key factor in the recent geomorphic evolution of the Polish Carpathians. Over the last two centuries deforestation, intensive agriculture, mining, housing developments on slopes, channelization of streams, and construction of reservoirs all have contributed to changes in the rate of the particular geomorphic processes. The intensity of sheet erosion depends on vegetation cover, land use, and cultivation techniques (terracing, contour tillage, crop rotation, etc.). Slopes bearing poorly constructed infrastructure have become susceptible to mass movements. At higher elevations debris flows, dirty avalanches,

Z. Rączkowska (✉)
Department of Geoenvironmental Research, Institute of Geography
and Spatial Organisation, Polish Academy of Sciences,
Św. Jana Str. 22, 31-018 Kraków, Poland
e-mail: raczk@zg.pan.krakow.pl

A. Łajczak
Institute of Geography, Jan Kochanowski University,
Świętokrzyska Str. 15, 25-406 Kielce, Poland
e-mail: alajczak@o2.pl

W. Margielewski
Institute of Nature Conservation, Polish Academy of Sciences,
A. Mickiewicza Av. 33, 31-120 Kraków, Poland
e-mail: margielewski@iop.krakow.pl

J. Święchowicz
Department of Geomorphology, Institute of Geography and Spatial Management,
Jagiellonian University, Gronostajowa Str. 7, 30-387 Kraków, Poland
e-mail: j.swiechowicz@geo.uj.edu.pl

and extreme floods are crucial in slope evolution. Over the last 20 years the ratio of agricultural land and the rate of sheet erosion and deflation have decreased, while gully erosion on slopes and the incision of river channels have intensified and the reactivation of shallow landslides has become more common. Increasingly frequent extreme weather events may reverse the stabilizing trend of landform evolution.

Keywords Mass movements • Sheet and rill erosion • Ephemeral gully erosion • Land use • Piping • Aeolian processes • Periglacial processes • Karst • Fluvial processes • Biogenic processes • Human impact • Polish Carpathians

5.1 Introduction

Adam Łajczak
Włodzimierz Margielewski
Zofia Rączkowska
Jolanta Święchowicz

The Polish Carpathians are the best-studied area in the entire Carpatho-Balkan-Dinaric Region. Geomorphic processes and landforms here have been the subject of intensive research since the beginning of the twentieth century (Margielewski et al. 2008; Rączkowska 2008).

The first studies of mass movements in the Outer Carpathians dealt with single *landslides* (Zuber and Blauth 1907; Łoziński 1909; Sawicki 1917). Complex landslide analysis in selected Carpathian regions began during the interwar period (Świderski 1932; Teisseyre 1929, 1936). In the second half of the twentieth century a novel attitude to mass movements emerged and researchers began to accept the view that mass movements effectively shaped the surface of the Carpathians during the Late Glacial and the Holocene (Flis 1958; Starkel 1960; Ziętara 1964). Research also addressed contemporary landslides as fundamental geomorphic processes and their negative impact on human economic activities (Jakubowski 1965; Ziętara 1968; Mrozek et al. 2000; Bajgier-Kowalska 2004; Gorczyca 2004). The Polish Geological Institute has been recording landslides in the Polish Carpathians for years (Poprawa and Rączkowski 2003; Rączkowski 2007). Researchers of the Research Station of the Institute of Geography and Spatial Organization, Polish Academy of Sciences (IGiPZ PAN), in Szymbark (Beskid Niski Mountains), identified precipitation thresholds that trigger mass movements (Gil 1997; Gil and Starkel 1979; Starkel 1996). In addition, selected Carpathian landforms are being closely monitored and allow researchers to identify complex hydrogeological triggering mechanisms and to predict future occurrences of landslides (Zabuski 2004).

Research on gravitational and other geomorphic processes in the Tatra Mountains began in the 1950s and led to a better understanding of the morphodynamics of high mountain slopes (Kotarba 1976, 1992; Kłapa 1980; Kotarba et al. 1983, 1987; Rączkowska 1995, 1999). *Debris flows* were added to research topics in the late

1980s with the identification of precipitation thresholds and changes of debris flow activity, especially for the past several centuries (Krzemień 1988; Kotarba 1992, 1997, 2004, 2005).

In the 1960s measurements of *sheet erosion* using special traps (Gerlach 1966, 1976b) as well as of *splash* (Gerlach 1976a) were performed in the Beskid Sądecki and Gorce Mountains, and both surface runoff (Słupik 1970) and sheet erosion (Gil 1976) were monitored at the IGiPZ PAN research station in Szymbark. Thus, the station, located on the boundary line between the Beskid Niski Mountains and the Ciężkowice Foothills, possesses the longest sheet and *rill erosion* measurement record on plots for the Polish Carpathians, continuous since 1968 (Gil and Słupik 1972; Słupik 1973, 1981; Gil 1976, 1986, 1994, 1999, 2009). Simultaneous measurements of soil splash, wash, and supply to fluvial transport were carried out at the IGiPZ PAN research station in Frycowa located in Beskid Sądecki Mountains (Froehlich and Słupik 1980b; Froehlich 1986), and splash and sheet wash were studied for different types of crops in the Beskid Niski Mountains (Śmietana 1987). Field measurements of splash, sheet, rill, and ephemeral gully erosion are also performed in the Wiśnicz Foothills (eastern part of the Wieliczka Foothills) as part of a research program run by the Institute of Geography and Spatial Management, Jagiellonian University (IGiGP UJ), at the research station in Łazy, near the town of Bochnia (Święchowicz 2002a, b, 2008, 2009, 2010).

The effects of *aeolian deposition* on snow were first studied in the Tatras (Kłapa 1963). Quantitative analyses of aeolian processes have been performed in the Jasło-Sanok Depression (Gerlach and Koszarski 1968, 1969; Gerlach 1986), the Western Bieszczady Mountains (Pękala 1969), the Beskid Niski Mountains (Janiga 1971, 1975), the Tatras (Izmaiłow 1984a, b), the Wiśnicz Foothills (Izmaiłow 1995a, b, c), as well as in the boundary zone between the Beskid Niski Mountains and the Ciężkowice Foothills (Welc 1977).

Less attention has been devoted to *piping*. The studies involved the mapping of landforms and the description of relief evolution, primarily in the Outer Eastern Carpathians in Poland (Czeppe 1960; Starkel 1965; Galarowski 1976).

Periglacial processes have been investigated in the Polish Tatra Mountains since the 1950s (Jahn 1958; Gerlach 1959; Kotarba 1976; Dobiński 1997; Gądek and Kędzia 2008; Gądek et al. 2009). This includes research performed at the IGiPZ PAN research station at Hala Gąsienicowa (Kłapa 1970, 1980; Rączkowska 1992, 1993, 1995, 1999, 2007; Kędzia et al. 1998; Mościcki and Kędzia 2001), where processes affecting slope evolution, especially nivation and solifluction, have been studied.

Large-scale research efforts on *fluvial processes* and landforms began in the 1960s and 1970s and have focused on the structure and dynamics of stream channel features (Kaszowski and Kotarba 1967; Kaszowski 1973; Niemirowski 1974; Kaszowski et al. 1976; Kaszowski and Krzemień 1977, 1979; Klimek 1979, 1987; Baumgart-Kotarba 1983; Krzemień 1984a, b, 1991) in light of the role of floods (Ziętara 1968; Froehlich 1975; Krzemień 1985; Malarz 2002; Izmaiłow et al. 2006) and river engineering (Wyżga 1993a, b; Krzemień 2003; Kościelniak 2004; Zawiejska and Krzemień 2004). Particular attention has been paid to the development of fluvial landforms affected by dams (Klimek et al. 1990; Łajczak 1996, 2006). The influence of accumulations of large woody debris is also being investigated with respect to stream channel evolution (Kaczka 2003, 2008; Wyżga et al. 2003). The Holocene evolution of stream channels and floodplains is also in the focus of research (Starkel 1965, 2001; Klimek and Trafas 1972; Klimek 1974; Froehlich et al. 1977). Fluvial geomorphologists are particularly interested in determining the rates of contemporary fluvial processes (Froehlich 1975, 1982; Krzemień 1984a, 1991; Krzemień and Sobiecki 1998; Łajczak 1999; Święchowicz 2002c; Wyżga et al. 2003).

Research on the formation and development of large *biogenic landforms* began during the second half of the twentieth century. Peat domes (Ralska-Jasiewiczowa 1972, 1980, 1989; Wójcikiewicz 1979; Żurek 1983, 1987; Obidowicz 1990) and human impact on the formation of peat bogs (Łajczak 2007b, 2011) are investigated in the Orawsko–Nowotarska Basin and in the Bieszczady Mountains.

5.2 Geological and Geomorphological Settings

Włodzimierz Margielewski

The Polish Carpathians is the northernmost section of the Alpine mountain belt. It is geologically divided into the *Inner (Central) Carpathians* (including the Tatras and the Podhale Basin) and the Outer (Flysch) Carpathians, separated by the Pieniny Klippen Belt (Fig. 5.1) (Książkiewicz 1972; Birkenmajer 1986; Oszczypko et al. 2006).

The *Tatra Mountains* are composed of an extensive crystalline core, including a Permian-Mesozoic sediment layer, the High-Tatric Nappe, and the north-verging

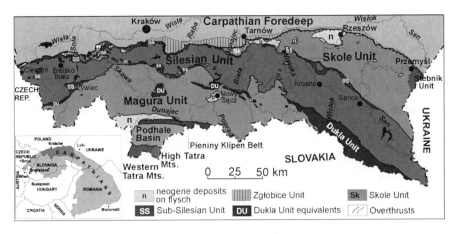

Fig. 5.1 Geological map of the Polish Carpathians (After Żytko et al. 1989)

Subtatric nappes (Książkiewicz 1972). The crystalline core is formed of intrusive Carboniferous granitoids of the High Tatras as well as metamorphic rocks (Paleozoic rocks: gneiss, amphibolite, metamorphic shale) found mainly in the Western Tatras. The crystalline core is unevenly covered with epicontinental autochthonous High-Tatra sediments such as quartzites, dolomites, limestones, marls, shales, conglomerates, and sandstones formed from the Triassic to the Middle Cretaceous. These High-Tatra rock series are covered from the south by the High-Tatric nappes of similar rock complexes. The following Subtatric nappes, formed between the Triassic and the Lower Cretaceous, are superimposed on the High-Tatric Nappe: the lower Križna Nappe, the middle Choč Nappe, and the upper Stražov Nappe. The Triassic formations developed here in a manner similar to that of the High-Tatric series. On the other hand, younger (Jurassic to Lower Cretaceous) formations are deep marine sediments (radiolarites, limestones – Książkiewicz 1972; Oszczypko 1995).

The *Podhale Basin* is filled with sediments of 2,500 m thickness from the Middle Eocene–Oligocene. Middle Eocene conglomerates are covered by nummulitic limestone, dominated by shale-sandstone sediments of the Podhale flysch type. The sediment fill is only slightly deformed and forms a wide syncline. Stronger deformations have only been found near the line of contact with the Pieniny Klippen Belt (Książkiewicz 1972; Oszczypko 1995).

The *Pieniny Klippen Belt* is a narrow geological structure, a suture formed in the collision zone between the North European Plate and the Adria-Alcapa terranes (microplates) (Oszczypko 1995; Oszczypko et al. 2006). Its constituent rocks form steep slide-folds and folds thrust in between the Podhale flysch formations in the south and between Magura-type nappes in the north. They consist of klippen belts built of resistant limestone and radiolarite as well as sheathing rocks such as less-resistant Upper Cretaceous to Paleogene marls, sandstones and shales (Birkenmajer 1986). Andesite intrusions associated with Miocene volcanic activity can be found along the northern edge of the klippen belt, along the line of contact with the Magura Nappe.

The *Outer (Flysch) Carpathians* are built primarily of Late Jurassic to Early Miocene flysch (turbidite) with occasional intercalations of marls and silicates (radiolarite, hornstones) (Fig. 5.1) (Książkiewicz 1972; Oszczypko 1995). Series of folded flysch sediments, disrupted from their substrate and displaced in a generally northerly direction, have formed discrete tectonic units (nappes) superimposed on one another and their foreland. These include the Magura Nappe, the Dukla Nappe (and its equivalents), the Silesian Nappe, the Sub-Silesian Nappe, and the Skole Nappe (Książkiewicz 1972; Oszczypko 1995). The Stebnik (Sambir) Unit in the northeast consists of folded Miocene formations (Fig. 5.1) (Oszczypko 1995; Oszczypko et al. 2006). The northern foreland of the Polish Carpathians contains Miocene Foredeep formations. The Zgłobice Unit is the folded part of these sediments, which appear in a narrow discontinuous belt in the Carpathian foreground (Fig. 5.1). The folded flysch formations of nappes are divided by a system of joints and numerous dislocations, while rock massifs are characterized by lithological variability.

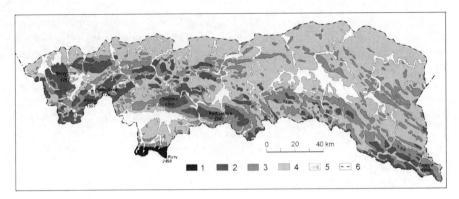

Fig. 5.2 Types of relief in the Carpathians (After Starkel 1980). 1, high mountains; 2, middle mountains; 3, low mountains and high foothills; 4, middle and low foothills; 5, bottom of valleys; 6, boundary of the Carpathians

Fig. 5.3 Geomorphological units of the Polish Carpathians (After Gilewska 1986; Starkel 1991). 1, Central Western Carpathians; 2, Outer Western Carpathians; 3, Outer Eastern Carpathians

The Carpathian relief is strongly linked to the lithology and tectonics of rock massifs and is subdivided into altitudinal belts (Starkel 1969). The Polish Carpathians encompass four dominant types of relief: high, middle and low mountains, foothills, and valley floors (Fig. 5.2 – Starkel 1972, 1991). *High mountains* (1,500–2,499 m a.s.l.) of postglacial and rocky slope type include the Tatra Range and the crest of Babia Góra Massif (1,725 m) in the Flysch Carpathians (Fig. 5.2 – Starkel 1991). *Middle mountains* (800–1500 m) with steep slopes (over 30%) on resistant sandstones include the compact Beskid Śląski, Beskid Żywiecki, and Beskid Sądecki mountain ranges and the isolated Beskid Wyspowy and Bieszczady ridges, ranging from 400 to 800 m in elevation (Figs. 5.2 and 5.3). *Low mountains* usually encircle medium-height mountains in the form of isolated ridges of resistant rocks, with steep slopes and a relative relief of 200–400 m, rising above depressions cut into less resistant rocks (like the Beskid Niski Mountains – Starkel 1991).

A common characteristic of relief in the *foothills* north of the Beskidy Mountains (see Fig. 5.3) is the presence of wide hills of various gradient (10–40%) with convex-concave slopes and flat valley floors. Three types of foothills are distinguished in the Flysch Carpathians: high foothills with crests at 200–300 m, middle foothills with crests at 120–200 m, and low foothills with crests at 40–100 m above valley bottoms, formed in the least resistant of rocks around the Polish Carpathians (Fig. 5.2). Protruding elements such as escarpments, associated with more resistant rocks on the margins of tectonic units, are remarkable features of the foothills in the Flysch Carpathians. The floors of valleys and basins are flat and often terraced (Starkel 1991).

Based on their diverse geological structure and dissected topography, the Polish Carpathians can be divided into three major morphostructural units (sub-provinces): the *Central Western Carpathians* (the Tatras and Podhale) (Fig. 5.3: 1), the *Outer Western Carpathians* (western part of Polish Flysch Carpathians) (Fig. 5.3: 2), and the *Outer Eastern Carpathians* (eastern part of Polish Flysch Carpathians) (Fig. 5.3: 3) (Gilewska 1986; Starkel 1991; Kondracki 2000).

Quaternary formations cover the slopes of the Polish Carpathians and the sides of valleys. They are formed of silt and debris clays from periglacial rock weathering, gravitational rock displacement on slopes, and accumulation in foothill areas (Stupnicka 1960; Cegła 1963; Starkel 1984; Henkiel 1972). In the medium-height mountain zone, the slope deposits on sandstones are relatively thin (70–80 cm) and only thicken out on flat footslopes (Kacprzak 2002–2003). In the foothill zone, the thickness of the slope deposits on flysch schist is greater (3–20 m), especially at lower elevations (Dziewański and Starkel 1967; Starkel 1969; Henkiel 1972). Plenivistulian (Last Glacial Maximum) loess patches, up to a dozen metres thick, locally occur in the foothills (Cegła 1963; Starkel 1984; Gerlach et al. 1991). Deep solifluctional-deluvial mantles cover the Polish Carpathian depressions and basins as a result of sheet erosion, gravitational displacement, and fluvial transport of slope material (Stupnicka 1960; Cegła 1963; Henkiel 1972).

5.3 Climate and Hydrology

Jolanta Święchowicz

A key characteristic of the Carpathian climate is both *vertical* (from 350 to 2,500 m elevation – Fig. 5.4) and horizontal *variability*. Each year maritime-polar and continental-polar air masses predominate in the area over 65% and 20% of the time, respectively. The foothills and the lower sections of the Beskidy Mountains, up to about 700 m elevation, are classified as moderately warm (Hess 1965). The boundary of the moderately warm climate zone decreases about 60 m in the easterly direction. The higher regions of the Beskidy Mountains are moderately cool, while the highest elevations of the Beskid Sądecki, the Gorce, the Beskid Śląski, and the Beskid Żywiecki Mountains fall within the cool climatic belt. The very cool climate zone

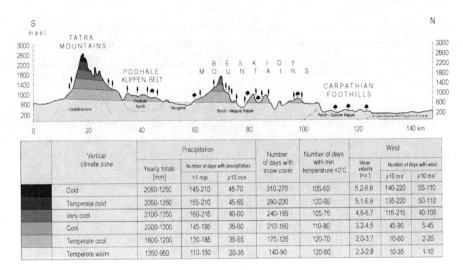

Fig. 5.4 Vertical climate zones in the Polish Western Carpathians (After Hess 1965)

is restricted to the Babia Góra Massif above the upper timberline. The Tatra Mountains show five climate zones (Fig. 5.4) with boundaries located 100–200 m higher on southern-facing slopes than on northern-facing slopes.

The average *precipitation* gradient in the Polish Carpathians is 60 mm per 100 m of elevation. The 700 mm isohyet runs along the edge of the Carpathian Foothills, while the 900 mm isohyet encircles the Beskidy Mountains. Strong rain shadow effect is typical of intramontane basins and N-S aligned valleys. Average annual precipitation drops towards the east from over 1,400 mm in the Beskid Śląski Mountains to 1,000–1,300 mm in the Bieszczady Mountains. In the Tatras, maximum annual precipitation is 1,500–1,700 mm (Niedźwiedź and Obrębska-Starklowa 1991; Obrębska-Starklowa et al. 1995). Intense continuous rainfall events are mainly characteristic in June and July and short local downpours in May. The predominance of continuous rainfalls in the west extends to the Dunajec and Biała Dunajcowa valleys (Cebulak 1992). Three types of precipitation determine the type and intensity of slope and fluvial processes: (1) short local downpours (intensity of 1–3 mm min^{-1}), which induce intense sheet erosion and mudflows; (2) continuous rainfalls (150–400 mm per 2–5 days), which lead to landslides, floods, river channel modifications, and sediment deposition on floodplains; (3) rainy seasons with monthly precipitation in the 100–500 mm range, which cause deep infiltration and rockslides (Starkel 1986, 1996; Froehlich and Starkel 1995).

Two *hydrological macroregions* are identified in the Polish Carpathians: the western and the eastern macroregion (Ziemońska 1973; Dobija 1981; Dynowska 1995). The drainage basins west of the Dunajec River are characterized by a rain-snow *precipitation type*, while those east of the river by a snow-rain type. The *rivers* of the Polish Carpathians show substantial seasonal and year-to-year variability of *discharge*. Low discharge occurs in the autumn-winter months (September to

February), while high discharge is typical in spring and summer (March to August) (Chełmicki et al. 1998–1999). The incidence of floods in different regions depends on precipitation type. Two discharge maxima occur during the spring-summer season in the western Polish Carpathians (in the Beskid Śląski and Beskid Żywiecki Mountains): (1) a spring maximum (mainly April) resulting from snowmelt and (2) a summer maximum (mainly June) resulting from rainfall. Only the Dunajec and Poprad rivers have an extended flood season, due to a combination of belated and prolonged snowmelt in the Tatras and summer rainfalls. Rivers east of the Dunajec River experience major snowmelt-induced spring floods and secondary rainfall-induced summer floods (Dynowska 1972, 1995; Ziemońska 1973; Chełmicki et al. 1998–1999). The former do not develop particularly rapidly as snow cover in the Carpathians melts gradually. Flash floods induced by downpours tend to be localized and occur mainly in June and July. Summer floods due to prolonged (3–5 days of) rainfall build up slowly, but usually affect large areas. Daily precipitation can total 200 mm in the Soła, Skawa, Raba, and Dunajec catchments, 150 mm in the Wisłoka basin and 100 mm in the San basin (Osuch 1991; Dynowska 1995; Punzet 1998–1999). During flood events, precipitation is normally higher in the western Polish Carpathians than in the east (Cebulak 1998–1999). For this reason, rivers from the western Carpathians (Soła, Skawa, Raba, Dunajec) contribute more substantially to flood waves on the Vistula River than do tributaries from the eastern part (Wisłoka, San). Fortunately, catastrophic floods never affect all Carpathian tributaries of the Vistula River simultaneously (Punzet 1998–1999).

5.4 Mass Movements

Włodzimierz Margielewski
Zofia Rączkowska

The shaping of slopes by mass movements in high mountains differs fundamentally from that taking place in medium-height and low mountains or in foothills. In the Tatras, mass movements are controlled by slope type and climate belt. Rockwalls and rocky slopes develop mainly as a result of weathering processes and rockfalls of various size. As confirmed by measurements of *rockwall retreat* and debris accumulation (Kotarba et al. 1983; Kotarba 1984), the conditions are most favorable in the cold climate zone and on western-facing walls (Klimaszewski 1971; Kotarba 1976). Debris accumulation rates on *talus slopes* range from 0.1 to 10 cm year^{-1} (Table 5.1), much higher on carbonates than on granite if elevation is the same. Debris mostly accumulates at the top of talus slopes and convex midslope sections (Kotarba et al. 1983). Twelve models of talus slope formation (Fig. 5.5) have been defined according to the morphometry of the rockwalls and slopes to the process which displaces debris (Kotarba et al. 1987).

Weathered surface material on debris-mantled slopes and Richter-denudation slopes stabilized by vegetation is displaced by slow mass movements, including

Table 5.1 Dynamics of present-day geomorphological processes in the Tatras

Geomorphological process	Rate of process in climatic belts					Study site	Author
	D	G	K	H	T		
Chemical denudation, $m^3 km^{-2} year^{-1}$	85.6	95.1	37.7	35.8		Limestones, dolomites, Mała Łąka valley	Kotarba (1971)
Output of mineral material, $m^3 km^{-2} year^{-1}$		12				Metamorphic rocks and granites, Starorobociański stream	Krzemień (1985)
Chemical denudation, $t km^{-2} year^{-1}$			10.3–10.9	10.2–10.4		Granites, upper Sucha Woda valley	Kot (1996)
Chemical denudation, $m^3 km^{-2} year^{-1}$		14				Metamorphic rocks and granites, Starorobociański stream	Krzemień (1984a)
Retreat of rockwalls, mm $year^{-1}$			0.028–0.26			Granites, Pięć Stawów Polskich valley	Rączkowski (1981)
			0.0004–0.26 mean 0.04			Metamorphic rocks and granites, Starorobociański stream	Koszyk (1977)
			0.7			Metamorphic rocks and granites, Starorobociański stream	Kotarba et al. (1987)
			0.1–3.0; mean 0.84	0.4		Granites, Chochołowska valley	Mamica (1984)
						Limestones, dolomites, Mała Łąka valley	Kotarba (1972)
			0.0004	2.5		Limestones, dolomites, Bělanske Tatry	Kotarba et al. (1987)
		0.01–0.43			0.003–0.019	Western Tatras	Midriak (1983)
Frost creep, cm $year^{-1}$	0.1–0.95	0.4–0.5	0.6–0.7	1.7		Mała Łąka valley	Midriak (1983) Kotarba (1976)
Frost creep and solifluction, cm $year^{-1}$			0.1–0.3	0.3–2.0		Metamorphic rocks, Goryczkowa Świńska valley	Baranowski et al. (2004)

5 Recent Landform Evolution in the Polish Carpathians

Process	Value	Location	Reference
Plowing boulders, cm year^{-1}	0.14–3.25 mean 1.3	Limestones, dolomites, Mała Łąka valley	Kotarba (1976)
Needle ice, cm year^{-1}	2.7	Granites, Sucha Woda valley	Kotarba et al. (1983) Gerlach (1959)
Deflation, g m^{-2} year^{-1}	38	Granites, Sucha Woda valley	Izmaiłow (1984a)
Eolian deposition, g m^{-2} year^{-1}	87.9; max. 163.7 1–265		Izmaiłow (1984b)
Denudation by tree uprooting, m^3 ha^{-1}	50,000	Mała Łąka valley	Kotarba (1970)
Max. accretion of talus slopes, cm year^{-1}	2–6 6–8 8–10 2–8	Mała Łąka valley	Kotarba (1976)
Accumulation on talus slopes around snow patches, cm year^{-1}	0.03–0.06	High Tatras	Rączkowska (1999)
Nival erosion, cm year^{-1}	0–5 0.05–4.2	Sucha Woda valley Belánske Tatry, Kopské Sedlo pass	Rączkowska (1993) Midriak (1996)
Water erosion, kg ha^{-1} year^{-1}	9–33 12–253		Kotarba (1976)
Length of bedload transport, m per flood	100–120 160 to few hundreds >20	Chochołowski stream Biały stream, tributary streams of the Chochołowski stream	Krzemień (1991) Kaszowski (1973), Krzemień (1991)

D temperate cool, (forest) 900–1,100 m above sea level, G cool (forest) 1,100–1,450 m, K very cool (subalpine) 1,450–1,850 m, H temperate cold (alpine) 1,850–2,200 m, T cold (seminival) >2,200 m

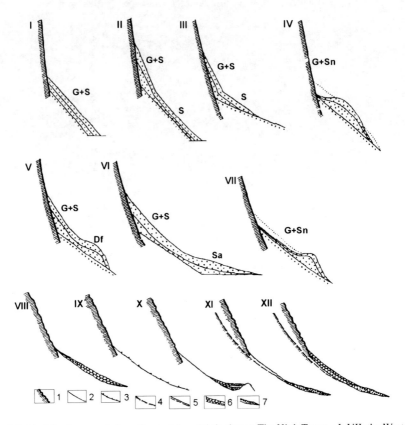

Fig. 5.5 Models of accumulation of material on debris slopes. The High Tatras – I–VII; the Western Tatras – VIII–XII; *G* supply of material by gravity, *S* sliding on debris surface, *Sn* sliding on snow surface, *Df* by debris flows, *Sa* slush avalanche. 1, rocky slope; 2, debris slope surface stabilized by vegetation; 3, sliding of single particles downslope; 4, rocky chute; 5, debris flow gully; 6, deposition of material by gravity; 7, deposition of metrial by debris flows (Kotarba et al. 1987)

solifluction mainly in the alpine and subalpine belts and by *creep* occurring overall. The highest rates of surface material displacement of 1.7–2.0 cm year^{-1} have been detected in the moderately cold zone (Table 5.1), decreasing in the forest belt (Kotarba 1976). Such processes smooth slopes with dome-shaped or step-type microtopography.

The most important geomorphic agent in the Tatra Mountains is abrupt mass movements, which displace and transport significant quantities of material (Midriak 1984; Krzemień 1988; Kotarba 1992, 1995, 2002; Rączkowska 2006, 2008). Such rapid and high-energy movements include debris flows, avalanches, and rockfalls (Kotarba 1992, 1995; Rączkowska 2006). *Debris flows* and *avalanches* take place across multiple altitudinal belts (Kotarba 2002) and often transport weathered material from crests to valley floors, thus connecting the slope system with the channel system (Kotarba et al. 1987). They create new landforms such as debris flow gullies, several meters deep and several hundred meters long, and debris flow levees, over 1 m in height or significantly change the morphometry of existing forms. The changes in relief are linear (debris flows, avalanches) or of point-like manner (rockfalls).

Among abrupt mass movements, debris flows are the most common and most influential in relief formation (Kotarba 1992, 1995, 1997). A single flow event can transport up to tens of thousands of cubic meters of material, often to the bottoms of cirques and glacial valleys, permanently transforming their topography and extending debris slopes (Midriak 1984; Kotarba et al. 1987; Krzemień 1988, 1991; Kotarba 1992; Rączkowska 1999). Debris flows are caused by short but powerful downpours. Precipitation events around 25 mm most often trigger debris flows (Krzemień 1988). They only affect the top of talus slopes. Precipitation intensities 35–40 mm h^{-1} generate debris flows that run across the full length of talus slopes (Kotarba 1992, 2002). The momentary intensity during such rainfalls is ≥1 mm min^{-1}.

The greatest changes in relief are caused by downpours during continuous and widespread rainfalls. In such cases, in addition to debris flows above the upper timberline, *mudflows* and *shallow landslides* occur in forests, affecting the entire length of the slope and removing all weathered material, leaving behind nothing but bare rock (Kotarba 1999, 2002). Lichenometric research has shown that debris flows and rockfalls used to be more active during the Little Ice Age (Kotarba et al. 1987; Krzemień 1988; Kotarba 1992, 1993–1994, 1995, 2004; Kotarba and Pech 2002). In the past 20 years precipitation intensity has increased and extreme geomorphic processes have become more widespread in the Tatra Mountains (Kotarba 1997, 2002, 2004; Kotarba and Pech 2002) than between 1958 and 1978 (Niedźwiedź 2003).

Surface mass movements commonly occur in the Outer Carpathians in the medium-height and low mountains and foothills (Fig. 5.6.: 1). In some mountain areas, they are of primary importance (Starkel 1960; Kotarba 1986; Ziętara 1988; Bober 1984). The major landslides date back to the Late Glacial and Holocene (Alexandrowicz 1997; Starkel 1997; Margielewski 2002, 2006a). Numerous contemporary mass movements are local phenomena and their intensity is associated with extreme precipitation events (Figs. 5.7: 1-4 and 5.8) (Ziętara 1968; Mrozek et al. 2000; Rączkowski and Mrozek 2002; Poprawa and Rączkowski 2003; Gorczyca 2004). Gravitational movements usually displace slope surface material in small and shallow *rotational landslides* (Fig. 5.7: 1–3) (Jakubowski 1965; Ziętara 1968; Poprawa and Rączkowski 2003; Mrozek et al. 2000; Gorczyca 2004). Mass movements that involve bedrock are usually shallow *translational* displacements over bedding planes, fractures and faults, called *structural landslides* (Fig. 5.7: 4 and 5.8) (Ziętara 1968; Bober 1984; Margielewski 2006b). Far less numerous are *deep-seated rockslides*, which tend to be composed of several types of gravitational movements (Fig. 5.7: 5) (Ziętara 1968). A common occurrence is the reactivation of old landslides by new generations of mass movements (Ziętara 1964; Jakubowski 1967; Bajgier-Kowalska and Ziętara 2002). In fact, some landslides exhibit signs of continuous development (Gil and Kotarba 1977). Less common types of landslides include mudflows and debris-and-mud flows (German 1998; Poprawa and Rączkowski 1998; Ziętara 1999; Gorczyca 2004). *Creep* is observed primarily in weathered (Starkel 1960; Jakubowski 1965; Gerlach 1966) and block-type material (Pękala 1969).

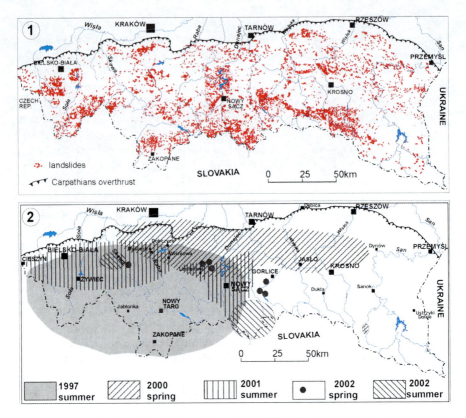

Fig. 5.6 1, Distribution of landslides registered by Polish Geological Institute in the Polish Carpathians (state in 2000) (After Poprawa and Rączkowski 2003, supplemented); 2, Occurrence and range of recent landslides in the Carpathian area, generated by various hydrometeorological factors (snow melting, downpours, heavy rainstorms) (After Poprawa and Rączkowski 2003, supplemented)

Local downpours, heavy continuous rainfalls, and abrupt snowmelt can exceed threshold values of gravitational processes and *trigger* mass movements. However, it is difficult to identify universal *precipitation thresholds,* which heavily depend on slope susceptibility to infiltration (Ziętara 1968; Gil and Starkel 1979; Starkel 1996, 2006; Gil 1997). It has been assumed that continuous and widespread rainfalls in the 150–400 mm range in 2–5 days generate earthflows and weathered material flows. It has also been assumed that rainy seasons with monthly totals of 100–500 mm lead to deep-seated landslides (Starkel 1996). Repeated hydrometeorological events are the most effective in geomorphological terms. Other very effective events include sudden downpours during continuous and widespread rainfalls (as observed in July, 1997 in the Beskid Wyspowy Mountains – Gorczyca 2004; Starkel 2006). The intensity of geomorphic transformation by mass movements depends on the incidence of intense precipitation events, which never affect the entire area of the Polish Flysch Carpathians. Such events can occur repeatedly

Fig. 5.7 Landslides formed and rejuvenated during current heavy rains (date shows the age of main stage and rejuvenation of each form): 1–3, shallow landslides formed in weathering (slope) mantles (forms 1–2) and fluvial deposits (form 3); 4, Kamionka Mała (Beskid Wyspowy): shallow weathering-rocky landslide (structural, slip consequence); 5, deep seated rocky landslide formed ca 150–200 years ago (see Pulinowa 1972) on Kicarz Mountain, Beskid Sądecki Mountains (Photo: W. Margielewski)

over the course of several consecutive years, as in 1958–1960 and 1997–2002 (Fig. 5.6: 2) (Mikulski 1954; Ziętara 1968; Cebulak 1992, 1998–1999; Poprawa and Rączkowski 2003; Rączkowski 2007) and cause pulses (uneven activity) in geomorphic processes (Ziętara 1968). The landforms produced by new and reactivated mass movements over the past 200 years are dated by lichenometry (Bajgier-Kowalska 2008) and dendrochronology (Krąpiec and Margielewski 2000). Both methods confirm the close correlation between mass movement activity and extreme precipitation events causing floods across the Carpathian region, while

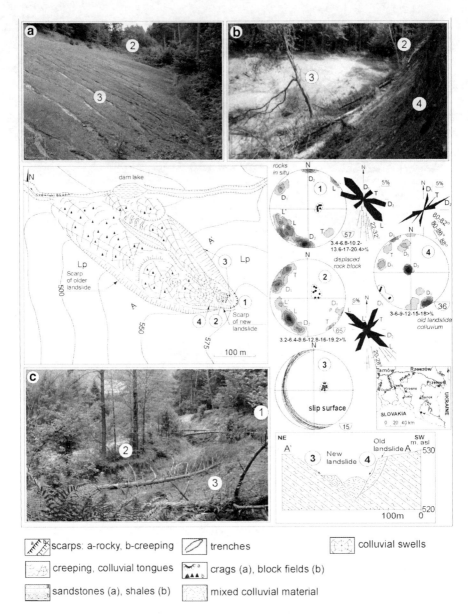

Fig. 5.8 Example of typical structural landslide, slip consequent, displaced along variegated shale (*Lp*) surface. Landslide in the Prełuki (Cygański stream), Bieszczady Mountains, formed in spring 2000 (After Margielewski 2004). Landslide successively damaged older form (signed on map and cross section). Map, cross section, and photographs of various parts of the landslide: *A* and *B* slip surface, *C* displaced rock mass. Joints on rose diagram, and contour diagram (equal area plot, pole projection on lower hemisphere, number of measurements and contour interval near each diagram). Bedding position (and slip surface) on pole point diagram (projection on lower hemisphere). Joints system; *L* longitudinal, *D1*, *D2* diagonal, *T* transversal (After Mastella et al. 1997) (Photo: W. Margielewski)

seismic tremors (Pagaczewski 1972; Schenk et al. 2001) triggering mass movements in the Polish Carpathians have not yet been properly documented (Gerlach et al. 1958; Poprawa and Rączkowski 2003). In addition to precipitation, other major factors of mass movements are slope susceptibility to landslides (geological structure), the thickness of slope deposits, vegetation cover, and human impact. Mass movements of different rate affect the sides of river valleys, valley heads, and slopes in the Beskidy Mountains and the Carpathian Foothills.

Flood-driven *lateral erosion* of valley sides is the most important trigger of mass movements, leading to abrupt displacements of slope cover and bedrock (Ziętara 1968; Kotarba 1986; Mrozek et al. 2000; Gorczyca 2004; Rączkowski 2007). Mass movements often reactivate old landslides (Fig. 5.7: 3). The resulting extensive landforms are composed of surface material and bedrock of a different age that remain subject to the influx of new material during subsequent flood events (Ziętara 1968). On a local scale, such processes lead to a gradual upslope extension and steepening of valley sides (Fig. 5.7: 3) (Ziętara 1968), accompanied by meander shifts (Starkel 1972). The colluvial material reaching the river channels is rapidly removed by floods. However, in the case of smaller valleys, an abrupt delivery of colluvia can agitate water to the point where it causes intensified fluvial erosion, local channel scour, widening of the valley floor, and steepening of valley sides (German 1998).

Mass movements can also be triggered by *headward erosion* (Starkel 1960; Ziętara 1968; Kotarba 1986). Intense precipitation leads to incision into slopes. Existing landslide surfaces become significantly larger as new material is brought down by repeated mass movements, resulting in concentric landforms (Ziętara 1968). In mountains of medium height, mass movements extend to headwater areas upslope, form stepped slope profiles, and, over thousands of years, lengthen valleys and create small tributary valleys (Ziętara 1968). In foothill areas, where valley heads are smaller, landslides help propagate them upslope. In effect, landslides shape the upper parts of slopes, and in the long term, cut through crests and form narrower ridges (Kotarba 1986).

Landslides occur as a result of the extra loading of masses of rock and regolith by rainwater. In the Carpathian Foothills they tend to be shallow and small in area (Fig. 5.7: 1–2) (Kotarba 1986), while in the Beskidy Mountains, extend high upslope and deep into the bedrock. Such landslides are called rockslides, sometimes also affecting weathered material. They can be quite extensive but of local occurrence (Fig. 5.7: 5) (Ziętara 1968; Pulinowa 1972; Oszczypko et al. 2002; Rączkowski and Mrozek 2002). In some cases, older landslides are reactivated by newer generations of mass movements and, in other cases, debris and mud flows occur in colluvia (Ziętara 1999). In the Flysch Carpathian region, under the current climatic conditions, shallow surface material and rock-weathered material landslides predominate. They take place primarily in the Carpathian Foothills and in medium-height mountains and are quite typical of deforested areas (Fig. 5.7: 1–2). Most contemporary mass movements develop in areas of previous landslide activity. They usually remove slope surface material, although there have been locally recorded cases of so strong a shearing action that bedrock became exposed (Poprawa and Rączkowski 1998). Such landslides normally do not produce remarkable topographic changes, but on a local scale, given their large number and density, even shallow movements have substantial effect: producing steeper (Fig. 5.7: 3) and also smoother valley sides (Fig. 5.7: 1). They eliminate or smoothe morphological irregularities like edges of old landslide scarps and road cuts and agricultural terraces (Ziętara 1968; Kotarba 1986; Mrozek et al. 2000; Gorczyca 2004). Repeated landslides can locally broaden and reshape valleys. An undulating or step-like slope profile with numerous depressions results. In effect, slopes affected by mass processes are characterized by an irregular shape both along their longitudinal and transversal sections (Starkel 1960; Ziętara 1968; Kotarba 1986; Gorczyca 2004; Rączkowski 2007). An important feature of landslides in regolith is their tendency to planate relief – even if the landslides in question tend to recur. Shallow landslides are moderately and occasionally influential in shaping the slopes of the Beskidy Mountains and Carpathian Foothills (Ziętara 1968; Mrozek et al. 2000; Gorczyca 2004).

Mass movements *in bedrock*, although only occuring locally, tend to be deeper and exert a more permanent impact on relief (Figs. 5.7: 5 and 5.8). Repeated landslide activity on slopes leads to a gradual fragmentation of straight slopes and the formation of side valleys. Side valleys in valley heads become elongated as a result of regression promoted by subsequent mass movements (Fig. 5.8) (Kotarba 1986). On the other hand, mass movements generate concave slope profiles (Starkel 1960). Both in the Beskidy Mountains and in the Carpathian Foothills, however, it takes thousands of years to produce perceptible topographic changes.

Rockslides can be traced for a very long time, particularly in resistant sandstones, where contemporary rockfalls of various dimensions accompany the retreat of landslide-created rock scarps (Jakubowski 1967; Ziętara 1968; Bajgier-Kowalska 2008). Large-scale landslide features produce distinct relief in resistant rocks. However, these types of deep-seated landslides have only been sporadically recorded in the past 200 years (e.g. Fig. 5.7: 5) (Zuber and Blauth 1907; Sawicki 1917; Pulinowa 1972).

5.5 Water Erosion

Jolanta Święchowicz

5.5.1 Sheet and Rill Erosion

For hydrological and geomorphological reasons, sheet and rill erosion in the flysch Carpathians is of a different nature than in the Beskidy Mountains and the Carpathian Foothills (Fig. 5.9) (Starkel 1972, 1991; Słupik 1978; Gil 1979). An *infiltration-evapotranspiration* water circulation pattern with a predominance of throughflow has been identified in middle mountains. In the case of straight and convex slopes covered by clays with abundant rock debris, rainwater tends to infiltrate, leach the regolith, and create piping tunnels. On complex slopes, steep on sandstones at the top and schist at the bottom, percolating waters into the fractured sandstone induce

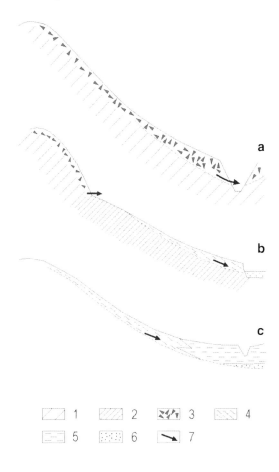

Fig. 5.9 Slope types and slope cover types in the Polish Carpathians (After Starkel 1991). (**a**) undercut mountain slope with accelerated throughflow, (**b**) bi-segmented mountain slope with limited throughflow, (**c**) foothill slope (upland); 1, sandstone series; 2, shale series; 3, debris slope cover (permeable); 4, solifluction clayey-skeletal cover; 5, alluvial clays/soils; 6, sand and pebbles; 7, principal groundwater flow directions

landslides and incision at footslopes. In mountain catchments, especially in upstream sections, slopes come into direct contact with stream channels, which directly receive water and material (Starkel 1972; Froehlich and Słupik 1977; Froehlich and Starkel 1995). In Carpathian Foothill catchments, infiltration-evapotranspiration is the dominant type of water circulation pattern with substantial input from surface runoff, which favors sheet erosion (Fig. 5.9) (Słupik 1973, 1978; Starkel 1991; Gil 1999). In the Carpathian Foothills, slopes are separated from stream channels by wide valley floors, which render direct delivery of weathered material impossible, and, thus most of it is deposited at footslopes (Święchowicz 2002c).

Vegetation cover is another key factor in the effectiveness of sheet and rill erosion. The forested slopes of the Beskidy Mountains and the Carpathian Foothills are not subject to significant sheet erosion (Table 5.2) (Gerlach 1976b; Gil 1976; Święchowicz 2002c). Forests, however, are dissected by a network of roads, footpaths, and logging tracks. While road density on forested slopes in the Beskidy Mountains is far lower than that in agricultural areas, roads play a key role in water and sediment transport. Roads facilitate flow and linear erosion and serve as delivery routes

Table 5.2 Rate of sheet and rill erosion (kg ha^{-1} year^{-1}) in different parts of the Polish Carpathians

Area	Author	Research period	Sheet and rill erosion [kg ha^{-1} year^{-1}]				
			Forest	Meadow	Fallow land	Potatoes	Cereal
Jaworki (Beskid Sądecki Mountains/ Pieniny Mountains)	Gerlach (1966, 1976a, b)	1956–1958 1968–1971		3.1–7.1 2.7–11.8		63,195.0	17,022.0
Jaszcze (Gorce Mountains)	Gerlach (1976b)	1969–1971	2.1–12.9				
Szymbark (Beskid Niski Mountains/ Ciężkowice Foothills)	Gil (1994, 1998, 1999, 2009)	1969–2000		3.5–466.1		371.5–99,139.9	10.5–8,495.1
Polanka Hallera (Wieliczka Foothills)	Drużkowski (1998)	–	25.0	50.0		75,000.0	1,250.0
Łazy (Wiśnicz Foothills)	Święchowicz (2010)	2007–2008	–	2.3–41.9	2.9–47,340.2^2	0–43,396.0	8.0–31.4

Based on Gerlach (1966, 1976b), Gil (1994, 1999, 2009), Drużkowski (1998), and Święchowicz (2010)

for suspended matter to stream channels (Froehlich and Słupik 1980a, 1986; Froehlich 1982; Soja and Prokop 1996).

Land use changes from season to season on arable lands by crop rotation and tillage operations in the Beskidy Mountains and the Carpathian Foothills. Along with changing weather conditions, this affects the rate of sheet and rill erosion (Table 5.2) (Gerlach 1966, 1976a, b; Gil 1976, 1979, 1986, 1998, 1999, 2009; Święchowicz 2002b, c, 2010). Areas of sparse or no vegetation are characterized by soil erosion several hundred times more intensive than areas with a dense vegetation cover (Table 5.2). In winter 30–40% of surfaces can be either free of vegetation or possess a sparse winter plant cover. Such conditions favor concentrated runoff and thaw facilitates rill erosion. Sheet erosion patterns are similar across fields planted with winter crops, and arable land in general. Intense water runoff takes place primarily during advective-type thaws often associated with rainfall (Słupik 1978; Gil 1999).

In summer runoff and soil erosion rates and variability are a function of land use, precipitation amounts, and rainfall intensity (Słupik 1973, 1978; Gil and Starkel 1979; Gil 1994, 1999, 2009; Starkel 1986, 1996; Święchowicz 2002a, b, 2008), as well as rainfall erosivity (EI_{30}) and 30-min maximum intensity (I_{30}) (Demczuk 2009; Święchowicz 2010). Intense sheet and rill erosion takes place on fields where plant cover is not dense enough to protect the ground surface from splash and sheet flow. Slopes and stream channels are primarily transformed by short local downpours and continuous rainfalls. Short but intense summer downpours produce precipitation amounts up to 100 mm at a rate of 1–3 mm min^{-1}. The infiltration capacity of the parent material is exceeded and heavy splashing ensues. The short-duration surface runoff transports soil over short distances to accumulate on densely vegetated slopes or on valley floors in the form of deluvial fans (Gil and Słupik 1972; Gerlach 1976b; Święchowicz 2002a, 2008). Stream channels only receive material transported by field roads from adjacent fields (Froehlich and Słupik 1980a, 1986; Gil 1994, 1999). Heavy sheet erosion mainly affects arable lands and may be catastrophic for the plowed layer (Gil 1998, 2009; Święchowicz 2008, 2009). Sheet erosion is light on fields with crops of dense surface cover and on meadows. Continuous rainfall amounting to 100 mm or more with a rate of 10 mm·h^{-1} saturates the parent material and surface runoff starts once the infiltration capacity of the soil is exceeded or precipitation intensity increases abruptly for a moment. The amount of surface runoff is then determined by the properties of the parent material and the water capacity of the slope material. At that point, land use and plant cover are less important (Słupik 1973, 1978). If the parent material becomes fully saturated with water, surface runoff restarts across affected surfaces. Surface runoff lasts significantly longer than precipitation. Sheet and rill erosion measured during continuous rainfall events is usually much more limited than during downpours (Gil 1986, 1999).

Measurements obtained on experimental fields suggest significant sheet and rill erosion on agricultural slopes. The mechanism of soil erosion is, however, complex and its high rates are not reflected directly in the *transport* and delivery *of material* from slopes to river channels. Material is transported in discrete steps over short distances and its amount varies along the length of the slope. Accumulation affects fields with dense vegetation, gentler slope sections, and footslopes (Gerlach 1966;

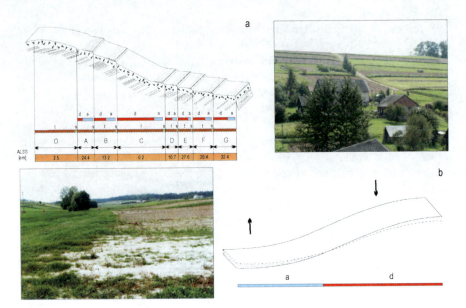

Fig. 5.10 (a) Terraced field pattern slope profile (Modified Gerlach 1966). *O* nonterraced slope, *A–G* agricultural slope, *t* terrace surface, *s* steep scarp (normally covered by grass or bush), *d* degradation surface, *a* aggradation surface, *ALSS* average loss of slope surface (cm), (b) Convex-concave slope profile: *d* degradation surface, *a* aggradation surface (Photo: J. Święchowicz)

Gil 1976; Święchowicz 2001, 2002c, 2008). Valley floors and river channels only receive material transported through (ephemeral) gullies during sudden downpours and continuous rainfall (Froehlich, Słupik 1986; Święchowicz 2008). The shape of slopes and their location with respect to local denudation zones influences material delivery to stream channels. Abundant material is transported down straight slopes and convex slopes located next to stream channels. Convex-concave slopes tend to deliver limited amounts of material. The longer the concave segment, the less material is delivered. When slopes deliver more material to valley floors, they themselves flatten out in downslope direction (downslope evolution). The opposite happens when more material is transported further in a stream channel than received (upslope evolution or stream-induced slope undercutting – Gerlach 1976b).

The transport and delivery of material from slopes to stream channels depend on the local *field pattern*. A mosaical field pattern characteristic of both the Beskidy Mountains and the Carpathian Foothills leads to different sheet and rill erosion rates on individual fields (Gil and Słupik 1972; Gerlach 1976b; Gil 1979; Święchowicz 2002a, c). In the Beskidy Mountains with high surface permeability and substantial variability in slope lengths and gradients, a field pattern parallel to contour lines prevails, and is stabilized by agricultural terraces (Gerlach 1976b; Gil 1979). In the marginal zone of the Carpathian Foothills, where permeability is lower and there is little diversity in terms of slope length and gradient, a longitudinal field pattern prevails with no elevated boundaries between fields (Święchowicz 2001). A permanent field boundary pattern and contour plowing significantly reduce soil erosion and lead to the formation of agricultural terraces. Water runoff and soil erosion take place on isolated fields that act as independent systems (Fig. 5.10) (Gerlach 1966; Słupik 1973; Gil 1976, 1979). Such fields are subject to short-distance transport and accumulation. Material is removed by a system of rills and field roads, which run downslope along field boundaries. It has been estimated that 40% less soil removal occurs on *terraced* agricultural *slopes* than on terrace-free slopes (Gerlach 1966).

Particle transfer systems are either linear-continuous or cascade-segmented. The former affects the entire length of the slope during snowmelt, most precipitation events, and is limited to roads, paths, and incisions such as rills, ephemeral gullies, and gullies. The latter affects terraced fields (Froehlich and Walling 1995). *Downslope plowing* intensifies erosion processes by orders of magnitude. If a slope is being used in a homogeneous manner from top to bottom, then sheet erosion intensity grows with the length of the slope and deluvial matter is accumulated at the foot of the slope (Fig. 5.10b) (Gil 1979; Święchowicz 2001, 2002c). Sheet and concentrated erosion are especially intense in large areas planted with the same type of crop (Święchowicz 2008). The absence of traditional erosion control measures such as elevated field boundaries and scarps intensify soil erosion. Deep rills or ephemeral gullies on slope surfaces, however, are only temporary phenomena and are usually destroyed by tillage operations (Święchowicz 2002c, 2004, 2008, 2009).

In Carpathian Foothill catchments, the delivery of material from slopes to stream channels happens locally and exclusively during extreme precipitation events. However, even then, most of the material is deposited on slope surfaces, deluvial flats at the foot of the slope, and on valley floors covered by permanent vegetation. The resulting relief is typical of the Carpathian Foothills: convex-concave, multi-segmented and stepped agricultural, and flat-bottom valleys with deluvial areas that constantly receive new slope material. The flats and flat valley floors constitute a zone that separates slopes from stream channels. It is also a zone where most of the material from slopes is deposited (Święchowicz 2002a, c).

In the Polish flysch Carpathians, the rate of slope transformation driven by sheet and rill erosion depends on vegetation cover and land use. Contour tillage, crop rotation, and all other tillage operations reduce soil erosion on arable land and lead to the formation of agricultural terraces and an irregular slope profile. On the other hand, on slopes with downslope plowing, the deposition of material at the foot of the slope or on the valley floor leads to the elongation of the concave part of the slope. This results in the elimination of clear morphological boundaries between slopes and valley floors (Fig. 5.10) (Święchowicz 2006).

5.5.2 Piping

In the Polish Carpathians the presence of debris, clayey-debris, and silt surfaces over impermeable bedrock favors piping (suffosion) (Czeppe 1960; Starkel 1960,

1965, 1972; Galarowski 1976). Piping occurs along fractures and channels produced mainly by small rodents in near-surface layers (40–60 cm). In deeper layers, living and dead tree roots serve as piping channels. Pipes also develop where slope clay comes into contact with the loam-debris layer or bedrock. Rainwater and meltwater flows under the compact turf layer along pipes and washes out fine particles. In most areas, piping is an episodic process during downpours and the melting of snow and ice (Starkel 1960). Piping features are created more frequently by meltwater than by rainwater (Czeppe 1960; Starkel 1960) and usually occur on permanent meadows and pasture lands. In tilled areas, plowing and related operations destroy existing channels and render impossible the development of typical piping landforms. Piping features are more common in the Eastern Polish Carpathians (Czeppe 1960; Starkel 1965; Galarowski 1976) than in the west (Starkel 1960). *Tunnels* are the only underground forms, as the development of shafts is limited by the shallow regolith. On the surface *piping pits*, foredeeps like cauldrons and funnels occur. Piping pits arise as a result of compaction and tend to be rather shallow (less than 1 m deep), covered with turf, and can be bowl-shaped, flat-bottom deeps, and ravines. The walls of foredeeps are usually steep while their bottoms tend to be flat and covered with turf. *Cauldrons* are of oval or rounded shape with steep sides, while their floors are parts of bottoms of piping channels. They are created when a ceiling caves in over a major tunnel. Cauldrons also develop due to landslides and the washing out of walls, eventually leading to the formation of a *piping fan* and dead-end valleys and gullies. Outlets of piping channels at the ends of valleys contribute to the upslope propagation of headwaters (Czeppe 1960; Starkel 1960; Galarowski 1976).

5.6 Aeolian Processes

Jolanta Święchowicz
Zofia Rączkowska

The rate of occurrence and the intensity of aeolian processes vary from year to year and with the individual landforms and also depend on elevation (Gerlach and Koszarski 1968; Janiga 1975; Gerlach 1976b, 1977; Welc 1977; Izmaiłow 1984a, b, 1995b; Drużkowski 1998).

Aeolian processes are induced by even moderately strong *winds* (>5 m s^{-1}), although they are the most effective with strong (>10 m s^{-1}) and very strong winds (>15 m s^{-1}). Under natural conditions, wind action is limited to vegetation-free surfaces (river banks, sandbars, levees), areas lying above the upper timberline, and in the case of larger forested areas, to the overturning of trees (Gerlach 1960, 1966, 1976a, b; Kotarba 1970). In the Tatra Mountains, as a result of a single wind event, 50,000 m^3 ha^{-1} of weathered and block-type material can be transferred on the roots of felled and uprooted trees. This produces a unique and relatively permanent slope micromorphology of mounds and depressions (Kotarba 1970).

Table 5.3 Rate of wind-driven soil particle deposition (t ha^{-1} year^{-1}) in different parts of the Polish Carpathians

Area	Author	Research period	Deposition (t ha^{-1} year^{-1})
Jasło-Sanok Depression	Gerlach T, Koszarski L (1969)	Winter 1964/1965	125.0–875.0
Bieszczady Mountains	Pękala K (1969)	Winters 1964/1965–1966/1967	1.1–9.0
Beskid Niski Mountains	Janiga S (1971)	Winters 1965/1966–1969/1970	70.0–241.0
Ciężkowice Foothills	Welc A (1977)	Winter 1968/1969	1.9–10.7
Tatra Mountains	Izmaiłow B (1984b)	1975–1980	0.01–2.65
Wieliczka Foothills	Izmaiłow B (1995a, b, c)	March 1994–February 1995	0.06–1.76

Based on Gerlach and Koszarski (1969), Pękala (1969), Janiga (1971), Welc (1977), Gerlach (1986), and Izmaiłow (1984b, 1995a, b, c)

Aeolian processes have been detected above the upper timberline in the Bieszczady Mountains (Pękala 1969) and the Tatra Mountains (Izmaiłow 1984a, b). The rate of aeolian *accumulation* on the leeward slopes of the Bieszczady Mountains ranges from 1.1 to 9.0 t ha^{-1} year^{-1}, while in the Tatra Mountains from 0.01 to 2.65 t ha^{-1} year^{-1} (Table 5.3). *Deflation niches* (Kotarba 1983) and *terracettes* develop on broad ridges in the Tatras, especially near mountain passes where wind speed can exceed 6.4 m s^{-1}, blowing away up to 1.63 t ha^{-1} year^{-1} of fine weathered particles (Izmaiłow 1984b). Aeolian processes are more decisive for relief evolution in environments affected by human impact. Arable land offers the most favorable conditions for aeolian processes. Deflation takes places mainly during winter (November–March) and is characterized by short episodes that do not recur year after year (Gerlach and Koszarski 1969; Janiga 1975; Welc 1977; Izmaiłow 1995a, b). The primary targets of the destructive power of the wind are flat surfaces and the upper parts of windward slopes, while leeward slopes are sites of accumulation. This reduces the heights of hills and slope gradients and causes footslope extension and accumulation in depressions (Gerlach 1986).

The most intensive deflation on windward slopes and the largest accumulation of aeolian sediment on leeward slopes, on the order of several hundred t ha^{-1} year^{-1}, has been detected in the Jasło-Sanok Depression (Gerlach and Koszarski 1968; 1969). Values recorded in the boundary zone between the Beskid Niski Mountains and the Carpathian Foothills do not exceed 10.7 t ha^{-1} year^{-1}, while at the edge of the Carpathian Foothills, despite the presence of silt formations, values on the order of several t ha^{-1} year^{-1} are found (Izmaiłow 1995a, c).

The intensity of deflation increases with increasing elevation above sea level (Janiga 1971, 1975). The average quantity of material deposited in the shallow valleys of the Jasło-Sanok Depression (320–350 m) separated by small hills reaches 70 t ha^{-1} year^{-1}. In the lower parts of the Beskid Niski Mountains (350–500 m), characterized by foothill relief, average deposition reached slightly over

128 t ha^{-1} year^{-1}, while in forests of higher elevation (550–600 m), it is slightly more than 241 t ha^{-1} year^{-1} (Table 5.3) (Janiga 1975).

In some parts of the Polish Carpathians, aeolian processes play a more significant part in geomorphic evolution than sheet erosion (Gerlach and Koszarski 1968). On the windward slopes of the Jasło-Sanok Depression, deflation was responsible for 60% of the soil removed over the course of the past 600 years, while sheet erosion was responsible for 40%. On the other hand, soil deposition on leeward slopes was far more common than sheet erosion (Gerlach and Koszarski 1968, 1969; Gerlach 1976b). The heights of rounded hilltops and the middle and upper parts of windward slopes have become reduced while their lower parts built up and extended. The reduction in height during the era of human activity has been about 60 cm while the amount of accumulation has reached 105 cm and extension about 50 m. Leeward slopes, except for their upper parts, have become longer and higher during the same period of time. Leeward slopes have become about 33 m longer and 15–115 cm higher. Where aeolian accumulation has been the greatest, an average of 3.5 m of new deposits have been recorded only in the last 100 years (Gerlach and Koszarski 1968, 1969; Gerlach 1976b, 1986). With the recent tendency to reduce the amount of cultivated land in the Polish Carpathians, the land area susceptible to deflation is steadily shrinking.

5.7 Periglacial Processes and Landforms

Zofia Rączkowska

In modern times, periglacial processes are significant only in high mountains such as the Tatras and Babia Góra Massif above the upper timberline. In the Tatra Mountains, they are predominant (Jahn 1958, 1975; Rączkowska 2007), although most of the active periglacial landforms can be found here at elevations ranging from 1,800 to 2,050 m (Kotarba 1976; Karrasch 1977; Rączkowska 2007). Only in the Tatras exist isolated patches of *permafrost* – above 2,300 m on southern-facing slopes and above 1,930 m on northern-facing slopes (Dobiński 1997, 2004; Mościcki and Kędzia 2001; Kędzia et al. 1998; Gądek et al. 2009; Gądek and Kędzia 2008).

Frost weathering is the principal type of weathering here. However, judging from the low rates of rockwall retreat, its intensity is rather low (Table 5.1). The intensity of weathering and rockfall was most likely the highest during the nineteenth century (Kotarba and Pech 2002), as a result of climatic conditions and earthquakes (Kotarba 2004). The lowering of ridges and slopes and the retreat of rockwalls in the Tatra Mountains as well as the cryoplanation terraces in the Babia Góra Massif are all effects of weathering.

Frost processes, such as *needle ice activity* and *solifluction*, participate in the transfer of weathered material on rock slopes with regolith mantle or on slopes of weathered material only. The rates of cryogenic processes are greater in the alpine belt than in the subalpine belt (Table 5.1). Both solifluction and creep show substantial spatial and temporal variability (Rączkowska 1999; Baranowski et al. 2004). Frost action and solifluction are primarily associated with daily and seasonal freeze-thaw cycles of the ground, mainly in spring and autumn (Baranowski et al. 2004; Rączkowska 2007). The resulting slope microtopography takes the form of *terracettes* and *solifluction garlands*. A much less common feature is the bound solifluction lobe. The processes of frost heaving and sorting produce *patterned grounds* including polygons and circles with a maximum diameter of 1 m but mostly a few dozen centimeters. In addition to patterned grounds, in the Tatras and Beskids, frost processes also create *thufurs*.

The role of *nivation* processes varies from case to case. In the Tatras, erosion processes associated with patches of snow lead to the fragmentation of slopes. *Nival niches* develop in areas occupied by snowpatches. The edges of nival niches retreat at a rate of 1–5 cm year^{-1}, while meltwater creates 20 cm deep gullies at the niche bottom or on the slope surface below. Accumulation nival niches develop on talus slopes (Rączkowska 1993, 1995, 2007). Protalus ramparts are rarely found and the height of active landforms does not exceed 1.5 m. The morphological role of nivation is limited to the Tatra Mountains and Babia Góra Massif.

5.8 Karst and Pseudokarst

Zofia Rączkowska
Włodzimierz Margielewski

Karst surfaces in the Polish Carpathians have a quite limited extension, as only fragments of the Tatras and the Pieniny Mountains are built of carbonate rocks. Given the low intensity of chemical denudation, their morphogenetic role is small under current climate conditions (Table 5.1). The highest rate of chemical denudation has been observed on carbonates in the forest belt of the Tatra Mountains (Kotarba 1972). All types of *karren* can be found on the northern-facing slopes of the Tatra Mountains, most commonly at 1,400–1,900 m elevation (Wrzosek 1933; Głazek and Wójcik 1963; Kotarba 1967, 1972). In addition, *dolines* can often become reflected in postglacial deposits in forested zones (Głazek 1964). Among the underground karst features in the Tatra massifs, foredeep-type *caves* and vertical caves with few speleothems can be mentioned (Kotarba 1972; Głazek and Grodzicki 1996). In the Pieniny Mountains, there are smaller caves less in number (Zarzycki 1982).

In the rock massifs of the Flysch Carpathians, non-karstifying rock cavities of any size are often called *pseudokarst* landforms (Vitek 1983). The largest of them can be over 20 m high and can exceed 1,500 m in length (Pulina 1997; Klassek and Mleczek 2008). They are quite common in the Polish Carpathian region. Until now, over 1,000 features of this type have been identified (Fig. 5.11: 1) (Klassek and Mleczek 2008). In most cases, they are partly of mass movement origin and can be called crevice-type (fissure caves) or talus-type caves (Fig. 5.11: 2a–c) (Vitek 1983). *Crevice-type caves* form as a result of the gradual release of shearing stress arising as a result of disturbances in slope equilibrium within rock massifs.

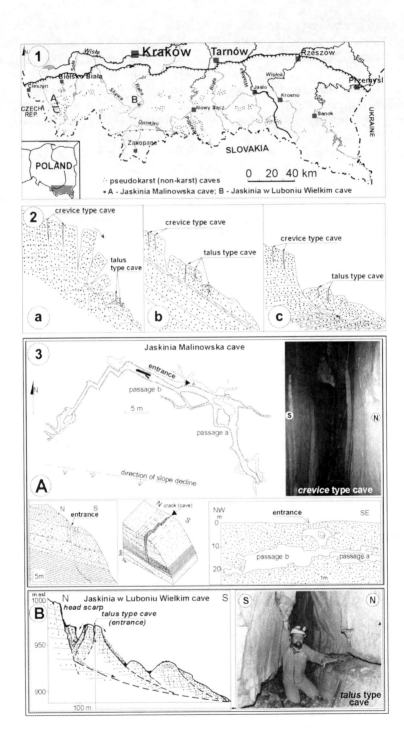

The relaxation of such stress can also take place along the surface of tectonic discontinuities such as fractures and faults, which slowly widen to cracks. This process continues until the massif becomes destroyed and gravity displaces the fragments of rock separated from the main rock mass by a system of cracks (Margielewski and Urban 2003, 2005). Crevice-type caves are an initial step in the triggering of mass movements on slopes that have not been subject to gravitational displacement up to this point (Fig. 5.11: 3A). They can also form unloading joints above landslide scarps (Fig. 5.11: 2). A unique kind of crevice-type landform are fissure caves featuring fractures along the slipping zones of landslides, called fissure macrodilatancy (Margielewski et al. 2007). *Talus-type caves* form within landslides (Fig. 5.11: 2) and can be found in gravitationally displaced rock packets (forming along fractures and faults) (Fig. 5.11: 3B), amidst chaotically arranged blocks of rock (Pulina 1997; Margielewski and Urban 2003). Speleothems are rarely found in fissure caves in flysch (Urban et al. 2007a, b). In addition, the flysch Carpathians also abound in *weathering-type caves* (fissure-type, bedding-type), partially of human and complex origin (Waga 1993; Urban and Otęska-Budzyn 1998).

5.9 Fluvial Processes

Adam Łajczak

The Polish part of the Carpathian Mountains is normally drained to the north by mountain and foothill tributaries of the Vistula River. At least 10% of this area is made up of valleys and basins. It contains Pleistocene and Holocene dry terraces that have remained intact to some extent, floodplains, river channels, old alluvial fans, and younger accumulating alluvial fans (Klimaszewski and Starkel 1972; Klimek 1979, 1991; Starkel 1991). The size of each fluvial landform depends on valley width and the sediment transport by the river flowing through the valley. The largest of fluvial landforms are found in mid-mountain basins and wider sections of large river valleys.

The rate of transformation of river channels, floodplains, and active alluvial fans depends on the frequency and magnitude of floods, the quantity of material transported

Fig. 5.11 Pseudokarst forms in the Polish Carpathians. 1, distribution of non-karst caves (pseudokarst caves) in the Polish flysch Carpathians (After Pulina 1997; compiled by Margielewski and Urban 2003, supplemented); 2, location of typical non-karst (pseudokarst) forms: crevice and talus type caves within mass movement forms, as: *a* topple, *b* translational slide, *c* rotational slide; 3, examples of non carst caves: *A* typical crevice type cave developed on slope not transformed by mass movements: Jaskinia Malinowska cave, Beskid Śląski Mountains (Map after Rachwaniec and Holek 1997; cross section after Ganszer and Pukowski 1997), *B* talus type cave in block landslide body of landslide (compound type) on Luboń Wielki Mountain, Beskid Wyspowy Mountains (After Margielewski 2006b). Position of both caves on the map above (Photo: W. Margielewski)

by rivers, and on human activity over the past few centuries. Older fluvial landforms found beyond the reach of peak floods are inactive. The construction of dams results in rapidly developing deltas, even large ones. These include channels, islands, levees, inter-levee basins, crevasses, and crevasse splays (Klimek et al. 1990; Łajczak 2006). Beaver activity is an additional factor that helps shape valley floor relief.

River channels are the most dynamic fluvial landforms. Their morphology is linked to variable lithological, orographic, and climatic conditions, vegetation cover, and past and present land use (Klimek 1979). Suspended sediment transport increases towards the east in large rivers, while bedload transport is reducing (Łajczak 1999). For this reason, west of the Dunajec River, most rivers have gravel beds, while rivers to the east tend to have long, wide sections of rock bed. Fluvial transport and fluvial landform dynamics are the most intense during summer flood events in the western part of the region. In the eastern part, on the other hand, this occurs during snowmelt. The delivery of material to channels is uneven along the lengths of major rivers depending on geological structure, relief, and land use in basins. Other factors of influence include channelization efforts and the construction of dams. The Polish part of the Carpathians constitutes 30% of the drainage basin of the upper Vistula but delivers 70% of the water and as much as 90% of the suspended sediment (Łajczak 1999). Spatially variable material delivery leads to the formation of eroded sections, transit-only sections, and accumulation sections that often alternate along the course of a river (Kaszowski et al. 1976; Froehlich et al. 1977; Kaszowski and Krzemień 1977; Klimek 1987, 1991).

Two periods, different in terms of the intensity and direction of fluvial landform development, have been identified for the period of human economic activity in the Polish Carpathians. The first period (from the seventeenth to the nineteenth century) is associated with agricultural colonization and logging activity, which have led to increased water flow dynamics in basins and increased material delivery to river channels. The second period, lasting since the late nineteenth century, has included river channelization measures and the construction of dams. The original purpose of human intervention was to limit the extent and intensity of floods as well as to affect fluvial landform dynamics.

Flood dynamics increased in the western part of the Polish Carpathians in the nineteenth century and involved a growing concentration of flood waves and increased discharge, which further enhanced sediment transport (Niemirowski 1974; Wyżga 1993a, b). Historical documents and hydrological research begun in the nineteenth century indicate that floods on the Soła River floodplain became ever more frequent from the fifteenth century to the end of the nineteenth century. The Soła is a typical gravel-bed river of the Polish Carpathians (Łajczak 2007a). The dominant tendency of floodplain accumulation and channel widening continued in the western Polish Carpathians (including the Dunajec River) until the end of the nineteenth century (Fig. 5.12: B). At times, the rivers in the region inundate the floodplain and form wider braided channels (maps by Mieg 1779–1782, Kummerer 1855, and "Die Spezialkarte..." 1894).

Large-scale, river-based transportation of logs led to the exposure of bedrock along upstream sections of rivers in the Beskidy Mountains in the nineteenth and the early twentieth centuries. Material removed along these sections of river would

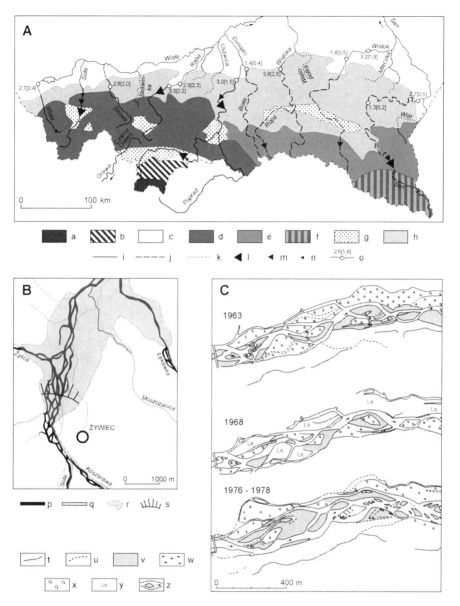

Fig. 5.12 Distribution of different types of river channels in Polish Carpathians (A), transformations of the Soła river channel in Żywiecka Basin since the 1850s (B) (After Starkel and Łajczak 2008, supplemented); transformations of the Białka river channel in Podhale, 1963–1978 (C) (After Baumgart-Kotarba 1983). Types of channels: a, Tatra; b, sub-Tatra; c, Pieniny; d, western Beskidy; e, eastern Beskidy; f, Bieszczady; g, intermontane basin; h, foothills; i, Vistula western Carpathian transit tributaries; j, Vistula eastern Carpathian transit tributaries; k, local foothill rivers. Dam reservoirs of the following water-storage capacity: l, >100 million m^3; m, 10–100 million m^3; n, <10 million m^3; o, deepening [m] of river channels in the selected water gauging stations during the twentieth century, a value in *brackets* concerns the second half of the twentieth century (After Wyżga and Lach 2002); p, a pattern of the Soła river channel according to the map by C. Kummerer (1855); q, present pattern; r, limit of backwater of the dam at Tresna on the Soła river (Żywiecki Reservoir) at the medium river stage; s, present limit of the delta front of the Soła river in the reservoir; t, clear escarpment; u, escarpments with gentle profile; v, fresh gravel bar; w, grass-covered bars; x, bars vegetated by willow and young forest; y, forest; z, channels with flow direction

build up alluvial fans and braided channels in the valleys of major rivers. The process of the *alluviation of valley floors* was associated with human activity and happened to coincide with climate change during the Little Ice Age. The gravel beds of rivers in the Polish Carpathians became shallower in the nineteenth century as a result of the introduction of potato cultivation (Klimek and Trafas 1972).

River engineering efforts commenced at the turn of the twentieth century started a trend of unprecedented river incision and this trend continues to this day. *Debris check dams* made of stone along the upstream sections of rivers and their tributaries halt bedload, allowing for the formation of small gravel-type alluvial fans. There is a tendency for these types of fluvial landforms to have coarser load towards the surface of sediment deposits. *Channelized streams* acquired a step-like longitudinal profile. Embankments made of stone or fascine stabilize the channel system. In the case of wide braided river channels, they create a narrow flow path and limit the amount of material being flushed out from the bases of riverbanks (Wyżga 1993a, b). The key reasons for Polish Carpathian rivers becoming deeper are the channelization of their downstream sections a century ago and of the Vistula River in the Polish Carpathian foreland (Klimek 1987; Wyżga 1993a; Łajczak 1999; Wyżga and Lach 2002). Water level measurements have shown that the downstream and middle sections of Carpathian tributaries of the Vistula River have become 2–3 m deeper during the twentieth century and often more than half of this change dates to the second half of the century. Land use change (that reduces suspended load transfer) has favorably affected upstream incision along eastern Polish Carpathian rivers since the 1940s. The same has been true of rivers across the entire Polish Carpathian region after 1989 (Łajczak 1999). The removal of coarse alluvia from river channels can lead to the exposure of bedrock (Soja 1977; Wyżga and Lach 2002; Krzemień 2003; Zawiejska and Krzemień 2004). The replacement of braided channels with single-thread channels, increasing channel depth, and riverbank gradient force a river to become more "efficient." Channelized rivers can transport material farther than ever before (Starkel and Łajczak 2008) with the exception of sections located behind dams with reverse flow. This does not favor the accumulation of *floodplains* in Polish Carpathian valleys, especially basins, and the part of the Carpathian Foothills where artificial levees raise riverbanks. Higher rates of accumulation have been observed across the full width of valley floors but only upstream from dams (Łajczak 1994, 1996), where large *deltas* continue to build of sand and silt. Deltas in far upstream sections of rivers could also build of coarse material. Deltas in reservoirs show three distinct zones: the normally exposed topset zone, the partially exposed foreset zone, and the usually submerged bottomset zone. In light of large fluctuations in the water levels of reservoirs up to 12 m, the longitudinal profiles of deltas are not consistent with the Gilbert delta model (Klimek et al. 1990; Łajczak 2006). Deltas extended towards the dam with sediment (mainly loam) accumulation at the bottom of the reservoir. An underwater wall, several meters high, forms upstream the dam and parallel to it, at the spot with a rising current. Downstream of dams, river channels incise as their water is not loaded with sediment. Fine particles are flushed out from coarse alluvium (Malarz 2002).

Despite intense human activity, there are still sections of river exhibiting natural flow dynamics in the Polish Carpathian region. Streams of I–IV order were usually dredged in the 1960s, while V–VII order streams tend to experience aggradation that has been moving upstream in non-forested areas (Kaszowski et al. 1976). In the eastern part of the Beskidy Mountains, an alternating pattern of erosion and accumulation sections can be observed. The following *types of river channel* have been identified in the Polish Carpathians (Fig. 5.12: A):

(a) Stream channels in the Tatras that cut into moraines, glacio-fluvial formations, and solid rock (carbonate and crystalline), with appreciable ion transport; most stable in middle sections of valleys (Kaszowski and Krzemień 1979; Krzemień 1985, 1991);
(b) River channels in the Tatra foreland flowing atop Podhale flysch, incising into glacio-fluvial and alluvial formations of the braided type. Channelization and removal of pebbles in some sections has exposed bedrock (Baumgart-Kotarba 1983; Krzemień 2003; Kościelniak 2004; Zawiejska and Krzemień 2004);
(c) The river channel of the Dunajec, unique for the Polish Carpathian region, flows through a deep gorge in the Pieniny Mountains, transporting less material downstream below the cascade;
(d) Stream and river channels in deep valleys in the western part of the Beskidy Mountains that tend to be rocky upstream and alluvial downstream (also in alluvial fans), continue as wide alluvial channels, and still braided in many places (Ziętara 1968; Krzemień 1976, 1984b);
(e) Slowly widening channels in the eastern Beskidy Mountains, which tend to alternate between gravel bars and rocky channels (Starkel 1965; Kaszowski, Kotarba 1967; Klimaszewski and Starkel 1972; Izmaiłow et al. 2006);
(f) Channels of rivers in the High Bieszczady Mountains that fit somewhere between the (d) and (e) channel types with alternating rocky channels and channels with thick layers of bottom gravel;
(g) Gravel-bed river channels in middle mountain basins, which exhibit aggradation caused by neotectonics; gravel extraction modifying their evolution; neighboring meandering streams and rivers that drain peat bogs (Baumgart-Kotarba 1983; Krzemień 2003; Kościelniak 2004; Zawiejska and Krzemień 2004; Łajczak 2007b); levee-bound floodplains in large basins with only local accumulation (Starkel and Łajczak 2008);
(h) Stream and river channels in the Carpathian Foothills which receive primarily fine-grained material from undercut banks (Krzemień and Sobiecki 1998; Święchowicz 2002c); floodplains of rivers with no artificial levees built up of suspended load;
(i) Channels of large rivers in valleys of varying width, which used to exhibit aggradation prior to channelization and now serve as transport routes for large amounts of suspended and bedload (still increasing downstream in the wake of channelization); local gravel bar building (Klimek and Trafas 1972; Wyżga 1993a; Malarz 2002) and floodplain alluviation.

The deepening of major river channels in the eastern Polish Carpathians in the twentieth century has altered their former floodplains near the edge of the Carpathians into (up to 9 m high) terraces. The narrow *floodplains* are quickly accumulating. Despite a larger gradient, the channels of gravel-bed rivers in the western Polish Carpathians have not incised considerably and their narrow floodplains below the flood terrace are not aggrading so rapidly (Starkel 2001).

5.10 Biogenic Processes

Adam Łajczak

Biogenic landforms in the Polish Carpathians include raised bogs, accumulations of large woody debris in channels and beaver habitats. The largest of peat-covered areas can be found in the Orawsko-Nowotarska Basin, in the Western Bieszczady Mountains in the upstream San and Wołosatka valleys, locally in the Tatra Mountain valleys, and on the slopes of the Beskidy Mountains (Żurek 1983, 1987). Human impact is transforming the landscape in this part of the region.

In the Orawsko-Nowotarska Basin more than 50% of the area was covered by *peat bogs* prior to its agricultural colonization (Łajczak 2007b). Areas of low peat occupied three times more space than the 18 large peat domes, two of which reached lengths of 4 and 6 km. As peat has been extracted, drained, and burned for agricultural purposes on a larger scale since the eighteenth century, total peat area in this basin dropped drastically. As a result of peat removal and fragmentation, remnants of peat domes, with numerous hollows and drainage canals, are bordered by scarps as high as 6 m. Large post-peat areas became exposed with their mineral parent material and the *paleo-channels* where peat formation had begun (Wójcikiewicz 1979; Obidowicz 1990). The removal of peat was restricted after 1990 and a recovery of peat domes and former peat bogs ensued. Currently, there are 30 peat bogs in the Polish part of the Orawsko-Nowotarska Basin with a total area of 1,312 ha. The existing peat bogs, low peat and former peat bogs included, occupy 70,000 ha or 14% of the basin. There are 18 raised bogs in the Polish part of the Bieszczady Mountains and 17 are found in a valley (Ralska-Jasiewiczowa 1972, 1980, 1989; Łajczak 2011). The bogs had been subject to human impact until the 1980s. Given the advanced state of regrowth in these areas, the geomorphological effects of this process are impossible to assess. Small peat bogs can be found in the Sucha Woda Valley and the Pańszczyca Valley in the Tatras. There are some minor peat bogs on slopes in headwater areas and in the landslide zones of the Polish Beskidy Mountains, mainly concentrated on Mount Barania Góra (1220), Pilsko Massif (1557), at the foot of the Babia Góra Massif (1725), and in the Beskid Niski Mountains (Gil et al. 1974; Obidowicz 2003; Margielewski 2006a).

Permanent accumulations of *large woody debris* in the stream and river channels of the Polish Carpathians have been noted in nature reserves and national parks (Kaczka 2003, 2008; Wyżga et al. 2003). *Beaver habitats* are expanding in Polish Carpathian valleys and, along with accumulations of woody debris, tend to cause fluvial deposition.

5.11 Human Impact

Adam Łajczak
Włodzimierz Margielewski
Zofia Rączkowska
Jolanta Święchowicz

Different forms of human impact that cause changes in relief and the intensity of geomorphological processes on a local and a regional scale have been intensifying in the Carpathian Foothills since the Neolithic (Pietrzak 2002). In the mountain interior, human impact has been a factor since the end of the Middle Ages and reached its peak intensity in the nineteenth and twentieth centuries. Various forms of human impact have been identified, in chronological order: settlement, construction, agriculture, logging, pastoral activity, mining, metal manufacturing, timber rafting, transportation, summer and winter tourism, and river engineering. Each of them developed and reached its peak at different dates.

Human impact has affected different elevations in different ways and to different degrees. High mountain areas have experienced minor human impact with the western part of the Polish Carpathians affected by more intense impact than the eastern part. Certain types of human activity still affect relief and geomorphological processes (logging, construction, river engineering, tourism) while others have been largely abandoned, however, their effects are still noticeable (mining, pastoral activity).

Human impact in high levels of the Tatras has been a fact of life for at least 200–300 years (Libelt 1988; Kaszowski et al. 1988; Mirek 1996), but intensified in the late nineteenth and the first half of the twentieth centuries (Jahn 1979; Libelt 1988; Hreško et al. 2005; Kotarba 2004, 2005), manifested in *mining* and *metal smelting,* which have left behind numerous traces: mine entrances, roads, and quarries. Furthermore, intense *sheep grazing*, which lasted until the 1960s, caused the destruction of subalpine shrubs and compact alpine meadows. The result has been intensified erosion as well as the formation of erosion niches and cattle terracettes on slopes. It took about 20 years for erosion to slow down and biological aggradation to take place following the end of sheep grazing (Jahn 1979; Kozłowska and Rączkowska 1999; Rączkowska and Kozłowska 2002). On the other hand, the homogeneous forests planted to replace natural woodland are still susceptible to the uprooting of trees by foehn winds (Kotarba 1970; Koreň 2005). Today, human impact primarily results from summer *tourism* (Kozłowska and Rączkowska 1999; Gorczyca and Krzemień 2002) and is limited to the 0.5–3.0 m wide zone along trails, where geomorphological processes are more active (Gerlach 1959; Kotarba 1976; Kłapa 1980; Krzemień 1991; Krusiec 1996; Czochański 2000; Gorczyca and Krzemień 2002; Balon 2002). These processes start in areas characterized by destroyed vegetation cover and stone pavement on trails. In addition to summer tourism, mountain climbing and skiing also impact the landscape to some extent. *Logging* in the forest zone also produces rills that result from the dragging of logs downslope. Once logging routes are abandoned, they are gradually obliterated by natural processes (Dudziak 1974). The geomorphological effects of skiing and other forms of human impact have been investigated in the Pilsko Massif area (Beskid Żywiecki Mountains).

Changes in land use in the Polish Flysch Carpathians have altered drainage patterns as well as reduced water retention depth, which favors accelerated runoff. As a result, extreme precipitation events often lead to the activation of shallow *landslides* (Jakubowski 1965, 1968; Gil 1997; Poprawa and Rączkowski 1998; Gorczyca 2004). In addition, poorly constructed hillside homes as well as road and rail construction that undercuts slopes induce pressure and landslides. The end results can be catastrophic in economic terms and as threats to human life (Fig. 5.7: 4) (Ziętara 1968; Poprawa and Rączkowski 2003; Oszczypko et al. 2002; Bajgier-Kowalska 2004). Intensive stone and gravel extraction in river valleys also upsets the equilibrium of slopes and valley sides, which can lead to mass movements (Pietrzyk-Sokulska 2005). Finally, *reservoirs* built along Polish Carpathian rivers help induce mass movements along riverbanks (Ziętara 1974; Bajgier 1992).

Intense deforestation in the eighteenth and nineteenth centuries led to an expansion of arable land. The beginning of the twentieth century saw a rapid division of farmland into smaller plots, which produced a dense network of *field roads* and *plot boundaries*. This plot mosaic survives to this day and affects the course and intensity of geomorphic processes on cultivated slopes. Consequently, weathering processes have changed significantly. *Sheet and rill erosion* is an important factor in slope evolution, as a result of which, mineral material is transported down-slope and accumulate at the base. Soil erosion is also a key process responsible for the delivery of suspended load to stream and river channels (Gerlach 1966, 1976a, b; Gil 1976, 1999; Froehlich and Słupik 1980a, 1986; Froehlich 1982; Lach 1984; Święchowicz 2002c). Intense *gully erosion* occurs on forested mountain and foothill slopes where logging is a key human activity. Agricultural slopes are characterized by terraces, plot boundaries, and wooded scarps that limit the amount of material transported downslope (cascade-fragmented system) and change their longitudinal profiles from convex or convex-concave to step-like. On shorter foothill slopes with no plot boundaries, eroded soil deposited at the slopefoot forms distinct deluvial flats, which act as a barrier to newer eroded material heading for the channel. The transfer of eroded material takes place primarily via a system of downslope ruts and field roads (linear-continuous system). In the light of the recent economic tendency to decrease the amount of arable land in the Polish Carpathians, the hazard of sheet and rill erosion and aeolian processes reduces remarkably.

Material transport and fluvial form dynamics in the river channels of the Polish Carpathians are undergoing changes as a result of human interference in the hydrology of drainage basins, a trend started as early as the seventeenth century (Klimek and Trafas 1972; Froehlich 1982; Wyżga 1993a, b; Łajczak 1999, 2007a; Krzemień 2003). The greatest *impact on fluvial processes* and landforms can be attributed to:

(a) Deforestation, agricultural colonization, the introduction of root crop cultivation on slopes (even at high elevations) with fields being plowed in downslope direction, as well as the creation of a dense network of roads separating fields, which

has resulted in substantially accelerated runoff and increased sediment delivery from slopes to stream and river channels (at least since the seventeenth century);
(b) Replacement of mixed forests in the lower forest belt with homogeneous spruce forests, another anthropogenic contribution to higher flood risk in the valleys of the Polish Carpathians and in the Vistula River valley of the Carpathian foreland as well as a key factor behind increased fluvial dynamics (nineteenth century);
(c) Creation of a dense network of forest roads, which leads to accelerated runoff and increased sediment delivery to channels (nineteenth and twentieth centuries);
(d) Channelization of streams and rivers designed to stabilize their banks; construction of debris control dams designed to halt the movement of bedload; and the construction of levees designed to decrease the size of the flooded zone in river valleys (twentieth century);
(e) Construction of dams, which permanently stop all bedload transport and up to 99% of suspended load (since 1932).

Changes in river channel morphology were rather gradual in the Polish Carpathians and have reflected the type of human impact predominant in the given period (Starkel and Łajczak 2008).

5.12 Regional Overview of Recent Landform Evolution

5.12.1 *Landform Evolution in High Mountains*

Zofia Rączkowska

Trends in relief transformation of high mountain slopes are dependent on slope type and elevation. *Rockwalls* and *rocky slopes* retreat by rockfalls and frost weathering and become fragmented and shortened by chutes and lowering (Kotarba et al. 1987). Chutes become deeper and wider through corrasion, avalanches, and debris flows (Kotarba 2002). *Debris slopes*, on the other hand, even at lower elevations with a denser plant cover, continue to accumulate. At the same time, they are fragmented by debris flow gullies and nival niches (Kotarba et al. 1987; Rączkowska 1995, 1999). Regolith slopes are shaped by cryonival processes, erosion, and mass movements, which lead to uniform transfer of slope deposits across the entire slope surface and finally to its degradation. Fresh debris flow gullies and erosion niches are rather rare.

Changes in the topography of the *valley floors* are usually limited, vary along the longitudinal profile, and are primarily restricted to stream channels (Kaszowski 1973; Baumgart-Kotarba and Kotarba 1979; Krzemień 1985, 1991; Kotarba 2002). Pleistocene glacial cirques are being filled with material delivered by debris flows and avalanches (Krzemień 1988; Kotarba 1992, 1995; Rączkowska 1999). The most significant changes occurring in valley floors, glaciated during the Pleistocene, affect footslopes through accumulation from debris flows and avalanches in the shape of poorly defined cones and tongues (Krzemień 1985, 1991; Kotarba 1992). Minor changes in stream *channel* morphology primarily arise from lateral erosion during flood events (Krzemień 1985, 1991). Major permanent changes along limited stretches of streams occur following extreme floods (e.g. in 1997) (Kotarba 1999). Despite relatively intense fluvial action, non-glacial valley floors are stable during event-free periods (Kaszowski 1973; Krzemień 1991). Bedrock channels exhibit a smoothing trend along their longitudinal profile (Kaszowski 1973). Major changes in channels are caused by extreme hydrometeorological events, which may completely flush out deposits (Kaszowski 1973; Kotarba 1999).

In general, debris flows, dirty avalanches, and extreme floods are crucial in the transformation of the relief. For this reason, the stabilizing trend after the Little Ice Age and with reduced human impact may reverse as the frequency of extreme events is increasing.

5.12.2 Landform Evolution in Middle Mountains

Włodzimierz Margielewski
Jolanta Święchowicz

In middle and low mountains, *linear (gully and stream) erosion* is central in the transformation of slopes during catastrophic precipitation events, leading to the incision of valleys, field roads, and logging roads, as well as to the initiation of surface mass movements (Ziętara 1968). Leaching, piping, and the uprooting of trees by wind are also very important, while sheet erosion is negligible on wooded and grassy slopes in the Beskids.

Numerous shallow landslides affect weathered bedrock (Fig. 5.7: 4 and 5. 8). Valley deepening and widening during floods induces mass movements, which transform the sides of valleys (Ziętara 1968). In some cases, landslides can shape large slope surfaces and their development is often stimulated by human activity (Ziętara 1968; Oszczypko et al. 2002; Bajgier-Kowalska 2004). However, given their discrete occurrence, their contribution to general slope evolution is limited and local. Furthermore, rock slides, deep-seated and extensive landslides, are rather rare in modern times (Fig. 5.7: 5). What happens most of the time can be termed a *reactivation of old landslides* (as observed in 1958–1960 and 1997–2002) involving only minor changes (local widening of valley sides and negligible retreats of valley heads – Ziętara 1968). Significant changes in the relief of the Beskidy Mountains only occur over the longer term (in thousands of years) and include the retreat of valley heads, the lengthening of river valleys, and the formation of tributary valleys and concave landslide slope profiles (Starkel 1960; Ziętara 1968; Kotarba 1986). In general, contemporary landslides are combined with mudflows, soil creep, sheet erosion, piping in unconsolidated material, erosion, and weathering (Fig. 5.8 upper left photo) (Ziętara 1968).

Terraced cultivated slopes are far less affected by sheet and rill erosion. Material is transported from slope surfaces directly to river channels only during heavy and lasting rainfalls. In such cases, natural rills, gullies, and field roads serve as main transport routes (Froehlich and Słupik 1980a, 1986; Froehlich 1982).

5.12.3 Landform Evolution in Foothills

Włodzimierz Margielewski
Jolanta Święchowicz

The Carpathian Foothills are used for agricultural purposes and shaped by soil erosion and mass movements, while in valley bottoms sediments transported downslope are deposited. In contrast to the Beskids, the Carpathian Foothills tend to be characterized by long fields perpendicular or diagonal to contour lines with no stabilized plot boundaries (Święchowicz 2001). Downslope plowing of the slope intensifies erosion processes quite substantially. *Sheet, rill,* and *ephemeral gully erosion* after heavy rainfalls particularly affects commercial farms with large acreage of crops (Święchowicz 2008, 2009). In foothill catchments, slope material is delivered to stream channels only in small amounts and during extreme precipitation events. Even then, most of the transported material is deposited close to footslopes, on deluvial plains, and across valley floors. This produces landforms typical of the Carpathian Foothills: *convex-concave, multi-segmented and step-like agricultural slopes* as well as *flat valley floors* with occasional deluvial surfaces (Święchowicz 2001, 2002a, b, 2008, 2009). Aeolian activity is only important in certain regions (e.g., in the Jasło-Sanok Depression), reducing slope inclination and transforming slope profiles (Gerlach and Koszarski 1968; Gerlach 1986).

Mass movements are important and sometimes predominant geomorphic processes. Although many *landslides* induced by extreme precipitation may be small (Fig. 5.7: 1–2), they are numerous. They make slopes steeper and smoother as well as help form wider river valleys. In general, however, they smooth out all morphological irregularities on slopes, creating *slopes of undulating profile* (Starkel 1960; Gorczyca 2004). Landslides in the heads of minor tributary valleys make them extend upslope, leading to crest fragmentation, overall crest lowering, and narrower ridges over the long term in the Beskids (Kotarba 1986).

Contemporary changes in land use (less arable land and more woods and meadows) reduce sheet erosion, but increase gully erosion and channel incision and also favor shallow landslides (Jakubowski 1964).

5.12.4 Valley Floor Evolution

Adam Łajczak

The longitudinal profiles of rivers in the Polish Carpathians and the Carpathian foreland are undergoing changes caused by *river engineering*, which include dam construction and other modifications of river channels and floodplains, which have reached unprecedented extents locally (Starkel and Łajczak 2008). Prior to river engineering, the gravel-bed braided rivers in the Polish Carpathians used to incise their channels upstream, aggrade, and widen downstream, allowing the inundation of floodplains. The erosional sections of rivers draining the eastern part of the Polish Carpathians used to be longer than those of rivers in the west. It was at that time that channels tended to become deeper downstream. Lower floodplains were starting to form, although aggradation also happened along the meandering sections of rivers. River engineering efforts and changes in land use led to the lengthening of erosional sections of river channels. Today, most rivers are eroding virtually along their entire length. Large-scale *deposition* of (mainly suspended) material occurs only in the backwaters of dams, especially *in long and deep reservoirs*. The suspended load accumulates in the flooded channel and across the former floodplain. *Channel incision* intensifies *below dams* and former floodplains are turned into dry terraces. Three dynamic types of longitudinal river channel and floodplain profiles have been identified in the Polish Carpathians in recent years, reflecting current land use trends (Starkel and Łajczak 2008):

(a) The gravel-bed type with an incised downstream channel and large growing deltas in reservoirs;
(b) Eastern Carpathian type with mainly suspended matter transports and currently incision virtually along the entire river length;
(c) Foothill river type with a floodplain undergoing rapid deposition.

5.13 Conclusions

Adam Łajczak
Włodzimierz Margielewski
Zofia Rączkowska
Jolanta Święchowicz

Deforestation and cultivation, particularly in the Carpathian Foothills, have made *sheet erosion* the predominant geomorphic process during the past 200 years. Material washed off fields builds up across deluvial flats at the base of slopes, which encroach onto flat valley floors. The steeper slopes of the Beskids show agricultural terraces that stabilize slope surfaces with fields that run perpendicular to contour lines, which results in limited sheet erosion. Gullies and field roads play key roles in the transport of surface material from slopes. In certain parts of the Polish Carpathians, intense aeolian processes lower crests and windward slopes, while leeward slopes build up and become longer.

Over the last 20 years the amount of *agricultural land* tends to *decrease* along with the rate of sheet erosion and deflation, while gully erosion is intensifying on slopes, river channels are incising, and shallow landslides are reactivating. The reduced delivery of fine-grained material from slopes to valley bottoms as well as the channelization of rivers since the early twentieth century has resulted in deeper river channels and reduced rates of sediment deposition on floodplains. The opposite is true in valley floors upstream from dams. On the other hand, rivers in the Carpathian Foothills that drain areas still dominated by agriculture are characterized by wide floodplains that continue to collect sediment during flood events.

Only high-energy precipitation can normally lead to simultaneous transformations of slopes and valley floors in high mountain, middle mountain, and foothill areas. Despite the creation of terraces and the construction of all types of dams on valley floors, localized *debris flows* overrun barriers and transport material along the full lengths of slopes and river channels, which carry suspended sediment beyond the Polish Carpathian region.

References

Alexandrowicz SW (1997) Holocene dated landslides in the Polish Carpathians. Palaeoclim Res 19:75–83

Bajgier M (1992) Rzeźba Pogórza Wielickiego ze szczególnym uwzględnieniem współczesnych procesów stokowych w otoczeniu zbiornika wodnego w Dobczycach (Relief of the Wieliczka Foothills with particular consideration of contemporary slope processes in the surroundings of the Dobczyce Reservoir). Rocz Nauk Dydakt Wyż Szk Pedagog Kraków Pr Geogr 14:33–61 (in Polish)

Bajgier-Kowalska M (2004) The role of the human activity in the incidence and reactivation of landslides in the flysch Carpathians. Folia Geogr Ser Geogr Phys 36:11–30

Bajgier-Kowalska M (2008) Lichenometric dating of landslide episodes in the Western part of the Polish Flysch Carpathians. Catena 72:224–234

Bajgier-Kowalska M, Ziętara T (2002) Sukcesja ruchów osuwiskowych w ostatnim 5-leciu w Karpatach fliszowych (Succession of the mass movements during last 5-years in the Flysch Carpathians). Prob Zagospod Ziem Górskich 48:31–42 (in Polish)

Balon J (2002) Górna granica kosodrzewiny jako wskaźnik stabilności geosystemu Tatr. In: Przemiany środowiska przyrodniczego Tatr (Upper boundary of the dwarf pine vegetation belt as the stability indicator of the Tatra Mts geosystem). Tatrzański Park Nar.-Pol. Tow. Przyjaciół Nauk o Ziemi, Kraków/Zakopane, pp 131–137 (in Polish with English summary)

Baranowski J, Rączkowska Z, Kędzia S (2004) Soil freezing and its relation to slow movements on alpine slopes (of the Tatra Mountains, Poland). Ann Univ de Vest din Timişoara Geogr 19:169–179

Baumgart-Kotarba M (1983) Kształtowanie koryt i teras rzecznych w warunkach zróżnicownaych ruchów tektonicznych (Channel and terrace formation due to differential tectonic movements – with the Eastern Podhale Basin as example). Pr Geogr Inst Geogr Przestrz Zagospod PAN 145:1–133 (in Polish with English summary)

Baumgart-Kotarba M, Kotarba A (1979) Wpływ rzeźby dna doliny i litologii utworów czwartorzędowych na wykształcenie koryta Białej Wody w Tatrach (Relation between the Quaternary relief and glacial deposits lithology and channel morphology). Folia Geogr Ser Geogr Phys 12:49–65 (in Polish with English summary)

Birkenmajer K (1986) Zarys ewolucji geologicznej pienińskiego pasa skałkowego (Outline of the geological evolution of the Pieniny Klippen Belt). Prz Geol 34:293–304 (in Polish)

Bober L (1984) Rejony osuwiskowe w polskich Karpatach fliszowych (Landslides in the Polish Flysch Carpathians and their relation to geology of the region). Biul Inst Geol 340:115–158 (in Polish with English summary)

Cebulak E (1992) Maksymalne opady dobowe w dorzeczu górnej Wisły (Maximum daily precipitation in the Upper Vistula Basin). Zesz Nauk Uniw Jagiell Pr Geogr 90:79–96 (in Polish with English summary)

Cebulak E (1998–1999) Charakterystyka wysokich opadów wywołujących wezbrania rzek karpackich (Characteristics of heavy precipitation causing increased flow in the Carpathian rivers). Folia Geogr Ser Geogr Phys 29–30:43–65 (in Polish with English summary)

Cegła J (1963) Porównanie utworów pyłowych kotlin karpackich z lessami Polski (On the origin of the Quaternary silts in the Carpathian Mountains). Ann UMCS Lublin Sec B 18:70–116 (in Polish with English summary)

Chełmicki W, Skąpski R, Soja R (1998–1999) Reżim hydrologiczny rzek karpackich w Polsce (The flow regime of Poland's Carpathian rivers). Folia Geogr Ser Geogr Phys 29–30:67–80 (in Polish)

Czeppe Z (1960) Zjawiska sufozyjne w glinach zboczowych górnej części dorzecza Sanu (Piping in the slope loams of the upper San drainage basin). Biul Państw Inst Geol 150:297–332 (in Polish)

Czochański JT (2000) Wpływ użytkowania turystycznego na rozwój procesów i form erozyjno-denudacyjnych w otoczeniu szlaków (Influence of tourism on erosional-denudational processes and forms in surrounding of tourist paths). In: Borowiak D, Czochański JT (eds) Z badań geograficznych w Tatrach Polskich, Wydaw. Uniw. Gdańskiego, Gdańsk, pp 331–348 (in Polish)

Demczuk P (2009) Wpływ erozyjności deszczu na wielkość erozji gleb w zlewni Bystrzanki w latach 1969–1993 (The influence of rainfall erosivity on soil erosion in Bystrzanka catchment in 1969–1993). In: Bochenek W, Kijowska M (eds) Zintegrowany Monitoring Środowiska Przyrodniczego. Funkcjonowanie środowiska przyrodniczego w okresie przemian gospodarczych w Polsce. PIOŚ, Warszawa, pp 231–238 (in Polish)

Die Spezialkarte der Österreichisch-Ungarischen Monarchie (1894) 1:75,000. Wien

Dobija A (1981) Sezonowa zmienność odpływu w zlewni górnej Wisły (po Zawichost) (Seasonal variability of river runoff in the upper Vistula Basin, Zawichost). Zesz Nauk Uniw Jagiell Pr Geogr 53:51–112 (in Polish with English summary)

Dobiński W (1997) Distribution of mountain permafrost in the High Tatras based on freezing and thawing indices. Biul Peryglac 36:29–37

Dobiński W (2004) Wieloletnia zmarzlina w Tatrach: geneza, cechy, ewolucja (Permafrost in the Tatra Mountains: genesis, features, evolution). Prz Geogr 76(3):327–343 (in Polish with English summary)

Drużkowski M (1998) Współczesna dynamika, funkcjonowanie i przemiany krajobrazu Pogórza Karpackiego (Current dynamics, functioning and transformations of the Carpathian Foothills landscape). Inst. Bot. Uniw. Jagiell, Kraków, 285 p (in Polish with English summary)

Dudziak J (1974) Obserwacje nad rozwojem rynien stokowych na polanach tatrzańskich (Development of slope gullies at glades in the Tatras). Czas Geogr 45(1):31–45 (in Polish with English summary)

Dynowska I (1972) Typy reżimów rzecznych w Polsce (Types of river regimes in Poland). Zesz Nauk Uniw Jagiell Pr Geogr 50:1–150 (in Polish with English summary)

Dynowska I (1995) Wody (Waters). In: Warszyńska J (ed) Karpaty Polskie. Przyroda, człowiek i jego działalność. Wyd. Uniw. Jagiell, Kraków, pp 49–67 (in Polish)

Dziewański J, Starkel L (1967) Slope covers on the middle Terrace at Zabrodzie upon the San. Stud Geomorphol Carpatho Balc 1:21–35

Flis J (1958) Formy terenu wywołane grawitacyjnymi ruchami mas skalnych w Sądecczyźnie (Morphological forms affected by gravitational mass movements in the Beskid Sądecki Mountains). Rocz Nauk Dydakt Wyż Szk Pedagog Kraków Geogr 8:35–54 (in Polish)

Froehlich W (1975) Dynamika transportu fluwialnego Kamienicy Nawojowskiej (The dynamics of fluvial transport in the Kamienica Nawojowska). Pr Geogr Inst Geogr Przestrz Zagospod PAN 114:1–122 (in Polish with English summary)

Froehlich W (1982) Mechanizm erozji i transportu fluwialnego w zlewniach beskidzkich (The mechanism of fluvial transport and the waste supply into the stream channel in a mountainous flysch catchment). Pr Geogr Inst Geogr Przestrz Zagospod PAN 143:1–144 (in Polish with English summary)

Froehlich W (1986) Influence of the slope gradient and supply area on splash: scope of the problem. Z Geomorph NF 60(Suppl):105–114

Froehlich W, Słupik J (1977) Metody badań transformacji opadu w odpływ oraz erozji na stoku w zlewni Homerki (Beskid Sądecki) (Methods of research on transformation of rainfall into runoff and slope erosion in Homerka catchment, Beskid Sądecki). In: Zasoby wodne w małych zlewniach. Ocena i zagospodarowanie. Seminarium Komisji Gosp. Wodnej i IMUZ-u, Falenty, pp 55–70 (in Polish)

Froehlich W, Słupik J (1980a) Drogi polne jako źródła dostawy wody i zwietrzelin do koryta cieku (Field roads as sources of water and sediment contribution to stream channel). Zesz Probl Post Nauk Rol 235:257–268 (in Polish with English summary)

Froehlich W, Słupik J (1980b) Importance of splash in erosion process with a small flysch catchment basin. Stud Geomorphol Carpatho Balc 14:77–112

Froehlich W, Słupik J (1986) Rola dróg w kształtowaniu spływu i erozji w karpackich zlewniach fliszowych (The role of roads in the generation of flow and erosion in the Carpathian flysch drainage basins). Prz Geogr 58(1–2):129–160 (in Polish with English summary)

Froehlich W, Starkel L (1995) The response of slope and channel systems to various types of extreme rainfall: a comparison between the temperate zone and humid tropics. Geomorphology 11:337–345

Froehlich W, Walling DE (1995) Ewolucja użytkowanego rolniczo stoku beskidzkiego w świetle badań metodami klasycznymi i radioizotopowymi (Evolution of agricultural Beskidian slope in view of classic and radioisotopic methods). In: Procesy geomorfologiczne. Zapis w rzeźbie i osadach. III Zjazd Geomorfol. Pol. Streszcz. komun. posterów i referatów, Wydz. Nauk o Ziemi Uniw. Śląski, Sosnowiec, pp 22–23 (in Polish)

Froehlich W, Kaszowski L, Starkel L (1977) Studies of present-day and past river activity in the Polish Carpathians. In: Gregory KJ (ed) River channel changes. Wiley, Chichester/New York, pp 418–428

Galarowski T (1976) New observations of the present-day suffosion (piping) processes in the Bereźnica catchment basin in the Bieszczady Mountains (Eastern Carpathians). Stud Geomorphol Carpatho Balc 1:115–124

Ganszer J, Pukowski J (1997) Jaskinia Malinowska – przekrój (Jaskinia Malinowska cave – cross section). In: Pulina M (ed) Jaskinie polskich Karpat fliszowych, vol 1. Pol. Tow. Przyjaciół Nauk o Ziemi, Warszawa (in Polish)

Gądek B, Kędzia S (2008) Winter ground surface temperature regimes in the zone of sporadic discontinuous permafrost, Tatra Mountains (Poland and Slovakia). Permafrost and Periglacial Proc 19(2):315–321

Gądek B, Kotyrba A (2003) Kopalny lód lodowcowy w Tatrach? (Fossil glacial ice?). Prz Geol 51(7):571 (in Polish)

Gądek B, Rączkowska Z, Żogała B (2009) Debris slope morphodynamics as a permafrost indicator in zone of sporadic permafrost, High-Tatras, Slovakia. Z Geomorph NF 53(Suppl 3):79–100

Gerlach T (1959) Needle ice and its role in the displacement of the cover of waste material in the Tatra Mountains. Prz Geogr 31:590–605

Gerlach T (1960) W sprawie kopczyków ziemnych na Hali Długiej w Gorcach (Report on the origin of small earth hillocks on the Hala Długa in the Gorce Mountains). Prz Geogr 32:1–2 (in Polish)

Gerlach T (1966) Współczesny rozwój stoków w dorzeczu górnego Grajcarka (Beskid Wysoki) (Developpement actuel des versants dans le Basin du Haut Grajcarek, Hautes Beskides, Carpates Occidentales). Pr Geogr Inst Geogr PAN 52:1–124 (in Polish with French summary)

Gerlach T (1976a) Bombardująca działalność kropel deszczu i jej znaczenie w przemieszczaniu gleby na stokach (L'importance de l'action des gouttes de pluie pour le transport du sol sur les versants). Stud Geomorphol Carpatho Balc 10:125–137 (in Polish with French summary)

Gerlach T (1976b) Współczesny rozwój stoków w polskich Karpatach Fliszowych (Present-day slope development in the Polish flysh Carpathians). Pr Geogr Inst Geogr Przestrz Zagospod PAN 122:1–116 (in Polish with French summary)

Gerlach T (1977) The role of wind in the present-day soil formation and fashioning of the Carpathian slopes. Folia Quat 49:93–113

Gerlach T (1986) Erozja wietrzna i jej udział w erozji gleb w Karpatach (Eolian erosion and its contribution to soil degradation in the Carpathians). Folia Geogr Ser Geogr Phys 18:59–72 (in Polish with English summary)

Gerlach T, Koszarski L (1968) Współczesna rola morfogenetyczna wiatru na przedpolu Beskidu Niskiego (The role of the wind in the contemporary morphogenesis of the Lower Beskid range, Flysh Carpathians). Stud Geomorphol Carpatho Balc 2:85–114 (in Polish with English summary)

Gerlach T, Koszarski L (1969) Badania nad pokrywami stokowymi w rejonie silnej współczesnej działalności wiatrów (Research on slope covers in areas affected by strong winds). Spraw z Posiedz Kom Oddz PAN w Krakowie 12(1):229–231 (in Polish)

Gerlach T, Pokorny J, Wolnik R (1958) Osuwisko w Lipowicy (The Landslide at Lipowica). Czas Geogr 30:685–700 (in Polish with English summary)

Gerlach T, Krysowska-Iwaszkiewicz M, Szczepanek K, Alexandrowicz SW (1991) Karpacka odmiana lessów w Humniskach koło Brzozowa na Pogórzu Dynowskim w polskich Karpatach fliszowych (The Carpathian variety of loesses at Humniska near Brzozów in the Polish flysch Carpathians). Kwart AGH Geol 17(1–2):193–219 (in Polish)

German K (1998) Przebieg wezbrania i powodzi 9 lipca 1997 roku w okolicach Żegociny oraz ich skutki w krajobrazie (Flood at 9 July 1997 in Żegocina and its effects in landscape). In: Starkel L, Grela J (eds) Powódź w dorzeczu górnej Wisły w lipcu 1997 roku. Wyd. Oddz. PAN, Kraków, pp 177–184 (in Polish)

Gil E (1976) Spłukiwanie gleby na stokach fliszowych w rejonie Szymbarku (Slopewash on flysch slopes in the region of Szymbark). Dok Geogr 2:1–65 (in Polish with English summary)

Gil E (1979) Typologia i ocena środowiska naturalnego okolic Szymbarku (Typology and evaluation of the natural environment in the region of Szymbark). Dok Geogr 5:1–91 (in Polish with English summary)

Gil E (1986) Rola użytkowania ziemi w przebiegu spływu powierzchniowego i spłukiwania na stokach fliszowych (The role of land use in the processes of the surface runoff and wash down in the flysch slopes). Prz Geogr 58:51–65 (in Polish with English summary)

Gil E (1994) Monitoring obiegu wody i spłukiwania na stokach (Monitoring of water circulation and slopewash). In: Starkel L, Bochenek W (eds) Zintegrowany Monitoring Środowiska Przyrodniczego. Stacja Bazowa Szymbark (Karpaty Fliszowe). PIOŚ, Warszawa, pp 66–87 (in Polish)

Gil E (1997) Meteorological and hydrological conditions of landslide, Polish Flysch Carpathians. Stud Geomorphol Carpatho Balc 31:143–157

Gil E (1998) Spływ wody i procesy geomorfologiczne w zlewniach fliszowych podczas gwałtownej ulewy w Szymbarku w dniu 7 czerwca 1985 roku (Runoff and geomorphic processes in the flysch catchments during heavy downpours in Szymbark on 7 June 1985). Dok Geogr 11:85–107 (in Polish with English summary)

Gil E (1999) Obieg wody i spłukiwanie na fliszowych stokach użytkowanych rolniczo w latach 1980–1990 (Water circulation and wash down on the flysch slopes used for farming purposes in 1980–1990). Zesz Inst Geogr Przestrz Zagosp PAN 60:1–78 (in Polish with English summary)

Gil E (2009) Ekstremalne wartości spłukiwania gleby na stokach użytkowanych rolniczo w Karpatach fliszowych (Extreme values of soil downwash on cultivated slopes in the Polish flysch Carpathians). In: Bochenek W, Kijowska M (eds) Zintegrowany Monitoring Środowiska Przyrodniczego. Funkcjonowanie środowiska przyrodniczego w okresie przemian gospodarczych w Polsce. PIOŚ, Warszawa, pp 191–218 (in Polish with English summary)

Gil E, Kotarba A (1977) Model of slide slope evolution in flysch mountains (an example drawn from the Polish Carpathians). Catena 4:233–248

Gil E, Słupik J (1972) The influence of plant cover and land use on the surface run-off and washdown during heavy rain. Stud Geomorphol Carpatho Balc 6:181–190

Gil E, Starkel L (1979) Long-term extreme rainfalls and their role in the modeling of the flysch slopes. Stud Geomorphol Carpatho Balc 13:207–220

Gil E, Gilot E, Szczepanek K, Kotarba A, Starkel L (1974) An Early Holocene landslide in the Beskid Niski and its significance for palaeogeographical reconstructions. Stud Geomorphol Carpatho Balc 8:69–83

Gilewska S (1986) Podział Polski na jednostki geomorfologiczne (Geomorphological subdivisions of Poland). Prz Geogr 58:15–42 (in Polish with English summary)

Głazek J (1964) Kras podmorenowy Doliny Pańszczycy w Tatrach (Karst under moraine in the Pańszczyca valley in the Tatras). Kwart Geol 8:161–170 (in Polish)

Głazek J, Grodzicki J (1996) Kras i jaskinie (Karst and caves). In: Mirek Z (ed) Przyroda Tatrzańskiego Parku Narodowego. Tatrzański Park Narodowy, Kraków/Zakopane, pp 139–168 (in Polish with English summary)

Głazek J, Wójcik Z (1963) Zjawiska krasowe wschodniej części Tatr Polskich (Karst phenomena in the Eastern part of the Polish Tatra Mountains). Acta Geol Pol 13:91–124 (in Polish with English summary)

Gorczyca E (2004) Przekształcanie stoków fliszowych przez ruchy masowe podczas katastrofalnych wezbrań (The transformation of flysch slopes by catastrophic rainfall-induced mass-processes, Łososina River catchment basin). Wyd. Uniw. Jagiell, Kraków, 101 p (in Polish with English summary)

Gorczyca E, Krzemień K (2002) Wpływ ruchu turystycznego na rzeźbę Tatrzańskiego Parku Narodowego (Impact of tourism on the relief of the Tatra National Park). In: Przemiany środowiska przyrodniczego Tatr. Tatrzański Park Nar.-Pol. Tow. Przyjaciół Nauk o Ziemi, Kraków/Zakopane, pp 389–394 (in Polish with English summary)

Henkiel A (1972) Solifukcja w polskich Karpatach (Solifluction in the Polish Carpathians). Czas Geogr 43:295–305 (in Polish with English summary)

Hess M (1965) Piętra klimatyczne w polskich Karpatach Zachodnich (Vertical climatic zones in the Polish Western Carpathians). Zesz Nauk Uniw Jagiell Pr Geogr 11:1–267 (in Polish with English summary)

Hreško J, Boltizar M, Bugar G (2005) The present-day development of landforms and landcover in alpine environment – Tatra Mts (Slovakia). Stud Geomorphol Carpatho Balc 39:23–48

Izmaiłow B (1984a) Eolian deposition above the upper timber line in the Gąsienicowa valley in the Tatra Mountains. Zesz Nauk Uniw Jagiell Pr Geogr 61:43–59

Izmaiłow B (1984b) Eolian process in alpine belt of the High Tatra Mountains, Poland. Earth Surf Process Landf 9:143–151

Izmaiłow B (1995a) Klimatyczne uwarunkowania morfologicznej działalności wiatru na progu Pogórza Wielickiego w rejonie Bochni (Climatic conditions of morphological activity of wind on the margins of the Wieliczka Foothills near Bochnia). In: Kaszowski L (ed) Dynamika i antropogeniczne przeobrażenia środowiska przyrodniczego Progu Karpat między Rabą a Uszwicą. Inst. Geogr. Uniw. Jagiell, Kraków, pp 195–203 (in Polish)

Izmaiłow B (1995b) Warunki anemologiczne Pogórza Karpackiego w rejonie Łazów w latach 1989–1992 i ich znaczenie dla procesów eolicznych (Anemological conditions of the Carpathian Foothills in the Bochnia environs during the period 1989–1992 and their importance for aeolian processes). Zesz Nauk Uniw Jagiell Pr Geogr 100:35–46 (in Polish with English summary)

Izmaiłow B (1995c) Wstępne wyniki badań nad eolicznym obiegiem materii w progowej części Pogórza Karpackiego koło Bochni (Preliminary results of aeolian matter circulation in the edge zone of the Carpathian Foothills near Bochnia). In: Kaszowski L (ed) Dynamika i antropogeniczne przeobrażenia środowiska przyrodniczego Progu Karpat między Rabą a Uszwicą. Inst. Geogr. Uniw. Jagiell, Kraków, pp 205–219 (in Polish)

Izmaiłow B, Kamykowska M, Krzemień K (2006) The geomorphological effects of flash floods in mountain river channels. The case study of the River Wilsznia (Western Carpathian Mountains). Pr Geogr Inst Geogr i Gospod Przestrz Uniw Jagiell 116:89–97

Jahn A (1958) Mikrorelief peryglacjalny Tatr i Babiej Góry (Periglacial microrelief in the Tatras and Babia Góra Massif). Biul Peryglac 4:227–249 (in Polish)

Jahn A (1975) Problems of the periglacial zone. Wyd. PWN, Warszawa, 223 p

Jahn A (1979) On Holocene and present-day morphogenetic processes in the Tatra Mountains. Stud Geomorphol Carpatho Balc 13:111–129

Jakubowski K (1964) Płytkie osuwiska zwietrzelinowe na Podhalu (Shallow weathering landslides in the Podhale region). Pr Muz Ziemi 6:113–145 (in Polish)

Jakubowski K (1965) Wpływ pokrycia roślinnego oraz opadów atmosferycznych na powstawanie osuwisk zwietrzelinowych (Influence of plant cover and precipitation on weathering landslides formation). Prz Geol 9:395–399 (in Polish)

Jakubowski K (1967) Badania nad przebiegiem wtórnych przeobrażeń form osuwiskowych na obszarze fliszu karpackiego (Investigations of secondary modifications of the landslide forms in the Carpathian Flysch). Pr Muz Ziemi 11:223–243 (in Polish with English summary)

Jakubowski K (1968) Rola płytkich ruchów osuwiskowych zwietrzeliny w procesach zboczowych na terenie wschodniego Podhala (Influence of shallow gravitational movement of the weathering cover on slope processes in the Eastern Podhale Region). Pr Muz Ziemi 13:137–314 (in Polish)

Janiga S (1971) Deflacyjna rola wiatru w kształtowaniu rzeźby Beskidu Niskiego (The deflanational effect of wind with shaping the relief of the Low Beskid Mountains). Prz Geogr 43:427–433 (in Polish with English summary)

Janiga S (1975) Deflacyjna i akumulacyjna rola wiatru w okresach zimowych na obszarze Beskidu Niskiego (Deflational and accumulational activities of wind during winter seasons in the Beskid Niski). Pr Monogr Wyż Szk Pedagog w Krakowie 15:1–51 (in Polish with English summary)

Kacprzak A (2002–2003) Pokrywy stokowe jako przedmiot badań geomorfologicznych i gleboznawczych (Slope deposits as a subject of geomorphological and pedological studies). Folia Geogr Ser Geogr Phys 33–34:27–37 (in Polish with English summary)

Kaczka RJ (2003) Coarse woody debris dams in mountain streams in Central Europe, structure and distribution. Stud Geomorphol Carpatho Balc 37:111–127

Kaczka RJ (2008) Dynamic of large woody derbis and wood dams in mountain Kamienica Stream, Polish Carpathians. In: Kaczka RJ, Malik I, Owczarek P, Gartner H, Helle G, Schleser G (eds) Tree rings in archaeology, climatology and ecology, vol 7. Association for Tree Ring Research, Potsdam, pp 133–137

Karrasch H (1977) Die klimatischen und aklimatischen Varianzfaktoren der periglazialen Höhenstufee in den Gebirgen West-und Mitteleuropas. Abh Akad Wiss Göttingen Math Phys Kl 31:157–177

Kaszowski L (1973) Morphological activity of the mountain streams. Zesz Nauk Uniw Jagiell Pr Geogr 31:1–100

Kaszowski L, Kotarba A (1967) Charkterystyka morfodynamiczna koryta Sanu koło Myczkowiec (Morphodynamic characteristics of the San river bed near Myczkowce). Stud Geomorphol Carpatho Balc 1:53–72 (in Polish with English summary)

Kaszowski L, Krzemień K (1977) Structure of mountain channel systems as exemplified by chosen Carpathian streams. Stud Geomorphol Carpatho Balc 11:111–125

Kaszowski L, Krzemień K (1979) Channel subsystems in the Polish Tatra Mountains. Stud Geomorphol Carpatho Balc 13:149–161

Kaszowski L, Niemirowski M, Trafas K (1976) Problems of the dynamic of river channels in the Carpathian part of the Vistula basin. Zesz Nauk Uniw Jagiell Pr Geogr 43:7–37

Kaszowski L, Krzemień K, Libelt P (1988) Postglacial modelling of glacial cirques in the Western Tatras. Zesz Nauk Uniw Jagiell Pr Geogr 71:121–141

Kędzia S, Mościcki J, Wróbel A (1998) Studies on occurence of permafrost in Kozia valley (High Tatra Mountains). In: Repelewska-Pękalowa J (ed) IV Zjazd Geomorfologów Polskich. Sesja Polarna. UMCS, Lublin, pp 51–57

Kłapa M (1963) Prace Stacji Badawczej Instytutu Geografii PAN na Hali Gąsienicowej w latach 1960 i 1961 (Report on Research Work Carried out at the Scientific Station of IG PAN in the Tatra Mountains at Hala Gąsienicowa in 1960–1961). Prz Geogr 35:221–237 (in Polish with English summary)

Kłapa M (1970) Problématique et méthodes de rechereches de la station scientifique de l'Institut de Géographie de l'Academie Polonaise des Sciences à Hala Gąsienicowa dans les Tatras. Stud Geomorphol Carpatho Balc 4:205–217

Kłapa M (1980) Procesy morfogenetyczne i ich związek z sezonowymi zmianami pogody w otoczeniu Hali Gąsienicowej w Tatrach (Morphogenetic processes and their connection with seasonal weather changes in the environment of Hala Gąsienicowa in the Tatra Mountains). Dok Geogr 4:1–55 (in Polish with English summary)

Klassek G, Mleczek T (2008) Eksploracja i inwentaryzacja jaskiń polskich Karpat fliszowych (wrzesień 2007-sierpień 2008) (Exploration and inventory of the caves in Polish Flysch Carpathians – September 2007–August 2008). In: Materiały 42 Sympozjum Speleologicznego. Tarnowskie Góry, 24–26.10.2008, pp 69–72 (in Polish)

Klimaszewski M (1971) A contribution to the theory of rockface development. Stud Geomorphol Carpatho Balc 5:139–151

Klimaszewski M, Starkel L (1972) Karpaty Polskie (Polish Carpathians). In: Klimaszewski M (ed) Geomorfologia Polski, vol. 1, Polska Południowa. Góry i wyżyny, vol 1. PWN, Warszawa, pp 21–115 (in Polish)

Klimek K (1974) The structure and mode of sedimentation of the floodplain deposits in the Wisłoka valley (South Poland). Stud Geomorphol Carpatho Balc 8:135–151

Klimek K (1979) Geomorfologiczne zróżnicowanie koryt karpackich dopływów Wisły (Morphodynamical channel types of the Carpathian tributaries to the Vistula). Folia Geogr Ser Geogr Phys 12:35–47 (in Polish with English summary)

Klimek K (1987) Man's impast on fluvial processes in the Polish Western Carpathians. Geogr Ann 69A:221–226

Klimek K (1991) Typy koryt rzecznych (Types of river chanel). In: Dynowska I, Maciejewski M (eds) Dorzecze górnej Wisły, vol 1. Wyd. PWN, Warszawa/Kraków, pp 231–259 (in Polish)

Klimek K, Trafas K (1972) Young-Holocene changes in the course of the Dunajec river in the Beskid Sądecki Mountains. Stud Geomorphol Carpatho Balc 6:85–92

Klimek K, Łajczak A, Zawilińska L (1990) Sedimentary environment of the modern Dunajec delta in artificial Lake Rożnów, Carpathian Mts., Southern Poland. Quest Geogr 11/12:81–92

Kondracki J (2000) Geografia regionalna Polski (Regional Geography of Poland). Wyd. PWN, Warszawa, 441 p (in Polish)

Koreň M (2005) Vetrová kalamita 19. novembra 2004: nové pohl'ady a konsekvencie (Wind disater 19 November 2004: new ideas and consequences). Tatry 44:6–28 (in Slovak)

Kościelniak J (2004) Influence of river training on functioning of the Biały Dunajec river channel system. Geomorphol Slovak 4:62–67

Koszyk H (1977) Zróżnicowanie procesów grawitacyjnych we współczesnym modelowaniu rzeźby krystalicznej części Tatr Zachodnich na przykładzie Doliny Starorobociańskiej (Differentiation of gravitational processes in present-day modelling of relief in the Western Tatras, example Starorobociańska valley). Manuscript. Inst. Geogr. Gospod. Przestrz. Uniw. Jagiell., Kraków (in Polish)

Kot M (1996) Denudacja chemiczna Tatr Wysokich (Chemical denudation in the High Tatra Mountains). In: Kotarba A (ed) Przyroda Tatrzańskiego Parku Narodowego a Człowiek. vol. 1. Nauki o Ziemi. Tatrzański Park Nar.-Pol. Tow. Przyjaciół Nauk o Ziemi, Kraków/Zakopane, pp 117–119 (in Polish with English summary)

Kotarba A (1967) Żłobki krasowe w Tatrach (Glaciated knobs in the Polish Tatra Mountains). Zesz Nauk Uniw Jagiell Pr Geogr 16:25–42 (in Polish with English summary)

Kotarba A (1970) The morphogenetic role of the foehn wind in the Tatra Mountains. Stud Geomorphol Carpatho Balc 4:171–187

Kotarba A (1971) Coarse and intensity of present-day superficial chemical denudation in the Western Tatra Mountains. Stud Geomorphol Carpatho Balc 5:111–127

Kotarba A (1972) Powierzchniowa denudacja chemiczna w wapienno-dolomitowych Tatrach Zachodnich (Superficial chemical denudation in the calcareous-dolomite Western Tatra Mountains). Pr Geogr Inst Geogr Przestrz Zagospod PAN 96:1–116 (in Polish with English summary)

Kotarba A (1976) Współczesne modelowanie węglanowych stoków wysokogórskich na przykładzie Czerwonych Wierchów w Tatrach Zachodnich (The role of morphogenetic processes in modelling high-mountain slopes). Pr Geogr Inst Geogr Przestrz Zagospod PAN 120:1–128 (in Polish with English summary)

Kotarba A (1983) Współczesne procesy eoliczne i stabilizacja zdegradowanych wierzchowin grzbietowych w piętrze halnym Tatr Polskich (Contemporary aeolian processes and stabilization of degraded dividing ranges in the alpine zone of the Polish Tatras). Prz Geogr 55:171–182 (in Polish with English summary)

Kotarba A (1984) Elevational differentiation of slope geomorphic processes in the Polish Tatra Mountains. Stud Geomorphol Carpatho Balc 18:117–133

Kotarba A (1986) Rola osuwisk w modelowaniu rzeźby Beskidzkiej i pogórskiej (The role of landslides in modelling of the Beskidian and Carpathian Foothills relief). Prz Geogr 58:119–129 (in Polish with English summary)

Kotarba A (1992) High energy geomorphic events in the Polish Tatra Mountains. Geogr Ann 74A:123–131

Kotarba A (1993–1994) Zapis małej epoki lodowej w osadach jeziornych Morskiego Oka w Tatrach Wysokich (Record of Little Ice Age in lacustrine sediments of the Morskie Oko Lake,

High Tatra Mountains). Stud Geomorphol Carpatho Balc 27–28:61–69 (in Polish with English summary)

Kotarba A (1995) Rapid mass wasting over the last 500 years in the High Tatra Mountains. Quest Geogr Spec Issue 4:177–183

Kotarba A (1997) Formation of high-mountain talus slopes related to debris-flow activity in the High Tatra Mountains. Permafr Periglac Process 8:191–204

Kotarba A (1999) Geomorphic effect of catastrophic summer flood of 1997 in the Polish Tatra Mountains. Stud Geomorphol Carpatho Balc 33:101–115

Kotarba A (2002) Współczesne przemiany przyrody nieożywionej w Tatrzańskim Parku Narodowym (Recent changes of abiotic components of natural environment of the Tatra National Park). In: Przemiany środowiska przyrodniczego Tatr. Tatrzański Park Nar.- Pol. Tow. Przyjaciół Nauk o Ziemi, Kraków/Zakopane, pp 13–20 (in Polish with English summary)

Kotarba A (2004) Zdarzenia geomorfologiczne w Tatrach Wysokich podczas małej epoki lodowej (Geomorphic events in the High Tatra Mountains during the Little Ice Age). Pr Geogr Inst Geogr Przestrz Zagospod PAN 197:9–55 (in Polish with English summary)

Kotarba A (2005) Geomorphic processes and vegetation pattern changes. Case study in the Zelene Pleso Valley, High Tatra. Stud Geomorphol Carpatho Balc 39:39–48

Kotarba A, Pech P (2002) The recent evolution of talus slopes in the High Tatra Mountains (with the Pańszczyca valley as example). Stud Geomorphol Carpatho Balc 36:69–76

Kotarba A, Kłapa M, Rączkowska Z (1983) Procesy morfogenetyczne kształtujące stoki Tatr Wysokich (Present-day transformation of alpine granite slopes in the Polish Tatra Mountains). Dok Geogr 1:1–82 (in Polish with English summary)

Kotarba A, Kaszowski L, Krzemień K (1987) High-mountain denudational system in the Polish Tatra Mountains. Pr Geogr Inst Geogr Przestrz Zagospod PAN Spec Issue 3:1–106

Kozłowska AB, Rączkowska Z (1999) Środowisko wysokogórskie jako system wzajemnie powiązanych elementów (The high-mountain environment as a system of interlinked elements). Pr Geogr Inst Geogr Przestrz Zagospod PAN 174:121–132 (in Polish with English summary)

Krąpiec M, Margielewski W (2000) Analiza dendrogeomorfologiczna ruchów masowych na obszarze polskich Karpat fliszowych (Dendrogeomorphological analysis of mass movements in the Polish Flysch Carpathians). Kwart AGH Geol 26(2):141–171 (in Polish with English summary)

Krusiec M (1996) Wpływ ruchu turystycznego na przekształcanie rzeźby Tatr Zachodnich na przykładzie Doliny Chochołowskiej (Influence of tourism on relief transformation in the Western Tatras: example Dolina Chochołowska). Czas Geogr 67:303–320 (in Polish with English summary)

Krzemień K (1976) Współczesna dynamika koryta potoku Konina w Gorcach (The recent dynamics of the Konina stream channel in the Gorce Mountains). Folia Geogr Ser Geogr Phys 10:87–122 (in Polish with English summary)

Krzemień K (1984a) Fluvial transport balance in high-mountain crystalline catchment. Pr Geogr Inst Geogr Uniw Jagiell 61:61–70

Krzemień K (1984b) Współczesne zmiany modelowania koryt w Gorcach (Contemporary changes in the modelling of the stream channels in the Gorce Mountain Range). Zesz Nauk Uniw Jagiell Pr Geogr 59:83–96 (in Polish with English summary)

Krzemień K (1985) Present-day activity of the high-mountain stream in the Western Tatra Mountains. Quest Geogr Spec Issue 1:139–146

Krzemień K (1988) The dynamics of debris flows in the upper part of the Starorobociańska Valley (Western Tatra Mountains). Stud Geomorphol Carpatho Balc 22:123–144

Krzemień K (1991) Dynamika wysokogórskiego systemu fluwialnego na przykładzie Tatr Zachodnich (Dynamics of the high-mountain fluvial system: example Western Tatra Mountains). Rozpr Hab UJ 215:1–160 (in Polish with English summary)

Krzemień K (2003) The Czarny Dunajec river, Poland, as example of human-induced development tendencies in a mountain river channel. Landf Anal 4:57–64

Krzemień K, Sobiecki K (1998) Transport of dissolved and suspended matter in small catchments of the Wieliczka Foothills near Łazy. Pr Geogr Inst Geogr Uniw Jagiell 103:83–100

Książkiewicz M (1972) Karpaty (The Carpathians). In: Pożaryski W (ed) Budowa Geologiczna Polski, vol 4, Tektonika, part 3. Wyd. Geol, Warszawa, 228 p (in Polish)

Kummerer Ritter von Kummersberg C (1855) Administrative Karte von den Königreichen Galizien und Lodomerien. 1:115,000. Wien

Lach J (1984) Geomorfologiczne skutki antropopresji rolniczej w wybranych częściach Karpat i ich Przedgórza (Geomorphological studies of agricultural pressure in some parts of the Carpathians and its Foreland). Pr Monogr Wyż Szk Pedagog w Krakowie 66:1–142 (in Polish with English summary)

Libelt P (1988) Warunki i przebieg sedymentacji osadów postglacjalnych w cyrkach lodowcowych Tatr Zachodnich na przykładzie Kotła Starorobociańskiego (Conditions and course of sedimentation of postglacial deposits by taking as example the Starorobociański Cirque, Western Tatra Mountains). Stud Geomorphol Carpatho Balc 22:63–82 (in Polish with English summary)

Łajczak A (1994) The rates of silting and the useful lifetime of dam reservoirs in the Polish Carpathians. Z Geomorph NF 38:129–150

Łajczak A (1996) Modelling the long-term course of non-flushed reservoir sedimentation and estimating the life of dams. Earth Surf Process Landf 21:1091–1107

Łajczak A (1999) Współczesny transport i sedymentacja materiału unoszonego (Contemporary transport and sedimentation of the suspended material in the Vistula river and its main tributaries). Monogr Komitetu Gospod Wod PAN 15:1–215 (in Polish with English summary)

Łajczak A (2006) Deltas in dam retained lakes in the Carpathian part of the Vistula drainage basin. Pr Geogr Inst Geogr Gospod Przestrz Uniw Jagiell 116:99–109

Łajczak A (2007a) River training vs. flood risk in the upper Vistula basin, Poland. Geogr Pol 80(2):79–96

Łajczak A (2007b) Natura 2000 in Poland, Area PLH120016 The Orawsko-Podhalańskie Peatlands. Inst. Bot. PAN, Kraków, 139 p

Łajczak A (2011) Contemporary changes in the relief of raised bogs on the example of the Polish Carpathians. Geogr Pol Spec. Vol. (in print)

Łoziński W (1909) O osuwaniu się gliny w Tymowej w brzeskim powiecie (About the gravitational movement of loam in Tymowa village, Brzesko district). Kom Fizyograficzna Akad Umiejętności 43:55–58 (in Polish)

Malarz R (2002) Powodziowa transformacja gruboklastycznych aluwiów w żwirodnnych rzekach zachodnich Karpat fliszowych (Flood transformation of coarse-grained alluvium in gravel-bed rivers of the West Flysch Carpathians: exemple Soła and Skawa rivers). Pr Monogr Akad Pedagog w Krakowie 335:1–146 (in Polish with English summary)

Mamica W (1984) Współczesna dynamika rzeźby wysokogórskiej na przykładzie doliny Wyżniej Chochołowskiej w Tatrach Zachodnich (Present-day dynamics of high-mountain relief of the Chochołowska Wyżnia Valley, Western Tatras). Manuscript. Inst. Geogr. Gospod. Przestrz. Uniw. Jagiell., Kraków (in Polish)

Margielewski W (2002) Late Glacial and Holocene climatic changes registered in landslide forms and their deposits in the Polish Flysch Carpathians. In: Rybař J, Stemberk J, Wagner P (eds) Landslides. AA. Balkema, Rotterdam, pp 399–404

Margielewski W (2004) Typy przemieszczeń grawitacyjnych mas skalnych w obrębie form osuwiskowych (Patterns of gravitational movements of rock masses in the Polish Flysch Carpathians). Prz Geol 52:603–614 (in Polish with English summary)

Margielewski W (2006a) Records of the Late Glacial-Holocene palaeoenvironmental changes in landslide forms and deposits of the Beskid Makowski and Beskid Wyspowy Mts. area (Polish Outer Carpathians). Folia Quat 76:1–149

Margielewski W (2006b) Structural control and types of movements of rock mass in anisotropic rocks: case studies in the Polish Flysch Carpathians. Geomorphology 77:47–68

Margielewski W, Urban J (2003) Crevice-type caves as initial forms of rock landslide development in the Flysch Carpathians. Geomorphology 54:325–338

Margielewski W, Urban J (2005) Pre-existing tectonic discontinuities in the rocky massifs as initial forms of deep-seated mass movement development: case studies of selected deep crevice-type caves in the Polish Flysch Carpathians. In: Senneset K, Flaate K, Larsen JO (eds) Landslides and Avalanches. 11th ICFL 2005 Norway, A. A. Balkema, London, pp 249–256

Margielewski W, Urban J, Szura C (2007) Jaskinia Miecharska Cave (Beskid Śląski Mountains, Outer Carpathians): case study of the crevice type cave developed on sliding surface. Nat Conserv 63(6):57–68

Margielewski W, Święchowicz J, Starkel L, Łajczak A, Pietrzak M (2008) Współczesna ewolucja rzeźby Karpat fliszowych (Recent evolution of the Flysch Carpathians relief). In: Starkel L, Kostrzewski A, Kotarba A, Krzemień K (eds) Współczesne przemiany rzeźby Polski. Inst. Geogr. Gospod. Przestrz. Uniw. Jagiell, Kraków, pp 57–133 (in Polish)

Mastella L, Zuchiewicz W, Tokarski AK, Rubinkiewicz J, Leonowicz P, Szczęsny R (1997) Application of joint analysis for paleostress reconstructions in structurally complicated settings: case study from Silesian Nappe, Outer Carpathians, Poland. Prz Geol 45:1064–1066

Midriak R (1983) Morfogenéza povrchu vysokých pohorí (Surface morphogenesis of high mountain chains). VEDA, Bratislava, 513 p (in Slovak with English summary)

Midriak R (1984) Debris flows and their occurrence in the Czechoslovak Carpathians. Stud Geomorphol Carpatho Balc 18:35–149

Midriak R (1996) Present-day processes and micro-landforms evaluation; case study of Kopske Sedlo, the Tatra Mountains, Slovakia. Stud Geomorphol Carpatho Balc 30:35–50

Mieg F (1779–1782) Karte des Königreiches Galizien und Lodomerien. 1:28,800. Wien

Mikulski Z (1954) Katastrofalne powodzie w Polsce (Catastrophic floods in Poland). Czas Geogr 25:380–395 (in Polish)

Mirek Z (1996) Antropogeniczne zagrożenia i przekształcenia środowiska przyrodniczego (Anthropogenic threats and changes of nature). In: Mirek Z (ed) Tatrzański Park Narodowy (Nature of the Tatra National Park). Tatrzański Park Narodowy, Kraków/Zakopane, pp 595–617 (in Polish with English summary)

Mościcki J, Kędzia S (2001) Investigation of mountains permafrost in the Kozia Dolinka valley, Tatra Mountains, Poland. Norsk Geogr Tidsskr 55:235–240

Mrozek T, Rączkowski W, Limanówka D (2000) Recent landslides and triggering climatic conditions in Laskowa and Pleśna Regions, Polish Carpathians. Stud Geomorphol Carpatho Balc 34:89–112

Niedźwiedź T (2003) Extreme precipitation events on the northern side of the Tatra Mountains. Geogr Pol 76(2):15–23

Niedźwiedź T, Obrębska-Starklowa B (1991) Klimat (Climate). In: Dynowska I, Maciejowski M (eds) Dorzecze górnej Wisły. Part 1. Wyd. PWN, Warszawa/Kraków, pp 68–84 (in Polish)

Niemirowski M (1974) The dynamics of contemporary river beds in the mountain streams. Zesz Nauk Uniw Jagiell Pr Geogr 34:1–97

Obidowicz A (1990) Eine Pollenanalytische und Moorkundliche Studie zur Vegetations-geschichte des Podhale-Gebietes (West-Karpaten). Acta Palaeobot 30(1–2):147–219

Obidowicz A (2003) The holocene development of forests in the Pilsko Mountain area (Beskid Żywiecki Range, South Poland). Folia Quat 74:7–15

Obrębska-Starklowa B, Hess M, Olecki Z, Trepińska J, Kowanetz L (1995) Klimat (Climate). In: Warszyńska J (ed) Karpaty Polskie. Przyroda, człowiek i jego działalność. Wyd. Uniw. Jagiell, Kraków, pp 31–47 (in Polish)

Osuch B (1991) Stany wód (Water levels). In: Dynowska I, Maciejowski M (eds) Dorzecze górnej Wisły. Part 1. Wyd. PWN, Warszawa/Kraków, pp 159–166 (in Polish)

Oszczypko N (1995) Budowa geologiczna (Geological structure). In: Warszyńska J (ed) Karpaty Polskie. Przyroda, człowiek i jego działalność. Wyd. Uniw. Jagiell, Kraków, pp 15–22 (in Polish)

Oszczypko N, Golonka J, Zuchiewicz W (2002) Osuwisko w Lachowicach (Beskidy Zachodnie): skutki powodzi z 2001 r. (The landslide at Lachowice, Western Outer Carpathians, Poland: effects of catastrophic flood in 2001). Prz Geol 50:893–898 (in Polish with English summary)

Oszczypko N, Uchman A, Malata E (eds) (2006) Rozwój paleotektoniczny basenów Karpat Zewnętrznych i pienińskiego pasa skałkowego (Palaeotectonic development of the Outer Carpathians and the Pieniny Klippen Belt basins). Inst. Nauk Geol. Uniw. Jagiell, Kraków, 199 p (in Polish with English summary)

Pagaczewski J (1972) Katalog trzęsień ziemi w Polsce (Catalogue of earthquakes in Poland), 1000–1970. Mater Pr Inst Geofiz PAN 51:3–36 (in Polish with English summary)

Pękala K (1969) Rumowiska skalne i współczesne procesy morfogenetyczne w Bieszczadach Zachodnich (Block fields and recent morphogenetic processes in the Western Bieszczady Mountains). Ann Uniw Maria Skłodowska Curie B 24(2):47–98

Pietrzak M (2002) Geomorfologiczne skutki zmian użytkowania ziemi na Pogórzu Wiśnickim (The impact of land-use change on ground relief in the Wiśnieckie Foothills, Southern Poland). Inst. Geogr. Gospod. Przestrz. Uniw. Jagiell, Kraków, 149 p (in Polish with English summary)

Pietrzyk-Sokulska E (2005) Kryteria i kierunki adaptacji terenów po eksploatacji surowców skalnych. Studium dla wybranych obszarów Polski (Criteria and adaptation trends at stone quarrying sites – study in selected areas of Poland). Stud. Rozpr. Monografie 131, IGMiE PAN, 171 p (in Polish with English summary)

Poprawa D, Rączkowski W (1998) Geologiczne skutki powodzi 1997 r na przykładzie osuwisk województwa nowosądeckiego (Geological effect of flood in 1997 on the base of landslides of the Nowy Sącz District). In: Starkel L, Grela J (eds) Powódź w dorzeczu Górnej Wisły w lipcu 1997r. Materiały Konf. Naukowej, 7–9.05.1998. Kraków. Wyd. PAN, Kraków, pp 19–131 (in Polish)

Poprawa D, Rączkowski W (2003) Osuwiska Karpat (Carpathian landslide in southern Poland). Prz Geol 51:685–692 (in Polish with English summary)

Pulina M (ed) (1997) Jaskinie polskich Karpat fliszowych (Caves of the Polish Flysch Carpathians), vols 1, 2. Pol. Tow. Przyjaciół Nauk o Ziemi, Warszawa, 261 p (in Polish)

Pulinowa M (1972) Osuwiska w środowisku naturalnym i sztucznym (Landslides in natural and artificial environment). Dok Geogr Inst Geogr PAN 4:1–112 (in Polish)

Punzet J (1998–1999) Występowanie wezbrań w karpackiej części dorzecza Wisły (Disastrous floods in the Carpathian part of the Vistula drainage area). Folia Geogr Ser Geogr Phys 29–30:81–111 (in Polish with English summary)

Rachwaniec M, Holek P (1997) Jaskinia Malinowska (Jama) (Cave map). In: Pulina M (ed) Jaskinie Polskich Karpat Fliszowych, vol 1 (Appendix). Pol. Tow. Przyjaciół Nauk o Ziemi, Warszawa

Rączkowska Z (1992) Niektóre aspekty niwacji w Tatrach Wysokich (Some aspects in nivation in the High Tatra Mountains). Pr Geogr Inst Geogr Przestrz Zagospod PAN 156:209–223 (in Polish with English summary)

Rączkowska Z (1993) Ilościowe wskaźniki niwacji w Tatrach Wysokich (Quantitative rates of nivation in the High Tatra Mountains). Dok Geogr Inst Geogr Przestrz Zagospod PAN 4–5:63–81 (in Polish with English summary)

Rączkowska Z (1995) Nivation in the High Tatras, Poland. Geogr Ann 77A(40):251–258

Rączkowska Z (1999) Slope dynamics in the periglacial zone of the Tatra Mountains. Biul Peryglac 38:127–133

Rączkowska Z (2006) Recent geomorphic hazards in the Tatra Mountains. Stud Geomorphol Carpatho Balc 40:45–60

Rączkowska Z (2007) Współczesna rzeźba peryglacjalna wysokich gór Europy (Present-day periglacial relief in the high mountains of Europe). Pr Geogr Inst Geogr Przestrz Zagospod PAN 212:1–252 (in Polish with English summary)

Rączkowska Z (2008) Współczesna ewolucja rzeźby Tatr (Recent evolution of the Tatra Mountains relief). In: Starkel L, Kostrzewski A, Kotarba A, Krzemień K (eds) Współczesne przemiany rzeźby Polski. Inst. Geogr. Gospod. Przestrz. Uniw. Jagiell, Kraków, pp 35–56 (in Polish)

Rączkowska Z, Kozłowska A (2002) Odzwierciedlenie wpływów antropogenicznych w wybranych elementach środowiska przyrodniczego otoczenia Kasprowego Wierchu (Anthropogenic transformations in the natural environment of the Kasprowy Wierch area). In: Przemiany środowiska przyrodniczego Tatr (Upper boundary of the dwarf pine vegetation belt as the stability indicator of the Tatra Mts geosystem). Tatrzański Park Nar.-Pol. Tow. Przyjaciół Nauk o Ziemi, Kraków/Zakopane, pp 403–406 (in Polish with English summary)

Rączkowski W (1981) Zróżnicowanie współczesnych procesów grawitacyjnych w Dolinie Pięciu Stawów Polskich (Tatry Wysokie) (Differentiation of present-day gravitational processes in the Dolina Pięciu Stawów Polskich valley, High Tatras). Biul Inst Geol 332:139–152 (in Polish)

Rączkowski W (2007) Landslide hazard in the Polish Flysch Carpathians. Stud Geomorphol Carpatho Balc 41:61–75

Rączkowski W, Mrozek T (2002) Activating of landsliding in the Polish Flysch Carpathians by the end of 20th century. Stud Geomorphol Carpatho Balc 36:91–101

Ralska-Jasiewiczowa M (1972) The forest of the Polish Carpathians in the late glacial and Holocene. Stud Geomorphol Carpatho Balc 6:5–19

Ralska-Jasiewiczowa M (1980) Late Glacial and Holocene of the Bieszczady Mountains (Polish Eastern Carpathians). Wyd. PWN, Warszawa/Kraków, 199 p

Ralska-Jasiewiczowa M (1989) Type region: the Bieszczady mountains. Acta Palaeobot 29:31–35

Sawicki L (1917) Osuwiska ziemne w Szymbarku i inne zsuwy powstałe w 1913 r. w Galicji Zachodniej (Die Szymbarker Erdrutschung und andere west-galizische Rutschungen des Jahres 1913). Rozp Wydz Mat Przyr Akad Umiejęt w Krakowie 56 A:227–313 (in Polish with German summary)

Schenk V, Schenkova Z, Kottnauer P, Guterch B, Labak P (2001) Earthquake hazards maps for the Czech Republic, Poland and Slovakia. Acta Geophys Pol 49:287–302

Słupik J (1970) Methods of investigating the water cycle within a slope. Stud Geomorphol Carpatho Balc 4:127–136

Słupik J (1973) Zróżnicowanie spływu powierzchniowego na fliszowych stokach górskich (Differerentiation of surface runoff on flysch mountain slopes). Dok Geogr 2:1–118 (in Polish with English summary)

Słupik J (1978) Obieg wody w glebie na stokach a rolnicze użytkowanie ziemi (Water circulation on cultivated slopes). In: Studia nad typologią i oceną środowiska geograficznego Karpat i Kotliny Sandomierskiej. Pr. Geogr. Inst. Geogr. PAN 125: 93–107 (in Polish with English summary)

Słupik J (1981) Rola stoku w kształtowaniu odpływu wody (Role of slope in generation of runoff in the Flysch Carpathian). Prace Geogr 142:1–98 (in Polish with English summary)

Słupik J (1986) Ocena metod badań roli użytkowania ziemi w przebiegu spływu wody i erozji gleb w Karpatach (Critical review of methods of studies on the influence of land use on runoff and soil erosion in the Carpathians). Przegl Geogr 58(1–2):41–50 (in Polish with English summary)

Śmietana M (1987) Zróżnicowanie rozbryzgu gleby na użytkowanych rolniczo stokach fliszowych (Differentiation of soil splash on agriculturally exploited flysch slopes). Stud Geomorphol Carpatho Balc 21:161–182 (in Polish with English summary)

Soja R (1977) Deepening of channel in the light of the cross profile analysis (Carpathian river as example). Stud Geomorphol Carpatho Balc 11:127–138

Soja R, Prokop P (1996) Drogi jako element antropogenicznego przekształcenia środowiska (Roads as elements of anthropogenic transformations of the environment). In: Soja R, Prokop P (eds) Zintegrowany monitoring środowiska przyrodniczego. Monitoring Geoekosyst. Górskich. Bibl. Monit. Środ., Warszawa, pp 91–98 (in Polish)

Starkel L (1960) Rozwój rzeźby Karpat fliszowych w holocenie (The development of the flysch Carpathians relief during the Holocene). Pr Geogr Inst Geogr PAN 22:1–239 (in Polish with English summary)

Starkel L (1965) Rozwój rzeźby polskiej części Karpat wschodnich (Geomorphological development of the Polish Eastern Carpathians). Pr Geogr Inst Geogr PAN 50:1–143 (in Polish with English summary)

Starkel L (1969) Odbicie struktury geologicznej w rzeźbie polskich Karpat fliszowych (Reflection of the geological structure on the relief of the Polish Flysch Carpathians). Stud Geomorphol Carpatho Balc 3:33–44 (in Polish with English summary)

Starkel L (1972) Charakterystyka rzeźby polskich Karpat i jej znaczenie dla gospodarki ludzkiej (An outline of the relief of the Polish Carpathians and its importance for human management). Probl Zagospod Ziem Górskich 10:75–150 (in Polish with English summary)

Starkel L (1980) Erozja gleb a gospodarka wodna w Karpatach (Soil erosion and water management in the Carpathians). Zesz Probl Postępu Nauk Rol 235:103–118 (in Polish)

Starkel L (1984) Karpaty i Kotliny Podkarpackie (Carpathians and Sub-Carpathian valleys). In: Sokołowski S, Mojski JE (eds) Budowa Geologiczna Polski, vol. 1. Stratygrafia, part 3b. Kenozoik. Czwartorzęd. Wyd. Geol, Warszawa, pp 292–308 (in Polish)

Starkel L (1986) Rola zjawisk ekstremalnych i procesów sekularnych w ewolucji rzeźby (na przykładzie fliszowych Karpat) (The role of extreme events and secular processes in the relief evolution of the flysch Carpathians). Czas Geogr 57:203–213 (in Polish with English summary)

Starkel L (1991) Rzeźba terenu (Relief), Geoekologiczne uwarunkowania obiegu wody (Geoecological conditions for water circulation). In: Dynowska I, Maciejewski M (eds) Dorzecze górnej Wisły, part 1. Wyd. PWN, Warszawa–Kraków. pp. 42–54, 91–95 (in Polish)

Starkel L (1996) Geomorphic role of extreme rainfalls in the Polish Carpathians. Stud Geomorphol Carpatho Balc 30:21–38

Starkel L (1997) Mass movement during the Holocene: Carpathian example and the European perspective. Palaeoclim Res 19:385–400

Starkel L (2001) Historia doliny Wisły (od ostatniego zlodowacenia do dziś) (Evolution of the Vistula river valley since the last glaciation till present). Monografie Geogr. Inst. Geogr. Przestrz. Zagospod. PAN 2:263 p (in Polish)

Starkel L (2006) Geomorphic hazards in the Polish Flysch Carpathians. Stud Geomorphol Carpatho Balc 40:7–19

Starkel L, Łajczak A (2008) Kształtowanie rzeźby den dolin w Karpatach (koryt i równin zalewowych) (Formation of relief of valley bottoms, channels and floodplains in the Carpathians). In: Starkel L, Kostrzewski A, Kotarba A, Krzemień K (eds) Współczesne przemiany rzeźby Polski. Inst. Geogr. Gospod. Przestrz. Uniw. Jagiell, Kraków, pp 95–114 (in Polish)

Stupnicka E (1960) Geneza glin lessowych Pogórza Cieszyńskiego i Beskidów Śląskich (Origin of the loess-like clays in the Cieszyn Upland and the Beskidy Śląskie Range). Acta Geol Pol 10(2):247–265 (in Polish with English summary)

Świderski B (1932) Przyczynki do badań nad osuwiskami karpackimi (Sur les éboulements dans le Carpates). Prz Geogr 12:96–111 (in Polish with French summary)

Święchowicz J (2001) Rola stoków i den dolin w odprowadzaniu zawiesiny ze zlewni pogórskiej (The role of slopes and valley floors in transporting suspended material from foothill catchments). In: Chełmicki W (ed) Przemiany środowiska na Pogórzu Karpackim. Procesy, Gospodarka, monitoring, vol 1. Inst. Geogr. UJ, Kraków, pp 31–49 (in Polish with English summary)

Święchowicz J (2002a) Linkage of slope wash and sediment and solute export from a foothill catchment in the Carpathian foothills of South Poland. Earth Surf Process Landf 27(12):1389–1413

Święchowicz J (2002b) The influence of plant cover and land use on slope-channel decoupling in a foothill catchment: a case study from the Carpathian Foothills, Southern Poland. Earth Surf Process Landf 27(5):463–479

Święchowicz J (2002c) Współdziałanie procesów stokowych i fluwialnych w odprowadzaniu materiału rozpuszczonego i zawiesiny ze zlewni pogórskiej (Linkage of slope and fluvial processes in sediment and solute export from a foothill catchment in the Carpathian Foothills of southern Poland). Przemiany środowiska na Pogórzu Karpackim, vol. 3, Inst. Geogr. Uniw. Jagiell., Kraków. 150 p (in Polish with English summary)

Święchowicz J (2004) Rola procesów ekstremalnych w transformacji stoków pogórskich (na przykładzie Dworskiego Potoku) (Role of extreme processes in slope transformation in the Carpathian Foothills). In: Izmaiłow B (ed) Przyroda – Człowiek – Bóg. Inst. Geogr. Gosp. Przestrz. Uniw. Jagiell, Kraków, pp 83–91 (in Polish with English summary)

Święchowicz J (2006) The role of precipitation on slope transformation in the Carpathian Foothills (Poland). Geomorphol Slovaca 2:5–13

Święchowicz J (2008) Soil erosion on cultivated foothill slopes during extreme rainfall events in Wiśnicz Foothills of southern Poland. Folia Geogr Ser Geogr Phys 39:80–93

Święchowicz J (2009) Geomorfologiczne i ekonomiczne skutki deszczu nawalnego z dnia 17 czerwca 2006 r. na terenie Rolniczego Zakładu Doświadczalnego UJ w Łazach (Pogórze Wiśnickie) (Geomorphic and economic effects of the heavy rainfall of 17 June 2006 at the Jagiellonian University's farmland in Łazy, Wiśnicz Foothills). In: Bochenek W, Kijowska M (eds) Zintegrowany Monitoring Środowiska Przyrodniczego. Funkcjonowanie środowiska przyrodniczego w okresie przemian gospodarczych w Polsce. PIOŚ, Warszawa, pp 219–230 (in Polish with English summary)

Święchowicz J (2010) Spłukiwanie gleby na użytkowanych rolniczo stokach pogórskich w latach hydrologicznych 2007–2008 w Łazach (Pogórze Wiśnickie) (Slopewash on agricultural foothill slopes in hydrological years 2007–2008 in Łazy, Wiśnicz Foothills). Pr Stud Geogr 45:243–263 (in Polish with English summary)

Teisseyre H (1929) Certaines observations morphologiques dans les Karpathes. Prz Geogr 9:330–347 (in Polish with French summary)

Teisseyre H (1936) Materiaux pour l'étude des éboulements dans quelques régions des Karpates et des Subkarpates. Rocz Pol Tow Geol 12:35–192 (in Polish with French summary)

Urban J, Otęska-Budzyn J (1998) Geodiversity of pseudokarst caves as the reason for their scientific importance and motive of protection. Geol Balc 28(3–4):163–166

Urban J, Margielewski W, Schejbal-Chwastek M, Szura C (2007a) Speleothems in some caves of the Beskidy Mountains, Poland. Nat Conserv 63(6):109–117

Urban J, Margielewski W, Żak K, Hercman H, Sujka G, Mleczek T (2007b) The calcareous speleothems in the pseudokarst Jaskinia Słowiańska-Drwali cave, Beskid Niski Mountains, Poland. Nat Conserv 63(6):119–128

Vitek J (1983) Classification of pseudokarst forms in Czechoslovakia. Int J Speleol 13:1–18

Waga JM (1993) Geneza Jaskinii Komonieckiej w Beskidzie Małym (Genesis of the Komoniecki Cave in the Beskid Mały). Kras Speleolog 7(16):92–101 (in Polish with an English summary)

Welc A (1977) Procesy eoliczne w zlewni Bystrzanki koło Szymbarku w latach 1969–1971 (Eolian processes in the Bystrzanka drainage basin near Szymbark, Polish Flysch Carpathians, in the years 1968–1971). Dok Geogr 6:67–85 (in Polish with English summary)

Wójcikiewicz M (1979) The stratigraphy of the Bór na Czerwonem peat bog with the consideration of sub-fossil associations and distribution and differentiation of contemporary plant communities. Part II. The characteristics of the peat bog plant cover. Zesz Nauk Akad Rol w Krakowie 153, Melioracja 10:133–193 (in Polish with English summary)

Wrzosek A (1933) Z badań nad zjawiskami krasowymi Tatr Polskich (Studies of karst pheonomena in the Polish Tatras). Wiad Służby Geogr 7(3):235–273

Wyżga B (1993a) Funkcjonowanie systemu rzecznego środkowej i dolnej Raby w ostatnich 200 latach (Evolution of the fluvial system of the middle and lower Raba river, Carpathians, Poland, in the last 200 years). Dok Geogr 6:1–92 (in Polish with English summary)

Wyżga B (1993b) River response to channel regulation: case study of the Raba river, Carpathians, Poland. Earth Surf Process Landf 18:541–556

Wyżga B, Lach J (2002) Współczesne wcinanie się karpackich dopływów Wisły – przyczyny, środowiskowe efekty oraz skutki zaradcze (Present-day downcutting of the Carpathian tributaries to the Vistula: causes, environmental effects and remedial measures). Probl Zagospod Ziem Górskich 48:23–29 (in Polish with English summary)

Wyżga B, Kaczka RJ, Zawiejska J (2003) Gruby rumosz drzewny w ciekach górskich – formy występowania, warunki depozycji i znaczenie środowiskowe (Large woody debris in mountain streams – accumulation types, depositional conditions and environmental significance). Folia Geogr Ser Geogr Phys 33–34:117–138 (in Polish with English summary)

Zabuski L (2004) Prediction of the slope movements on the base of inclinometric measurements and numerical calculations. Polish Geol Inst Spec Issue 15:29–38

Zarzycki K (ed) (1982) Pieniny. Przyroda w obliczu zmian (The nature of the Pieniny Mountains, West Carpathians, in face of the coming changes). PWN, Warszawa/Kraków, 575 p (in Polish)

Zawiejska J, Krzemień K (2004) Man-induced changes in the structure and dynamics of the upper Dunajec river channel. Geogr Čas 56(2):111–124

Ziemońska Z (1973) Stosunki wodne w polskich Karpatach Zachodnich (Hydrographic conditions in the Polish West Carpathians). Pr Geogr Inst Geogr PAN 103:1–127 (in Polish with English summary)

Ziętara T (1964) O odmładzaniu osuwisk w Beskidach Zachodnich (On the rejuvenation of landslides in the West Beskids). Rocz Nauk Dydakt Wyż Szk Pedagog Kraków Pr Geogr 22:55–86 (in Polish with English summary)

Ziętara T (1968) Rola gwałtownych ulew i powodzi w modelowaniu rzeźby Beskidów (Impact of torrential rains and floods on the relief of the Beskidy Mountains). Pr Geogr Inst Geogr PAN 60:1–116 (in Polish with English summary)

Ziętara T (1974) Rola osuwisk w modelowaniu pogórza Rożnowskiego (Zachodnie Karpaty fliszowe) (Le rôle des glissements dans le modelage du Pogórze de Rożnów). Stud Geomorphol Carpatho Balc 8:115–133 (in Polish with French summary)

Ziętara T (1988) Landslide areas in the Polish Flysch Carpathians. Folia Geogr Ser Geogr Phys 20:21–67

Ziętara T (1999) The role of mud and debris flows modelling of the Flysch Carpathians relief, Poland. Stud Geomorphol Carpatho Balc 33:81–100

Zuber K, Blauth J (1907) Katastrofa w Duszatynie (Disaster in Duszatyn village). Czas Techn Lwów 25:218–221 (in Polish)

Żurek S (1983) Stan inwentaryzacji torfowisk w Polsce (State of stock-taking of peat bogs in Poland). Wiad Melioracyjne Łąkarskie 7:210–215 (in Polish)

Żurek S (1987) Złoża torfowe Polski na tle stref torfowych Europy (The peat deposits of Poland in European comparison). Dok Geogr Inst Geogr Przestrz Zagospd PAN 4:1–84 (in Polish with English summary)

Żytko K, Zając R, Gucik S, Ryłko W, Oszczypko N, Garlicka I, Nemčok J, Eliaš M, Menčik E, Stranik Z (1989) Map of the tectonic elements of the Western Outer Carpathians and their foreland 1:5,000,000. In: Poprawa D, Nemčok J (eds) Geological atlas of the western outer Carpathians and their foreland. Państ. Inst. Geol, Warszawa

Chapter 6
Recent Landform Evolution in the Moravian–Silesian Carpathians (Czech Republic)

Jaromir Demek, Jan Hradecký, Karel Kirchner, Tomáš Pánek, Aleš Létal, and Irena Smolová

Abstract Topographic changes in the Moravian-Silesian Carpathians have been due to variations of natural conditions (climatic changes, accelerated rates of exogenic geomorphic processes during the Little Ice Age), but mainly to the growing intensity of human activities (tillage, deforestation, accelerated soil erosion, urban sprawl). In this geomorphologically highly sensitive region, land-use changes exerted a great influence on the intensity and type of exogenic geomorphological processes in the last millennium. Their impact is studied on archive maps. The density of slope deformations (like deep-seated slope failures, lateral spreading, toppling, sackung, translational and rotational landslides, earthflows, debris flows, and rockfalls) in the study area is the highest in the Czech Republic. Other geomorphic

---

Professor Dr. J. Demek, DrSc
The Silva Tarouca Research Institute for Landscape and Ornamental Gardening,
Průhonice, Czech Republic

Department of Landscape Ecology, Lidická 25/27, 602 00 Brno,
Czech Republic
e-mail: DemekJ@seznam.cz

J. Hradecký (✉) • T. Pánek
Department of Physical Geography and Geoecology, Faculty of Science,
University of Ostrava, Chittussiho 10, 710 00 Ostrava – Slezská Ostrava, Czech Republic
e-mail: jan.hradecky@osu.cz; tomas.panek@osu.cz

K. Kirchner
Institute of Geonics, Academy of Sciences of the Czech Republic,
Branch Brno, Drobného 28, 602 00 Brno, Czech Republic
e-mail: kirchner@geonika.cz

A. Létal • I. Smolová
Department of Geography, Faculty of Science, Palacký University in Olomouc,
17. listopadu 12, 771 46 Olomouc, Czech Republic
e-mail: ales.letal@upol.cz; irena.smolova@upol.cz

processes presented in this overview are erosion by water on the surface and underground (piping), wind erosion, and a range of anthropogenic processes (urbanization, mining, industry, water management, and transport).

Keywords Land use changes • Archive maps • Slope deformations • Water erosion • Piping • Wind erosion • Anthropogenic processes • Western Carpathians

6.1 Introduction

Jaromír Demek

The Moravian–Silesian Carpathians (MSC) are the westernmost member of the Western Carpathians in the southeastern part of the Czech Republic. The young mountains and lowlands are a very sensitive relief with neotectonic movements. Built of Cretaceous and Cainozoic rocks, the relief of the MSC changed substantially not only due to variations of *natural conditions*, but mainly due to *activities of human society* during the last millennium. Around 1000 AD the warm and dry climate turned wetter and colder. Soil erosion in ancient core settlement areas accelerated and abandoned settlements of the Great Moravian Empire in floodplains were buried under flood loam deposits. Slavonic villages were relocated from frequently inundated floodplains further away from river channels onto lower river terraces. In the twelfth century Carpathian flysch highlands, mountains, and floodplains were still forested, while arable land prevailed only in ancient core settlements of lowlands. Soil erosion processes were further accelerated by the medieval colonization and deforestation of highlands of the West European platform in the twelfth and thirteenth centuries. Simultaneously the studied area experienced a general deterioration of the climate between the years 1150 and 1460 and a very cold and wet climate between 1560 and 1850 transformed geomorphic processes, vegetation and land use (Little Ice Age). Two climatic minima about 1650 and about 1770 correspond with intensification of soil erosion and slope deformations. Sediment transport by watercourses was influenced by the construction of a large number of fishponds since the thirteenth century.

The character of flysch mountain landscape radically changed with the Walachian colonization at the end of fifteenth and beginning sixteenth century when Walachian pastoral tribes arrived into the MSC (Macůrek 1959). Deforestation and grazing accelerated erosion and landslides on steep mountain slopes as well as accumulation on valleys bottoms. In the eighteenth century the largest landscape user was agriculture and changes in the agriculture (e.g. the Austrian agricultural reform) significantly influenced both the type and intensity of geomorphic processes. The first stage of scientific-technical revolution in agriculture starting around 1750 resulted in the extension of arable land at the expense of permanent grassland and drying of fishponds systems. Open-cast hard coal mining in the Ostravská pánev Basin began shortly after the discovery of coal seams in the late eighteenth century.

River regulation began in nineteenth century. The deep man-made channels and dykes retained floodwater and disturbed the connectivity of valley slopes, floodplains, and stream channels. The regulation thus resulted in the fragmentation of floodplains and changes in the lengths and sinuosity of watercourses. The urban landscape was expanding in the second half of the nineteenth century. The cultural landscape in the MSC experienced essential changes in the second half of the twentieth century due to the impact of the 2nd Czechoslovak agrarian reform on the landscape and the ensuing collectivization of Czechoslovak agriculture, merging private plots into large fields.

6.2 Geomorphological Conditions

Jaromír Demek

6.2.1 Highlands and Mountains of the Outer Western Flysch Carpathians

The mountain ridges are composed of massive flysch sandstones and depressions developed mostly in shales. The system of sandstone mountain ridges and depressions shows a zonal arrangement (Fig. 6.1). Several subsystems can be distinguished: the South Moravian Carpathians, the Central Moravian Carpathians, the Moravian–Slovakian

Fig. 6.1 The zonal structure of the MSC on digital elevation model

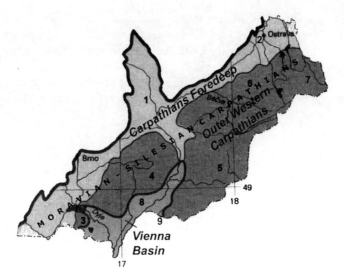

Fig. 6.2 Geomorphological divisions of the Moravian–Silesian Carpathians and environs on the territory of the Czech Republic. *1* Western Outer Carpathian depressions, *2* Eastern Outer Carpathian depressions, *3* South Moravian Carpathians, *4* Central Moravian Carpathians, *5* Moravian–Slovakian Carpathians, *6* Western Beskids Foothills, *7* Western Beskids, *8* Lower Moravian Graben, *9* Záhorská nížina Lowland. Source: Demek and Mackovčin (2006)

Carpathians, the Podbeskydská pahorkatina Hills (piedmont of the Western Beskids), and the Western Beskids on the Slovak border (see Fig. 6.2).

The *South Moravian Carpathians* (Fig. 6.2: 3) are characterized by their white limestone klippen forming the highest peaks in the Pavlovské vrchy Hills (Mt. Děvín 548.7 m elevation). Steep Jurassic and Cretaceous limestone klippen rise above the piedmont hills of the Ždánice Flysch Unit.

The *Central Moravian Carpathians* (Fig. 6.2: 4) are a strip of flysch hills and highlands extending between the Věstonická brána Gate on the Dyje River in the west and the Napajedelská brána Gate on the Morava River in the east. Their axis is formed by the Ždánický les Highland (U slepice 437.4 m.) and the Chřiby Highland (Brdo 586.7 m) surrounded by foothills. The central ridge of the Ždánický les Highland has a planation surface at 300–430 m elevation. Typical are the Tertiary pediments and Quaternary cryopediments for the hilly rim of the central ridge of the Ždánický les. The Chřiby Highland, the eastern continuation of the Ždánický les Highland, reaches as far as the Napajedelská brána Gate to the east. The central higher part of the Highland consists of two parallel forested ridges with flysch sandstone tors and castle coppies. On the watersheds expressive flats can be found truncating the surface of folded rocks (predominantly by Paleogene sandstones and claystones). The elevation of these flats is usually 350–500 m. In the north and south a lower and less dissected hills strip of variable width joins the central range: the Litenčická pahorkatina in the north and the Kyjovská pahorkatina in the south (Fig. 6.2). The Litenčická pahorkatina Hills are rounded interfluvial ridges with broad dry valleys and cryopediments on Neogene deposits.

The *Moravian–Slovakian Carpathians* (Fig. 6.2: 5) consist of two flysch ranges extending along the Czech-Slovak border (the Bílé Karpaty and Javorníky Mountains) and the Vizovická vrchovina Highland. A characteristic feature of the Bílé Karpaty Mountains (White Carpathians) is the high massive Magura flysch sandstone ridge (Velká Javořina 970.0 m) with smooth and less dissected slopes. The Bílé Karpaty Mountains are dissected by the valleys of the right tributaries of the Váh River into a series of isolated mountain groups. Numerous landslides developed on the slopes of the mountain ridge. The Bílé Karpaty continues in the central ridge of the Javorníky Mountains (the highest point is Malý Javorník 1019.2 m) built of flysch sandstones of the Rača unit. On Hradisko Mountain (773.1 m), near the village of Pulčín, the largest "rock city" in the Czech Republic developed in flysch sandstones and conglomerates of the Outer Western Flysch Carpathians with castellated rocks, gorges, and pseudokarst caves. The Vizovická vrchovina Highland extends from the foot of the Bílé Karpaty and Javorníky Mountains up to the Morava River valley in the west. The main axis is the central anticlinal Klášťovský hřbet Ridge composed of resistant sandstones (Klášťov 752.9 m). Extensive pediments and cryopediments developed on the less resistant shales of the surrounding hills.

The subsystem of the *Západní Beskydy* (*Western Beskids* – Fig. 6.2: 7) represents a mountain range extending on the territory of the Czech Republic from the Morava River in the west to the Slovak and Polish border in the east. The Západní Beskydy Mountains are divided by the Jablunkovský průsmyk Pass into a Moravian–Silesian part and a Polish-Slovakian part. With the predominance of resistant Cretaceous and Paleogene flysch sandstones, they are characterized by considerable elevation (Lysá hora 1323.3 m), massive ridges, and relatively steep slopes. The Moravian–Silesian part has its core in the Moravskoslezské Beskydy Mountains and also includes the Hostýnské vrchy and Vsetínské vrchy Hills, the Jablunkovské mezihoří Highlands and the foothills of the Podbeskydská pahorkatina. To the Polish-Slovakian part belong the Slezské Beskydy (Silesian Beskids) and the tectonic depression Jablunkovská brázda Furrow. The relief of the Slezské Beskydy Mountains is influenced by complex tectonics, deep-seated movements, and numerous block landslides (e.g. Velký Šošov 885.6 m). The topography of the Moravskoslezské Beskydy Mountains is controlled by their structure. The thick layers of the Godula and Istebná flysch sandstones are slightly inclined to the south and control the appearance of flat summits (Novosad 1966). The northern boundary of the mountains forms steep and high structural slopes on heads of dipping resistant sandstone strata. Pseudokarst features are typical on sandstone summits and due to gravitational tectonics there are numerous landslides on the slopes. The depression of Rožnovská brázda Furrow divides the Moravskoslezské Beskydy from the Vsetínské vrchy Hills. The axis of the depression forms the valley of the Rožnovská Bečva River bordered by fluvial terraces and valley pediments. The Vsetínské vrchy Hills are situated at the head of the flysch Magura nappe. The highest point is Vysoká (1,024 m) in the Soláňský hřbet sandstone ridge. There are many extensive landslides on its slopes. The mountain range of the Hostýnské vrchy Hills begins near the town of Holešov in the west and reaches to the Bečva River valley in the east. The flysch highland is limited by a steep structural slope in the north bordered by rock pediments.

The Hostýnské vrchy Hills and the Moravskoslezské Beskydy Mountains are accompanied on the NW side by the continuous belt of the *Podbeskydská pahorkatina Hills* (Fig. 6.2: 6). The Maleník Ridge between towns of Hranice na Moravě and Lipník, built of Paleozoic rocks, is partly thrust on the Neogene deposits of the Moravian Gate.

6.2.2 Outer Carpathian Depressions

The Carpathian Foredeep developed in the Neogene as a continuous depression bordering the Flysch Carpathian arc on the outside (Fig. 6.2: 1). The depression was filled with Neogene marine and lacustrine deposits, which were partly overridden by the marginal flysch nappes of the Carpathians during the Neogene orogenies. The Outer Carpathian Depressions are followed by the foredeep and represented on the territory of the Czech Republic by three basins connected by two gates, forming a pronounced boundary between the Bohemian Highlands and ranges of Moravian–Silesian Flysch Carpathians (Fig. 6.2). The outer border line of the Moravian–Silesian Carpathians is marked by an expressive fault scarp between the Bohemian Massif and Outer Carpathian Depressions along the line Znojmo–Brno–Litovel–Přerov–Hranice na Moravě and Ostrava.

The *Dyjsko-svratecký úval Graben* is the first westernmost link in the chain of the Outer Carpathian Depressions (Fig. 6.2: 8). It stretches south and southwest from the city of Brno in the north up to Czechia–Austria border in the south. With its subsidence character and unconsolidated Neogene and Quaternary deposits, its relief is flat or slightly undulating. The narrow graben around the town of Vyškov, called *Vyškovská brána Gate*, connects the grabens of Dyjsko-svratecký úval and Hornomoravský úval. The Hornomoravský úval (*Upper Moravian Graben*) represents the central link of the chain of the Outer Carpathian Depressions, filled by Neogene sediments overlain by Quaternary fluvial and aeolian deposits, along the Morava River. It is the last northeastern link of the Outer Carpathian Depressions with the Basin of Ostrava through the narrow elongated Moravská brána (*Moravian Gate*). The Ostravská pánev (*Basin of Ostrava* – Fig. 6.2: 2) in the northeast is a Quaternary accumulation plain with a lowland relief along the Odra and Olše Rivers, glaciated during the Pleistocene. The landscape is substantially changed by human activities (hard coal mining).

6.2.3 The Inner Carpathian Depressions

On the territory of the Czech Republic the Inner Carpathian depressions are the northern extensions of the Vienna Basin, such as the *Dolnomoravský úval Graben* (Fig. 6.2: 8) with lowland relief and the wide *floodplains* of the *Morava and Dyje* (Thaya) Rivers.

6.3 History of Geomorphological Research in the Czech Republic

Jaromír Demek
Karel Kirchner

The systematic study of present-day geomorphological processes started at the beginning of the twentieth century with the study of slope deformation, especially *landslides*. H. Götzinger (1909) called attention to relation of landslides on the Neogene/Quaternary boundary. More papers were published in the interwar period, mostly on slope deformations related to road and railway construction in the Flysh Mountains between Moravia and Slovakia and on establishing water reservoirs in this sensitive terrain (Záruba 1938). Q. Záruba (1922) described landslides in the vicinity of the town of Vsetín and Stejskal (1931) in the Pavlovské vrchy Hills. During World War II J. Krejčí studied landslides around Zlín, while S. Novosad (1956) described deep-seated movements in the Moravskoslezské Beskydy Mountains in. Attention was also paid to local landslides in the MSC – e.g., in Neogene deposits on Výhon Hill near Židlochovice (Quitt 1960). A catalogue of historical landslides was compiled by M. Špůrek (1972).

Through the systematic research of landslides a wealth of new knowledge was obtained on slope processes, particularly in the years of 1962 and 1963 from the landslide hazard studies in Handlová in the Slovak Republic in December 1960 (Záruba and Mencl 1969). The database of the Geofond in Prague was compiled and supplied with further data by the Czech Geological Survey in the following years resulting in an *inventory* of landslides and other dangerous slope deformations. The constructions of dams on rivers in the MSC (Šance Dam on the Ostravice River, Karolinka Dam on the Stanovnice Stream) induced slope deformations that were monitored (Krejčí 2004; Novosad 2006). The events of extreme precipitation and flash floods in July 1997 drew attention again to the research and monitoring of slope deformations, the reactivation of landslides and other hazardous slope processes (mud flows, gully formation, etc.) with extensive damage and loss of lives. Detailed investigations of landslides, led by the Czech Geological Survey, started in the most endangered regions. In 1999 further research was initiated by the Ministry of Environment of the Czech Republic (for more details see Brázdil and Kirchner 2007). Recent slow slope movements were measured by J. Demek (1986, 1988) in the Javorníky Mountains. K. Šilhán (2010) applied dendrogeomorphology for the study of the periodicity of rockfalls in the Moravskoslezské Beskydy Mountains.

The geomorphological research of fluvial processes in Czechia is closely connected with *archeological and historical investigations*. Especially after the emergence of the concept of the Great Moravian Empire, the interests of geomorphologists and Quaternary geologists focused on problems of floodplains evolution

(Opravil 1983). The study of river processes concentrated on studies of the impacts of floods and of suspension load and bedload transport (Brosch 2005; Buzek 1977, 1991), particularly the geomorphological impact of floods in 1980 (Gargulákova 1983), 1960 (Kříž 1963), and 2007 (channel configuration changes, Křížek 2008). Changes in the courses of rivers were studied by M. Havrlant and V. Kříž (1984).

Accelerated sheet and gully erosion in the Dyjsko-svratecký úval Graben, the Vyškovská brána Gate and the Central Moravian Carpathians in the eighteenth and nineteenth centuries were studied by Z. Láznička (1959) from historical data. Gully erosion in the Moravian Carpathians was described by K. Gam and O. Stehlík (1956), who also indicated extreme gully erosion after flash floods in the Kyjovská pahorkatina Hills near Bzenec (Stehlík 1954). Estimations of soil loss by water erosion in the Kyjovská pahorkatina Hills were published by V. Vaníček (1963) and by Demek and Stehlík (1972). Related deposits in the Outer Carpathians depressions deposited in connection with soil erosion after 1000 AD were studied by R. Netopil (1954), J. Demek (1955), and E. Opravil (1973, 1983). A general overview of soil erosion was published by O. Stehlík (1981), while the impact of hollow roads on gully development was revealed by J. Demek (1960). Interesting quantitative studies of the importance of wood felling and forest roads for soil erosion in the Moravskoslezské Beskydy Mountains were published by L. Buzek (1977, 1981, 1986, 1993, 1996).

Wind erosion in the Vizovická vrchovina Highland was studied in several papers, especially by R. Švehlík (1978, 1983, 1985). *Piping* in the Moravian-Silesian Carpathians was described by L. Buzek (1969), K. Kirchner (1981, 1982, 1987a, b), V. Cílek (2000), and P. Kos et al. (2000). The investigation of *pseudokarst* phenomena in the Silesian and Moravian-Silesian Beskids, Hostýnské vrchy Hills, Vizovická vrchovina Highland, and Javorníky Mountains substantially contributed to knowledge on deep-seated deformations in the Flysh Carpathians (Wagner et al. 1990; J. Wagner 1994).

Anthropogenic geomorphic processes and landforms were studied in the Ostrava Region (Havrlant 1979; 1980). From the geomorphological point of view, the Moravian-Silesian Carpathians are a very sensitive area. The intensity of natural endogenic and exogenic geomorphological processes is very high. In the last millennium the intensity of many natural exogenic geomorphological processes was accelerated by human activities.

6.4 Land-Use Changes and Their Impact in the Last Millennium

Jaromír Demek

Land-use changes influenced the intensity and type of exogenic geomorphological processes in the last millennium. The map *Landscape in the 12th century* compiled by B. Nováková at the scale of 1:500,000 published in the *Landscape atlas of the*

Czech Republic (Hrnčiarova et al. 2009) shows ancient *core settlement areas* of Outer Carpathians depressios, the Dolnomoravský úval Graben and flat foothills of Carpathians (e.g., Milovická pahorkatina, Kyjovská pahorkatina, Hlucká pahorkatina, and the western Podbeskydská pahorkatina) as agricultural landscapes with small isolated forests. Floodplains are mostly depicted as wetlands, but the extent of floodplain *forests* was probably greater than shown on the map. On the other hand, highlands and mountains of the MSC are shown as continuous forest landscapes.

Important administrative centers of the Great Moravian Empire (as Mikulčice, Staré Město, and Uherské Hradiště) in the floodplain of the Morava River and Pohansko of the Dyje floodplain were situated in the ancient core settlement areas of the grabens Dyjsko-svratecký úval and Dolnomoravský úval. These centers were buried under floodplain loams due to climatic changes and simultaneous accelerated soil erosion during floods becoming more frequent after the end of the Empire in the tenth century AD. Major international *trade routes* ran through the Moravian-Silesian Carpathians, along the Morava, Bečva, and Odra Rivers (e.g., the Amber Road from the Baltic to the Mediterranean).

Changes in landscape structure reflected the introduction of a new land administration in the late twelfth and early thirteenth centuries. The beginning of the Little Ice Age in the twelfth century brought higher precipitations causing *floods* and accelerated accumulation in floodplains. Most of the settlements located at riverbanks had to move from the floodplains up to higher ground not regularly inundated by floods (e.g., onto low river terraces). Settlements gradually spread into the higher parts of Carpathians from the ancient core areas.

Less fertile and forested flysch highlands and mountains of Eastern Moravia and Silesia were sparsely inhabited for a long time. The *colonization of forested landscapes* of Moravian-Silesian Carpathians started in the thirteenth century when settlers began to occupy mountain valleys. Towns were also founded (e.g., Rožnov pod Radhoštěm in 1267). The character of mountain landscapes radically changed only by the *Walachian colonization* in the late fifteenth and early sixteenth centuries, when Walachian pastoral tribes arrived into Moravian-Silesian Carpathians migrating westwards with their herds of shaggy sheep and goats from southern Romania along the Carpathian ranges. The local feudal lords did not profit from those forested mountain areas until the arrival of these mountain shepherds. The local Czech population was involved in agriculture in the valleys and did not have skills in mountain economy. Throughout the summer the Walachians' sheep and goats grazed in the mountains, where at the same location milk, wool, and other products were processed (alpine cottage production). The mountain grazing season started usually in May, when sheep and goats were driven from valleys to the cottages on mountain slopes and ranges. New land was gained mainly from clearing the woods for grazing cattle and growing crops. *Deforestation* and *grazing* on steep slopes naturally caused acceleration of geomorphological processes on less permeable flysh deposits, especially soil erosion due to distortion of grass cover, landsliding, and mudflows. Eroded material accumulated on valley bottoms. This situation is evidenced by deep accumulations on the floodplain of the Bečva River in the town of Vsetín (0.7–0.8 m of flood loam upon Late Holocene gravels). The base of the

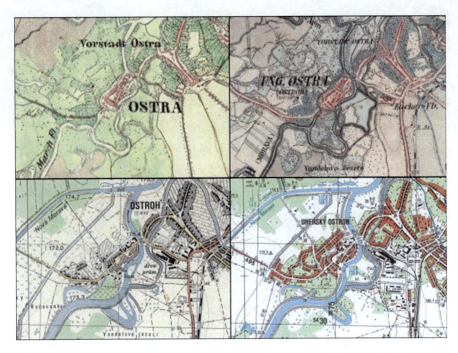

Fig. 6.3 Examples of topographic maps used for land-use evaluation. *Upper left*: map of the first Military Survey – Morava floodplain, 1768, *upper right*: map of the second Military Survey, 1836, *lower left*: map S52 of the Czechoslovak Military Survey, 1955, *lower right*: map of the Czech mapping, 2006

flood loam layer is dated 410 ± 30 ^{14}C years BP (calibrated years 1430–1510 AD), evidencing major environmental changes after Walachian colonization (Baroň et al. 2008). Deforestation also caused more frequent *flash floods*. Catastrophic processes occurring on slopes were the reason for a Decree of Empress Maria Theresa, which made it the duty of local citizens to plant trees. A similar decree was issued in 1796.

The construction of *fishponds* and entire fishpond systems changed the character of agricultural landscapes in the period of flourishing pisciculture in the fifteenth and sixteenth centuries and also influenced the water and sediment discharges of lowland watercourses. Fishponds changed runoff conditions and intercepted suspended load. An extensive fishpond system was established in the Dyjsko-svratecký úval Graben near the town of Pohořelice in the sixteenth century and near the village Lednice in the Dolnomoravský úval Graben, including the largest Moravian fishpond, Nesyt (296 ha). Many fishponds occur in the Dyjsko-moravská pahorkatina Hills in the catchment of the Kyjovka River.

Useful information on land-use changes is contained on detailed *topographical maps* drawn since the second half of the eighteenth century (Fig. 6.3). Land-use changes in the Moravian-Silesian Carpathians and their impact on the type and intensity of geomorphological processes are studied at the Department of Landscape Ecology of The Silva Tarouca Research Institute for Landscape and Ornamental

Gardening, using manual and computer supported analyses of topographical maps from the periods 1764–1836, 1836–1875, 1875–1950, 1950–1990, and 1990–2006 (Mackovčin 2009). The topographic maps from the firtst Austrian Military Survey of Moravia in 1764–1768 (at a scale of 1:28,800) provide a unique picture of the landscape at the beginning of the agricultural revolution. In the period of the first Survey, Europe was mainly agricultural. The largest landscape user was agriculture and its changes (e.g., Austrian agricultural reform) significantly influenced both the type and intensity of geomorphological processes. The agricultural revolution was followed by the first stage of scientific-technical revolution in agriculture that resulted in the extension of arable land at the expense of permanent grassland and in changes of runoff. In this period major elements of the agricultural landscape were wet floodplains with prevailing floodplain forests, meadows, and fishponds. Some fields appear in the warm floodplain Středodyjská niva on the boundary to Austria. In the floodplains, rivers freely meandered and formed anastomosing channels. The Morava River was markedly anastomosing in the vicinity of the town of Strážnice in the Dolnomoravský úval Graben (Culek et al. 1999). An elaborate network of river arms formed in floodplain forests in the region of Soutok near the confluence of the Dyje River with the Morava River on the border with Slovakia and Austria. Typical landscape features were fishpond systems. The low hills were typically rural landscapes dominated by arable fields, vineyards, and rural settlements. The landscape to the south of the town of Bzenec (Dúbrava in older maps, Doubrava in modern maps) still featured freely moving sand dunes. Rivers, their floodplains, and adjacent low hills were mutually linked in the landscape. Meadows and pastures locally reached up to the watersheds in highlands and deforested mountains on the boundary to Slovakia. Especially in regions of Walachian mountain farming, farms were also built on higher mountain slopes (called *kopanice* in Czech). Forests were affected by forest grazing.

The fact that the maps from the first Austrian Military Survey were not based on a triangulation network did not allow computer-aided evaluation, only manual analysis.

Landscape structure and geomorphological processes in the first half of the nineteenth century are well depicted on topographical maps from the second Austrian Military Survey, carried out in Moravia between 1836 and 1841, at 1:28,800 scale. For the survey a triangulation network was already employed and thus the maps can be georeferenced and computer processed. The 1:200,000-scale land-use map sheet M-33-XXIX (Brno) was compiled by the Silva Tarouca Research Institute for Landscape and Ornamental Gardening in Průhonice. The map includes the Dyjsko-svratecký úval Graben, the Výškovaká brána Gate, the Southern Moravian and Central Moravian Carpathians, and part of the Dolnomoravský úval Graben. Results of the computer-aided quantitative analysis of land use on the 1:200,000-scale map sheets M-33-XXX (Zlín) and the Czech part of sheet M-34-XXV (Žilina) showing the Carpathians in southeastern Moravia are presented in Table 6.1.

Rural landscape still dominated in the Moravian-Silesian Flysch Carpathians. In 1823 reforestation started in the Doubrava blown sand area (Vitásek 1942, p. 1 – Fig. 6.4) near the town of Bzenec in the Dolnomoravský úval Graben. A number of

Table 6.1 Land use changes on map sheets M-33-XXX (Zlín) and M-34-XXV (Žilina, Moravian part) in the period 1836–2006

Code	Categories	1836 sq.km	1836 %	1876 sq.km	1876 %	1956 sq.km	1956 %	1996 sq.km	1996 %	2006 sq.km	2006 %
1	Arable land	1,907.47	45.55	2,215.93	52.92	2,291.87	54.73	1,889.79	45.13	1,727.13	41.26
2	Permanent grassland	1,010.16	24.12	640.30	15.29	291.96	8.97	373.38	8.92	476.95	11.39
3	Garden and orchard	19.73	0.47	24.53	0.59	37.34	0.89	67.99	1.62	38.11	1.62
4	Vineyard and hop field	53.73	1.28	41.28	0.99	27.72	0.66	54.49	1.30	1,475.11	0.91
5	Forest	1,089.64	26.02	1,153.18	27.54	1,317.70	31.47	1,430.07	34.15	1,475.11	35.23
6	Water surface	5.27	0.13	0.61	0.01	6.64	0.16	19.33	0.46	20.88	0.50
7	Built-up area	101.20	2.42	111.10	2.65	209.82	5.01	331.16	7.91	354.93	8.48
8	Recreation area	0.00	0.00	0.00	0.00	2.12	0.05	17.48	0.42	22.44	0.54
0	Other areas	0.21	0.01	0.38	0.01	2.24	0.05	3.62	0.09	3.80	0.09
	Total	4,187.31	100	4,187.31	100	4,187.31	100	4,187.31	100	4,187.31	100

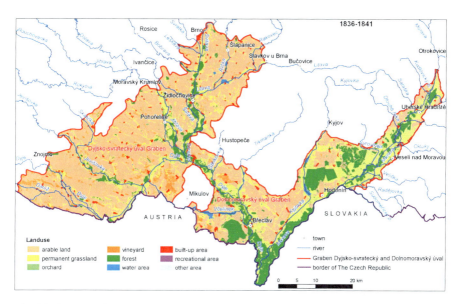

Fig. 6.4 Example of a land-use map of the Dyjsko-svratecký and Dolnomoravský úval Grabens based on computer-aided analysis of topographic maps of the second Austrian Military Survey in 1836. Rivers are not trained and in wet floodplains there are floodplain forests and permanent grasslands

fishponds were drained and replaced by country estates (Hlavinka and Noháč 1926, p.11) in this period. The river patterns experienced considerable changes. The main channel of the Olšava River shifted from the direction of the town of Uherské Hradiště to the west of Kunovice and then meandered as a yazoo towards the south parallel with the main channel of the Morava River, reaching it at the town of Ostroh. To the south of the town of Hodonín, the Morava River was braided. The degree of connectivity, both longitudinal and lateral, was still considerable in the Moravian-Silesian Carpathians. The map shows a dense network of gullies, especially in loess, many of them developed from abandoned medieval hollow roads that concentrated runoff. A group of new parallel hollow roads from this period were used up to the twentieth century.

For the assessment of landscape pattern and geomorphic processes in the second half of the nineteenth century maps from the third Austrian Military Survey, prepared at a scale of 1:25,000 for Moravia in 1875–1877, are useful. The period between the second and the third military surveys was a period of a very rapid spreading of the cultural landscape. Land use changed over 20.37% of the flat area of the Dolnomoravský úval Graben (Table 6.2).

The agricultural revolution leading to higher intensity of agricultural production was responsible for the increasing proportion of arable land (Table 6.1) – again at the expense of permanent grassland. In lowlands and flat hills the cultivation of sugar beet expanded and led to soil exhaustion and the application of manure as well

Table 6.2 Land use changes in the Dolnomoravský úval Graben, 1836–2006 (Demek et al. 2009)

Code	Categories	1836	1876	1955	1992	2006
1	Arable land	38.06	46.46	49.61	49.26	49.52
2	Permanent grassland	32.00	22.96	14.95	6.92	5.70
3	Orchard	0.67	0.50	0.87	1.44	1.78
4	Vineyard and hopfield	1.62	1.24	1.26	2.83	2.36
5	Forest	23.86	25.42	26.46	26.51	27.09
6	Water surface	1.42	0.74	1.21	3.41	3.57
7	Built-up area	2.35	2.67	5.42	9.14	9.47
8	Recreational area	0.00	0.00	0.03	0.26	0.31
0	Other area	0.01	0.01	0.19	0.23	0.18
	Total	100.00	100.00	100.00	100.00	100.00

Table 6.3 River channel length changes (in km) in the Dyjsko-svratecký úval and Dolnomoravský úval Graben in the course of time (Demek et al. 2008a,b)

River	1836	1876	1944	1954	1991	2007[a]
Svratka	52.50	44.62	40.25	40.21	35.33	36.75
Cézava	28.24	26.48	n.a.	24.46	24.41	24.48
Jihlava	26.32	25.20	25.36	25.55	24.52	24.97
Dyje	68.39	67.14	n.a.	61.46	59.11	60.15
Jevišovka	31.82	32.36	n.a.	31.79	31.46	31.49
Morava[b]	145.53	144.51	n.a.	112.12	97.29[c]	n.a.

[a]Mapping scale 1:10,000
[b]From Napajedla up to the confluence with the Dyje River (Kilianová 2001)
[c]Data for 1999 (Kilianová 2001)

as mineral fertilizers. Agriculture, however, remained backward in highlands and mountains in this period.

The floodplains showed a reduced share of forests and arable land sprawled from the adjacent hills into the floodplains. The number of fishponds in the landscape was dramatically reduced (e.g., in the Dyjsko-svratecký úval Graben from 0.21% to 0.003%). The development called for the regulation of the Svratka and Svitava river channels, launched in 1848. The original confluence of both rivers was artificially moved (Demek et al. 2007). A new deep channel was excavated and embankments built between the city of Brno in the north and the village of Přízřenice in the south. Channelization disturbed the connectivity of valley slopes, floodplains, and stream channels. The regulation thus resulted in the fragmentation of floodplains; also changing river length and sinuosity (see Table 6.3).

During the extreme flood in Moravia (1877) the dykes of the Bečva River breached (Peřinka 1911, p. 397) and the floodplain Středomoravská niva was inundated. A large flood also occurred in 1894 (Peřinka 1912, p. 577). Urban growth in floodplains accelerated after river regulations (Table 6.1). The construction of transport networks and extension of settlements required the extraction of buildings materials. Gravels and sand pits were opened in the floodplains and raw materials for brick making (namely loess and clay) were extracted in the adjacent hills.

No integrated set of large-scale topographic maps exists for the first half of the twentieth century that would make it possible to follow land-use changes continually. Only the maps from the third Austrian Military Survey were gradually revised. The first Czechoslovak agricultural reform after World War I led to the reduction of large estates, allotting land to small farmers and creating a more conspicuous mosaic of small fields in the rural landscape. The Morava River in the Carpathians was regulated as late as the twentieth century. At the beginning of the twentieth century, river training started between the towns of Napajedla and Lanžhot and the main channel was shortened (Table 6.3). The topographic map at 1:75,000 scale dating from about 1930 already shows a new navigation canal named Morávka excavated between the towns of Veselí nad Moravou and Vnorovy in Slovakia. The longitudinal connectivity of rivers was greatly disrupted by weirs. The area of blown sands in the Doubrava was already fully reforested and the dunes stabilized. Settlements were sprawling, also on floodplains, and the share of urban landscape was increasing (Table 6.1).

In the second half of the twentieth century, the cultural landscape in the Moravian-Silesian Carpathians experienced essential changes and geomorphological processes were accelerated. After a long break of about 75 years, another integrated set of Czechoslovak military topographic maps was published in 1952–1955 (S52). The post-war period saw the second Czechoslovak agricultural reform, and the industrialization and collectivization of Czech agriculture after 1955. The maps document the re-establishment of several large fishponds (e.g., in the vicinity of Pohořelice in the Dyjsko-svratecký úval Graben). A network of three large water reservoirs (Nové Mlýny) was constructed at the confluence of the Dyje and Svratka Rivers in the Dyjsko-svratecký úval and Dolnomoravský úval Grabens in 1975–1982.

Renewed Czechoslovak color military topographic maps (S42) from 1990–1992 document the impact of the second Czechoslovak *agrarian reform* on the landscape as well as the subsequent *collectivization*. The private field boundaries were obliterated and large fields were created. A system of shelter belts was planted in the dry and warm Jaroslavická pahorkatina and Drnholecká pahorkatina Hills in Southern Moravia. The maps S42 also document a considerable expansion of urban landscapes and *suburbanization*. At the same time the connectivity of ecologically more stable landscape segments and patterns was decreasing. An unfavorable fact was the further spreading of built-up areas into floodplains (resulting in major damage during the 1997 catastrophic flood).

The contemporary landscape structures are shown on the digital grid Basic Map of the Czech Republic at the scale 1:10,000 and on recent aerial and satellite images. The share of arable land slightly decreased after 1989 (Table 6.1), while large fields still prevail. The share of forests increased to its historical peak (Table 6.1). The digital maps also represent a rapid growth in built-up land (Table 6.1). The number of anthropogenic landforms in the landscapes has extremely grown.

6.5 Hillslope Processes and Landforms

Tomáš Pánek
Karel Kirchner

The extensive occurrence of hillslope processes in the MSC is explained by both suitable *preparatory* and *triggering factors*. To the first category belongs (1) the susceptible anisotropic flysch substrate, (2) dissected mountain terrain with river incision, (3) large portions of slopes weakened by deep-seated slope failures or old landslides, and (4) intensive anthropogenic activity – mainly deforestation during the past centuries. Main triggering factors especially include extreme hydrometeorological events, i.e., local heavy downpours, incessant rains with regional extent followed by heavy downpours or intensive snowmelt (often enhanced by rainfall events). Some slope failures have been induced by active undercutting of slopes. The MSC are the most hazardous landslide terrain on the territory of Czech Republic with 9100 inventoried active and dormant slope failures until 2007 (Krejčí et al. 2008).

According to the classification of R. Dikau et al. (1996), in the last millennium the territory of the MSC has been affected by almost a full range of slope processes, including deep-seated slope failures (lateral gravitational spreading, toppling, and sackung), translational and rotational landslides, earthflows, debris flows, and rockfalls. The highest ridges formed by most resistant sandstones of the Silesian Unit (e.g., Moravskoslezské Beskydy Mountains) have been mostly modeled by slow and massive deep-seated slope failures and debris flows, less dissected uplands on the fine-rhythmical (clay-rich) flysch of the Magura Unit (e.g. Hostýnsko-vsetínská hornatina Mountains or Javorníky Mountains) are extremely susceptible to landslides and earthflows (Krejčí 2005).

Deep-seated slope failures (mainly lateral gravitational spreading) have been described from numerous places situated especially in the most elevated parts of the Silesian and Magura Units (Baroň et al. 2004; Krejčí et al. 2004; Hradecký et al. 2007; Hradecký and Pánek 2008). Suitable conditions for such failures involve mainly high ridges formed by jointed and/or faulted sandstones underlying by weak (plastic) fine-rhythmical flysch. Many manifestations of deep-seated failures are connected with widespread landslide terrains; some of them present incipient stages of landslide evolution (Baroň et al. 2004; Hradecký and Pánek 2008). Morphological expressions of deep-seated failures involve double (or multiple) ridges, counter-slope scarps, and crevice-type caves (Fig. 6.5). Dilatometric or extensometric measurements performed within crevice-type caves and on surface ruptures in the last decade have revealed very slow, often reversible movements do not exceeding 1 mm year^{-1} (Rybář et al. 2006). For instance, maximum amplitude of movements within the longest Cyrilka cave in the mountain Ridge Radhošťský hřbet in the Mountain Moravskoslezské Beskydy (>370 m long) in the period 2000–2007 was 0.03 mm year^{-1} (Klimeš and Stemberk 2007). Higher amplitudes were measured by J. Stemberk and J. Rybář (2005) in the Kněhyňská cave (shaft) (0.3 mm year^{-1}), and in the

Fig. 6.5 Morphological expressions of deep-seated slope failure (Lukšinec ridge, Moravian–Silesian Beskids). Some of the opened trenches (not visible on this picture) originated after July 1997 event (Photo T. Pánek)

Lukšinec hills (0.5 mm year^{-1}). Most of these results indicate pulse-like movements with seasonal shortening and dilatation also indicating external (climatic) influence. Despite the fact that monitoring of slow movements of deep-seated failures brings valuable information, due to the limited number of measurements, it cannot cover the full range of movement intensities. Periods of much faster opening of some gravitational trenches were observed, e.g., just after high-intensity rainfalls in July 1997. J. Wagner (2004) described that some new tensional cracks on the Lukšinec ridge (Lysá hora mountain, the Moravskoslezské Beskydy Mountains) opened within several days/week after mentioned extreme hydrometeorological event.

The most abundant and geomorphologically effective slope processes are *landslides* involving *translational, rotational,* or *compound* movements. They are often associated with and/or gradually evolve to the earthflows with protruding lobes. Nearly 40% of recently activated landslides are nested on slopes weakened by older landslides or deep-seated failures (Krejčí et al. 2002, 2008). Historic sources (e.g., Špůrek 1972) and case studies (e.g., Adámek, 1973; Pečeňová-Žďárská and Buzek 1990; Krejčí et al. 2002; Baroň et al. 2004, etc.) indicate more than 50 landslide events in the MSC that have taken place since 1770. While some of these events were isolated phenomena (e.g., Burkhardt et al. 1972), some phases had catastrophic dimensions with genesis and reactivation of several tenths to thousands slope failures (Krejčí et al. 2002; Bíl and Müller 2009). Four types of hydrometeorological regimes in the study area can be recognized as leading to landslides: (1) prolonged rainfalls enhanced by heavy downpours; (2) rapid snowmelt accompanied by rainfalls; (3) localized short-term heavy downpours exceeding 100 mm day^{-1}, and (4) prolonged (several years) abnormally humid periods. Of the above-mentioned triggers only the first two categories usually induce landslides over vast areas, whereas the two last types cause rather spatially localized and isolated failures.

The most pronounced landslide event arose during *extreme rainfalls* in July 1997 when 5-day (from 4th to 8th July) precipitation amounts in the Moravskoslezské Beskydy, Hostýnsko-vsetínská hornatina, and Javorníky Mountains reached 200–400 mm at meteorological stations, whereas a 1-day precipitation record was measured on the Lysá hora (234 mm on 7th July 1997). The hydrometeorological events of July 1997 caused or reactivated more than 1,500 landslides with the highest concentration in the clay-rich substrate of the Magura Unit in the Vsetín and Zlín districts (Krejčí et al. 2002) (Fig. 6.6a).

The melting of an anomalously deep snow cover coupled with the intensive rainfalls of March/April 2006 caused another major landslide event (Bíl and Müller 2009). The cumulative precipitation (both rainwater and snowmelt) reached 143 mm during the thaw season and led to the formation of more than 90 rather shallow landslides situated preferentially on the deforested slopes in the Bílé Karpaty Mountains and Vizovická vrchovina Highland (Bíl and Müller 2009; Klimeš et al. 2009) (Fig. 6.6b). Similar snowmelt combined with rainfalls induced the catastrophic January 1919 Hošťálková landslide in the Hostýnské vrchy Hills (Záruba 1922).

Localized heavy downpours (usually summer rainstorms) could also trigger landslides, especially if their intensity (or sum of several events during 1 day) exceeds 100 mm day^{-1} (Hradecký and Pánek 2008). If catchments are not saturated by antecedent rainfalls, such events are capable of activating only a limited number of shallow landslides (e.g., debris slides) concentrated at the sites of rainstorms, as the rainstorm event on 24th August 2005 in the Lysá hora. Several shallow debris slides (up to 100×40 m) developed within one day during the rainstorm. Their preferential location was on the deforested steepest slopes, e.g., headscarps of older deep-seated landslides or roadcuts (Fig. 6.6c).

Rather isolated but often pronounced deep-seated landslides occur after prolonged anomalously wet periods, not directly associated with particular rainfall events. Deformation in this category was described by Burkhardt et al. (1972) from the Oznice site in the Hostýnské vrchy Hills (Fig. 6.6d). A structurally preconditioned rotational landslide evolved in August 1967 as a short-time event (observed directly by a local resident) during which large, internally almost non-deformed block moved ca 40 m downslope in the direction of the brachysynclinal axis. A landslide emerged suddenly after a period of two abnormally wet years (121% of average annual precipitation) and just after a wet spring/summer period.

Detailed geomorphological mapping, geophysical survey, and dating reveal that most of the largest recent landslides (e.g., during July 1997 or March/April 2006

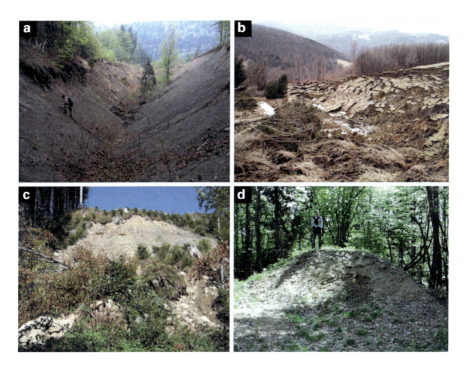

Fig. 6.6 Landslides triggered by different types of hydrometeorological events. (**a**) Brodská landslide with protruding lobe evolved in older, structurally preconditioned landslide area during the extreme rainfalls in July 2007 (Photo T. Pánek); (**b**) *Upper part* of the Hluboče landslide, evolved after massive snowmelt and rainfalls of March/April 2006 (Photo T. Kosačík); (**c**) An example of shallow landslide in the Jestřábí Stream (eastern slopes of Lysá hora), happened after localized downpours on 24th August 2005 (Photo T. Pánek). (**d**) Middle part of the Oznice landslide with small activated area (Photo I. Baroň)

landslide events) are in fact reactivated movements in older landslide bodies (Fig. 6.7). For instance, the Hluboče landslide (Bílé Karpaty Mountains) of March/April 2006, the largest and most catastrophic landslide, is situated in an older failure that was AMS dated to the $1,337 \pm 22$ cal year BP (Klimeš et al. 2009). The analysis of historical aerial photographs reveals that the catastrophic movement in early April, 2006, was anticipated by more than 50 years long period of creeping and gradual evolution of the headscarp area (Fig. 6.8).

Similar observations of recurrent behavior of landslides were obtained from other parts of the region (e.g., Baroň 2007, 2009; Pánek et al. 2009b), like the Hlavatá ridge (Moravskoslezské Beskydy Mountains), where a landslide reactivated at least three times over 1,500 years (Pánek et al. 2009b). The catastrophic January 1919 Hošťálková landslide (Hostýnské vrchy Hills) that destroyed several houses and caused blockage of a valley is situated on a fossil landslide of a minimum age 1.4 cal ka BP (Baroň 2007, 2009). In the past 100 years, on the Skalice hill (Podbeskydská pahorkatina Hills near Frýdek-Místek) numerous activations of

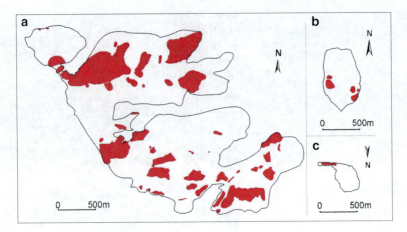

Fig. 6.7 Spatial connection between old deep-seated landslides in the Magura Unit (around Vsetín) and failures activated during July 1997 event (*red patches*). All displayed landslides show a complex Holocene history. The Vaculov–Sedlo landslide (**a**) has a minimal age 6.1 ka BP, the Kobylská landslide (**b**) is older than 9.0 ka BP and the Kopce failure (**c**) is older than 2.5–3.0 ka BP (After Baroň et al. 2004)

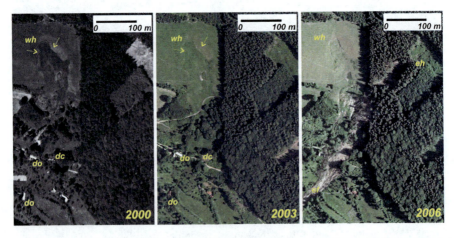

Fig. 6.8 Evolution of the Hluboče landslide. Note the very pronounced, but not fully developed, western branch of the headscarp (*wh*) already 6 years before catastrophic movement. At the turn of March/April 2006 catastrophic earthflow (*ef*) originated after intersection of western (*wh*) and eastern headscarps (*eh*). Several buildings (*do*) and roads were destroyed. A disastrous earthflow affected the ca. 1.3 ka old landslide body (Klimeš et al. 2009)

frontal landslide area were associated with extreme events and undercutting of slopes by the anabranching the Morávka River (Adámek 1973; Pečeňová-Žďárská and Buzek 1990; Pánek and Hradecký 2000). *Slow creep*-like movements (inclinometric measurements provided by (Kovář et al. 2006), show cumulative movements in 2005 up to 6 mm) are disrupted once in 10–20 years by catastrophic landslides causing significant economic losses at each time (e.g., destroyed road in August 1972 or five recreational objects in July 1997).

Less pronounced hillslope processes in the MSC are *debris flows* and *rockfalls*. Their low abundance in the area is due to the lack of major rock faces and steep hillslopes above the timberline. They only occur in the highest regions, mainly in the Moravskoslezské Beskydy and Hostýnsko-vsetínská hornatina Mountains. A higher activity of debris flows and rockfalls is assumed for the Pleistocene/Early Holocene periods with different climatic conditions. A rare example of Late Holocene catastrophic rock-avalanche/debris flow was described from the northern slope of Mt. Ropice (Moravskoslezské Beskydy Mountains) by T. Pánek et al. (2009a). An extensive rock avalanche originated ca 1.4 cal ka BP as a consequence of the collapse of vast rock packet situated in the unstable part of the deep-seated landslide (Pánek et al. 2009a). A large number of debris flows were described by K. Šilhán and T. Pánek (2006) from the highest parts of the Moravskoslezské Beskydy Mountains, the majority of them, however, being prehistoric landforms. Historic events only comprise accumulations of 10^2–10^4 m^3 volume. The analysis of historical sources and dendrochronological record points to 33 minor debris flow events since 1939, mostly after daily rainfalls above 100 mm (Šilhán and Pánek 2010). More than 20 years history of active debris flow fan at the outlet of a high-gradient valley on the steep western slopes of Mt. Travný was described by K. Šilhán (2008) (Fig. 6.9). Rockfalls are relatively more common processes in the study area (associated with failure scarps, structural and undercut slopes), but, due to their small dimensions, their geomorphic impact is quite negligible. Preliminary dendrochronological investigations in the highest parts of the Moravskoslezské Beskydy Mountains indicate that triggering factors for rockfall activity could be both extreme rainfalls and freeze-thaw cycles (Šilhán 2009). On the western slopes of Smrk Mountain (Moravskoslezské Beskydy Mountains) average rockfall frequency is up to two events year^{-1} (Hradecký and Pánek 2000; Šilhán 2009). A rainfall-triggered rockfall from an active landslide area took place in the Bystřička valley (Vsetínské vrchy Hills) in July 1997 (Kirchner and Krejčí 1998).

6.6 Water Erosion and Landforms

Jan Hradecký
Aleš Létal

About 25% of the MSC territory is forested and about 60% is arable land. As in other areas of Central Europe, human activity is the predominant cause of water erosion. In the last millennium, streams of the Czech Carpathians recorded a few fundamental changes consequent to global climatic changes and regional environmental processes, many of which were induced by *human action*. The rate of erosion is directly proportional to man-induced changes in the landscape.

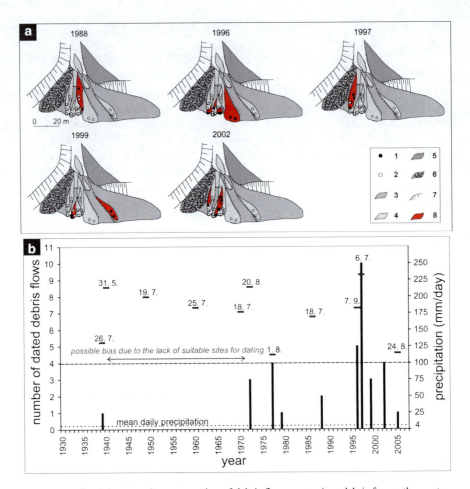

Fig. 6.9 (a) Spatial-temporal reconstruction of debris flows on a minor debris fan on the western slope of the Travný Mountain. The present-day fan developed gradually over several tens of years due to low-magnitude events: *1* tree affected by debris flows in particular years, *2* sampled tree, *3* area of the debris fan, *4* lobe of debris flow deposits, *5* old generation of fan, *6* accumulation of hyperconcentrated flow, *7* scarp, *8* active part of the fan in particular years. (b) Dendrochronological reconstruction of debris flows in the Moravian–Silesian Beskids for the last 80 years. Mean daily precipitation is shown by the *lower dashed line*. *Numbers* above indicate dates of extreme daily rainfalls exceeding 100 mm (*higher dashed line*) (Both figures after Šilhán and Pánek 2010)

O. Stehlík (1981) found four periods of increased erosion activity in the Czech Republic (750–850, 1300–1400, 1750–1850, and 1952-). The last two periods have been particularly effective. The MSC is an area of increased erosion activity and the central and southern parts present the highest erosion hazard in the Czech Republic (Fig. 6.10). Anthropogenic pressure in relation to erosion has evolved over time and ordered in importance as

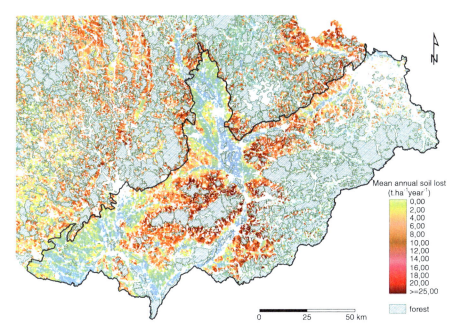

Fig. 6.10 Map of soil loss (Dostál 2007)

- Deforestation;
- Intensification of agriculture (agrarian revolution – change to fallow system – three-field system, change to Norfolk crop rotation, industrialization and collectivization of agriculture – land consolidation);
- Forestry (forest grazing, heavy mechanization in forestry).

Climatic fluctuation of the last 1,000 years has led to a wide range of flood phases accompanied by *increased geomorphologic activity of streams*. Studies focusing on the reconstruction of the historic climate of the last millennium assume the alternation of warmer and colder episodes and also of wetter and drier intervals (Brázdil and Kotyza 1998). In wetter phases increased *flood frequency* (Brázdil et al. 1999, 2005; Brázdil and Kirchner 2007) changed channel and floodplain morphology. Extreme discharge values are related to climatic periods such as the Little Ice Age between 1300 and 1850 AD in Europe. According to Brázdil and Kotyza (1998), it commenced in the Czech Republic between 1400 and 1419 AD. The highest frequency of floods on the Morava River is observed in the following periods: 1591–1610, 1711–1720, 1831–1880, and 1871–1880 (Brázdil et al. 2005). The period between 1501 and 1880 witnessed 75 historic floods, 48 of which is documented in our study area, while the rest of them are related to upper reaches of the Morava River. Less information is available for floods in the Odra River basin (Brázdil et al. 2005). Historic sources contain records of 31 floods (Brázdil et al. 2005), 20 of which are related directly to the wider Carpathian territory (1501,

1531, 1533, 1542, 1571, 1584, 1593, 1649, 1713, 1723, 1740, 1741, 1831, 1845, 1871, 1878, 1879, 1880, 1891, and 1892). For example, the Bečva River basin (Brázdil and Kirchner 2007), which was affected by 83 flood events in the years between 1494 and 1899. The data are even more valuable considering the fact that the Bečva River is a typical gravel-bed Carpathian stream with channel branching. An even higher number of floods (151) was recorded in the Dyje River basin (the Morava River right tributary), namely in the period of 1519–1900 (Brázdil and Kirchner 2007). However, it must be stressed here that the discharge of the Dyje River is largely influenced by precipitation events outside the area of interest. The twentieth century, on the contrary, was much poorer in flood events. According to records, in July 1903 the studied area was hit by a flood (Řehánek 2002; Brázdil and Kirchner 2007) that significantly affected the character of gravel-bed Carpathian streams that flow into the Ostravská pánev basin, streams springing in Nízký and Hrubý Jeseník Mountains, or into the above mentioned Bečva River. Another larger flood hit the Bečva, Morava, and Dyje Rivers and streams in the Odra River basin in early September 1938. The last extreme flood events affecting Carpathian streams and streams of the Carpathian Foreland with visible impact on channel morphology were the floods of the 6th and 8th July 1997 (Demek et al. 2006). Many streams within the Odra River basin experienced 50- or 100-year floods and in some places even higher discharges (Řehánek et al. 1998). As a result, the Bečva River considerably widened and recorded renaturalization of its gravel bed and developed branches (Klečka 2004). The Morávka River, by contrast, was affected by extensive deep incision (Hradecký and Pánek 2008). Similarly large floods also occurred in the basins of the Morava and Dyje Rivers, whereas in the case of the Dyje River others followed in the spring of 2006. Particularly strong influence on stream morphology and transport of large amounts of sediments is exerted by the *flash floods* that occurred in the MSC area a few times during the twentieth century. The latest extreme floods of this type took place in the area of the Štramberská vrchovina Highland (e.g., Jičínka and Zrzávka Rivers) in June 2009.

Important information on stream fluvial activity not only in the last millennium but also in the preceding phases has been obtained from the study of *floodplains*. Research on the Morava and Dyje floodplains revealed a relatively long time span of the beginning of flood loam sedimentation: from 965 ± 95 BP to 3,720 ± 60 BP (Havlíček and Smolíková 1994). Floodplain accretion is particularly related to deforestation and the advent of agricultural activities, which led to the loss of a large amount of fine-grained soil that had deposited in the floodplain zone during the floods (Opravil 1983; Havlíček 1991). In areas of high relative relief increased runoff from deforested areas might have led to voluminous coarse bedload transport, which, consequently, resulted in increased channel sedimentation (Hradecký and Škarpich 2009).

The most complex research was performed on part of the Morava River floodplain (Strážnické Pomoraví) over the last few years (Kadlec and Beske-Diehl 2005; Kadlec et al. 2009; Grygar et al. 2009), applying modern scientific *tools* (e.g., radiocarbon dating, clay mineral identification and quantitative analysis of expandable clay minerals, micromorphology of sediments, magnetic susceptibility, particle size distribution, ICP/MS including isotopic analysis, X-ray fluorescence analysis, xylotomy analysis, ^{137}Cs measurement, etc.). The fluvial archive of the Morava River, preserved in the study reach from the last millennium, consists of up to 5-m-thick formations of clayey or silty overbank sediments exposed in the eroded banks (Kadlec et al. 2009).

The *rate of deposition* (channel aggradation and lateral accretion) is a very important indicator of human influence on floodplain evolution (mainly the impact of deforestation and arable farming) (Grygar et al. 2009). The rate of deposition for the topmost sediment layers of the analyzed profiles was probably influenced by past land-use changes and flow regulations (e.g., flood defense constructions and channelization) of the river upstream during the twentieth century (Grygar et al. 2009). Different rates and quality of overbank aggradation are evident within the whole sediment sequence. The lowest rates are typical of distal floodplain sediments (0.2–0.3 cm year^{-1}) for the periods of 720–1320 AD and 1140–1900. The floodplain aggradation during the first half of the last millennium is a consequence of medieval farming boom. Similar rates are identified with upward coarsening clayey sediments grading from distal to proximal floodplain deposits (Grygar et al. 2009). Kadlec et al. (2009) recognized a pronounced change in the *grain size composition* of the sediment that took place in the sixteenth century when a shift from more clayey deposits (assigned here to the distal floodplain) to more silty and sandy deposits (assigned here to the proximal floodplain) occurred (probably as an effect of the LIA).

From the complex analyses of the sediment record it was possible to identify three simultaneous *changes in depositional conditions*: one in ~1200 AD, another in ~1600 AD and the last one in ~1900 AD. Faster accumulation (0.42–0.55 cm year^{-1}) is accompanied with upward coarsening (Kadlec and Beske-Diehl 2005), mostly silty and proximal floodplain deposition in the period from 1535 AD to 1900 AD (Grygar et al. 2009). The topmost 50 cm of the flood sediments reveal anthropogenic pollution caused by leaded gasoline, DDTs, and PCBs. The rate of deposition within this industrial layer is estimated at 0.8 cm per year (Kadlec et al. 2009). The upper 12-cm layer of sediments deposited after the Chernobyl nuclear power plant accident in April 1986 (Kadlec et al. 2009). High sedimentation rates in the topmost part of the floodplain sequence are attributed to increased erosion in response to intensive agricultural activities observed in the Morava River basin over the last 50 years (Kadlec et al. 2009) and due to intensive collectivization after 1950 (which caused higher rates of soil and gully erosion in the upper parts of the river basins and accelerated accumulation along the lower sections).

The percentage of arable land in the MSC is relatively high and, therefore, *sheet erosion* is of decisive significance. The rate of sheet erosion has only been studied in a few localities of the Central Moravian Carpathians, where long-term erosion loss from the loess watershed was estimated at 17.6 t ha^{-1} year^{-1} (Mařan 1958), and in the southwestern part of the Central Moravian Carpathians (in the Trkmanka River catchment) it was found to be 43.5 t ha^{-1} year^{-1} (Vaníček 1963; Demek and Stehlík 1972). Due to the industrialization, collectivization, and intensification of agricultural production (farming on large fields, use of heavy machinery), the rate significantly increased in the second half of the twentieth century.

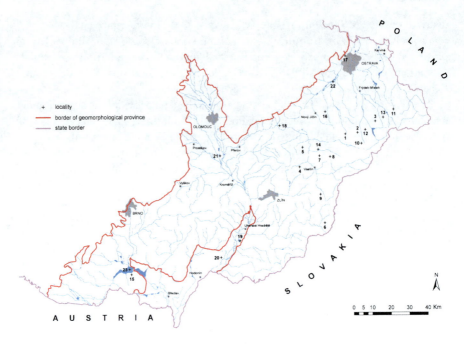

Fig. 6.11 The main localities described in the paper

Permanent gullies are important indicators of the historical evolution of anthropogenic pressure on the landscape. Rill and gully erosion spring from human activity and directly contribute to land degradation. The present permanent gully system in the MSC is mostly older than 200 years (Láznička 1957). Gully density does not exceed 0.5 km km^{-2} with maximum values of 3–5 km km^{-2} (Gam and Stehlík 1956). Increased erosion was caused by a combination of several factors: the massive colonization of forested areas, spreading arable land, cold and wet climate periods (Little Ice Age) (Havlíček 1994; Stehlík 1981), and extreme precipitation events. Gully erosion is triggered by extreme downpours, flash floods, and/or continuous rainfall. These extreme precipitation events have strictly local effect and, therefore, eroded sediments only appear in the third or fourth level subcatchments.

6.6.1 Recent Soil Erosion

Water erosion is one of the major problems of agriculture in the Czech Republic. Over 40% of arable land in the CR is potentially at risk by water erosion, mainly *sheet wash* and *ephemeral rill erosion*. Ephemeral rills have a limited duration because of tillage, but their occurrence is evident throughout the cultivated area. Ephemeral rills are directly linked to row crops (maize, root crops). The most recent episodes were documented in 1997, June and July 2009, and August 2010 floods. The floods in the Luha River catchment (tributary of the Odra River) involved the development of numerous ephemeral rills and gullies, with soil loss from the largest gully (474 m length, width at the mouth 4 m, depth 1.25 m) being 264 m^3. Recent erosion is due to compaction by soil tillage and the use of heavy machinery, which affect about 45% of agricultural land in the Czech Republic (Javůrek and Vach 2008). Rill and gully erosion do not only occur on arable land, but also in forests – directly influenced by human action (logging in the dormancy period, mechanization and forest roads). The relationship between logging and soil erosion has long been studied by L. Buzek (1981, 1986). Erosion induced by human activity in the forests of the MSC is mainly observed in the past 50 years. It could be associated with overexploitation of forests (forest grazing) around settlements in some areas even before the twentieth century. Gullies in the Carpathians often form on landslides of the Carpathian flysch (Vizovická vrchovina Highland, Bílé Karpaty Mountains, and Chřiby Highland). A new phenomenon emerging in the recent erosion processes springs from the disruption of drainage system during amelioration. *Amelioration drainage* systems were frequently built in the 1960s and 1970s to increase arable land. This technique is quite often applied for erosion control on slopes as well. Without technological and financial conditions of maintenance, combined with the decline of intensive agriculture after 1989, it may be a serious problem in the near future, especially in areas with steeper slopes. Rapid erosion associated with growing crops (maize, root crops) on graded areas remains as a major issue. Given the present constraints on agricultural activities, erosion rates are likely to have reached their peak between 1960 and 1990. Rainfall activity is also a decisive factor in determining the type and extent of erosion processes. If land use pressure is coupled with humid weather, as it probably happened in some past periods, we can expect even higher rates of water erosion of all forms.

6.6.2 Piping Landforms and Processes

Piping (suffosion) landforms and processes in the MSC were first described by Buzek (1969) and Kirchner (1981). Most of the piping forms studied in the MSC are underground features with some manifestations on the surface (Kirchner 1987a, b). The *piping channels* and *depressions* studied are several meters long and 2–3 m deep. Minor features preferentially occur along rodent pathways, root systems of trees, or even on the surface of block fields. In the Štramberská vrchovina Highland present piping forms under cliffs are 2–5 m deep and are up to 8 m long, on the surface of debris fans reach 2 m length and 0.5 m depth (Buzek 1969). A piping depression of 4.6 m long, 1.9 m wide, and 1.1 m deep was described in the clay-loamy sediments of the Vizovická vrchovina Highland (Kirchner 1987a, b) and a piping *tunnel* 4.5 m long and 1.4 m deep from the Hostýnské vrchy Hills (Kirchner 1981). Smaller forms are found on loess and loess-loamy soils in the Chřiby Highland (Hořáková 2007). The maximum depth of monitored forms was 1.5 m with maximum lengths of 4.5 m. Piping processes are mostly associated with spring snowmelt and extreme summer rainfall events. V. Cílek (2000) and P. Kos et al. (2000) described a pseudokarst system of chimneys, piping wells, and two suffosion caves in loess at the foot of the Pavlovské vrchy Hills in the South Moravian Carpathians.

Piping induced by *burrowing* (tunnels less than 10 cm diameter) is significant in the reactivation of permanent gullies in the area of Chřiby Highland and Vsetínské vrchy Hills. Piping processes directly induce the formation of new lateral branches of gullies and transformation of gully heads. The various piping features may survive for several hours or even several years under favorable conditions. Interesting permanent gully development was detected in the Vsetínské vrchy Hills, where sediments flushed out from piping tunnels accumulated around trees on gully slopes. Lateral tunnel branches longer than 3 m may develop during short but heavy rainfall events. In general, however, piping processes and forms are not common in the MSC and they are of limited geomorphological significance.

6.7 Wind Erosion and Landforms

Aleš Létal

Wind erosion in the MSC occur in the Dyjsko-svratecký úval Graben, the Kyjovská pahorkatina and Hlucká pahorkatina Hills, and in the Bílé Karpaty Mountains (Dufková 2007). The occurrence of wind erosion and its manifestations are directly proportional to the intensity of land cultivation. The intensity of wind erosion has increased markedly in the second half of the twentieth century through the processes of collectivization (land consolidation). About 90% of wind erosion occurs on cultivated land (Pasák 1970). The most endangered areas lie on lighter soils (light sandy soils, loamy-sandy soils) (Pasák 1970), but in the Hlucká pahorkatina Hills and Bílé Karpaty Mountains wind erosion occurs in areas with heavier clay-loamy soils. While the maximum size of soil particles blown away by wind erosion is generally 0.8 mm, in these specific regions soil particles of size 1.12–2 mm, in exceptional cases about 4 mm, are also entrained (Janeček 2002) by strong and dry southern, southeastern, and eastern winds in autumn, winter, or spring. The long-term systematic research of wind erosion in that area has been carried out by R. Švehlík, who found that the annual removal of soil is about 4–5 mm, in dust storms about 1–2 cm. Higher rates of wind erosion were identified in the Hlucká pahorkatina Hills in 1972 (193 m^3 ha^{-1} $year^{-1}$) (Švehlík 2002). The long-term monitoring of wind erosion is coupled with a detailed classification of aeolian forms (Švehlík 2002).

6.8 Anthropogenic Impact on Topography

Irena Smolová

Anthropogenic features in the Czech Carpathian region are dominated by mining, aquacultural (fishponds), and transportation features, markedly concentrated in the zone of the Outer Carpathian depressions and the northern extension of the Vienna

Basin (Kirchner and Smolová 2010). Human impact usually reaches down to hundreds of meters and sometimes even more than a kilometer in the case of underground mining. The deepest mines in operation are those in the Ostrava Basin (Doubrava III is 1,176 m deep with shaft top at 281 m above, and bottom at 895 m below sea level). Direct influence in the form of exploration and test borings usually reaches the depths of 1–2 km. The deepest borehole in the Czech Carpathian Mountains is Jablůnka 1 (near the town of Vsetín) with a depth of 6,506 m, drilled in 1982 (closed at present).

From the *historical aspect*, the first known anthropogenic impact on the relief of the Czech Carpathian Mountains dates back to the Lower Paleolithic period, well documented by archaeological finds and closely connected to the colonization of the Moravian Grabens. The oldest documented settlement was in karst areas. Caves utilized as shelters were not heavily transformed (as shown by the archeological finds in the Šipka Cave in the Štramberk Karst). The core residential area was the Vienna Basin and the Outer Carpathian depressions, where the most important Upper Paleolithic cultures evolved (Gravettian and Pavlovian), and some Moravian locations became world famous (e.g., Dolní Věstonice and Pavlov). Early *residential communities* (settlements on hills, rondels typical of the Neolithic Age, strongholds or *grad*s), the regulation of watercourses and the cultivation of land involved erosion processes. Burial sites (barrows) are preserved from the early historical period. A historically important Roman settlement is the Mušov stronghold in a strategically important site at the confluence of the Svratka and Dyje Rivers (preserved aqueduct, defence ditches, etc.) and the sites Pasohlávky, Olomouc-Neředín, and Hulín.

Along with human settlement in the Moravian Grabens, flood-control dykes were constructed in numerous locations and houses were often built on artificial mounds. (In Otrokovice, for instance, the waterlogged ground at the confluence of the Morava and the Dřevnice Rivers was elevated artificially by 1–4 m in the 1930s using material from the nearby Tresný Hill.

Human impact on the relief culminated in the period of the Industrial Revolution (second half of the nineteenth century) and towards the end of the socialist industrialization (second half of the twentieth century), characterized by extensive extraction of minerals for heavy industry in the city of Ostrava, especially underground mining of hard coal in the Ostrava Basin, one of the areas most affected by human activities in the Carpathian Mountains with abundant *underground* as well as *surface features of mining*. At the initial stage, shortly after coal deposits had been discovered in the late eighteenth century, opencast mining prevailed. The first underground mine in the Ostrava Basin (Anselm Mine) was established in Landek in 1782. At present, the mines in operation reach depths of ca 1 km, and after 1990 are concentrated in the eastern part of the Ostrava Basin, around Karviná (mines Československá armáda, Lazy, ČSM, Darkov, Paskov). In the area of the Moravian-Silesian Beskyds there is the now conserved mine Doubrava in the town of Frenštát pod Radhoštěm, excavated in the 1970s. Underground mining in the Ostrava Basin resulted in a number of spoil heaps, disposal, and subsidence areas. At present, more than 300 *spoil heaps* are registered, 5 of them are still active, one of the largest being Ema (82 ha area, 315 m altitude, the highest peak of the Ostrava district). The heap includes more than 4 million m³ of mine waste from the already closed mine Trojice. The creation of anthropogenic surface features, especially *subsidence areas*, also heavily influenced the drainage conditions within the Ostrava Basin. J. Maníček (2003) describes the *influence upon the watercourses* of the Ostrava Basin in the overall length of almost 120 km, including 23 km along the Ostravice River, 15 km along the Odra River, and more than 20 km along the Olše River. Because of the subsidence, artificial watercourses were established in the Ostrava Basin, primarily for the *drainage* of subsidence areas (e.g., the Černý příkop Canal, constructed to drain the area around the mine Odra in the 1950s). An example of markedly influenced watercourse is the Karvinský potok Stream, originally a tributary of the Stonávka River, but connected, to the parallel lateral canal along the Olše River in the 1960s to allow the drainage of the most intensively subsiding areas.

The subsidence in the undermined section of the Ostrava Basin influences the statics of buildings and induces *seismic activity*. Since 1989, disturbances have been recorded by a network of seismic stations situated directly in the mines areas. According to the statistics, 20,000–50,000 disturbances of various energy levels are recorded annually by the Czech Geological Survey. The strongest of them often cause mining accidents and reach the local magnitude of 2–4 (e.g., tremor at mine Doubrava on 13 June, 2002, 3.9 M; at mine Lazy on 11 March, 2004, 3.1 M, 7 victims).

Materials mined underground in the Western Carpathians include lignite, oil, and natural gas in the Neogene deposits in the South Moravia. The first prospect *oil and natural gas wells* were drilled in 1899 and reached the depth of 649 m. The first exploited well was 529 m deep near Slavkov, which, in 1908, reached gas reserves. More than 10,000 prospect borings were realized along the Moravian-Slovak border to date aimed at the extraction of oil and natural gas. The deepest exploitation well is near the village of Jarošov (5,587 m) near Uherské Hradiště. The well near the village of Hrušky at the town of Břeclav is 3,885 m deep. The Outer Carpathian Depressios have the highest concentration of *exhausted reservoirs* in the Czech Republic. The oldest was established in 1965 (aquifer-type underground gas storage in the village of Lobodice), followed by other reservoirs, e.g., at village Tvrdonice (formerly Hrušky) near Břeclav (at depths of 1.1–2.5 km), at Třanovice (formerly Žukov) near the town of Český Těšín, at Štramberk (Příbor) on the boundary between the Bohemian Massif and the Carpathians, at Uhřice, at Dolní Dunajovice in the Dyjsko-svratecký úval Graben, or at Dolní Bojanovice near the town of Hodonín (in the depth of 750–2,070 m). The impact on the Earth's crust is documented by repeated surveying. (For instance, at the gas storage site Hrušky documents periodic oscillations and an annual increase of tilt by 0.4–0.5 mm since 1978 was found with subsidence towards the center of the depression.)

For building materials production in the Carpathian area, the largest *quarries* are a limestone quarry in the town of Štramberk (1.18 km²) and a stone quarry near Hranice (0.17 km²). The largest yield is produced by the stone quarry Podhůra near the town Lipník nad Bečvou where Culm greywacke is quarried. The Outer Carpathian Depressions and the Lower Moravian Graben are also heavily influenced by the extraction of gravel and sand, mostly by wet process from the floodplain, creating new water surfaces. The largest sand pits are found at Ostrožská Nová Ves (working area 5.17 km^2), Vracov-Bzenec (2.66 km^2), and Žabčice near the town of Břeclav (2.41 km^2).

Important *industrial anthropogenic features* in the Carpathian area include industrial spoil heaps, industrial platforms, especially in new industrial zones (e.g., the Nošovice Zone with an area of 260 ha, from where 2.2 million m^3 of soil was removed), and large industrial sites, from where considerable volumes of material were removed during construction. One of the largest industrial spoil heaps is at Třinec (on the left bank of the Olše River), which consists of clinker and waste from the local iron works piled up from 1839 to 1995. At present, the spoil heap area exceeds 65 ha and its relative height is more than 40 m.

The water regime is being influenced by the construction of *water-management structures*. The oldest interferences with the natural regime derive from agriculture: ponds, flumes, and canals for irrigation, land drainage or supply of water to mills (mill races). The largest pond is Nesyt (296 ha) near Břeclav; other extensive pond systems are in the surroundings of towns of Hodonín, Tovačov and along the Odra River. Artificial canal systems are found around Břeclav and Hodonín. In numerous locations, the stream flow was trained by modification of riverbeds and straightening of watercourses, resulting in changes of longitudinal profiles, intensified erosion, and reduced natural infiltration caused by the modification of banks. H. Kilianová (2001) claims that in the period from 1836 to 1999, the course of the Morava River was shortened by 67.34 km, most fundamentally along the lower course of the Morava River in the Lower Moravian Graben (shortening by 48 km between 1836 and 1999), while the length of the middle course in the Upper Moravian Graben was cut by 13 km and by 6 km in the Mohelnická brázda Furrow. The upper course of the Morava River remains relatively unaffected by regulation, excluding head and mill races. The most prominent water-management structure is the three reservoirs of the Nové Mlýny barrage system at the confluence area of the Svratka, Dyje, and the Jihlava Rivers, completed in the late 1980s. Numerous barrages were constructed for the regulation of the upper courses of rivers in the Bečva River basin and in the Moravian–Silesian Beskids. To provide sufficient volume of drinking and industrial water for the dynamically developing Ostrava region, a water-supply system in the Odra River basin was built in the 1970s and 1980s. After repeated floods in 1997 and subsequent years, new dry polders (emergency reservoirs) were constructed in order to provide flood protection for the municipalities.

The Outer Carpathian Depressions are an important *transport corridor*. The most important and oldest route crossing here was the Amber Road between the Baltic Sea coast and the Adriatic Sea via the Moravian Gate (Moravská brána). As transportation developed, *railways and roads* were constructed. Their *embankments* lie in the axis of the Outer Moravian Depressions, vast areas are occupied by transportation platforms (including the Mošnov Airport in the Moravian Gate) and other kinds of transportation infrastructure.

References

Adámek J (1973) Sesuvné území ve Skalici u Frýdku (Landslides in the Skalice near the town of Rrýdek). Těšínsko 3:1–4 (in Czech)

Baroň I (2007) Výsledky datování hlubokých svahových deformací v oblasti Vsetínska a Frýdeckomístecka (Results of dating of deep seated slope deformations in the Vesín and Frýdek Místek Region). Geologické výzkumy na Moravě a ve Slezsku 14:10–12 (in Czech)

Baroň I (2009) Svahová deformace v Hošťálkové – Uvezeném (Landslides in the village Hošťálková – Uvezené). In: Baroň I, Klimeš J (eds) Svahové deformace a pseudokras. Elektronický sborník referátů z konference, 13–15.5.2009 ve Vsetíně. ČGS & ÚSMH AV ČR (in Czech)

Baroň I, Cílek V, Krejčí O, Melichar R, Hubatka F (2004) Structure and dynamics of deep seated slope failures in the Magura Flysch Nappe, Outer Western Carpathians (Czech Republic). Nat Hazards Earth Syst Sci 4:549–562

Baroň I, Kirchner K, Nehyba S, Jankovská V, Nývlt D (2008) Fluviální sedimenty řeky Bečvy ve Vsetíně (Fluvial deposits of the Bečva River at Vsetín). In: Roszková A, Vlačik M, Ivanov M (eds) 14. Kvartér 2008. Sborník abstrakt. s. 1, Ústav geologických věd PřF MU, Česká geologická společnost, Brno (in Czech)

Bíl M, Müller I (2009) The origin of shallow landslides in Moravia (Czech Republic) in the spring 2006. Geomorphology 99:246–253

Brázdil R, Kirchner K (eds) (2007) Vybrané přírodní extrémy a jejich dopady na Moravě a ve Slezsku (Selected natural extremes and its impacts in Moravia and Silesia). Masarykova univerzita, Brno – Český hydrometeorologický ústav, Praha – Ústav geoniky Akademie věd České republiky, Brno, 431 p (in Czech with English summary)

Brázdil R, Kotyza O (1998) Kolísání klimatu v Českých zemích v první polovině našeho tisíciletí (Climatic changes in the Czech lands during the first half of our millenium). Archeologické rozhledy 49:663–669

Brázdil R et al (1999) Flood events of selected European rivers in the sixteenth century. Clim Chang 43:239–285

Brázdil R et al (2005) Historické a současné povodně (Historical and contemporary floods). Masarykova univerzita v Brně, Český hydrometeorologický ústav v, Praze, 369 p

Brosch O (2005) Povodí Odry (Odra River Watershed). Anagram, Ostrava, 232 p (in Czech)

Burkhardt E, Liškutínová D, Plička M (1972) Význačný sesuv u Oznice v Hostýnských vrších (Imporant landslide near Oznice in the Hostýnské vrchy Mts.). Sborník Československé společnosti zeměpisné 3:219–225 (in Czech)

Buzek L (1969) Geomorfologie Štramberské vrchoviny (Geomorphology of the Štramberská vrchovina Highland). Spisy Pedagogické fakulty v Ostravě 11:1–91 (in Czech)

Buzek L (1977) Příspěvek ke studiu současných morfogenetických procesů v povodí Morávky v Moravskoslezských Beskydech (Contribution to the study of present-day morphogenetic processes in the Moravka catchment in the Moravian–Silesian Beskids). Sborník prací Pedagogické fakulty v Ostravě 51(E-7):97–109 (in Czech)

Buzek L (1981) Eroze proudící vodou v centrální části Moravskoslezských Beskyd (Water erosion in the central part of Moravian–Silesian Beskids). Spisy Pedagogické fakulty v Ostravě 45:1–109 (in Czech)

Buzek L (1986) Degradace lesní půdy vodní erozí v centrální části Moravskoslezských Beskyd (Degradation of forest soils by water erosion in the central part of Moravian–Silesian Beskids). Sborník Československé geografické společnosti, Praha 91:112–126 (in Czech)

Buzek L (1993) Vliv dešťových srážek a tání sněhu na intenzitu eroze půdy v Moravskoslezských Beskydech v letech 1976–1990 (Influence of precipitation and snow melting on intensity of soil erosion in Moravian–Silesian Beskids in years 1976–1990). Acta Fac Rer Nat Univ Ostraviensis Geogr-Geol Ostrava 1(136):33–45 (in Czech)

Buzek L (1996) Faktory urychlené eroze v jižním horském zázemí Ostravské průmyslové aglomerace (Factors of accelerated erosion in the southern mountainous vicinity of Ostrava industrial agglomeration). Sborník České geografické společnosti, Praha 101(3):211–224 (in Czech)

Cílek V (2000) Sprašové jeskyně u Dolních Věstonic pod Pavlovskými vrchy (Loess caves near Dolní Věstonice at the foot of the Plavlovské vrchy Hills). Speleo 30:14–19 (in Czech)

Culek M, Ivan A, Kirchner K (1999) Geomorphologie der Talaue der March zwischen der Napajedla Pforte und dem Zusammenfluß mit der Thaya (Zum Naturmillieu in der Umgebung von Mikulčice und Staré Město). In: Poláček L, Dvorská J (eds) Internationale Tagungen in Mikulčice, Band V: Probleme der Mitteleuropäischen Dendrochronologie und Naturwissenschaftliche Beiträge zur Talaue der March. Archeologický ústav AV ČR, Brno, pp 199–221

Demek J (1955) Vznik a stáří tzv. povodňových kalů našich údolních niv (Origin and age of flood deposits in our floodplains). Sborník Československé společnosti zeměpisné, Praha 60:1–20 (in Czech)

Demek J (1960) Problémy eroze půdy a ochrany krajiny v území jihozápadně od Brna (Problems of soil erosion and landscape erosion southwest of Brno). Ochrana přírody 15:132–136 (in Czech)

Demek J (1986) Kvantitativní výzkum svahových pohybů ve Vnějších Západních Karpatech (Quantitative research of slope movements in Outer West Carpathians). Geografický časopis 38(2–3):178–185 (in Czech)

Demek J (1988) Quantitative study of debris movements on slopes in Western Carpathians. Studia geomorphologica Carpatho-Balcanica 22:83–90

Demek J, Mackovčin P (eds) (2006) Zeměpisný lexikon ČR. Hory a nížiny (Geographical encyclopedia of the Czech Republic. Mountains and Lowlands), 2nd edn. Agentura ochrany přírody a krajiny ČR, Brno, 531 p (in Czech)

Demek J, Stehlík O (1972) Urychlená eroze půdy – zdroj devastace krajiny a životního prostředí (Accelerated soil erosion – the source of landscape and environmental devastation). Životné prostredie 4:186–191 (in Czech)

Demek J, Kalvoda J, Kirchner K, Vilímek V (2006) Geomorphological aspects of natural hazards and risks in the Czech Republic. Studia Geomorphologica Carpatho-Balcanica 40:79–92

Demek J, Havlíček M, Mackovčin P, Stránská T (2007) Brno and its surroundings a landscape-ecological study. In: Dreslerová J (ed) Ekologie krajiny. J Landsc Ecol, Prague:32–53

Demek J, Havlíček M, Chrudina Z, Mackovčin P (2008a) Changes in land-use and the river network of the Graben Dyjsko-svratecký úval (Czech Republic) in the last 242 years. J Landsc Ecol 1(2):22–51

Demek J et al (2008b) Mezinárodní konference Antropogenní stres v horách a podhůří – zápis změn v tvarech terénu i v sedimentech (Anthropogenic stress in landscape – traces of changes in relief and deposits). Výzkumný ústav Silva Taroucy pro krajinu a okrasné zahradnictví, Brno, 18 p (in Czech)

Demek J, Havlíček M, Mackovčin P (2009) Landscape changes in the Dyjsko-svratecký úval Graben and Dolnomoravský úval Graben in the period 1764–2009 (Czech Republic). Acta Pruhoniciana 91:23–30

Dikau R, Brunsden D, Schrott L, Ibsen ML (eds) (1996) Landslide recognition: identification, movement and causes. Wiley, Chichester

Dostál T (ed) (2007) Metody a zpusoby predikce povrchoveho odtoku, eroznich a transportnich procesu v krajine (The methods of surface runoff, erosion and transport processes in a landscape prediction). CTU Prague Research report of the project COST 634, ČVUT v Praze, Fakulta stavební, Katedra hydromeliorací a krajinného inženýrství, Praha (in Czech)

Dufková J (2007) Comparison of potential and real erodibility of soil by wind. Acta Universitatis Agriculturae et Silviculturae Mendelianae Brunensis, Brno 55(4):15–21

Gam K, Stehlík O (1956) Příspěvek k poznání stržové erose na Moravě a ve Slezsku (Contribution to the knowledge of gully erosion in Moravia and Silesia). Sborník Československé společnosti zeměpisné 61(3):270–273 (in Czech)

Garguláková Z (1983) Povodeň v Ostravě a okolí v roce 1880 – největší známá živelní pohroma na Ostravsku (Flood in Ostrava and its vicinity in the year 1880 – the greatest geohazard in the Ostrava region). In: Ostrava. Sborník příspěvků k dějinám a výstavbě města, 12, Ostrava (in Czech)

Götziger H (1909) Geologische Studien im subbeskidischen Vorland auf Blatt Freistadt in Schlesien. Jahrbuch der k.k. Geol. Reichsanstalt, Wien

Grygar T et al (2009) Geochemical tools for the stratigraphic correlation of floodplain deposits of the Morava River in Strážnické Pomoraví, Czech Republic from the last millennium. Catena 80:106–121

Havlíček P (1991) The Morava River basin during the last 15 000 years. In: Starkel L, Gregory KJ, Thornes JB (eds) Temperate palaeohydrology. Wiley, Chichester, pp 319–341

Havlíček P (1994) State of research of the Moravian rivers in Holocene time. In: Růžičková Eliška ZA (ed) Holocene flood plain of the Labe river. Geological Institute of Academy of Sciences of the Czech Republic, Prague, pp 98–99

Havlíček P, Smolíková L (1994) Vývoj jihomoravských niv (Origin of Moravian floodplains). Bull Czech Geol Surv 69:23–40 (in Czech)

Havrlant M (1979) Antropogenní formy reliéfu a životní prostředí v příměstské zóně ostravské průmyslové oblasti. (Anthropogenic landforms and environment in the suburban zone of Ostrava industrial region). Sborník prací pedagogické fakulty v Ostravě, řada E-10 67:55–78, SPN Praha (in Czech)

Havrlant M, Kříž V (1984) Antropogenní hydrografické změny Odry, Opavy a Ostravice v širším územním prostoru Ostravy (Anthropogenic hydrographic changes of the rivers Odra, Opava and Ostravice in the Ostrava region). Sborník prací Českého hydrometeorologického ústavu, Praha 29:45–59 (in Czech)

Hlavinka K, Noháč J (1926) Hodonský okres (District Hodonín). Vlastivěda Moravská II. Místopis. Musejní spolek v Brně, Brno, 259 p (in Czech)

Hořáková M (2007) Rozšíření sufoze ve vybrané oblasti Vnějších Západních Karpat (Piping in some regions of the Outer Western Carpathians). PhD thesis, Palacký University, Olomouc, 114 p (in Czech)

Hradecký J, Pánek T (2000) Geomorphology of the Smrk Mt. area in the Moravskoslezské Beskydy Mts. (Czech Republic). Moravian Geographical Reports, Brno 8:45–54

Hradecký J, Pánek T (2008) Deep-seated gravitational slope deformations and their influence on consequent mass movements (case studies from the highest part of the Czech Carpathians). Nat Hazards 45:235–253

Hradecký J, Pánek T, Klimová R (2007) Landslide complex in the northern part of the Silesian Beskydy Mountains (Czech Republic). Landslides 4:53–62

Hradecký J, Škarpich V (2009) General scheme of the Western Carpathian stream channel behaviour. Geomorfologický sborník 8: 8, Česká asociace geomorfologů, Kašperské hory

Hrnčiarova T, Mackovčin P, Zvara I (eds) (2009) Landscape atlas of the Czech Republic. Ministry of Environment of the Czech Republic, Prague – Silva Tarouca Research Institute for Landscape and Ornamental Gardening, Průhonice, 331 p

Janeček M (2002) Ochrana zemědělské půdy před erozí (Protection of agricultural soils from erosion). Nakladatelství, ISV Praha, 201 p (in Czech)

Javůrek M, Vach M (2008) Negativní vlivy zhutnění půd a soustava opatření k jejich odstranění (Negative impact of soil compression and preventive measures). Výzkumný ústav rostlinné výroby, Praha-Ruzyně, 24 p (in Czech)

Kadlec J, Beske-Diehl S (2005) Magnetic properties of floodplain deposits along the banks of the Morava River (Czech Republic). The IRM Q 15:2–3

Kadlec J et al (2009) Morava River floodplain development during the last millennium, Strážnické Pomoraví, Czech Republic. Holocene 19:499–509

Kilianová H (2001) Hodnocení změn lesních geobiocenóz v nivě řeky Moravy v průběhu 19. a 20. století (Evaluation of changes of forest geobiocenoses in the Morava floodplain during the 19th and 20th centuries AD). PhD thesis, Mendlova zemědělská a lesnická univerzita v Brně. 118 p (in Czech)

Kirchner K (1981) Příspěvek k poznání sufoze v Hostýnských vrších (východní Morava) (Contribution to the knowledge of piping in the Hostýnské vrchy Mountains (Eastern Moravia). Zprávy Geografického ústavu ČSAV v Brně 18(2):119–125 (in Czech)

Kirchner K (1982) K rozšíření a vývoj sufozních tvarů na východní Moravě (Distribution and development of suffosion landforms on the Eastern Moravia). Stalagmit 41–42 (in Czech)

Kirchner K (1987a) Výskyt sufozních tvarů v okolí Vsetína (Piping landforms in the surroundings of the town of Vsetín). Československý kras 38:129–132 (in Czech)

Kirchner K (1987b) Sledování vývoje sufozních tvarů v oblasti Vsetína (Studies of piping in the vicinity of the town of Vsetín). Sborník prací, Geografický ústav ČSAV v Brně 14:135–143 (in Czech)

Kirchner K, Krejčí O (1998) Slope movements in the flysch Carpathians of eastern Moravia (Vsetín district) triggered by extreme rainfalls in 1997. Moravian Geographical Reports, 6(1): 43–52, Brno

Kirchner K, Smolová I (2010) Základy antropogenní geomorfologie (Anthropogenic Geomorphology). Univerzita Palackého, Olomouc, 287 p (in Czech)

Klečka J (2004) Early stadiums of floodplain forest succession in a wide river beds upon an example of Bečva. J Forest Sci 50:338–352

Klimeš J, Stemberk J (2007) Svahové deformace v okolí sedla Pustevny (Slope deformations near the saddle Pustevna). Geomorfologický sborník 6:72–73 (in Czech)

Klimeš J, Baroň I, Pánek T, Kosačík T, Burda J, Kresta F, Hradecký J (2009) Investigation of recent catastrophic landslides in the flysch belt of Outer Western Carpathians (Czech Republic): progress towards better hazard assessment. Nat Hazards Earth Syst Sci 9:119–128

Kos P et al (2000) Sprašové jeskyně u Dolních Věstonic pod Pavlovskými vrchy (Loess caves near Dolní Věstonice at the foot of Pavlovské vrchy Hills). Speleo 30:14–19 (in Czech)

Kovář L, Marschalko M, Liberda A (2006) Významná svahová deformace „U hřbitova" ve Skalici u Frýdku-Místku. (Important slope deformation "U hřbitova" in the village Skalice near the town of Frýdek Místek). Workshop Svahové deformace a pseudokras 2006 - 10. - 13. května 2006, Ostravice. Sborník referátů a prezentací z odborného semináře. ČGS, ÚSMH AVČR, Ostravice. CD-ROM (in Czech)

Krejčí O (2004) Tektogeneze oblasti styku Českého masívu a Západních Karpat (Tectogenesis of the contact zone between the Bohemian massif and the Outer Western Carpathians). PhD thesis, Czech Geological Survey, Brno, 55 p (in Czech)

Krejčí O (2005) Slope instability hazard evaluation in the Flysch Western Carpathians (Czech Republic) in comparison with standards of European Commission. Mineralia Slovaca 37(3):409–412

Krejčí O, Baroň I, Bíl M, Hubatka F, Jurová Z, Kirchner K (2002) Slope movements in the Flysch Carpathians of Eastern Czech Republic triggered by extreme rainfalls in 1997: a case study. Phys Chem Earth Parts A/B/C 27:1567–1576

Krejčí O, Hubatka F, Švancara J (2004) Gravitational spreading of the elevated mountain ridges in the Moravian–Silesian Beskids. Acta Geodyn Geomater 3:1–13

Krejčí O et al (2008) Podprogram "ISPROFIN č. 215124–1 Dokumentace a mapování svahových pohybů v ČR" – Přehled provedených prací (Documentation and mapping of slope movements). Výzkumná zpráva. Česká geologická služba, Praha, 114 p (in Czech)

Kříž V (1963) Hydrologické vyhodnocení povodně z července 1960 v povodí Odry (Hydrological evaluation of flood in July 1960 in the Odra catchment). Český hydrometeorologický ústav, Praha (in Czech)

Křížek M (2008) Configuration of channel bars in the Černá Opava River as a consequence of flooding in September 2007. Acta Universitatis Carolinae, Geographica 42(1–2):145–162

Láznička Z (1957) Stržová erose v údolí Jihlavy nad Ivančicemi (Gully erosion in the Jihlava R. valley upstream of the town Ivančice). Práce Brněnské základny ČSAV 29(9):393–415 (in Czech)

Láznička Z (1959) Historické zprávy o erosi půdy v Brněnském kraji (Historical reports on soil erosion in the Brno region). Sborník ČSSZ, Praha 64(1):13–28 (in Czech)

Mackovčin P (2009) Land use categorization based on topographic maps. Acta Pruhoniciana, Průhonice 91:5–13

Macůrek J (1959) Valaši v Západních Karpatech v 15.-18. století (Walachians in the Western Carpathians in 15– 18th century AD). Krajské nakladatelství v Ostravě (in Czech)

Maníček J (2003) Mapa ovlivnění vodních toků a vodních ploch dobýváním černého uhlí v české části hornoslezské pánve se zákresem záplav povodně v roce 1997 (Map of influence of watercourses by hard coal mining in the Czech part of the Upper Silesian Basin with boundaries of area flooded in 1997). Documenta Geonica, Milan Čermák Publishers, Ostrava, pp 63–79 (in Czech)

Mařan B (1958) Výskum erose a protierosních opatření na zemědělských a lesních půdách (Study of erosion and its control on agricultural and forest soils). Výskumný ústav zemědělských a lesnických meliorací ČSAZV v Praze (in Czech)

Netopil R (1954) Vztah mezi geomorfologickým vývojem aluviální nivy Jihlavy u Ivánĕ a odtokovými poměry (Relation between geomorphological evolution of the Jihlava floodplain near Iváň and discharge conditions). Vodní hospodářství 4:156–160 (in Czech)

Novosad S (1956) Fosilní rozeklání hřbetu Lukšinec u Lysé hory (Fossil spreading of Lukšinec Ridge near Lysá hora). Časopis pro mineralogii a geologii 1(2):126–131 (in Czech)

Novosad S (1966) Porušení svahů v godulských vrstvách Moravskoslezských Beskyd (Slope deformations in Godula formation of the Moravian–Silesian Beskids). Sborník geol Věd HIG 5:71–86 (in Czech)

Novosad S (2006) Výsledky detailního monitoringu pohybu sesuvů na příkladech ze Vsetínska a Beskyd (Results of detailed monitoring of landslide movements on examples from District Vsetín and Moravian–Silesian Beskids). In: Baroň I, Klimeš J, Kašperáková D (eds) Svahové deformace a pseudokras. Elektronický sborník referátů a prezentací z odborného semináře. Česká geologická služba a Ústav struktury a mechaniky hornin AV ČR, Praha (in Czech)

Opravil E (1973) Předběžné výsledky analýzy rostlinných zbytků z výplně říčního koryta u Mikulčic (Preliminary results of plant remnants analysis from the oxbow near Mikulčice). Archeologický ústav ČSAV, Brno, pp 43–52 (in Czech)

Opravil E (1983) Údolní niva v době hradištní (The floodplain in the medieval period). Studie Archeologického ústavu ČSAV v Brně 11(2):1–77 (in Czech)

Pánek T, Hradecký J (2000) Současný geomorfologický výzkum v Západních Beskydech a Podbeskydské pahorkatině (Contemporary geomorphological research in the Western Beskids and in the Podbeskycká pahorkatina Hills). Geologické výzkumy na Moravě a ve Slezsku v roce 1999. Roč. 7, Masarykova univerzita, Brno, pp 44–47 (in Czech)

Pánek T, Hradecký J, Minár J, Hungr O, Dušek R (2009a) Late Holocene catastrophic slope collapse affected by deep-seated gravitational deformation in flysch: Ropice Mountain, Czech Republic. Geomorphology 103:414–429

Pánek T, Smolková V, Hradecký J, Šilhán K (2009b) Late Holocene evolution of landslides in the frontal part of the Magura Nappe: Hlavatá Ridge, Moravskoslezské Beskydy Mountains. Moravian Geographical Reports 4:2–11

Pasák V (1970) Wind erosion on soils. Research Institute of Soil and Water Conservation, Prague. Scientific monographs, Výzkumný ústav meliorací, Praha 3, 187 p

Pečeňová-Žďárská M, Buzek L (1990) Svážný terén ve Skalici, místní část Frýdku-Místku (Landslides in Skalice, part of Frýdek-Místek). Acta Facultatis Pedagogicae Ostraviensis 122:173–185 (in Czech)

Peřinka F (1911–1912) Vlastivěda Moravská II. Místopis, Kroměřížský okres (District Kroměříž). I-II část. Nákladem Musejního spolku, Brno, 604 p (in Czech)

Quitt E (1960) Sesuvné území na Výhoně u Židlochovic (Landslides on Výhon Hill near Židlochovice). Geografický časopis 12:189–197 (in Czech)

Řehánek T (2002) ovodeň na řece Odře v červenci 1997 (Flood on the Odra River in July 1997). Práce a studie, seš. 31, Český hydrometeorologický ústav v Praze, Praha, 41 p

Řehánek T et al (1998) Hydrometeorologická zpráva o odtokové situaci v červenci 1997 v povodí Odry (Discharge situation in July 1997 in the Odra River Catchment: hydrometeorological report). Český hydrometeorologický ústav, Ostrava

Rybář J, Jánoš V, Klimeš J, Nýdl T (2006) Rozpad synklinálního hřbetu Ondřejníku v Podbeskydské pahorkatině (Spreading of the syncline ridge Ondřejník in the Podbeskyská pahorkatina Hills). Zprávy o geologických výzkumech v roce 2006. Česká geologická služba, Praha, pp 92–96

Šilhán K (2008) Dendrochronologie kuželu blokovobahenních proudů na západním svahu Travného (1203 m) (Moravskoslezské Beskydy) (Dendrochronology of the mus cones on the western slope of Travný (1203 m) (Moravian–Silesian Beskids). Časopis Slezského Muzea Opava (A) 57:265–270 (in Czech)

Šilhán K (2009) Morfologie a morfometrie gravitačních proudů v Moravskoslezských Beskydech. (Morphology and morphometry of gravitational streams in the Moravian-Silesian Beskids). Zprávy o geologických výzkumech v roce 2008. 75–78 (in Czech)

Šilhán K (2010) Dendrogeomorphology of spatio-temporal activity of rockfall in the Flysch Carpathians: a case study on the western slope of Mt. Smrk, Moravskoslezské Beskydy Mountains (Czech Republic). Moravian Geograph Rec 189(3):33–42

Šilhán K, Pánek T (2006) Mury v kulminační části Moravskoslezských Beskyd (Debris flows in the Moravian–Silesian Beskids). In: Geomorfologické výzkumy v roce 2006. Univerzita Palackého, Olomouc, pp 260–265 (in Czech)

Šilhán K, Pánek T (2010) Fossil and recent debris flows in medium-high mountains (Moravskoslezské Beskydy, Czech Republic). Geomorphology 124:238–249

Špůrek M (1972) Katalog historických sesuvů na území ČR (Catalogue of historical landslides in the Czech Republic). Studia Geographica 19:1–180

Stehlík O (1954) Stržová eroze na Moravě (Gully erosion in Moravia). Práce Brněnské základny ČSAV 26(9):201–234 (in Czech)

Stehlík O (1981) Vývoj eroze půdy v ČR (Soil erosion in the Czech Republic). Studia Geographica 72:1–72 (in Czech)

Stejskal J (1931) Svážná území na Pavlovských vrších (Landslides in the Pavlovské vrch Hills). Sborník Čsl akademie zemědělských věd 6:55–94 (in Czech)

Stemberk J, Rybář J (2005) Risk assessment of deep-seated slope failures in the Czech Republic. In: Hungr O (ed) Landslide risk management. Taylor & Francis, London, pp 497–502

Švehlík R (1978) Kategorizace orné půdy ohrožené větrnou erozí v jihovýchodní části okresu Uherské Hradiště. (Classification of the arable land dangered by wind erosion in the southeastern part of the district Uherské Hradište). Sborník Československé společnosti zeměpisné 83(3):163–169 (in Czech)

Švehlík R (1983) Vítr jako geomorfologický činitel. (Wind as a geomorphological factor). Sborník prací 1:75–80, Geografický ústav ČSAV, Brno (in Czech)

Švehlík R (1985) Větrná eroze půdy na jihovýchodní Moravě (Wind erosion in southeastern Moravia). Česká státní pojišťovna – Zabraňujeme škodám. Svazek 20. SZN, Praha, 80 p (in Czech)

Švehlík R (2002) Větrná eroze na jihovýchodní Moravě v obrazech (Wind erosion on southeastern Moravia in pictures). Sborník Přírodovědného klubu v Uherském Hradišti, Supplementum 8, 78 p (in Czech)

Vaníček V (1963) Příspěvek k poznání vodní eroze na zemědělských pozemcích (Contribution to soil erosion studies on agriculture land). Sborník Vysoké školy zemědělské, Brno, pp 191–195 (in Czech)

Vitásek F (1942) Dolnomoravské přesypy (Sand dunes in the Lower Moravian Graben). Práce Moravské přírodovědecké společnosti 14(9):1–2 (in Czech)

Wagner J (1994) Objevy nových jeskyních ve Vizovických vrších (Česká republika) (Discovery of new caves in the Vizovická vrchovina Highland, Czech Republic). Proceedings of the 5th Pseudokarst Symposium, Szczyrk (in Czech)

Wagner J (2004) Pseudokrasové jeskyně v Moravskoslezských Beskydech (Pseudokarst cave in Moravian–Silesian Beskids). In: Baroň I, Klimeš J, Wagner J (Eds), Svahové deformace a pseudokras. Elektronický sborník referátů z konference, 1–3 Apr 2004 v Hutisku-Solanci. ČGS & ÚSMH AV ČR (in Czech)

Wagner J, Demek J, Stráník Z (1990) Jeskyně Moravskoslezských Beskyd a okolí (Caves of Moravian–Silesian Beskids and their surroundings). Knihovna České speleologické společnosti 17:1–130 (in Czech)

Záruba Q (1922) Studie o sesuvných terénech na Vsatsku a Valašsku (Study of landslides in District Vsetín and Walachia). Práce z geologického ústavu čes. Vysokého učení technického, Praha, pp 170–177 (in Czech)

Záruba Q (1938) Sesuvy v Lyském průsmyku a jejich význam pro komunikační stavby (Landslides on the Lyský průsmyk Saddle and their importance for road construction). Technický obzor 73:1–6, 26–29 (in Czech)

Záruba Q, Mencl V (1969) Landslides and their control. Elsevier/Academia, Amsterdam/Praha, 205 p

Chapter 7
Recent Landform Evolution in Slovakia

Miloš Stankoviansky, Ivan Barka, Pavel Bella, Martin Boltižiar,
Anna Grešková (†), Jozef Hók, Pavol Ištok, Milan Lehotský,
Monika Michalková, Jozef Minár, Martin Ondrášik, Rudolf Ondrášik,
Jozef Pecho, Peter Pišút, Milan Trizna, and Ján Urbánek

Abstract In addition to the morphostructures and climatic conditions, landform evolution on the territory of Slovakia has been heavily influenced by the history of human settlement. Human impact on geomorphic processes has a longer history in the sub-Carpathian lowlands (embracing millennia) than in the Carpathian mountains (measured in centuries). In addition to the predominant gravitational processes (first of all landslides), major natural exogenic geomorphic agents in Slovakia include fluvial processes, soil erosion by water, karst and pseudokarst evolution, aeolian, cryogenic and nival actions. The slopes of agricultural areas were completely remodeled by tillage and water erosion processes. Increased frequency of extreme meteorological-hydrological events during the Little Ice Age resulted in

---

†Deceased

M. Stankoviansky (✉) • M. Michalková • P. Pišút • M. Trizna
Department of Physical Geography and Geoecology, Faculty of Natural Sciences,
Comenius University in Bratislava, Mlynská dolina, 842 15 Bratislava 4, Slovakia
e-mail: milos.Stankoviansky@test.fns.uniba.sk; michalkovam@fns.uniba.sk;
pisut@fns.uniba.sk; trizna@fns.uniba.sk

I. Barka
Department of Ecology and Biodiversity of Forest Ecosystems, Forest Research Institute,
National Forest Centre, T.G. Masaryka 22, 960 92 Zvolen, Slovakia
e-mail: barka@nlcsk.org

P. Bella
State Nature Conservancy of the Slovak Republic, Slovak Caves Administration,
Hodžova 11, 031 01 Liptovský Mikuláš, Slovakia

Department of Geography, Pedagogical Faculty, Catholic University,
Hrabovská cesta 1, 034 01 Ružomberok, Slovakia
e-mail: bella@ssj.sk; bella@ku.sk

M. Boltižiar
Department of Geography and Regional Development, Faculty of Natural Sciences,
Constantine the Philosopher University in Nitra, Trieda A. Hlinku 1, 949 74 Nitra, Slovakia
e-mail: mboltiziar@ukf.sk

enormous activation of gullying, floods and other precipitation-induced processes. Drastic terrain adjustments, including the levelling of terraced plots, were done due to collectivization in agriculture. The predicted climate change could contribute to an increase in natural hazards, particularly those related to precipitation extremes (floods and landslides).

Keywords Extreme precipitation • Human occupation • Mass movements • Fluvial processes • Water erosion • Karst and pseudokarst • Aeolian processes • Cryogenic processes

7.1 Geological and Geomorphological Settings

The whole territory of Slovakia (49,035 km^2) is part of the Alpine–Himalayan mountain system. Its northern and central parts belong to the subsystem of the Carpathians (36,491 km^2, i.e., more than 74% of the area of Slovakia) and its southwestern and

---

A. Grešková (†) • M. Lehotský • J. Urbánek
Institute of Geography, Slovak Academy of Sciences, Štefánikova 49, 814 73 Bratislava, Slovakia
e-mail: geogleho@savba.sk; jan.urbanek@zoznam.sk

J. Hók
Department of Geology and Paleontology, Faculty of Natural Sciences, Comenius University in Bratislava, Mlynská dolina, 842 15 Bratislava 4, Slovakia
e-mail: hok@fns.uniba.sk

P. Ištok
D. Štúra 758/3, 926 01 Sereď, Slovakia
e-mail: pavol.istok@gmail.com

J. Minár
Department of Physical Geography and Geoecology, Faculty of Natural Sciences, Comenius University in Bratislava, Mlynská dolina, 842 15 Bratislava 4, Slovakia

Department of Physical Geography and Geoecology, Faculty of Science, University of Ostrava, Chittussiho 10, 710 00 Ostrava, Czech Republic
e-mail: minar@fns.uniba.sk; josef.minar@osu.cz

M. Ondrášik
Department of Geotechnics, Slovak University of Technology in Bratislava, Radlinského, 11 813 68 Bratislava, Slovakia
e-mail: martin.ondrasik@stuba.sk

R. Ondrášik
Department of Engineering Geology, Faculty of Natural Sciences, Comenius University in Bratislava, Mlynská dolina, 842 15 Bratislava 4, Slovakia
e-mail: ondrasik@fns.uniba.sk

J. Pecho
Institute of Atmospheric Physics, Academy of Sciences of the Czech Republic, Boční II 1401, 141 31 Praha 4, Czech Republic
e-mail: pecho@ufa.cas.cz

southeastern parts to the subsystem of the Pannonian Basin (12,544 km^2). The predominant section of the Slovak Carpathians (32,007 km^2, 88%) forms the province of the Western Carpathians, the rest the Eastern Carpathians. The Western Carpathians are also higher and culminate in the Gerlachovský štít Peak in the Tatra Mountains (2,655 m above sea level). The highest mountain of the Eastern Carpathians, built mostly of Paleogene flysch and Neogene volcanic complexes, is Kremenec (1,221 m) in the Bukovské vrchy Mountains. Both Carpathian provinces are divided into the subprovinces of Outer and Inner Western/Eastern Carpathians (Fig. 7.1).

The *Western Carpathians* form the northernmost segment of the European Alps of a general west to east strike and are linked to the Eastern Alps in the west and to the Eastern Carpathians in the east. The northern boundary of the Western Carpathians is marked by the erosional margins of Alpine nappes thrust over the North-European platform, while its inner limit is less distinct because it is covered by sediments of the Pannonian Basin. The Western Carpathians are characterized by a predominant nappe structure with zonal arrangement and orogenetic polarity of processes migrating in time from south to north. They are classified according to the age of development of the Alpine nappe structure as the *Outer Western Carpathians* (Externids, OWC) with the Neo-Alpine (Miocene) nappes and the *Inner Western Carpathians* (Internids, IWC) with the Paleo-Alpine (Pre-Tertiary) nappe structures. The IWC also contain relicts of an older (Hercynian) tectogenesis, transformed and incorporated into Alpine structures. This subdivision partly corresponds to the geomorphological subdivision of the mountain range.

The OWC predominantly comprise the broad Flysch belt (Fig. 7.2), representing the accretion wedge with rootless nappe structure built of Jurassic and Cretaceous sediments but above all of Paleogene flysch sediments. The Flysch belt is generally divided into two groups, namely the Silesian-Krosno nappes in the north and the Magura thrust system in the south.

The boundary between the OWC (the Flysch belt) and the IWC is formed by the Klippen belt – the most complicated structure of the Western Carpathians. Its designation was influenced by the characteristic morphotectonic and geomorphic features, rocky blocks with steep slopes – klippen, built of Jurassic and Lower Cretaceous limestones, rising above the surrounding gentle relief on softer rock complexes of the Late Cretaceous and Paleogene.

The IWC consist of units that were tectonically separated before the Late Cretaceous and were thrust generally from the south to north (in accordance with the current geographical coordinates). From the north to south these units are as follows: the Core mountains belt, the Vepor belt, and the Gemer belt.

The northernmost of them is the *Core mountains belt*. The structure of individual core mountains is formed by a crystalline core and cover sediments of Late Paleozoic and mainly Mesozoic age, jointly named the Tatricum. Superimposed above the Tatricum, there are two nappe structures, namely the lower Fatricum and the higher Hronicum nappes.

The *Vepor belt* structure, constituted mainly of the Veporicum, is thrust from the south over the Tatricum. Similarly to the Tatricum, the Veporicum comprises crystalline basement and cover sequences of Late Paleozoic to Mesozoic age. The Veporicum is overlain by the Hronicum and Silicicum nappes.

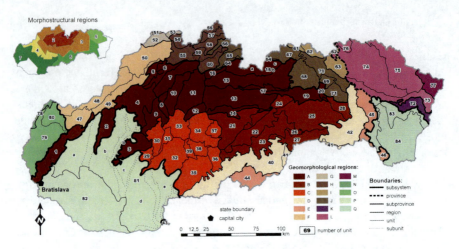

Fig 7.1 Geomorphological divisions of Slovakia (Mazúr and Lukniš 1978). Hierarchical levels: Subsystems, Provinces, Subprovinces, Regions, Units (and selected subunits). CARPATHIANS: *Western Carpathians: Inner Western Carpathians:* A. Fatra–Tatra Region: 1, Malé Karpaty Mts; 2, Považský Inovec Mts; 3, Tribeč Mts; 4, Strážovské vrchy Mts; 5, Súľovské vrchy Mts; 6, Žilinská kotlina Basin; 7, Malá Fatra Mts; 8, Žiar Mts; 9, Hornonitrianska kotlina Basin; 10, Turčianska kotlina Basin; 11, Veľká Fatra Mts; 12, Starohorské vrchy Mts; 13, Nízke Tatry (Low Tatra Mts); 14, Horehronské podolie Basin; 15, Podtatranská kotlina Basin; 16, Chočské vrchy Mts; 17, Kozie chrbty Mts; 18, Tatry: a, Západné Tatry (Western Tatra Mts); b, Vysoké Tatry (High Tatra Mts); c, Belianske Tatry Mts; 19, Hornádska kotlina Basin; 20, Branisko Mts; B. Slovenské rudohorie: 21, Veporské vrchy Mts; 22, Stolické vrchy Mts; 23, Revúcka vrchovina Mts; 24, Spišsko-gemerský kras Karst; 25, Volovské vrchy Mts; 26, Rožňavská kotlina Basin; 27, Slovenský kras Karst; 28, Čierna hora Mts; C. Slovenské stredohorie: 29, Pohronský Inovec Mts; 30, Vtáčnik Mts; 31, Žiarska kotlina Basin; 32, Štiavnické vrchy Mts; 33, Kremnické vrchy Mts; 34, Zvolenská kotlina Basin; 35, Krupinská planina Mts; 36, Ostrôžky Mts; 37, Poľana Mts; 38, Javorie Mts; 39, Pliešovská kotlina Basin; D. Lučenec–Košice Depression: 40, Juhoslovenská kotlina Basin; 41, Bodvianska pahorkatina Hills; 42, Košická kotlina Basin; E. Mátra–Slanské vrchy Region: 43, Burda Mts; 44, Cerová vrchovina Mts; 45, Slanské vrchy Mts; 46, Zemplínske vrchy Mts; *Outer Western Carpathians:* F. Slovak-Moravian Carpathians: 47, Myjavská pahorkatina (Myjava Hills); 48, Biele Karpaty Mts; 49, Považské podolie Basin; 50, Javorníky Mts; G. Západné Beskydy (Western Beskids): 51, Moravsko-sliezske Beskydy Mts; 52, Turzovská vrchovina Mts; 53, Jablunkovské medzihorie Mts; H. Stredné Beskydy (Central Beskids): 54, Kysucké Beskydy Mts; 55, Kysucká vrchovina Mts; 56, Oravské Beskydy Mts; 57, Podbeskydská brázda Furrow; 58, Podbeskydská vrchovina Mts; 59, Oravská Magura Mts; 60, Oravská vrchovina Mts; I. Východné Beskydy (Eastern Beskids): 61, Pieniny Mts; 62, Ľubovnianska vrchovina Mts; 63, Čergov Mts; J. Podhale–Magura Region: 64, Podtatranská brázda Furrow; 65, Skorušinské vrchy Mts; 66, Oravská kotlina Basin; 67, Spišská Magura Mts; 68, Levočské vrchy Mts; 69, Bachureň Mts; 70, Spišsko-šarišské medzihorie Mts; 71, Šarišská vrchovina Mts; *Eastern Carpathians: Inner Eastern Carpathians:* K. Vihorlat–Gutin Region: 72, Vihorlatské vrchy Mts; *Outer Eastern Carpathians:* L. Nízke Beskydy (Low Beskids): 73, Beskydské predhorie Foothills; 74, Ondavská vrchovina Mts; 75, Laborecká vrchovina Mts; 76, Busov Mts; M – Poloniny: 77, Bukovské vrchy Mts PANNONIAN BASIN: *Western Pannonian Basin: Vienna Basin:* N. South-Moravian Basin: 78, Dolnomoravský úval Graben, O. Záhorská nížina Lowland: 79, Borská nížina Lowland; 80, Chvojnická pahorkatina Hills; *Little Danube Basin:* P – Podunajská nížina (Danube Lowland): 81, Podunajská pahorkatina Hills: a, Trnavská pahorkatina Hills; b, Nitrianska pahorkatina Hills; c, Žitavská pahorkatina Hills; d, Hronská pahorkatina Hills; e, Ipeľská pahorkatina Hills; 82, Podunajská rovina (Danube Plain); *Eastern Pannonian Basin: Great Danube Basin:* Q – Východoslovenská nížina Lowland: 83, Východoslovenská pahorkatina Hills; 84, Východoslovenská rovina Plain. Top left: Morphostructural regions of the Western Carpathians and their foreland (Minár et al., 2011). 1, Tatra region; 2, Central region; 3, Transitional region; 4, West marginal region; 5, Northeastern marginal region; 6, Southeastern region; 7, Southwestern foreland

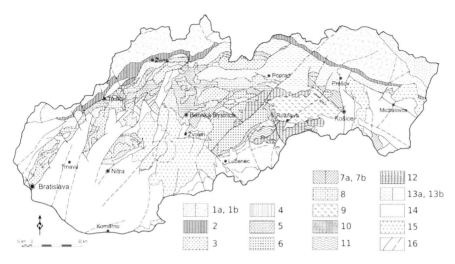

Fig. 7.2 Geological map of Slovakia (J. Hók after Biely et al., 1996). 1, Flysch Belt: 1a, Krosno units; 1b, Magura units; 2, Pieniny Klippen Belt; 3, Tatricum crystalline basement; 4, Tatricum sedimentary cover; 5, Fatricum; 6, Veporicum crystalline basement; 7, Sedimentary covers: 7a, Veporicum; 7b, Zemplinicum; 8, Hronicum; 9, Gemericum; 10, Meliaticum; 11, Turnaicum; 12, Silicicum; 13a, sediments of the Central Carpathian Paleogene Basin; 13b, Eocene to Early Miocene sedimentary complexes of the Buda Basin; 14, Neogene to Quaternary sediments of the Pannonian Basin and intra-Carpathian basins; 15, Neogene to Quaternary volcanics; 16, main faults

The vergence of the *Gemer belt* structure is from the south to the north over the Veporicum. This belt represents the innermost and structurally highest zone of the Alpine nappes in the Western Carpathians. Its crystalline basement – the Gemericum – is covered by the Meliaticum, Turnaicum, and Silicicum nappes. As opposed to the Tatricum and Veporicum, the Gemericum is built mainly of low-grade metamorphosed rocks predominantly of Early Paleozoic age. The Meliaticum is a tectonic unit of oceanic origin.

The postnappe formations of the IWC are built of Late Cretaceous and principally of Tertiary sediments. The transgressional sediments are discordant upon the nappe structure. The Paleogene sediments in the IWC can be divided into two major lithological types: the Inner Carpathian Paleogene, which is by far the prevailing Paleogene rock group in this part of Slovakia, and the Buda Basin group. Neogene clays, sands, and gravels (interbedded with Neogene volcanites, volcanoclastic formations, and coal seams) fill the intra-Carpathian basins indicating marine, brackish, lacustrine and fluvial (deltaic) environments. The Neogene volcanics, mainly andesites and also rhyolites, and basalts, are associated with the subduction in the OWC and back-arc extension. The youngest volcanic activity is dated to the Pleistocene.

The extensive *Pannonian Basin* penetrates onto Slovak territory in the form of three separated sub-Carpathian lowlands, namely the Záhorská nížina Lowland and the Danube Lowland in the south-west and the Východoslovenská nížina Lowland in the southeast, filled by Neogene sediments of up to 8 km thickness in the Gabčíkovo depression (Danube Plain).

The Quaternary cover sediments increase in thickness from the north to south, i.e., from the Carpathians to areas of young sub-Carpathian tectonic depressions along the margin of the Pannonian Basin, where mostly thick layers of fluvial sandy gravels, loess, and blown sands are deposited. Maximum thickness is ca 550 m of fluvial to fluvio-limnic sands and gravels in the Gabčíkovo depression.

A detailed historical overview of geomorphological investigation in Slovakia indicating a development of conceptions on relief evolution with a special regard to planation surfaces (L. Sawicki, F. Machatschek, J. Hromádka, L. Dinev, M. Lukniš, E. Mazúr) in the context of fundamental geological works (D. Štúr, F. Hauer, A. Rehman, V. Uhlig, H. Stille, D. Andrusov, T. Buday) is provided by Mazúr (1965). He was the first to use the morphostructural approach in the assessment of the relief in Slovakia.

The Western Carpathians represent a morphostructural megaform: a relatively flat elliptical dome, comprising multiple morphostructures of lower hierarchical levels. Its surface is irregular, but a mosaic of discrete mountains (mainly horsts and domes as positive morphostructures) and basins (mainly grabens and flexures as negative morphostructures) creates a distinctive pattern, which results from Neogene tectonic evolution (Mazúr 1965). However, various geological and geomorphological markers indicate the crucial role of the Late Miocene, Pliocene, and even Quaternary influences on the morphostructural pattern (Minár et al. 2011).

Mazúr (1976) proposed the first morphostructurally based subdivision of the Western Carpathians and adjacent areas. His units became the foundations for a geomorphological division of Slovakia (cf. Mazúr and Lukniš 1978) (Fig. 7.1). Lacika and Urbánek (1998) proposed another morphostructural subdivision of the Slovak territory, while the most recent morphostructural division is based on targeted morphometric and cluster analyses and enables a more precise delimitation of nine specific active morphostructural units (regions) (Minár et al. 2011), seven of them lie predominantly or at least partially on Slovak territory: the Tatra region, Central region, Transitional region, West marginal region, Northeastern marginal region, Southeastern region, and Southwestern foreland (Fig. 7.1, top left).

In the Pleistocene glacials the lithological-structural properties of rock complexes (passive geological structure) controlled geomorphic evolution and created some specific, rock-controlled relief types: relief in the flysch and klippen belts, crystalline cores, folded Late Paleozoic and Mesozoic rocks, Neogene volcanic structures, and on Neogene sedimentary rocks (Lukniš 1972).

7.2 Climatic and Hydrological Settings

Climate in Slovakia is transitional between Atlantic and continental with mild summers and cold winters. The coldest month is January, the warmest is July. The warmest region is the Danube Lowland with an average July temperature of 20–21 °C and the January average of −2 °C to −1 °C. The coldest region is in the High Tatra Mountains, where the yearly average is −3.9 °C (−11.3 °C in February, rising to 3.6 °C in August).

Precipitation conditions are obviously significantly influenced by the location of Slovakia and also by characteristic atmospheric circulation patterns in Central Europe. In general, four factors control the precipitation regime in Slovakia, namely the Atlantic Ocean, the Mediterranean Sea, the Eurasian continent, and convection (Lapin and Melo 2005). Average annual rainfall varies from about 520 mm in the eastern part of the Danube Plain, to 2,000 mm in the High Tatras. Rainfall generally increases with altitude at the rate of about 50–60 mm for every 100 m. About 40% of the yearly rainfall comes in summer (June to August), 25% in spring, 20% in autumn, and 15% in winter, the wettest months being June and July, the driest January and February.

The highest activity and effectiveness of the precipitation-induced geomorphic processes are related to *extreme rainfalls* and the resulting runoff. The highest rainfall intensity is regularly observed in southern and western Slovakia and during the summer even in the eastern Slovakia. The catchments of the upper Nitra, Myjava, Slaná, upper Váh, Hron, Hornád, Ondava, and Topľa Rivers are among the most susceptible regions to extreme precipitation event incidence. The most torrential and catastrophic rainfalls occurred in Salka in the southern part of the Ipeľská pahorkatina Hills in July 1957 (228 mm in 65 min) and in Trnava, Trnavská pahorkatina Hills, in June 1951 (163 mm within only 2 h).

Extreme precipitation-runoff situations in Slovakia fall into two principal groups. To the first group belong events with *lower intensity* and *long duration* (several days or weeks). Precipitation intensity per hour is less than 20% that of a convective event. Much precipitation associated with cyclonic systems occurs in the winter half-year, especially during autumn and early spring. Precipitation is caused by individual centers of low pressure and by prolonged series of low pressure systems. The longer a cyclonic situation lasts, the larger is the area affected (individual river systems or entire regions). Cyclonic situations occurring during the summer are particularly dangerous. After the progressive saturation of soils, huge runoff and mostly regional floods are generated (e.g., floods of the Danube River in 1954, 1965, 2002, Morava 1997, Hron and Hornád 1974, Bodrog 1924).

The second group of extreme precipitation-runoff situations is induced by convective rainfalls with *short duration* (from several minutes to hours) and *high intensity* (more than 1 mm min^{-1}) affecting a limited area. Convection mainly occurs in the summer or late spring, most intensely along the southern and western edges of the Carpathians. High rainfall intensity and extreme runoff result in flash floods. During flash floods both runoff and fluvial processes are strongly concentrated in time and space, and this markedly increases their efficiency and has devastating effect. Disastrous examples are events of July 1998 in the Malá Svinka catchment, Šarišská vrchovina Mountains, and of July 1999 in the Zolná River, Zvolenská kotlina Basin.

The analysis of *trends* in daily and monthly precipitation has revealed a significant decrease in precipitation recorded for southern Slovakia (−15%) and a slight increase (up to 5%) in the northern mountainous regions over the past 110 years. The change in precipitation regime is connected with longer and more severe drought spells and short periods (several days) with heavy precipitation mainly in the growing season. Also the distribution of heavy rains and maximum daily precipitation have changed in Slovakia. The most significant increase has been registered in eastern and southern Slovakia (Faško et al. 2009b) – probably due to the intensification of the hydrologic cycle resulting from global warming in the twentieth century (Faško et al. 2009a).

Recently, growing climatic instability springs from the combination of the above two types of extreme situations. Almost every year during the first decade of the twenty-first century there have been recorded severe floods, particularly in the eastern Slovakia, confirmed by the statistical analysis of 5-day precipitation totals, which show a positive trend for the majority of meteorological stations.

Most of the Slovak Carpathians are part of the Danube River basin and drained to the Black Sea. Only smaller part of northern Slovakia with the Poprad and Dunajec Rivers belongs to the Baltic Sea catchment. Slovakia is a source area and majority of streams is fed by rainfall. Only the border rivers Danube and Morava bring water from abroad.

As to the *hydrological regime*, the typical major Slovak rivers, such as Váh, Hron, Nitra, and Hornád, collecting waters mainly from mid-altitude areas, have maximum discharge in March/April. The eastern Slovakian streams on flysch catchments have mostly uneven discharge and are prone to flash floods induced by torrential rainfalls. The streams of the highest mountains of the Carpathians – the Tatras and Low Tatras – running in materials derived from crystalline rocks, often build coarse alluvial fans. They are of high gradient and may form series of rapids, cascades, and waterfalls. Riverbeds are primarily composed of boulders with cobbles and gravel. Along foothill sections they abound in bedload and form braided or wandering channels. Receiving most water from snowmelt at high altitudes, they have maximum discharge in May or June. The typical third to sixth order Carpathian streams are confined or at least partly confined with coarser alluvial deposits and younger alluvial fans on the foothills. Along non-channelized river stretches riparian forests and woody debris still have significant impact on the behavior and morphology of river channels (Grešková and Lehotský 2007).

7.3 Environmental Changes and Their Influence on Landform Evolution

In the youngest Holocene, the pattern and dynamics of natural geomorphic processes were modified by the human occupation of the landscape and also by climatic fluctuations over most of the Slovak territory. The beginning of a gradual transformation of the original natural landscape into a cultural landscape coincides with the last phases of the *Mesolithic* when the first farming communities settled on the territory of the present Slovakia. Human interventions peaked in the *Bronze Age*, followed gradually by interferences associated with the people of central Danubian urn fields culture, Lusatian culture, the Celts, Germanic tribes, Slavs, Avars, and Hungarians. The most important human intervention since the arrival of the first farmers was *forest clearance*. Settlement started in lowlands but penetrated gradually into the Carpathian foothills and mountains. However, the Carpathians above 400–500 m had not been settled permanently before the end of the twelfth century (Demo et al. 2001).

More significant human interventions in the Carpathian landscape can be dated to the *Great colonization* (feudal and German) in the thirteenth and fourteenth centuries. Towns were founded and central Slovakia even became the land of mining and metallurgy. Another important intervention in the Carpathian landscape was the *Walachian* shepherd *colonization*, which reached here in the fifteenth century and peaked in the sixteenth and seventeenth centuries. The Walachian colonization proceeded along the ridges of the Carpathians from the east and southeast mainly to regions of the upper Hron River basin, Orava River basin, upper and middle Váh River basin with the Biele Karpaty Mountains and the Myjava Hills as the westernmost locations. The next (so-called Goral) shepherd colonization took place in the seventeenth to eighteenth century. *Gorals* settled the territory in the upper reaches of the Kysuca and Orava Rivers and a part of the northern Spiš region. The so-called "*kopanice*" *colonization* began in the sixteenth century and culminated in the eighteenth and the first half of the nineteenth centuries.

In the course of settlement, farmer settlers from below and shepherd settlers from above attacked primeval forest cover and gradually reduced its spatial extent. Intense forest clearance was also performed by *miners* and *smelters* who needed wood for charcoal in the surroundings of the sites of ore extraction and processing. Through deforestation colonists promoted the expansion of accelerated geomorphic processes and processes emerged that previously had had negligible or no influence on the natural landscape. The devastation of extensive cultivated areas by a complex of anthropogenically accelerated geomorphic processes called for local *afforestation*. The artificial restoration expanded to the majority of the exploited forests after 1770, it was intensified in the nineteenth century and especially after World War II (Stankoviansky and Barka 2007).

Long-term farming resulted in a typical *land use* pattern of a *mosaic* of small, narrow plots, as we know it from the first half of the twentieth century. After World War II (namely after the advent of socialism in 1948), the original land use pattern was changed. The collectivization of land fundamentally changed land use pattern, merged former small private plots into *large cooperative fields* and removed the dense network of man-made linear landscape elements. The mosaic of small plots was replaced by vast blocks of fields.

The restoration of capitalism after 1989 did not result in such a marked change in land use. The areal extent of arable land was reduced but large-scale farming survived until now. The main land cover types in the current landscape are represented by agricultural land (45%; including 34% arable land), occurring predominantly in lowlands and intramountain basins, and forests (38%), distributed mostly in mountains (Feranec and Oťaheľ 2001).

However, within the last millennium land use changes were coupled with important *climate changes*. Paleoclimatologists recognize four periods – the Medieval Warm Epoch (MWE – between about 1150 and 1300), the transitional period of climate deterioration, the Little Ice Age (LIA – 1550–1850), and the period of global warming.

An increased frequency of extreme meteorological-hydrological events in the LIA induced an acceleration of some geomorphic processes. The most extensive and also the most damaging geomorphic response to climatic fluctuations occurred precisely in the areas most affected by human interference (Stankoviansky and Barka 2007).

The above outline of environmental changes in Slovakia suggests that human impact on geomorphic processes has longer history in the sub-Carpathian lowlands than in the Carpathian mountains. It is explained by the history of settlement in both areas, measured in Carpathians in centuries and in lowlands in millennia. As the massive settling of spatially dominating Carpathians is connected with the Great colonization, this Slovak chapter focuses on the period of the past eight centuries.

7.4 Exogenic Geomorphic Processes: Spatial Distribution and History of Investigation

The best known exogenic geomorphic processes operating on the territory of Slovakia in recent times fall into the groups of:

– Gravitational processes (mass movements),
– Fluvial processes,
– Runoff processes,
– Processes modeling karst and pseudokarst landforms,
– Eolian processes,
– Cryogenic processes,
– Nival processes.

The first five groups are *azonal* processes, while cryogenic and nival actions are *zonal processes*. In addition, two other azonal groups are organogenic and anthropogenic processes. Considerable amounts of material are transported by the activities of plants and animals, especially within the root system when trees are uprooted by wind calamities (cf. Stankoviansky and Barka 2007). Unfortunately, only little attention has been dedicated so far to this phenomenon. More attention was devoted to anthropogenic processes. *Direct human interventions* into topography were studied mostly on the territory of the Slovak capital (e.g., Mazúrová 1985; Stankoviansky 1992) and in the mining region of the Hornonitrianska kotlina Basin (cf. Jakál 1998) – not treated here. The effects of tillage on erosion, however, are detailed below, along with the impact of human activities accelerating natural processes (*indirect anthropogenic influence*).

The spatial distribution of the above groups of processes is depicted on the map of *morphodynamic units* (Fig. 7.3). Three types of criteria were used to refer morphodynamic units into three hierarchic levels: morphoclimatic zones (systems), degree of human impact (subsystems), and predominant genetic types of recent exogenic processes. The main criteria used were *altitude* (systems and regions with predominant linear fluvial processes, i.e., vertical erosion), *land cover* (subsystems

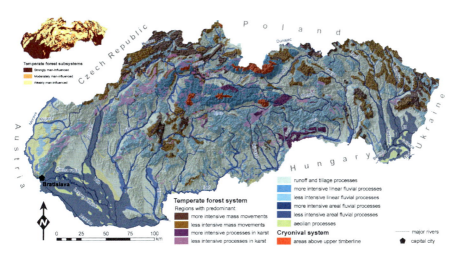

Fig. 7.3 Map of morphodynamic units of Slovakia (By J. Minár, J. Urbánek and I. Barka)

and regions with predominant runoff and tillage processes), *rock* and sediment properties (regions with predominant processes in karst, mass movements or aeolian processes), and *tectonic subsidence* and *inundation* (regions with predominant areal fluvial processes, mainly lateral erosion and overbank accumulation).

Research into exogenic geomorphic processes only began after World War II – although landslides and other gravitational deformations are mentioned as early as the second half of the nineteenth century and pioneer studies of fluvial processes even date back to the late eighteenth century. Early investigations are summarized in Slovakia-wide well-known monographs by Zachar (1970, 1982) on soil erosion by water, Nemčok (1982) on mass movements and Midriak (1983) on geomorphic processes operating in high mountains. An overview on recent and current exogenic morphogenesis is given in works by Stankoviansky and Midriak (1998) and Stankoviansky and Barka (2007), while an assessment of natural hazards in Slovakia is provided by Minár et al. (2009).

7.5 Recent Landform Evolution in the Cryonival System

The occurrence of exogenic geomorphic processes does not depend only on geology and relief but also on vertical zonation of climate and vegetation. Vertical climatic and vegetation zonation is reflected in the present-day altitudinal morphoclimatic zones (systems) characterized by sets of processes that involve specific types of landform development (Stankoviansky and Barka 2007). Kotarba and Starkel (1972) recognize two climatic-morphogenetic systems in the Carpathians: the cryonival system and the temperate forest system.

The *cryonival morphoclimatic system* comprises the highest parts of the Western Carpathians above the upper timberline. These areas are typified by cold and very cold mountain climate and the occurrence of zonal (cryogenic and nival) processes. There are six high mountains in Slovakia (Lukniš and Plesník 1961): the Tatras and Low Tatras, Veľká Fatra and Malá Fatra Mountains, the Oravské Beskydy Mountains, and the Chočské vrchy Mountains. The total area above the upper timberline in these mountains is ca 49,000 ha (1% of Slovakia's territory) but within them it ranges from 2% of the area of the Chočské vrchy Mountains to 50% of the Tatra Mountains. Although the characteristic landforms have been created by frost action and nivation, runoff processes dominate in intensity (Midriak 1999).

The environment of the Western Carpathian high mountains above the upper timberline can be divided into several geoecological belts, characterized by relief type, climate, and prevailing geomorphic processes. The scheme of geoecological belts was defined in the Polish Tatras by Kotarba et al. (1987) but it is also applicable to the Slovak high mountains. According to this scheme, there are three belts in the Tatra Mountains within the cryonival system: subalpine, alpine, and seminival belts, the latter corresponding to the nival belt in the Alps. An idealized and simplified model of the actual morphodynamics was based on the characteristics of the prevailing processes within each belt.

The *highest (seminival) belt* is situated above 2,150–2,300 m elevation. It comprises ridges and peaks with steep cliffs. As weathering mantles are thin or absent, physical weathering, rockfalls, corrasion by sliding debris, and cryogenic processes are the main geomorphic processes, while gelifluction only occurs locally. Snow avalanches are frequent, but have little morphological significance.

The *alpine belt*, where cirques, troughs, moraines, and other glacial landforms resulting from Pleistocene glaciations are common, starts from 1,650 to 1,800 m. The slopes and ridges are mostly covered by soil-weathering mantle and dissected by chutes. Microrelief is highly variable. Soil cover is discontinuous, interrupted by rock debris tali. Considerably large parts of alpine belt lacks vegetation cover; moreover during winters a protecting snow cover is highly affected by wind and thus it is discontinuous. Bare, unprotected surface is prone to intensive cryogenic processes. Gravitational processes and frost weathering on cliffs and rocky slopes, frost creep, gelisaltation, gelifluction and creep of debris covers on smooth slopes, wind erosion in exposed locations and nivation in depressions are typical of the alpine belt (Midriak 1983). Debris flows and snow avalanches are very dynamic. Several smaller glacier lakes in the Tatra Mountains were filled up by material from tributary streams or from talus cones. Some moraines and block glaciers were destroyed by fluvial processes.

The *subalpine belt* is situated above the upper timberline. The presence or absence of dwarf pine is decisive for the occurrence and intensity of geomorphic processes. Dwarf pine reaches the highest coverage in the lower part of subalpine belt, preventing intensive geomorphic processes. Avalanche activity (mainly of wet slab avalanches) and nivation are the most important local processes (Barka 2005). Above the dwarf pine zone, in the upper part of the subalpine belt, diverse geomorphic processes operate: cryogenic processes create vegetation-soil forms (girland soils, thufurs), runoff and fluvial processes, debris flows, and creep or piping are also common. The subalpine belt is also important as a source or transport zone of snow avalanches affecting other belts (Midriak 1983; Boltižiar 2009). The deposition of the material from higher geoecological belts is an important process in the Tatra Mountains.

It is only in the Tatra Mountains that all belts are present; the alpine and subalpine belts are found in the Low Tatra Mountains; while in other high mountains only the subalpine belt occurs. The character of each geoecological belt in the individual mountains markedly depends on the presence of a higher belt and on the potential occurrence of allochtonous processes and, therefore, varies considerably between these mountain ranges.

Describing the high-mountain system of sediment transfer, Kotarba et al. (1987) identified two basic morphogenetic subsystems: slopes and channels. Allochtonous processes can affect lower belts mainly within channel subsystems. The *upper timberline* (especially the natural, climatic one) is not an abrupt discontinuity for geomorphic processes but processes cross it from above along channel subsystems. This "avalanche-type timberline" in the sense of Plesník (1971) is interrupted and locally squeezed to lower positions. An anthropogenically lowered timberline (typical of the Malá and Veľká Fatra Mountains) is a sharper boundary. According to Midriak's (1999) survey, the mean elevation of the present timberline in the Western Carpathians ranges from 1,185 m (the northern part of the Malá Fatra) to almost 1,440 m (the High Tatras and the eastern part of the Low Tatras). Human action lowered the timberline by 280 m and locally even by 350–400 m. The rate of lowering reached its maximum during the Walachian colonization in areas with rich soils on calcareous substrate (limestones, calcareous marls) and more subdued relief. Dwarf pine and forest stands had been sporadically destroyed until the 1950s. Ensuing intensive overgrazing led to the rapid spreading of geomorphic processes (e.g. runoff and wind erosion, avalanche activity). Due to the systematic afforestation of selected slopes and the natural overgrowing of abandoned pastures, the altitude of the timberline has been rising recently.

The most dynamic, high-energy gravitational processes in the cryonival system of the Slovak Carpathians and in particular in the High Tatra Mountains are *debris flows*. The main cause of debris flow formation is extreme precipitation of high intensity, reaching great efficiency during relatively short spells. Debris flows operate exclusively above the upper timberline and damage rather extensive areas. Their geomorphic effects are most marked on talus cones.

Detailed studies of Kotarba (2004, 2005, 2007a, b) in the Dolina Zeleného plesa Valley in the Slovak part of the High Tatras revealed the changing intensity of debris flows in the past with changing climate. Rates of debris flows varied during and after the LIA (in the Tatras: the period 1400–1925). Marked climatic anomalies during the LIA resulted in much more frequent and intensive debris flows than in later periods. Analysis of lacustrine sediments confirmed that the high frequency of rapid slope movements in the Tatra Mountains took place between 1400 and 1860 and continued with reduced intensity until 1925, as confirmed by lichenometric dating. Perhaps the best-known debris flow event in the Slovak high mountains during the period of the LIA happened in the Malá Fatra Mountains in 1848. It destroyed the village of Štefanová under Veľký Rozsutec Mountain (1,610 m) (Stankoviansky and Barka 2007).

According to Kotarba (2004), in the High Tatras a calmer period followed the LIA till the 1970s. However, also during this period marked debris flows occurred from time to time. Záruba and Mencl (1987) document the response to the extreme precipitation event of July 15, 1933, when the meteorological station Starý Smokovec recorded 45.8 mm rainfall with a maximum intensity of 26 mm h^{-1}. This torrential rain resulted in a sequence of major debris flows that destroyed fully vegetated talus cones in some valleys. Ingr and Šarík (1970) refer to a large debris flow in July 19, 1970, in the Roháčska dolina Valley, Western Tatra Mountains. Since the 1970s, debris flow activity increased in the High Tatra again (cf. Kotarba 2004). This statement is also supported by Kapusta et al. (2010), who studied the long-term development of debris flows in the Velická dolina Valley and Dolina Zeleného plesa Valley in the same mountains between 1949 and 2006. Multitemporal trend analyses of remote sensing data show a growing number of debris flows in both valleys since the 1980s.

An increased rate of snow avalanches is also observable since the winter of 2000, mainly in the forest belt (Hreško et al. 2005). Avalanches were triggered even in dense broadleaf forests, and in the Malá and Veľká Fatra Mountains they reached valley bottoms below 500 m. The extraordinary huge snow avalanche (recurrence interval 100 years) happened in the Žiarska dolina Valley in the Western Tatra Mountains in March 20, 2009 (Minár et al. 2009). However, the current activity of snow avalanches falls behind the LIA level.

7.6 Exogenic Processes in the Temperate Forest System

7.6.1 Gravitational Processes

Gravitational processes represent the most significant geomorphic processes in mountain and submontane areas of the Carpathians. Most of slope deformations were caused by gravitational movements in prehistoric times, probably since the end of Pliocene until the beginning of the Holocene. Unfortunately, there are only two works dealing with dating *old landslides*. Harčár and Krippel (1994) dated landslide generations in the neovolcanic Vihorlatské vrchy Mountains in the Eastern Carpathians, while Pánek et al. (2010) dated a large long-runout landslide on the flysch slope of Kykula Mountain in the Kysucké Beskydy Mountains (the northwestern corner of Slovakia). Landslides from both regions derive from the transition between Late Glacial and Holocene times. Most of the prehistoric landslides became stabilized due to the Holocene climate change accompanied by forest development. In more humid spells, especially during the Holocene Climatic Optimum, and more recently during the Little Ice Age, movements partly reactivated. The idea of reactivation of older landslides was raised early by Andrusov (1931), who studied these phenomena in the Orava River basin. According to him they were formed partly in the Pleistocene and partly in the Holocene, but some of them has been active until the present. According to Paul (1868), numerous landslides in this area are of historical age.

Unfortunately, with very few exceptions there is no information on the exact date of landslide reactivation. The first written note about the landslide is in the travel book by Mednyanský (1844; the first edition in 1826) who described a devastating landslide next to the village of Vlašky in the upper Váh River valley, Podtatranská kotlina Basin, which was initiated in 1813 by an extreme rainfall event bringing about the worst known flood on this river. Further information about landslides is from the late nineteenth century, related to *research* undertaken for the purpose of railway building in the mountain valleys of the Hron and Orava Rivers (Nemčok 1982). The first written record on landslides, including their map documentation, is the work of Koutek (1935), describing his research in the Chočské vrchy Mountains and the adjacent part of the Podtatranská kotlina Basin. The detailed studies of landslides in Slovakia were started by Záruba (1938), who investigated the stability of slopes during construction of railways across the Flysch belt on the southern edge of the Javorníky Mountains. An increased interest to study slope movements was attracted by natural landslide disasters, especially in Handlová in the Hornonitrianska kotlina Basin at the turn of 1960 and 1961. This event stimulated regional research and inventory of slope deformations in the former Czechoslovakia (Matula et al. 1963). Nemčok and Pašek (1969) and Nemčok (1972) explained the origin of high-mountain doubled crests (which had been considered earlier as a product of cryogenic processes) as result of gravitational movements. This research was succeeded by Mahr (1977) in the area of crystalline rocks and by Baliak (Baliak and Mahr 1979) in the area of Mesozoic complexes. Malgot (1975) investigated slope deformations at the edges of neovolcanic mountains. Nemčok (1982) evaluated results of numerous authors, as well as of his own research on landslides in Slovak Carpathians in his monograph.

Current research of landslides is summarized in the Atlas of Slope Stability in Slovak Republic (cf. Šimeková et al. 2006; Kopecký et al. 2008), where 21,190 slope deformations affecting a total area of 257,591 ha (5.25% of Slovakia), mostly of Paleogene complexes (60%) are recorded. Up to 94.5% of the total number of registered slope deformations are landslides and earthflows. At present 30 selected slope deformations are monitored.

Most of the movements (63%) were active in the recent past and can resume activity again; 24.9% of slope deformations are stabilized and 11.6% are active, the rest are combined movements. Active deformations are mostly landslides (94.9%) and flows (3.5%), a similar share applies to potential deformations. The most common direct natural causes of movement are climatic factors (mostly extreme precipitation) and indirectly rainfall-induced phenomena (e.g. gully formation, fluvial erosion, effects of subsurface water), which activated 89% of them. Man-made causes of slope deformations are especially undercutting of slopes (49%) and their overloading (16.3%).

In the territory of Slovakia we find all main types of slope deformations as classified by Nemčok et al. (1974), which is mandatory and still used in this country, namely creeping, sliding, flowing, and falling. *Deep-seated creep* occurs on favorable geological structures: brittle rock complexes moving slowly along plastic strata. Creeping involves several millions or tens of millions m^3 of Mesozoic and Paleogene sedimentary rocks, Neogene volcanic rocks, and also crystalline Paleozoic rocks. During creeping the slowly moving parts of massifs are disintegrated to blocks separated by crevasses and slope deposits along their front are prone to sliding. The existence of creep was experimentally proved by surveying of ten fixed points located on morphologically striking neovolcanic blocks on the slopes of the Kremnické vrchy Mountains and adjacent slopes of the Hornonitrianska kotlina Basin. The measurements since 1907 proved slow movement (1–4 cm year^{-1} – Malgot 1975).

The most common type of slope movements in Slovakia are *landslides*. Regarding the volume of replaced material, the landslides have a wide range of forms, from small slides with the volume of several tens of m^3 to large, several tens of meters thick sliding bodies with volume of several millions of m^3. The largest active landslide was located on the northern margin of the Turčianska kotlina Basin at the foot of the Malá Fatra Mountains close to Sučany (maximum depth of the slip plane: 80 m, volume: ca 40 million m^3 – Záruba and Mencl 1987). It was developed in the fine-grained Neogene sediments through slope undercutting by the Váh River. The regulation of the Váh River (construction of hydropower plants and derivation canals) stabilized the landslide. During the top recent activity in the period 1947–1957 the rate of movement was 8–10 cm per year. The above-mentioned Handlová landslide had a different character: sliding of slope deposits in front of creeping blocks at a rate of 6.3 m day^{-1} (26 cm h^{-1}). In the course of several weeks ca 20 million m^3 of earth moved downslope. The last decade is the period of the first attempts to assess the *impact of climate*, primarily precipitation, on the development of slope movements (Fussgänger and Jadroň 2001), especially landslides (Kopecký 2004). Research conducted in the Kysuce flysch region (Northwest-Slovakia) points to the role of extreme precipitation in landslide generation. Most important is the long-term balance of rainfall because it directly influences the saturation of the ground. Precipitation infiltrating to groundwater and affecting its free or piezometric levels is most influential ("effective precipitation" by Kopecký 2004). However, landslides can also be activated by sudden melting of thick snow cover in combination with a rainfall, as it happened near the village of Bukovec in the Myjava Hills on March 29, 2006 (Ondrášik et al. 2006).

Flows occur above all in flysch mountains. Shallow earth flows are typical here usually during the spring months, in enduring rainy periods, or after torrential rains. Sometimes, however, also large-scale earth flows of volume more than 100,000 m^3 are generated – as it happened on the slopes of Brusová Mountain, Javorníky Mountains, at a maximum rate of 16 m h^{-1} in 1937 (Záruba 1938). Another huge earthflow of the Flysch belt was activated in 1962 in the village of Riečnica (today Nová Bystrica) in the Kysucká vrchovina Mountains, also Northwest-Slovakia, and reached a maximum rate of 20 m h^{-1} (Záruba and Mencl 1987).

Rockfalls, either in the form of falling small fragments or large blocks, mostly occur in high mountains, especially in the Tatra and Fatra Mountains. The oldest rockfalls, known from historical records, affected peaks of Gerlachovský štít (1590), Lomnický štít (1662) (cf. Špůrek 1972), and Slavkovský štít (1662) (cf. Kotarba 2004) in the High Tatras. Rockfalls in other regions are limited to rocky slopes and cliffs, common in the Klippen belt, producing huge talus cones mostly below rock walls. A huge rockfall with a volume of ca 22,000 m^3 from Jurassic and Cretaceous limestone klippen affected a railway track and the bank of the Orava River near the village of Podbiel, Oravská vrchovina Mountains, in 1975 (Slivovský 1985).

In the twentieth century the *activation* of landslides was often caused *by human action*, construction of roads or expansion of settlements reducing slope stability. A considerable part of such slope deformations had to be stabilized by complex technical remediation works. Slope stability was also improved by river regulations. On the other hand, construction of the Oravská priehrada dam on the Orava River (the largest Slovak dam with an area of 35 km^2) and the Domaša reservoir on the Ondava River generated extensive landslides in flysch, reaching from tens to hundreds of meters away from the reservoir banks (e.g., Harčár and Trávniček 1992).

7.6.2 Fluvial Processes

Fluvial processes in Slovakia and sediment input are highly variable along the longitudinal profiles of streams, dependent on the river hydrological regime, the geology of drainage basins, soils, vegetation and land use. Fluvial transport and accumulation prevail in the intra-Carpathian basins, in wider valleys and alluvial plains passing into sub-Carpathian lowlands. In the past, sediment delivery generally maintained an active *meandering* character of these stretches. Larger rivers have also accumulated pronounced *levees*, causing some tributaries to flow parallel to them (e.g., the Váh and Dudváh Rivers). In past decades, the fluvial regimes of most major rivers have been strongly affected by channel straightening, training works and construction of dams. Bed degradation downstream of hydroelectric plants and reservoirs has reduced river and groundwater levels. On the contrary, bed aggradation at the head of artificial reservoirs builds prograding deltas.

The *study of fluvial processes* and landforms dates back to the 1770s, when the cross-sections and longitudinal profiles of the Danube and Váh River channels were first surveyed. A modern chapter in fluvial geomorphic research began in the twentieth century. First systematic studies were represented by continuous measurements of suspended and bedload launched in 1952, motivated by dam planning and construction. The characteristics of the regime of sediment load on the Danube, Váh, Morava, Nitra, Hron, Laborec and Uh Rivers were published by Szolgay and Náther (1954), Almer (1955), Almer and Szolgay (1955), and several others. The results of comprehensive studies carried out by hydrologists and engineers until 1972 were published in 18 volumes of erosion-sedimentation atlases of Slovak rivers (Holubová 1997). In 1961, estimated 2,800,000 tons of total sediment was transported from the territory of Slovakia by rivers (Kozlík et al. 1961). Pašek (1958) and Kňazovický (1962) described channel character, planform, and changing morphology in response to total water erosion and catastrophic floods in the catchments of the Laborec and some other rivers in eastern Slovakia. Lehotský and Grešková (2007) studied fluvial response of an anastomosing mountain stream, the Studený potok, High Tatras, to large-scale forest destruction. The catastrophic windstorm of November 2004 initiated lateral and bed erosion, increased transport of bedload and reactivated old avulsion channels.

The most important fluvial system of Slovak Carpathians is the gravel-bed *Váh River*, the longest river (403 km) of Slovakia with $Q_m = 196$ m^3 s^{-1} at its confluence with the Danube River. At present, most of the Váh River valley is regulated with canals, artificial dams, and 22 hydropower stations (known as the Váh Cascade). Prior to regulation the Váh was a high-energy, wandering river along its middle and upper reaches, while along its lower reaches a laterally unconfined, actively meandering river, undercutting the footslopes of the Nitrianska pahorkatina Hills. Heavy bank erosion caused numerous and often catastrophic landslides along a 17-km long stretch (Lukniš 1951). In the past the Váh River played a key role in rafting timber and other products from the mountains of northern Slovakia. Most serious geomorphic effects were linked to the greatest flood disaster on record in Slovakia (500–1,000-year flood) in August 23–26, 1813. This flood, caused by heavy rains lasting for 56 h, claimed at least 243 victims and destroyed over 50 villages. It remodeled the whole Váh River floodplain and also hit the basins of the Poprad, Slaná, Hornád, and Hron Rivers. Other major floods that affected basins of several Carpathian rivers occurred in 1662, 1683, 1725, 1845, and 1899 (Horváthová 2003).

The best-documented examples of most recent human impact on fluvial systems are from the Morava and Hron Rivers (second order streams), which were straightened between 1930 and 1960 (Holubová et al. 2005). The *Morava River* has a discharge $Q_m = 120$ m^3 s^{-1} (flood discharge $Q_{100} = 1,500$ m^3 s^{-1}) at its confluence with the Danube River. Along its Slovakian–Austrian reach, the Morava is a characteristic medium-energy river with fine-grained bed and relatively cohesive banks. Prior to regulation, the Morava at its lowermost reach (river km 16–21) was an actively meandering, single-thread river (Grešková 2002) at least since 1807, with the mean rate of lateral erosion 1.75 m year^{-1} (Pišút 2006b). In the 1930s–1960s, the lower Morava was shortened by more than 10 km by cutting off 23 meanders. The present-day channelized Morava is a low sinuosity river with lateral migration mostly prevented by bank revetments. The synergistic effect of increased slope, sediment discharge deficit and gravel extraction induced bed degradation, locally up to 2 m (Holubová 1999). Mean annual bedload transport is ca 45,000 m^3. Intensive sedimentation reduced floodplain capacity and caused degradation of the oxbow system (Holubová and Lisický 2002).

The *Hron* is the second longest Slovak river (279.5 km). The Lower Hron is flanked by higher terrain (locally over 12 m high) and has no continuous flood dykes. It was artificially straightened from 80 to 74.5 km length. In contrast to the Morava River, the channel still maintains a certain degree of freedom to migrate, although flow dynamics and sediment transport are influenced by small hydroelectric power stations. The Hron River has a coarser bed material and bed armoring. The high loess pseudoterraces along the river show major bank failures, resulting in higher concentrations of suspended load and rapid sedimentation in the cut-off meanders (Holubová et al. 2005).

Fluvial processes and their effects also attracted much attention on the *Danube River*. Downstream of the Devín Gate Gorge, Malé Karpaty Mountains, in the subsiding Danube Plain the river built the largest alluvial fan in Slovakia. At this reach, the gravel-bed Danube River has a mean annual discharge of 2024 m^3 s^{-1}, still maintaining its Alpine character of water regime. Fluvial processes show great dynamics, although it is currently largely determined by engineering structures and dams. Between 1378 and 1528 AD, large avulsions on the Danube River resulted in the abandonment of the 24 km long lowermost stretch of the Dudváh River (Pišút 2006a). Frequency and magnitude of floods on the Danube also increased from 1550s onward, along with lateral shifting of the river and sediment transport (Pišút 2002, 2006a).

In the past the key mechanisms of the Danube channel change were meander development through progression, neck, and chute cutoffs, abandonment of secondary channels and a tendency for channel switching. In the second half of the eighteenth century the frequency of high-magnitude flood events increased. At Bratislava the floods of the 1760s–1770s triggered a series of channel adjustments and subsequent human interventions, leading to permanent instability of the river channel (Pišút 2002) and rapid formation of new bends downstream (Pišút and Timár 2007). Maximum rates of lateral erosion at the village of Čunovo (to the east of the Slovak–Hungarian–Austrian frontier point) between 1783 and 1794 up to 72 m year^{-1} are relatively high for Central European conditions (Pišút 2008). This period was also specific as to the extreme floods, including the second largest Danube flood of the past millennium in November 1787 (Pišút 2011). Downstream of Čunovo, channel adjustments brought about a dramatic planform change from meandering to a high-energy, multiple braided channel in the 1780s–1790s (Pišút 2006a). The shallow channels with many bars enhanced the geomorphic effects of frequent ice-jamfloods, typical of the peaking LIA (the most catastrophic floods occurred in 1809, 1830, 1838, and 1850).

The modern Danube is the result of the mid-flow channelization in 1886–1896. Prior to the completion of upstream dams, which reduced sediment load to about one-quarter, an average of ca 600,000 m^3 of pebbles and gravel and ca seven million tons of suspended load were annually transported through Bratislava. A substantial part of this material was deposited within the system of side channels, which were cut off during low-flow regulation. Since the 1960s natural bed aggradation has been alternating by degradation (Holubová et al. 2004). Present-day fluvial processes of Danube are restricted to the riverbed and floodplain area between the embankments (Szmańda et al. 2008).

The comparison of bedload measurements in 1952–1960 with those in 1992–1998 has shown that sediment transport through the Slovak section of the Danube has been recently significantly affected by the Freudenau (at Vienna, Austria) and Gabčíkovo hydropower plants (Holubová 2000). Downstream of Gabčíkovo the deficit in sediment supply and altered flow dynamics has caused bed degradation and deposition along lower reaches. Since 1992 maximum drop in low water level reached 1.29 m compared with pre-dam conditions (Holubová et al. 2004).

Direct systematic *human interventions* into the channels of major Slovak rivers date back to the 1770s, primarily in order to improve navigability and facilitate river transport. The earliest structures of erosion control and flow diversion (wicker works, fascines, cut trees serving as breakwaters, sporns, groynes, bank revetments) often proved ineffective during the extreme floods of the LIA. Since medieval times, several riverside settlements fell victim to lateral erosion and floods or had to be relocated several times (Lukniš 1951; Horváthová 2003; Pišút 2006a). Natural changes and human interventions resulted in substantial simplification of floodplains and water concentration into main channels (Pišút 2002).

In general, several pronounced stages of river activity, correlated to the colder and wetter spells of the LIA, can be recognized on major Slovak rivers and best documented on the Danube River. Channel instability and increased sediment supply occurred between 1378 and 1528 AD, in the 1550s–1570s and 1770s–1790s (so far the most fundamental change in river behavior, which persisted until the midflow channelization in 1886–1896).

Damming of rivers and *construction of valley reservoirs* significantly affected sediment transport. The annual deposition of suspended load in Slovak reservoirs is 8–10 times higher than that of bedload. The problem of reservoir siltation is particularly observed for the middle and upper Váh River. Sediments are mostly eroded from the flysch parts of the OWC and transported by right-bank tributaries into the reservoirs Krpeľany, Hričov, and Nosice, which have lost 58, 25, and 22% of their original volume, respectively. In total, over 12.7 million m^3 of deposits have been accumulated behind dams since they were put into operation (35% reduction of total volume – Holubová 1997).

Janský (1992) assessed the rate of siltation in 27 small reservoirs for irrigation both in mountains and lowlands. The area of individual reservoirs ranges from 1 to 20 ha and the amount of sediments was 4.8–83.6% of their total storage capacity. Annual deposition ranged from 188 to 7,554 m^3 (with a weighted average of 2,897 m^3), i.e., storage volume is reduced by 0.32–9.30%.

Larger reservoirs also face a problem of *wave erosion*. According to Harčár (1986) and Lukáč (1988), on flysch rocks bank erosion may form up to 6 m high undercut banks and cliffs, such as at the Domaša reservoir on the Ondava River.

7.6.3 Runoff Processes

In general, the Slovak territory is profoundly susceptible to runoff processes, referring to water erosion and accumulation induced by heavy rainfalls or snowmelt, operating almost exclusively in deforested areas. Human transformation of the natural landscape along with past climatic oscillations controlled the areal extent, frequency and magnitude of runoff erosion events in time and space (Stankoviansky et al. 2006). Runoff processes were most effective when human interference was combined with a higher frequency of extreme rainfall events during colder and wetter climatic fluctuations. According to Bučko (1980), the Slovak territory suffered the consequences of four main *historical phases* of accelerated erosion associated with the Late Bronze Age (1260 to 750 BC), the Great Moravian Empire (the ninth century), and, more importantly, the Great colonization (the thirteenth and fourteenth centuries) and especially the period from the sixteenth to nineteenth centuries. The last period of extreme erosion was connected with the combined influence of the dispersed settlement in the wake of the youngest (Walachian, Goral, and kopanice) colonization waves and the LIA (Stankoviansky 2003).

In the past, extensive areas were affected by *linear (gully) erosion* resulting in the formation of relatively dense network of permanent gullies. The spatial distribution of gullies in Slovakia according to their density is depicted on the unique map by Bučko and Mazúrová (1958) at a scale of 1:400,000. The densest gully network is linked to uplands and hills where above all two factors influenced an activity of concentrated runoff: a deep regolith in areas built of less resistant (flysch and volcanic) rocks and land use (mostly pastures and arable land). The locally high density of gullies permits to classify the most affected areas as badlands.

A key area for the assessment of geomorphic effect of the historical gully erosion is the low, flat-topped *Myjava Hills* in the western Slovakia, built predominantly of flysch-like rocks of medium to low resistance. Average gully density in the Myjava Hills as a whole is ca 1.2 km km^{-2}, but field research revealed 11 km km^{-2} maximum density locally. Gully depths are commonly 10–15 m, but exceptionally more than 20 m. The field reconnaissance and the analysis of old cadastral maps suggests that the pattern and density of these gullies have been primarily controlled by artificial linear landscape elements typical for land use before collectivization (access roads, field boundaries, drainage furrows, etc.). The old military survey maps from years 1782, 1837 (1:28,800) and 1882 (1:25,000) and both regional and local historical sources indicate at least *two main periods of gully formation*: between the mid-sixteenth century and the 1730s and between the 1780s and the mid-nineteenth century. Conditions for gully erosion were created by extensive forest clearance and expansion of farmland due to the "kopanice" settlers and to overgrazing by Walachian shepherds. Gully formation was very probably triggered by extreme rainfalls and snowmelts during the LIA. All gullies were later afforested. Naturally, the areas that were settled earlier could also be affected by gully formation. The abandonment of some villages in the fourteenth century in the western margin of the study area may indicate such an older phase. These oldest gullies were probably later filled, but reactivated locally in the LIA. Gully erosion was accompanied with *muddy floods*. This phenomenon is not connected with permanent watercourses, but exclusively with cultivated slopes. The more intense erosion is, the more extreme manifestations of muddy floods occur. Repeated muddy floods from gullies left thick colluvial cones prograding onto the floodplains of the valleys where the gullies emptied (Stankoviansky 2003).

The operation of runoff processes was significantly changed by the collectivization of agriculture. Large-scale land use changes brought about their distinct acceleration. According to Jambor (1997), water (runoff) erosion has increased by 4–5 times, according to Juráň (1990) even 10 times. With the formation of large cooperative fields and the removal of a dense network of artificial linear landscape elements the predominant gully erosion was replaced by intense sheet wash, rill, and inter-rill erosion.

The lowering of the surface of hillslopes affected by *sheet wash* is in fact unnoticeable to the naked eye. However, if shortly after sowing the soil layer is washed away from the whole field together with seeds, it is possible to estimate the thickness of removed soil. Minimum slope lowering of 2–3 cm per downpour was observed in the southern part of the Trnavská pahorkatina Hills (Bučko 1963), during two consecutive cloudbursts by at least 3–4 cm in the southern part of the Myjava Hills (Stankoviansky 2003).

More visible are the geomorphic effects of *rill erosion*. In general, the depth of rills did not exceed tillage depth. Most of these microforms were deep from some centimeters to 30 cm (Stankoviansky 2003). The highest average soil loss due to rill erosion, 623 m^3 ha^{-1}, was recorded on a potato field near Lučatín in the northern margin of the Zvolenská kotlina Basin during a disastrous rainfall on May 23, 1958 (Zachar 1970).

Land use changes also influenced gully erosion. Contrary to the past, mostly topographically controlled *ephemeral gullies* are formed on large sloping cooperative fields. Two types of ephemeral gullies were recorded in the Myjava Hills: wide (up to 5–6 m) and shallow gullies, deepened only in the plowed layer and narrow and deep gullies, incised in the plow pan with less than 1 m depth over the whole post-collectivization period (Stankoviansky 2003). This is in accordance with Lukniš (1954), who claimed that under Slovak conditions these forms are rarely deeper than 1 m. Gullies on fields are always erased by the subsequent tillage operation but are re-formed in the same places during the next extreme events.

In contrast to arable land, fresh gullies on *pastures* can survive and grow gradually into permanent gullies. Gully erosion is also linked with linear artificial landscape features, mostly with *dirt roads*. Zachar (1970) documents the formation of a 2 m deep gully during the only extreme rainfall event in 1958 in the Zvolenská kotlina Basin, Košťálik (1965) records creation of a 3 m deep gully in the period 1959–1961 in the Nitrianska pahorkatina Hills. However, despite such records, gully erosion at present is hardly comparable with the disastrous gullying in the LIA (Stankoviansky et al. 2006).

The increased intensity of runoff processes after collectivization is also confirmed by deep *correlative sediments*. A maximum total amount of ca 1 m colluvia of post-collectivization deposition is observed at some footslopes and on dry valley bottoms. The depth of fill in some narrow valley cuts or gullies locally exceeds 2 m (Stankoviansky 2003). The thick accumulation of eroded and transported material is due to muddy floods from agricultural fields. They accompany runoff erosion during extreme rainfall events in times when the soil is bare or poorly protected by vegetation, above all in spring months. Muddy floods are off-site effects of runoff and erosion resulting mostly from ephemeral gully erosion along thalwegs of dry valleys (cf. Stankoviansky et al. 2010).

The documented manifestations of extreme rainfall events that affected neighboring villages Kunov and Prietrž on the western margin of the Myjava Hills in 2009 indicated that under specific circumstances the current combination of gully formation and muddy floods can show similar features to their more intense operation in the past. The material carried away by muddy floods arrived not only from fresh ephemeral gullies but also from old permanent gullies that were deepened during these events, while the fresh incision in the bottoms of these historical gullies, partially in underlying Miocene sandstones, ranged from 0.5 to 3 m (Stankoviansky and Ondrčka 2011).

A totally different course of runoff processes is observed in *woodlands*. Erosion under natural conditions is negligible, but any human intervention in the forest landscape markedly changes the situation. It is evident that overall erosion in forest has increased considerably in the last decades as a result of mechanization of timber harvesting. In the flysch area of the Bukovské vrchy Mountains, Midriak (1995) found soil loss caused by gully erosion nearly hundred times exceeding the loss caused by sheetwash (4.4–14 m^3 ha^{-1} year^{-1}). The average soil removal from one meter of length of forestry skidding roads ranges from 0.13 to 0.61 m^3 year^{-1} due to both mechanical scraping by logged trees and the subsequent concentrated flow.

According to Stankoviansky (2003), the post-collectivization period represents the fifth period of accelerated erosion and also the first erosion phase exclusively induced by human interference.

7.6.4 Combined Tillage and Runoff Processes

Since the conversion of woodland into farmland regularly cultivated hillslopes are under the influence of combined sheet wash and tillage erosion. Sheet wash accompanied by rill and inter-rill erosion occurred irregularly during extreme meteorological-hydrological events while tillage erosion took place regularly during temporally fixed tillage operations.

The first attempts to point out the role of tillage in soil degradation and geomorphic evolution dates back to the 1950s. According to Lobotka (1955), on slopes with contour cultivation, tillage erosion clearly predominated over runoff erosion and unintentionally created terraced fields. Before collectivization, terraced fields in many areas occupied up to 90% of all arable land (e.g., in the Kysucká vrchovina, Javorníky, or Vtáčnik Mountains). On the other hand, there were regions with maximum extent of terraced fields not higher than 4–6% of arable (e.g., in the Krupinská planina Mountains). In the case of lots of contour parcels arranged one above the other, an interesting, *step-like slope profile* was often formed (Lobotka 1958). The most marked terraces with the highest steps up to 4–5 m have originated on the steepest slopes in areas with a high relief. Up-and-down tillage together with runoff erosion resulted locally in a complete loss of soil, abandonment of cultivation, and use as poor pastures (Lobotka 1958). In the uppermost slope segments and especially on ridge tops tillage erosion clearly predominated, influenced by the spatial pattern of plots with different land use types and the techniques of tillage employed.

The above scheme was totally changed by collectivization. The most drastic terrain adjustment was *leveling* of the terraced plots, and numerous smaller gullies were planated. The resulting effect was return to a smoothly shaped relief, similar to that inherited from the times of periglacial morphogenesis. However, the present-day surface on arable land is markedly lowered, not only in comparison with the surface from the onset of cultivation, but also with respect to the surface from the beginnings of collectivization (Stankoviansky 2003).

Naturally, such relief evolution was characteristic mostly for the lower parts of the Carpathians. The morphogenesis of the agricultural landscape in higher regions was markedly different. Thus for example, most of the terraced fields around the village of Liptovská Teplička in the Low Tatras were grassed and used as pastures. The narrow and long fields were only merged on gentler slopes and on valley bottoms and the step-like character of slopes was preserved until now. Reviewing the role of tillage erosion in geomorphic evolution, Stankoviansky (2008) considers it the predominant process of relief modeling on sloping arable land. The total geomorphic effect of tillage erosion in a given area depends on the duration and frequency of tillage.

In *lowlands*, the combined operation of tillage and runoff processes (accompanied by aeolian processes) has a longer history than in the Carpathians, lasting millennia. Eroded material from convexities is deposited gradually on footslopes or in various depressions. Thus, the long-term operation of mentioned processes results in the decrease of relative relief (Stankoviansky 2008). Landform changes due to these processes are distinct especially in areas of loess hills with chernozems. Their surface is spangled with bright patches indicating the incorporation of loess into the tillage horizon after partial or complete erosional removal of the overlying humous horizon. Detailed investigation in broader surroundings of the village of Voderady in the Trnavská pahorkatina Hills showed their great diversity. Patches linked with convex forms and upper portions of slopes indicate an origin from tillage, while other positions suggest polygenetic origin (Smetanová 2009). Studies of the spatio-temporal evolution of bright patches in the selected dry valley basin revealed their significant (almost fourfold) spatial growth in the post-collectivization period (Smetanová et al. 2009).

7.6.5 Karst and Pseudokarst Evolution

Surface and subsurface karst landforms are formed by corrosion of soluble rocks in dependence on autochthonous or allochthonous position of an area, degree of structural-tectonic disturbance of rocks, slope, hydrological and climatic conditions, the presence of weathering mantles, soils, and vegetation. Karst extends over 2,700 km^2 in Slovakia (Jakál 1993).

In the allochthonous karst of the northern Low Tatra Mountains Droppa (1976, 1978) found the following values for the *chemical denudation* in karst systems drained by two local streams: for the Demänovka 46.8 m^3 km^{-2} year^{-1} (for 1974), for the Štiavnica in the Jánska dolina Valley 28.7–42.8 m^3 km^{-2} year^{-1} (for 1974–1976). Locally *new ponors* are opening in surface channels, mostly in allochthonous streams of mountain regions, and are later completely filled by fluvial sediments, mainly by granite gravels. For the last time a new ponor was created in May 2006 in the channel of the stream of Štiavnica, which completely drained the surface watercourse (Holúbek 2006). In the Slovenský kras Karst marginal *karst lakes* appear and vanish. Jašteričie (Silické) jazero lake with the area of 1.22 ha vanished due to the formation of a new ponor, filling the lake by soil sediment washed from surrounding shales (Cílek 1996). The marginal karst lake originated at Silica in 1953 due to clogging the ponor in the depression called Papverme (Turkota 1972). The soil sediment washed in from nearby fields is also transported into some inflow caves of the contact karst. In the Domica Cave, Slovenský kras Karst, there are masses of sediment that partially filled underground channels and adjacent parts of corridors during disastrous floods (Jakál 2006). Floods, extremely rich in sediment, were connected with improper agricultural land use in the drainage area of the cave in the first decades of the post-collectivization period.

In karst areas block sinking and *collapses* of *cave ceilings* are also observable. According to Jakál (2000), a significant precondition is a high degree of underground karstification at a near-surface part of the karst massif what is typical of extensive cave systems with an existence of domed spaces. The hazard of collapsing increases in areas where carbonate massifs are strongly affected by tectonics. Both phenomena are caused by corrosive broadening of fissures, by congelifraction acting along them, and by water percolation in the vadose zone. Their most frequent occurrence is in the Silicicum nappe, mostly in the central part of Silická and Plešivecká planina Plateaus, Slovenský kras Karst, where the surface is deformed more markedly by block subsidence and collapse of chasms of light hole type.

More caves have originated in limestone bodies of nappe outliers, disintegrated by gravitation processes (e.g., Nešvara 1974; Hochmuth 1976). The rate of movement of rock blocks in one of monitored caves – the Driny Cave, Malé Karpaty Mountains – ranges from 0.05 to 0.08 mm year^{-1} (Briestenský and Stemberk 2008). In some caves, recent movements of rock blocks are indicated by shift of separated parts of oval corridors along open fissure, for example, in the Harmanecká jaskyňa Cave, Veľká Fatra Mountains (Bella 2000).

Gravitational movements in caves are indicated mostly by broken and chapped stalagnate or flowstones formed on destabilized rock surfaces. Compaction and collapsing of fine-grained clastic sediments in intermittently flooded caves due to changes of their mechanical properties under repeated moistening and dessication result in chapping and deformation of overlain stalagnates, for instance in the Jasovská jaskyňa Cave in the Slovenský kras Karst (Zacharov 1984).

Crevice caves are often formed in travertine bodies but also in flysch sandstones and in volcanic rocks (e.g., Cebecauer and Liška 1972; Hochmuth 1995; Gaál 2003). The part of monitoring of gravitational disintegration of the travertine body with Spišský hrad Castle on top in the Hornádska kotlina Basin is also the measurement of fissure-controlled block movements in the Podhradská jaskyňa Cave that revealed average annual rate of fissure opening since 1992 ca 1 mm (Vlčko et al. 2002).

Mechanical erosion by subsurface water in noncarbonatic complexes is mainly caused by piping of permeable layers of the regolith. It is frequent in glacial and glaciofluvial accumulations. In unstable areas piping favors landslides and gully formation. Mazúr (1963) and Stankoviansky (2003) document hollows and pipes created in the gully bottom fills in the Žilinská kotlina Basin and the Myjava Hills, respectively. Holúbek (2008) introduces a 12-m-long cave formed due to piping in silty sediment of flysch complex in the northwestern part of the Podtatranská kotlina Basin. Maglay (2010) mentions up to 6–8 m deep pits formed by collapses of the loess series underlain by sandy gravels from the southern and eastern parts of the Trnavská pahorkatina Hills. Leaching in combination with piping in lower layers of loess is responsible for the formation of cavities with possible collapse of loess. The origin of observed phenomena was probably linked to older drillings of hydrocarbon exploration.

7.6.6 Aeolian Processes

Though aeolian processes do not belong to the most important processes of the recent morphogenesis, under favorable circumstances they can have significant geomorphic effect. Dry lowlands with soils light in texture are regularly affected by this phenomenon. The most extensive area covered by aeolian sands is in the Borská nížina Lowland lying in the westernmost part of Slovakia. Sand dunes not afforested by secondary pine forests and still used as arable land often suffer from sandy storms. Other smaller sandy areas very prone to wind erosion are scattered in the Danube Plain and in the Východoslovenská rovina Plain. Apart from major hot spots with sandy soils, wind erosion affects also areas with loamy soils in hilly parts of lowlands covered by loess, naturally with lesser intensity. In dry summer periods after harvest, however, dust storms can be observed occasionally (Stankoviansky et al. 2006). According to Streďanský (1993), the most favorable conditions for the highest intensity of wind erosion occurs in the winter and early spring months when a frozen, dusty soil surface is without vegetation cover. Janšák (1950) admits a possibility of wind erosion also at the beginning of autumn, if October is dry.

Ištok and Ižóf (1990) refer to a considerable dynamics of aeolian processes both in the past and today in the northern and central parts of the Danube Plain. Wind has accumulated and currently accumulates blown sand and dust more or less evenly throughout the plain. However, older aeolian deposits were preserved on the surface only in places where they were not washed away or were protected under younger fluvial sediments. A significant lowering of local relief was observed on the old aggradation ridge between the Váh and Dudváh Rivers. The terrain was leveled through the silting of paleochannels, lowering of older dunes by water and wind erosion, and the accumulation of young sand covers and low dunes. In the site Szilasi dülő the authors have recorded 25–40 cm eolian accumulation of sandy loess accumulated since the eleventh to twelfth centuries (i.e. since the abandonment of medieval village).

In the area between Lesser Danube and the Váh River Ištok and Ižóf (1990) ascribe a relatively young age to ca 80–100 elevations of aeolian origin of small areal extent, built of sandy and dusty sediments, when their formation date to the Middle Ages or later. Their common denominator is an absence of any archaeological finds typical of older dunes.

The destruction of sand dunes in this area was documented for example at the village of Dolná Streda where the surface was lowered by the removal of ca 1 m thick layer of the dune body. Along with tillage the predominant role in the destruction was played by wind while its final stage took place in the twentieth century. The contours of the original dune are indicated by archaeological finds from the High Middle Ages preserved in a ring along its former perimeter (Ištok and Ižóf 1990).

The first reports on current aeolian processes relate to the landscape after World War II, still unaffected by the collectivization. Janšák (1950), who documented the spatial distribution of sand dunes in detail for all the three sub-Carpathian lowlands, also perceived their contemporary activity. He registered the formation of wind furrows and the short distance transport of sand on dunes in the southernmost part of the Hronská pahorkatina Hills where the surface was not sufficiently fixed by crops. Similar present-day microforms created by wind erosion were identified by Lukniš and Bučko (1953) in the contact zone of the Danube Plain and the Hronská pahorkatina Hills. Shallow linear depressions and low linear elevations were formed mostly in bottoms of old, Pleistocene deflation troughs. Lukniš and Mazúr (1959) refer to the fact that paleochannels in the central part of the "island" between the Danube and Lesser Danube Rivers, Danube Plain, are obliterated by both eolian and human activity and minor dunes were also formed.

After collectivization wind erosion significantly intensified. According to Streďanský (1993), there was some tenfold increase locally connected with the shape and large size of contemporary fields, reaching in the most affected lowland parts 100–200 ha. Streďanský (1977) recorded a triangular body of aeolian deposits with thicknesses of 50–70 cm and lengths of 4–6 m accumulated behind the barrier formed by a railway embankment during 7-h of intense deflation from a 250 m-long cooperative field in the village of Kľačany in the Nitrianska pahorkatina Hills.

According to Ištok (2007), in the period 1980–2004 the area between the Váh and Dudváh Rivers to the SW from Sereď, Danube Plain, was affected by frequent intense aeolian transport and accumulation in a 100–150 m wide belt of NW-SE direction in the continuation of the axis of the Trnávka River valley, deepened in the neighboring Trnavská pahorkatina Hills. The wind accelerated by the "jet effect" of the valley resulted both in erosional obliteration of relief and areal or point-like accumulation of transported material in this belt.

An extreme event was recorded in 1986 in the area to the south from the confluence of the Dudváh and Čierna voda Rivers. In the course of one day of very intense west wind, the tilled 1.5–2 m high sand dune of oval shape with longer axis of NW-SE direction and an area ca 450–600 m^2 was almost totally removed. The dune material was scattered and accumulated on the adjacent leeward area, mostly in microdepressions between aggradation ridges (Ištok 2007).

7.7 Perspectives: Impact of Climate Change

One of the consequences of the predicted climate change could be an increase in natural hazards, particularly those that relate to precipitation extremes. Daily and multi-day extreme precipitation events can cause major flooding of profound geomorphic effect. Over the last few years, research has provided an improved understanding of the processes leading to extreme precipitation. According to climatic models, the intensification of the hydrologic cycle due to global warming is very likely. Historical climate records from the meteorological network of Slovakia indicate that extreme precipitation events could occur at a far greater scale than they have in recent years. It is highly probable that the expected rise in their frequency will further increase their geomorphic effectiveness.

Some studies have already referred to an increasing frequency and intensity of precipitation-induced processes within the last two to three decades: landslides (Fussgänger and Jadroň 2001), debris flows (Kapusta et al. 2010), and muddy floods (Stankoviansky and Ondrčka 2011). The authors suggest possible linkage of these developments with global climate change.

Naturally, the marked increase of the frequency of extreme precipitation event is not a new phenomenon. However, so far it was characteristic mostly for colder and at the same time wetter climatic fluctuations (the last such a period was the LIA). The present-day increase of their frequency is associated with climate change of opposite character, namely with global warming.

In any case, regardless of a character of climate change, we have to highlight the need to continue research on practical issues, improving our capacity to learn from the mistakes of the past in planning future land management strategies.

References

Almer D (1955) Pozorovania plavenín na niektorých slovenských tokoch (Observation of suspended load in some Slovak rivers). Vodohospodarsky časopis 3(1–2):95–103 (in Slovak)

Almer D, Szolgay J (1955) Štúdium splavenín a plavenín na Dunaji, Laborci a Uhu a výustných tratiach riek Moravy, Váhu, Nitry a Hrona: Záverečná správa (Study of bedload and suspended load in the Danube, Laborec and Uh rivers and along the lower stretches of Morava, Váh, Nitra and Hron rivers). Final Report, VÚV, Bratislava (in Slovak)

Andrusov D (1931) Poznámka o sesuvech v povodí Oravy na Slovensku (Remarque sur les glissements de sol dans la vallée de l' Orava en Slovaquie). Věstník Státního geologického ústavu ČSR, Praha 7:172–174 (in Czech with French summary)

Baliak F, Mahr T (1979) Svahové deformácie v oblasti vysokých pohorí Západných karpát (Slope deformations in the high mountains of the Slovak Carpathians). In: Matula M (ed) Inžinierskogeologické štúdium horninového prostredia a geodynamických procesov (Engineering geological studies of rocks and geodynamic phenomena). Slovak Academy of Sciences, Bratislava, pp 133–145 (in Slovak)

Barka I (2005) Niektoré metodické postupy pri mapovaní vybraných geomorfologických procesov (Selected methods for mapping geomorphological processes). Faculty of Natural Sciences, Comenius University in Bratislava, 108 p (in Slovak)

Bella P (2000) Harmanecká jaskyňa – názory a problémy genézy, základné morfologické a genetické znaky (Harmanecká jaskyňa Cave – opinions and problems of genesis, basic morphological and genetic features). In: Výskum, využívanie a ochrana jaskýň (Demänovská dolina, 16–19. 11. 1999). Proceedings. Správa slov. jaskýň, Liptovský Mikuláš. 2:71–81 (in Slovak)

Biely A (ed) (1996) Geological map of the Slovak Republic 1:500,000. Ministry of Environment of the Slovak Republic, Geological Survey of the Slovak Republic, Bratislava

Boltižiar M (2009) Vplyv georeliéfu a morfodynamických procesov na štruktúru vysokohorskej krajiny Tatier (Influence of the relief and morphodynamic processes to the spatial structure of high-mountain landscape, Tatra Mountains). Univ. Konšt. Filoz.; Ústav krajinnej ekológie SAV, Nitra, 158 p

Briestenský M, Stemberk J (2008) Monitoring mikropohybov v jaskyniach západného Slovenska (Micromovements monitoring in the Western Slovakia caves). Slovenský kras 46(2):333–339 (in Slovak)

Bučko Š (1963) Erózia pôdy v dolnom povodí Váhu (Soil erosion in the Lower Váh River Basin). Sborník ČSSZ 68:72–76 (in Slovak)

Bučko Š (1980) Vznik a vývoj eróznych procesov v ČSSR (Rise and development of erosion processes in Czechoslovakia). In: Protierózna ochrana (Soil conservation). Proceedings. Dom techniky ČSVTS, Banská Bystrica, pp. 1–14 (in Slovak)

Bučko Š, Mazúrová V (1958) Výmoľová erózia na Slovensku (Gully erosion in Slovakia). In: Zachar D (ed) Vodná erózia na Slovensku (Water erosion in Slovakia). Slovak Academy of Sciences, Bratislava, pp 68–101 (in Slovak)

Cebecauer I, Liška M (1972) Príspevok k poznaniu krasových foriem spišských travertínov a ich kryhových zosuvov (Contribution to the knowledge of karst forms of the Spiš travertines and their block slidings). Slovenský kras 10:47–60 (in Slovak)

Cílek V (1996) Silica – zrození a smrt jezera (Silica – birth and death of a lake). Speleofórum 15:36–42 (in Czech)

Demo M, Bielek P, Čanigová M, Ďuďák J, Fehér A, Hraška Š, Hričovský I, Hubinský J, Ižáková V, Látečka M, Lednár Š, Martuliak P, Molnárová J, Muchová Z, Paška I, Rataj V, Štefanovič M, Valšíková M, Vereš A, Vontorčík E, Vontorčík J (2001) Dejiny poľnohospodárstva na Slovensku (History of agriculture in Slovakia). Slovak University of Agriculture in Nitra – Soil Science and Conservation Research Institute, Bratislava, 662 p (in Slovak)

Droppa A (1976) Intenzita korózie krasových tokov v Demänovskej doline (Corrosion intensity of karst streams in the Demänovská dolina Valley). Slovenský kras 14:3–30 (in Slovak)

Droppa A (1978) Intenzita korózie tokov v Jánskej doline (The intensity of stream corrosion in the Jánska dolina Valley). Slovenský kras 16:39–67 (in Slovak)

Faško P, Lapin P, Melo M, Pecho J (2009a) Changes in precipitation regime in Slovakia – past, present and future. In: A changing climate for biology and soil hydrology interactions. Proceedings of the 2nd international conference on bioclimatology, Bratislava, 21–24 Sept 2009. Institute of Hydrology SAS, Bratislava

Faško P, Pecho J, Mikulová K, Nejedlík P (2009b) Trends of selected characteristics of precipitation in the Northern Carpathians in the light of water supply for agriculture. In: Impact of climate change and adaptation in agriculture. Extended abstracts of the international symposium, Vienna, 22–23 June 2009. BOKU-met report 17, pp 106–109 http://www.boku.ac.at/met/report

Feranec J, Oťaheľ J (2001) Land cover of Slovakia. Veda, Bratislava, 124 p

Fussgänger E, Jadroň D (2001) Vplyv súčasných klimatických pomerov na vývoj svahových gravitačných pohybov (Influence of recent climatic conditions on slope gravitational movement evolution). In: Proceedings of the 2nd conference: Geológia a životné prostredie (Geology and the environment), Bratislava, 24–25 Feb 2001. Dionýz Štúr State Geological Institute, Bratislava, pp 15–18 (in Slovak)

Gaál Ľ (2003) Genetické typy rozsadlinových jaskýň na Slovensku (Genetic types of crevice caves in Slovakia). Slovenský kras 41:29–45 (in Slovak)

Grešková A (2002) Dynamika a transformácia nivy rieky Moravy študovaná pomocou historických máp a leteckých snímok (Dynamics and transformation of river Morava floodplain studied in the historical maps and aerial photographs). Geomorphologia Slovaca 2(2):40–44 (in Slovak)

Grešková A, Lehotský M (2007) Vplyv lesných brehových porastov na správanie a morfológiu riečneho koryta (Impacts of riparian forest on behaviour and morphology of river channel). Geomorphologia Slovaca et Bohemica 7(1):36–42 (in Slovak)

Harčár J (1986) Abrázia vodnej nádrže Domaša v Nízkych Beskydách (Abrasion along the fringes of the water reservoir Domaša in the Nízke Beskydy mountains). Geografický časopis 38(4):322–341 (in Slovak)

high-mountain landscape, Tatra Mountains). Univ. Konšt. Filoz.; Ústav krajinnej ekológie SAV, Nitra, 158 p

Briestenský M, Stemberk J (2008) Monitoring mikropohybov v jaskyniach západného Slovenska (Micromovements monitoring in the Western Slovakia caves). Slovenský kras 46(2):333–339 (in Slovak)

Bučko Š (1963) Erózia pôdy v dolnom povodí Váhu (Soil erosion in the Lower Váh River Basin). Sborník ČSSZ 68:72–76 (in Slovak)

Bučko Š (1980) Vznik a vývoj eróznych procesov v ČSSR (Rise and development of erosion processes in Czechoslovakia). In: Protierózna ochrana (Soil conservation). Proceedings. Dom techniky ČSVTS, Banská Bystrica, pp. 1–14 (in Slovak)

Bučko Š, Mazúrová V (1958) Výmoľová erózia na Slovensku (Gully erosion in Slovakia). In: Zachar D (ed) Vodná erózia na Slovensku (Water erosion in Slovakia). Slovak Academy of Sciences, Bratislava, pp 68–101 (in Slovak)

Cebecauer I, Liška M (1972) Príspevok k poznaniu krasových foriem spišských travertínov a ich kryhových zosuvov (Contribution to the knowledge of karst forms of the Spiš travertines and their block slidings). Slovenský kras 10:47–60 (in Slovak)

Cílek V (1996) Silica – zrození a smrt jezera (Silica – birth and death of a lake). Speleofórum 15:36–42 (in Czech)

Demo M, Bielek P, Čanigová M, Duďák J, Fehér A, Hraška Š, Hričovský I, Hubinský J, Ižáková V, Látečka M, Lednár Š, Martuliak P, Molnárová J, Muchová Z, Paška I, Rataj V, Štefanovič M, Valšíková M, Vereš A, Vontorčík E, Vontorčík J (2001) Dejiny poľnohospodárstva na Slovensku (History of agriculture in Slovakia). Slovak University of Agriculture in Nitra – Soil Science and Conservation Research Institute, Bratislava, 662 p (in Slovak)

Droppa A (1976) Intenzita korózie krasových tokov v Demänovskej doline (Corrosion intensity of karst streams in the Demänovská dolina Valley). Slovenský kras 14:3–30 (in Slovak)

Droppa A (1978) Intenzita korózie tokov v Jánskej doline (The intensity of stream corrosion in the Jánska dolina Valley). Slovenský kras 16:39–67 (in Slovak)

Faško P, Lapin P, Melo M, Pecho J (2009a) Changes in precipitation regime in Slovakia – past, present and future. In: A changing climate for biology and soil hydrology interactions. Proceedings of the 2nd international conference on bioclimatology, Bratislava, 21–24 Sept 2009. Institute of Hydrology SAS, Bratislava

Faško P, Pecho J, Mikulová K, Nejedlík P (2009b) Trends of selected characteristics of precipitation in the Northern Carpathians in the light of water supply for agriculture. In: Impact of climate change and adaptation in agriculture. Extended abstracts of the international symposium, Vienna, 22–23 June 2009. BOKU-met report 17, pp 106–109 http://www.boku.ac.at/met/report

Feranec J, Oťaheľ J (2001) Land cover of Slovakia. Veda, Bratislava, 124 p

Fussgänger E, Jadroň D (2001) Vplyv súčasných klimatických pomerov na vývoj svahových gravitačných pohybov (Influence of recent climatic conditions on slope gravitational movement evolution). In: Proceedings of the 2nd conference: Geológia a životné prostredie (Geology and the environment), Bratislava, 24–25 Feb 2001. Dionyz Štúr State Geological Institute, Bratislava, pp 15–18 (in Slovak)

Gaál Ľ (2003) Genetické typy rozsadlinových jaskýň na Slovensku (Genetic types of crevice caves in Slovakia). Slovenský kras 41:29–45 (in Slovak)

Grešková A (2002) Dynamika a transformácia nivy rieky Moravy študovaná pomocou historických máp a leteckých snímok (Dynamics and transformation of river Morava floodplain studied in the historical maps and aerial photographs). Geomorphologia Slovaca 2(2):40–44 (in Slovak)

Grešková A, Lehotský M (2007) Vplyv lesných brehových porastov na správanie a morfológiu riečneho koryta (Impacts of riparian forest on behaviour and morphology of river channel). Geomorphologia Slovaca et Bohemica 7(1):36–42 (in Slovak)

Harčár J (1986) Abrázia vodnej nádrže Domaša v Nízkych Beskydách (Abrasion along the fringes of the water reservoir Domaša in the Nízke Beskydy mountains). Geografický časopis 38(4):322–341 (in Slovak)

Harčár J, Krippel E (1994) K morfogenéze gravitačných svahových deformácií vo Vihorlate (založené na palynologickom datovaní) (To the morphogenesis of gravitational slope deformation in the Vihorlat mountains based on palynological dating). Geografický časopis 49(2):283–290 (in Slovak)

Harčár J, Trávniček D (1992) Genéza a vývoj svahových pohybov v Novej Kelči v Nízkych Beskydách (Genesis and development of the slope movement in Nová Kelča in the Nízke Beskydy mountains). Geografický časopis 44(3):259–272 (in Slovak)

Hochmuth Z (1976) Priepasťové jaskyne na Bielej skale vo Veľkej Fatre (Abyss-like caves at the white rock in the Veľká Fatra mountains). Československý kras, Praha 27:113–116 (in Slovak)

Hochmuth Z (1995) Pseudokrasové gravitačné jaskyne v pohorí Bachureň (centrálnokarpatský flyš) (Pseudokarst gravitation caves in the Bachureň mountains, Central Carpathian Flysch zone). In: Preserving of pseudokarst caves. Proceedings of the international meeting. Slovak Environmental Agency, Banská Bystrica, pp 61–67 (in Slovak)

Holúbek P (2006) Nový ponor v Jánskej doline (New ponor in the Jánska dolina Valley). Bull Slovak Speleol Soc 37(2):1–24 (in Slovak)

Holúbek P (2008) Suffosion (piping) cave in the foothill of the West Tatras in Slovakia. In: Zacisk, Special Issue: 9th international symposium on pseudokarst, Bielsko-Biala, pp 18–19

Holubová K (1997) Problémy systematického sledovania erózno-sedimentačných procesov v oblasti vodných diel (Problems of systematic observations of erosion-sedimentation processes in the area of hydroelectric reservoirs). In: Práce a štúdie 135. VÚV, Bratislava, 62 p (in Slovak)

Holubová K (1999) Hydrológia a úpravy rieky (Hydrology and river regulation). In: Šeffer J, Stanová V (eds), Aluviálne lúky rieky Moravy – význam, obnova a manažment. (Morava river floodplain meadows – importance, restoration and management). Daphne, Bratislava, pp 7–23 (in Slovak)

Holubová K (2000) Some aspects of the bedload transport regime in the Slovak section of the Danube River. In: Proceedings of 20th conference of the Danubian countries on hydrological forecasting and hydrological bases of water management, Bratislava, 4–8 September 2000, CD-ROM. Institute of Hydrology SAS, Bratislava

Holubová K, Lisický MJ (2002) Morava river floodplain: the need for river restoration and management strategy. In: Proceedings of Gemenc international workshop on side arm rehabilitation. Eötvös József College, Baja, pp 1–12

Holubová K, Capeková Z, Szolgay J (2004) Impact of hydropower schemes on bedload regime and channel morphology of the Danube River. In: Greco M, Carravetta A, Della MR (eds) River flow. Taylor & Francis Group, London, pp 135–142

Holubová K, Hey RD, Lisický MJ (2005) Middle Danube tributaries: constraints and opportunities in lowland river restoration. Arch Hydrobiol Suppl Large Rivers 15(1–4):507–519

Horváthová B (2003) Povodeň to nie je len veľká voda (Flooding is not just big water). Veda, Bratislava, 224 p (in Slovak)

Hreško J, Boltižiar M, Bugár G (2005) The present-day development of landforms and landcover in alpine environment – Tatra Mountains (Slovakia). Studia Geomorphologica Carpatho-Balcanica 39:23–38

Ingr M, Šarík I (1970) Suťový prúd v Roháčoch (Debris current in Roháče). Mineralia Slovaca 2(8):309–313 (in Slovak)

Ištok P (2007) Zonalita osídlenia južnej časti Dolnovážskej nivy a priľahlých území (Zonality of settlement in the southern part of the Dolnovážska niva Floodplain and adjacent areas). Manuscript (in Slovak)

Ištok P, Ižóf J (1990) Podmienky vzniku a vývoja osídlenia krajiny dolného toku Váhu vo svetle geografických a archeologických prieskumov (Conditions of the rise and development of settlement in the Lower Váh River landscape in the light of geographical and archeological survey), vol 26. Študijné zvesti Archeologického ústavu SAV, Nitra, pp 145–170 (in Slovak)

Jakál J (1993) Karst geomorphology of Slovakia. Typology. Map at scale 1:500,000. Geographia Slovaca 4:38 p

Jakál J (1998) Antropická transformácia reliéfu a jej odraz v krajine Hornej Nitry (Anthropic transformation of relief and its reflection in the landscape of Horná Nitra). Geografický časopis 50(1):4–20 (in Slovak)

Jakál J (2000) Extrémne geomorfologické procesy v krase (Extreme geomorphological processes in karst). Geografický časopis 52(3):211–219 (in Slovak)

Jakál J (2006) Geomorfologické hrozby a riziká v krase (Geomorphological hazards and risks in karst of Slovakia). Slovenský kras 44:5–22 (in Slovak)

Jambor P (1997) Effective erosion control is fully in hands of land user. In: Proceedings, Soil Fertility Research Institute, Bratislava 20, pp 249–254

Janšák Š (1950) Eolické formácie na Slovensku (Les formations éoliennes en Slovaquie). Geographica Slovaca 2(1–2):5–48; (3–4):7–31 (in Slovak with French summary)

Janský L (1992) Sediment accumulation in small water reservoirs utilized for irrigation. In: Younos T, Diplas P, Mostaghimi S (eds), Land reclamation – advances in research and technology: proceedings of the international symposium, Nashville, 14–15 Dec 1992. American Society of Agricultural Engineering, St Joseph, pp 76–82

Juráň C (1990) Erózne procesy na Slovensku a perspektíva protieróznej ochrany poľnohospodárskej pôdy (Erosion processes in the territory of Slovakia and perspectives of agricultural soil conservation). In: Proceedings, Pôda – najcennejší zdroj (Soil – a most valuable resource). Soil Fertility Research Centre, Bratislava, pp 60–74 (in Slovak)

Kapusta J, Stankoviansky M, Boltižiar M (2010) Changes of debris flow activity in the High Tatra Mountains within the last six decades. Studia Geomorphologica Carpatho-Balcanica 44:5–33

Kňazovický L (1962) Les – voda – pôda (Forest, water, soil). Slovak Publishers of Agricultural Literature, Bratislava, 224 p (in Slovak)

Kopecký M (2004) Možnosti prognózovania vzniku zosuvov v SR na základe analýzy klimatických a hydrogeologických pomerov (Possibilities of prognosis of landslide formation in Slovak Republic based on analysis of climatic and hydrogeological conditions). In: Proceedings of scientific survey of VSB – Technical University Ostrava 50(2):63–72 (in Slovak)

Kopecký M, Martinčeková T, Šimeková J, Ondrášik M (2008) Atlas zosuvov – výsledky riešenia geologickej úlohy (Landslide atlas – results of the geological project). In: Proceedings of the 6th conference: Geológia a životné prostredie (Geology and the environment). Dionyz Štúr State Geological Institute, Bratislava, pp 115–121 (in Slovak)

Košťálik J (1965) Príspevok ku štúdiu erózie pôd v katastrálnom území Bojničky a Dvorníky (Beitrag zum Studium der Bodenerosion im Katastralgebiet Bojničky und Dvorníky). Geografický časopis 17(4):301–318 (in Slovak with German summary)

Kotarba A (2004) Zdarzenia geomorfologiczne w Tatrach Wysokich podczas małej epoki lodowej (Geomorphic events in the High Tatra Mountains during the Little Ice Age). In: Kotarba A (ed) Rola małej epoki lodowej w przekształcaniu środowiska przyrodniczego Tatr (Effect of the Little Ice Age on transformations of natural environment of the Tatra Mountains), vol 197. Prace Geograficzne IG i PZ PAN, Kraków, pp 9–55 (in Polish)

Kotarba A (2005) Geomorphic processes and vegetation pattern changes. Case study in the Zelené pleso Valley, High Tatras, Slovakia. Studia Geomorphologica Carpatho-Balcanica 39:39–48

Kotarba A (2007a) Extreme geomorphic events in the High Tatras during the Little Ice Age. In: André M-F (ed) From continent to catchment: theories and practices in physical geography. Presses Universitaires Blaise Pascal, Clermont–Ferrand, pp 325–334

Kotarba A (2007b) Geomorphic activity of debris flows in the Tatra Mountains and in other European mountains. Geographia Polonica 80(2):137–150

Kotarba A, Starkel A (1972) Holocene morphogenetic altitudinal zones in the Carpathians. Studia Geomorphologica Carpatho-Balcanica 6:21–35

Kotarba A, Kaszowski L, Krzemień K (1987) High-mountain denudational system of the Polish Tatra Mountains. Geographical Studies, Ossolineum, Wrocław, Special Issue 3, 106 p

Koutek J (1935) Geologická mapa Prosečnických hor a přilehlých oblastí flyšových (Carte géologique des Montagnes de Prosečno (Carpathes Occidentales) et de la region adjacente). Věstník du

Service Géologique de la République Tchécoslovaque 11:115–127 (in Czech with French summary)

Kozlík V, Mališ O, Alena F (1961) Ochrana pôdy pred vodnou eróziou (Soil conservation against water erosion). Slovak Publishers of Agricultural Literature, Bratislava, 228 p (in Slovak)

Lacika J, Urbánek J (1998) New morphostructural division of Slovakia. Slovak Geological Magazine 4(1):17–28

Lapin M, Melo M (2005) Priestorová interpretácia výstupov klimatických scenárov v povodí Hrona a Váhu geoštatistickými metódami (Spatial interpretation of climate scenario outputs in the Hron and Váh River basins using geostatistical methods). In: Pekárová P, Szolgay J (eds) Scenáre zmien vybraných zložiek hydrosféry a biosféry v povodí Hrona a Váhu v dôsledku klimatickej zmeny (Assessment of climate change impact on selected components of the hydrosphere and biosphere in Hron and Váh River basins). Veda, Bratislava, pp 49–80 (in Slovak)

Lehotský M, Grešková A (2007) Morphological response of the high-gradient river to the windblown forest – ecological aspect. Geomorphologia Slovaca et Bohemica 7(2):79–84 (in Slovak)

Lobotka V (1955) Terasové polia na Slovensku (Terraced fields in Slovakia). Poľnohospodárstvo 2(6):539–549 (in Slovak)

Lobotka V (1958) Príspevok k problému erózie z orania (Beitrag zu den Problemen der Erosion durch das Acker). Poľnohospodárstvo 5(6):1172–1191 (in Slovak with German summary)

Lukáč M (1988) Influence of wind wave motion on the transformation of reservoir banks in Slovakia. In: Volker A, Henry JC (eds) Side Effects of Water Resources Development, vol 172. IAHS Publ, Wallingford, pp 51–78

Lukniš M (1951) Sosunové územie na ľavom brehu Váhu medzi Hlohovcom a Šintavou (Landslide area on the left bank of the Váh River between Hlohovec and Šintava). Geographica Slovaca, Academia Scientiarum et Artium Slovaca 3:53–77 (in Slovak)

Lukniš M (1954) Všeobecná geomorfológia (General geomorphology), vol 1. Slovak Pedagogical Publishers, Bratislava, 341 p (in Slovak)

Lukniš M (1972) Relief. In: Lukniš M (ed) Slovensko (Slovakia) 2: Príroda (Nature). Obzor, Bratislava, pp 124–203 (in Slovak)

Lukniš M, Bučko Š (1953) Geomorfologické pomery Podunajskej nížiny v oblasti medzi Novými Zámkami a Komárnom (Geomorphological conditions of the Danube Lowland in the area between Nové Zámky and Komárno). Geografický časopis 5(3–4):131–168 (in Slovak)

Lukniš M, Mazúr E (1959) Geomorfologické regióny Žitného ostrova (Die Geomorphologischen Regionen des Žitný ostrov (Schüttinsel)). Geografický časopis 11(3):161–206 (in Slovak with German summary)

Lukniš M, Plesník P (1961) Nížiny, kotliny a pohoria Slovenska (Lowlands, basins and mountains of Slovakia). Osveta, Bratislava, 140 p (in Slovak)

Maglay J (2010) Geomorfológia Trnavskej pahorkatiny v závislosti na štruktúrnom vývoji a charakteristike usadenín vrchného pliocénu a kvartéru (Geomorphology of the Trnavská pahorkatina Hills in relation with structural development and characteristics of Upper Pliocene and Quaternary deposits). Manuscript PhD thesis, Faculty of Natural Sciences, Comenius University in Bratislava, 210 p (in Slovak)

Mahr T (1977) Deep-reaching gravitational deformations of high mountain slopes. Bull Int Assoc Engin Geol Krefeld 16(1):121–127

Malgot J (1975) Gravitačné deformácie na okrajoch vulkanických pohorí Slovenska (Gravitationsdeformationen der Abhänge an Rändern der Vulkangebirge der Slowakei). Geografický časopis 27(3):216–226 (in Slovak with German summary)

Matula M, Nemčok A, Pašek, JM, Řepka L, Špůrek M (1963) Sesuvná území ČSSR (Slide areas of Czechoslovakia). Summary final report, Geological survey, Geofond, Prague, 55 p (in Czech)

Mazúr E (1963) Žilinská kotlina a priľahlé pohoria (The Žilina Basin and the adjacent mountains: Geomophology and Quaternary era). Slovak Academy of Sciences, Bratislava, 185 p (in Slovak)

Mazúr E (1965) Major features of the West Carpathians in Slovakia as a result of young tectonic movements. In: Mazúr E, Stehlík O (eds) Geomorphological Problems of Carpathians. Slovak Academy of Sciences, Bratislava, pp 9–54

Mazúr E (1976) Morphostructural features of the West Carpathians. Geografický časopis 28(2):101–111

Mazúr E, Lukniš M (1978) Regionálne geomorfologické členenie SSR (Regional geomorphological division of Slovakia). Geografický časopis 30(2):101–125 (in Slovak)

Mazúrová V (1985) Antropogénne zmeny reliéfu v oblasti Bratislavy (Anthropogenic changes in the relief within the Bratislava area). Geografický časopis 37(4):380–393 (in Slovak)

Mednyanský A (1844) Malerische Reise auf dem Waagflusse in Ungarn. Zweite Ausgabe, Pesth

Midriak R (1983) Morfogenéza povrchu vysokých pohorí (Surface morphogenesis of high-mountain chains). Veda, Bratislava, 516 p (in Slovak)

Midriak R (1995) Povrchový odtok a erózne pôdne straty v lesných porastoch flyšovej oblasti CHKO – Biosférickej rezervácie Východné Karpaty (Surface runoff and erosive soil losses in forest stands of the flysch zone of the Landscape pretection area – The East Carpathians biosphere reserve). In: Hochmuth Z (ed), Reliéf a integrovaný výskum krajiny (Relief and integrated landscape research). Proceedings, Pedagogical Faculty, P. J. Šafárik University, Prešov, pp 58–63 (in Slovak)

Midriak R (1999) Geomorfologické procesy a krajinné formy na rozličných povrchoch v kryoniválnom morfogenetickom systéme Západných Karpát (Geomorphological processes and landforms on different landscape surfaces in cryonival morphogenetic system of the West Carpathians). Acta Facultatis Ecologiae Techn Univ Zvolen 6:9–21 (in Slovak)

Minár J, Barka I, Hofierka J, Hreško J, Jakál J, Stankoviansky M, Trizna M, Urbánek J (2009) Natural geomorphological hazards in Slovakia. In: Ira V, Lacika J (eds) Slovak geography at the beginning of the 21th century. Geographia Slovaca 26:201–220

Minár J, Bielik M, Kováč M, Plašienka D, Barka I, Stankoviansky M, Zeyen H (2011) New morphostructural subdivision of the Western Carpathians: an approach integrating geodynamics into targeted morphometric analysis. Tectonophysics 502:158–174

Nemčok A (1972) Gravitačné svahové deformácie vo vysokých pohoriach Slovenských Karpát (Gravitational slope deformations in the high mountains in the Slovak Carpathians). Sborník geologických věd, Řada HIG, ÚÚG, Praha 10:7–38 (in Slovak)

Nemčok A (1982) Zosuvy v slovenských Karpatoch (Landslides in the Slovak Carpathians). Veda, Bratislava, 319 p (in Slovak)

Nemčok A, Pašek J (1969) Deformácie horských svahov (Mountain slope deformations). Geologické práce, Správy, Dionyz Štúr State Geological Survey 50:5–28 (in Slovak)

Nemčok A, Pašek J, Rybář J (1974) Dělení svahových pohybů (Slope movement classification). Sborník geologických věd, Řada HIG, ÚÚG, Praha 11:77–97 (in Czech)

Nešvara J (1974) Deformace krasových masívů Západních Karpat (Deformation of karst massifs in the West Carpathians). Sborník geologických věd, Řada HIG, ÚÚG, Praha 11:177–192 (in Czech)

Ondrášik M, Wagner P, Ondrejka P (2006) Zosuvy na Slovensku sa prebúdzajú (Landslides in Slovakia are reactivating). Enviromagazín 11:13–14 (in Slovak)

Pánek T, Hradecký J, Smolková V, Šilhán K, Minár J, Zernitskaya V (2010) The largest prehistoric landslide in northwestern Slovakia: Chronological constrains of the Kykula long-runout landslide and related dammed lakes. Geomorphology 120(3–4):233–247

Pašek J (1958) Některé formy erose v povodí Laborce (Some forms of erosion in the Laborec River Basin). Ochrana přírody 4:91–96 (in Czech)

Paul CM (1868) Die nördliche Arva. Jahrbuch der k-k geologischen Reichsanstalt 18(2):27–46

Pišút P (2002) Channel evolution of the pre-channelized Danube river in Bratislava, Slovakia (1712–1886). Earth Surface Processes and Landforms 27(4):369–390

Pišút P (2006a) Changes in the Danube riverbed from Bratislava to Komárno in the period prior to its regulation for medium water (1886–1896). In: Mucha I, Lisický MJ (eds) Slovak–Hungarian environmental monitoring on the Danube. Groundwater Consulting, Bratislava, pp 186–190

Pišút P (2006b) Evolution of meandering Lower Morava River (West Slovakia) during the first half of 20th Century. Geomorphologia Slovaca 6(1):55–68

Pišút P (2008) Endangerment of the village Čunovo (Slovakia) by lateral erosion of the Danube river in the 18th century. Moravian Geographical Reports 16(4):33–44

Pišút P, Timár G (2007) History of the Ostrov Kopáč. In: Majzlan O (ed) Nature of Kopáč island. Fytoterapia Citizens Association, Bratislava, pp 7–30 (in Slovak)

Pišút P (2011) Dunajska povoden roku 1787 a Bratislava (The 1787 flood of the River Danube in Bratislava). Geograficky casopis, 63(1):87–109 (in Slovak)

Plesník P (1971) Horná hranica lesa vo Vysokých a Belanských Tatrách (Die obere Waldgrenze in der Hohen und Belauer Tatra). Slovak Academy of Sciences, Bratislava, 240 p (in Slovak with German summary)

Šimeková J, Martinčeková T, Abrahám P, Gejdoš T, Grenčíkova A, Grman D, Hrašna M, Jadroň D, Záthurecký A, Kotrčová E, Liščák P, Malgot J, Masný M, Mokrá M, Petro Ľ, Polaščinová E, Solčiansky R, Kopecký M, Žabková E, Wanieková D (2006) Atlas of slope stability in Slovak Republic 1:50,000. INGEO – IGHP, Žilina (in Slovak)

Slivovský M (1985) Podbiel – rock fall. In: A guide to the 25th nationwide geological congress of the Slovak geological society. Dionyz Štúr State Geological Institute, Bratislava, pp 123–127 (in Slovak)

Smetanová A (2009) Bright patches in Chernozems areas and their relationship to the relief. Geografický časopis 61(3):215–227

Smetanová A, Kožuch M, Čerňanský J (2009) The land use changes in the 20th century and their geomorphological implications in lowland agricultural area (Voderady, Trnavská tabuľa Table Plain, Slovakia). Geomorphologia Slovaca et Bohemica 9(2):57–63

Špůrek M (1972) Historical Catalogue of Slide Phenomena. Studia Geographica, vol 19. Brno, 178 p

Stankoviansky M (1992) Hodnotenie stavu prírodných a prírodno-antropogénnych morfolitosystémov (na príklade vybranej časti Bratislavy) (Evaluation of state of the natural and anthropogenic morpholithosystems on the example of selected part of Bratislava). Geografický časopis 44(2):174–187 (in Slovak)

Stankoviansky M (2003) Geomorfologická odozva environmentálnych zmien na území Myjavskej pahorkatiny (Geomorphic response to environmental changes in the territory of the Myjava Hill Land). Comenius University, Bratislava, 152 p (in Slovak)

Stankoviansky M (2008) Vplyv dlhodobého obrábania pôdy na vývoj reliéfu slovenských Karpát (Influence of long-term tillage on landform evolution in the Slovak Carpathians). Acta Geographica Universitatis Comenianae 50:95–116 (in Slovak)

Stankoviansky M, Barka I (2007) Geomorphic response to environmental changes in the Slovak Carpathians. Studia Geomorphologica Carpatho-Balcanica 41:5–28

Stankoviansky M, Midriak R (1998) The recent and present-day geomorphic processes in Slovak Carpathians. State of the art review. Studia Geomorphologica Carpatho-Balcanica 32:69–87

Stankoviansky M, Ondrčka J (2011) Current and historical gully erosion and accompanying muddy floods in Slovakia. Landscape Analysis (in press)

Stankoviansky M, Fulajtár E, Jambor P (2006) Slovakia. In: Boardman J, Poesen J (eds) Soil erosion in Europe. Wiley, Chichester, pp 117–138

Stankoviansky M, Minár J, Barka I, Bonk R, Trizna M (2010) Muddy floods in Slovakia. Land Degrad Dev 21(4):336–345

Streďanský J (1977) Kritické rýchlosti vetrov z hľadiska erodovateľnosti pôd na južnom Slovensku (Critical wind velocities from the viewpoint of the soil erodibility in the southern Slovakia). Slovak University of Agriculture, Nitra, 138 p (in Slovak)

Streďanský J (1993) Veterná erózia pôdy (Wind erosion of soil). Slovak University of Agriculture, Nitra, 66 p (in Slovak)

Szmańda JB, Lehotský M, Novotný J (2008) Sedimentological record of flood events from year 2002 and 2007 in the Danube River overbank deposits in Bratislava (Slovakia). Moravian Geogr Rep 16(4):25–32

Szolgay J, Náther B (1954) Niektoré poznatky zo štúdia splaveninového režimu Dunaja (Some data from the study of the Danube bedload regime). Vodohospodársky časopis 2(1):28–37 (in Slovak)

Turkota J (1972) Exkurzia po Silicko-Gombaseckej jaskynnej sústave (Excursion through the Silica-Gombasek cave system). In: Prírodné vedy – Geografia (Natural Sciences – Geography) 2. Universitas Comeniana, Facultas Peadagogica Tyrnaviensis, Bratislava, pp 27–55 (in Slovak)

Vlčko J, Petro Ľ, Bašková L, Polaščinová E (2002) Stabilita horninových masívov pod historickými objektami (Stability of the rock massifs under historic structures). Geologické práce, Správy, State Geological Institute Dionyz Štúr, Bratislava 106:89–96 (in Slovak)

Zachar D (1970) Erózia pôdy (Soil erosion). Vydavateľstvo SAV, Bratislava, 527 p (in Slovak)

Zachar D (1982) Soil erosion. Elsevier, Amsterdam, 545 p

Zacharov M (1984) Výskum geologicko-štruktúrnych pomerov a deformácií v Jasovskej jaskyni (Investigation of geological-structural conditions and deformations in the Jasovská jaskyňa cave). Slovenský kras 22:69–94

Záruba Q (1938) Sesuvy v Lyském průsmyku a jejich význam pro komunikační stavby (Landslides in the Lyský průsmyk Pass and their significance for transport structures). Technický obzor 46(1):1–6; (2):26–29; (3):37–39 (in Czech)

Záruba Q, Mencl V (1987) Sesuvy a zabezpečováni svahů (Landslides and their control). Academia, Praha, 338 p (in Czech)

Chapter 8
Recent Landform Evolution in the Ukrainian Carpathians

Ivan Kovalchuk, Yaroslav Kravchuk, Andriy Mykhnovych, and Olha Pylypovych

Abstract In the Ukrainian Carpathians the boundary between the forest and subalpine belts (at about 1,400–1,500 m) divides different systems of geomorphic processes. In the subalpine belt mass movements, gully erosion and nivation dominate and affect the slopes of extensive erosional surfaces. Avalanches occur on slopes with inclinations of 20–45° at elevations of 300–2,000 m. Neotectonic uplift and denudation rates are in approximate equilibrium (both 1.5–2.5 mm year^{-1}). Gullies and ravines are the most widespread form of erosion at lower elevations. There are three types of karst in the Ukrainian Carpathians: carbonate, sulfate, and salt karsts. In recent decades human impact on geomorphic processes has intensified and landslides, debris flows, and extreme floods have become more frequent. The alterations in the water and sediment discharges and channel morphology of rivers are good indicators of environmental changes.

Keywords Water erosion • Debris flows • Avalanches • Sheet erosion • Gully erosion • Floods • Channel evolution • Nivation • Karst • Human impact • Ukrainian Carpathians

---

I. Kovalchuk (✉)
Department of Geodesy and Cartography, Faculty of Land Use, National University of Nature Use and Life Sciences of Ukraine, Vasylkivs'ka Str. 17/114, 03040 Kyiv, Ukraine
e-mail: kovalchukip@ukr.net

Y. Kravchuk
Department of Geomorphology, Faculty of Geography, Ivan Franko National University of Lviv, Doroshenko Str. 41, 79000, Lviv, Ukraine
e-mail: 2andira@ukr.net

A. Mykhnovych • O. Pylypovych
Department of Applied Geography and Cartography, Faculty of Geography, Ivan Franko National University of Lviv, Doroshenko Str. 41, 79000 Lviv, Ukraine
e-mail: 2andira@ukr.net; olha@tdb.com.ua

8.1 The History of Geomorphological Investigations in the Ukrainian Carpathians

The history of research into recent geomorphological processes in the Ukrainian Carpathians is summarized in a range of papers (Krawczuk 1999, 2003; Kovalchuk 1997; Kovalchuk and Kravchuk 2001; Shushniak 2004). One of the first scientific papers dedicated to the Ukrainian Carpathians was written by the Polish geographer W. Pol (1807–1872) on the Tatra, Beskyds, Petros (Horhany), Polonynas, Chornohora, Pokuttia, and Bukovyna regions (Pol 1851). In the subsequent period research was conducted by the State Geological Institute of Austro-Hungary, established in Vienna in 1849.

The *Department of Geography at Lviv University* was founded in 1882. Its first head, A. Reman, was the author of a regional physico-geographical atlas of Poland and the Carpathians. H. Velychko (1863–1932) published the monograph "Relief of Polish–Ukrainian lands with special attention to the Carpathians" (1889) and was the first to regionalize the Carpathian Mountains. E. Romer (1871–1954) and S. Rudnyts'kyi (1877–1937) studied the origins and geomorphic development of mountains ridges, river valleys, and mountain glacial features (Shushniak 2004; Romer 1909). Rudnyts'kyi ascertained the existence of denudation surfaces in the Beskids and their absence in the Horhany, and revealed the different character of river valleys formed in different regions of the Carpathians. Based on expeditions, E. Romer carried out detailed morphostructural analyses of the Eastern Carpathians and elaborated a genetical-chronological approach to the Dnister river valley. After World War I experts of the Polish State Geological Institute and the Warsaw University were also involved in research. B. Sviders'kyi (1892–1943) is known for his monograph "Geomorphology of Czarnogora (Chornohora)" (1937), supplemented with a pioneering large-scale geomorphological map of Chornohora at a scale of 1:25,000. In 1922, during the 13th International Geological Congress in Bruxelles, the Carpathian Geological Association was founded. Its meetings promoted the study of Carpathian geomorphology (Shushniak 2004).

After World War II the government of the *Soviet* Union launched several scientific *expeditions* to the region. The best-known researchers were H. Alferyev and I. Hofstein from the State Carpathian Scientific Research Institute in Boryslav. Morphotectonical and geomorphological investigations were carried out by collaborators at the Lviv (P. Tsys', S. Subbotin), Chernivtsi (B. Ivanov, K. Herenchuk) and Moscow (A. Spiridonov) universities, the Moscow Geological Survey (A. Bogdanov, M. Muratov, G. Raskatov), the Lviv Branch of Ukrainian Geological Survey (M. Yermakov, M. Zhukov), the Leningrad Oil Institute (O. Vyalov) (Shushniak 2004; Professor… 2004).

In 1950 the *Department of Geomorphology* was founded at the Lviv University by Petro Tsys', a representative of regional geomorphological analysis. His achievements were the first detailed geomorphological regionalization and morphostructural analysis of the Ukrainian Carpathians (1956), the study of neotectonic (Tsys' 1956) movements, and the determination of the main phases of the geomorphic evolution of the Carpathians with special attention to the development of river valleys and denudational-accumulational surfaces. During next 20 years a set of papers were published on the problems of morphotectonics (P. Tsys', K. Herenchuk), river valleys development (M. Kozhurina), and denudation surfaces (Ya. Kravchuk) (Kravchuk 2001; Professor... 2004).

The Department of Geomorphology was engaged in regional geomorphology (initiated by P. Tsys'), engineering geomorphology (Ya. Kravchuk), experimental and environmental geomorphology (I. Kovalchuk), and paleogeography of the Pleistocene (A. Bohuts'kyi). Experimental and field investigations of recent geomorphologic processes of the Ukrainian Carpathians were carried out by O. Boliukh, Ya. Kravchuk, M. Kit, D. Stadnyts'kyi, I. Kovalchuk, M. Aizenberg, A. Oliferov, Ya. Khomyn, R. Slyvka, and V. Shushniak. The formation of denudational surfaces, the problems of geomorphologic regionalization and the morphosculpture of the Ukrainian Carpathians have been studied by Yaroslav Kravchuk. Engineering geomorphology and nature conservation planning appear in Ya. Kravchuk's, Yu. Zin'ko's, V. Brusak's, and N. Karpenko's works. Geomorphologic mapping projects have been carried out by Ya. Kravchuk, R. Hnatiuk, and V. Shushniak. Problems of paleogeography are studied by A. Bohuts'kyi and his students A. Yatsyshyn and R. Dmytruk. Relief genesis and the history of river valley development in Bukovyna at that period were also studied by geographers of Chernivtsi University (B. Ivanov, M. Kozhurina, M. Kunytsia, V. Lebediev, N. Krasuts'ka). In the 1980s and 1990s a new direction of geomorphological research, regional environmental-geomorphological analyses emerged (encouraged by Ivan Kovalchuk) and practiced by M. Symonovs'ka, L. Dubis, A. Mykhnovych, O. Pylypovych, N. Yedynak, S. Volos, and others.

An increased interest of the Lviv geomorphologists in the investigation of *recent exogenic processes* is observed since the mid-1960s (Tsys' 1964; Tsys' et al. 1968). In addition to field surveys and experiments, recently remote sensing methods are also used in geomorphological mapping, dynamic geomorphology, planning of landform protection and monitoring. In 1966–1969 a working group on recent exogenic processes in Carpathians was formed at the Lviv University under the coordination of Oleh Boliukh. In 1972 on the initiative of O. Boliukh and O. Skvarchevs'ka stationary and semistationary investigations of erosional processes began in the Precarpathians. The results were published in numerous monographs, dissertations, and papers (Skvarchevs'ka 1962; Boliukh et al. 1976; Holoyad et al. 1995; Kovalchuk 1997; Kravchuk 1999, 2005; Slyvka 2001). The results of observations on erosional and accumulative processes at *research stations* (runoff experiments, rainfall simulations) were presented in numerous papers (Boliukh et al. 1976; Tsys' et al. 1968; Kovalchuk et al. 1978; Professor... 2004; Problems... 2008; Relief... 2010). Karst processes, mudflows, and landslides have been investigated at stations over the past 30–40 years under the coordination of G. Rud'ko.

Since the 1990s, on the initiative and under coordination of I. Kovalchuk, river systems and their changes due to the *impact of* natural and *human factors* have been intensively studied (Kovalchuk 1997; Mykhnovych 1998; Kovalchuk et al. 2000, 2008; Bogacki et al. 2000; Kowalczuk and Mychnowicz 2009). Hydrogeomorphological hazards (Kovalchuk et al. 2008; Romashchenko and Savchuk 2002; Lymans'ka 2001) and various types of exogenous geomorphic processes are further investigated (Kovalchuk 1997; Slyvka 1994; Holoyad et al. 1995; Kravchuk 1999, 2005; Rud'ko and Kravchuk 2002; Kovalchuk and Mykhnovych 2004; Dubis et al. 2006), also at the Kyiv (Rud'ko et al. 1997; Obodovs'kyi 2001), Chernivtsi (Yushchenko 2009), Tavrian (Oliferov 2007), Rivne (Budz and Kovalchuk 2008), and Transcarpathian universities (Habchak 2010).

8.2 Geomorphological Regions

Most of the scientists mention the marked *vertical and horizontal zonation* of the Ukrainian Carpathians. Typical geomorphological surfaces (Tsys' 1968) are the denudation surfaces of summit levels, gently denuded upper terraced slopes at medium levels, and clearly discernible terraces at lower elevations of the mountains. The geomorphologic regions distinguished are the Precarpathian and Transcarpathian plains, Lump (Skybovi in Ukrainian), the Dividing Verkhovyna, Polonyna-Chornohora, Volcanic Carpathians, and Marmarosh Massif (Fig. 8.1, Table 8.1).

Fig. 8.1 Geomorphological units of the Ukrainian Carpathians

8 Recent Landform Evolution in the Ukrainian Carpathians

Table 8.1 Geomorphological units, their genetic types, and main geomorphic processes in the Ukrainian Carpathians

Geomorph. province	Geomorph. region	Geomorph. subregion	Geomorphological district	Genetic type of relief	Exogenic geomorphic processes
Eastern Carpathians (subprovince Forested or Ukrainian Carpathians)	Precarpathian premountain highland (1)	Pre-Beskid (Northwestern) Precarpathians (1.1)	Sian Plain (1.1.1)	Moraine-outwash-alluvial plain	Relict landslides, karst, activation due to human activity
			Sian-Dnister Highland (1.1.2)	Hills with glacial and fluvioglacial landforms	Erosion and landslides
			Upper Dnister Plain (1.1.3)	Alluvial	Bank erosion, floods
			Stryvihor Highland (1.1.4)	Erosional-accumulational with glacial landforms	Sheet and bank erosion
			Drohobych Highland (1.1.5)	Erosional-accumulational	Sheet and gully erosion
			Morshyn Highland (1.1.6)	Erosional-accumulational	Sheet and gully erosion
				Accumulational plain	Bank erosion, karst
		Pre-Horhany (Central) Precarpathians (1.2)	Zalissia (Svicha) Highland (1.2.1)	Erosional-accumulational	Sheet and gully erosion
			Kalush Valley (1.2.2)	Erosional-accumulational	Sheet and gully erosion
			Voinyliv Highland (1.2.3)	Structural-erosional	Erosion
			Lukva Highland (1.2.4)	Accumulational plain	Bank erosion
			Maidan Low Mountains (1.2.5)	Denudational	Sheet and gully erosion, landslides, debris flows
			Bystrytsia Valley (1.2.6)		
			Bystrytsia Highland (1.2.7)		
			Prut–Bystrytsia Highland (1.2.8)	Erosional-accumulational	Sheet and gully erosion, landslides.

(continued)

Table 8.1 (continued)

Geomorph. province	Geomorph. region	Geomorph. subregion	Geomorphological district	Genetic type of relief	Exogenic geomorphic processes
		Pokuttia-Bukovyna (Southeastern) Precarpathians (1.3)	Prut–Liuts'ka Highland (1.2.9)	Erosional-accumulational	Erosion, landslides
			Sloboda–Runhur Low Mountains (1.2.10)	Structural-erosional	Sheet and gully erosion, landslides, defluction
			Kolomyia–Chernivtsi Plain (1.3.1)	Alluvial plain	Bank erosion
			Pokuttia Highland (1.3.2)	Sculptural	Erosion, landslides
			Bahnens'ka Plain (1.3.3)	Alluvial plain	Sheet and gully erosion, bog processes
			Brusnyts'ka Highland (1.3.4)	Structural-erosional	Erosion, landslides
			Chernivtsi Highland (1.3.5)	Structural-sculptural	Erosion, landslides
			Siret Highland (1.3.6)	Erosional-accumulational	Sheet and gully erosion, landslides
			Krasnoil highland (1.3.7)	Erosional-accumulational	Sheet and gully erosion, landslides
			Tarashan Highland (1.3.8)	Structural-erosional	Erosion, landslides
			Hertsaiv Highland (1.3.9)	Structural-erosional-accumulational	Erosion, landslides

Lump Carpathians (2)	Beskids (2.1)	Upper Dnister Beskids (2.1.1)	Anticlinal low mountains	Erosion, mudflows, landslides
		Skole Beskids (2.1.2)	Monoclinal-lump low mountains	Erosion, mudflows, debris flows, landslides
	Horhany Mountains (2.2)	Low mountains of the Lump Horhany (2.2.1)	Monoclinal-lump low mountains	Erosion, mudflows, landslides, debris flows
		Middle mountains of the Lump Horhany (2.2.2)	Monoclinal-lump middle mountains	Erosion, mudflows, landslides, debris flows
	Pokuttia–Bukovyna Carpathians (2.3)	Pokuttia–Bukovyna Low Mountains (2.3.1)	Structural-denudational low mountains	Erosion, landslides
		Pokuttia–Bukovyna Middle Mountains (2.3.2)	Anticlinal denudational middle mountains	Erosion, landslides, debris flows
Dividing-Verkhovyna Carpathians (3)		Stryi–Sian Verkhovyna (3.0.1)	Structural-denudational low mountains	Mudflows, landslides, structural-tectonic landslides
		Volovets'–Mizhhirya Verkhovyna (3.0.2)	Structural-denudational low mountains	Mudflows, landslides, structural-tectonic landslides
		Dividing Verkhovyna (3.0.3)	Structural-denudational low mountains	Mudflows, landslides
		Deviding Horhany (3.0.4)	Anticlinal-lump middle mountains	Stone fields, debris flows, mudflows
		Vorokhta–Putyla Low Mountains (3.0.5)	Relict longitudinal valley	Erosion, landslides
		Yasinia Valley (3.0.6)	Terraced erosional low mountains	Erosion, landslides

(continued)

Table 8.1 (continued)

Geomorph. province	Geomorph. region	Geomorph. subregion	Geomorphological district	Genetic type of relief	Exogenic geomorphic processes
	Polonyna-Chornohora Carpathians (4)	Polonyna Ridge (4.1)	Polonyna Rivna (4.1.1)	Lump middle mountains with denudational surfaces	Erosion, landslides, mudflows, debris flows
			Polonyna Borzhava (4.1.2)	Lump middle mountains with denudational surfaces	Erosion, debris flows, landslides, mudflows
			Polonyna Krasna (4.1.3)	Lump middle mountains with denudational surfaces	Erosion, debris flows, landslides, mudflows
		Svydivets'–Chornohora massif (4.2)	Svydivets' Massif (4.2.1)	Lump Alpine and middle mountains	Erosion, debris flows, avalanches
			Chornohora Massif (4.2.2)	Lump Alpine and middle mountains	Mudflows, avalanches
			Hryniava and Losova (4.2.3)	Lump middle mountains with denudational surfaces	Erosion, gravitational, mudflows
	Marmarosh Crystalline Massif (5)		Rakhiv Mountains (5.0.1)	Vaulted lump Alpine and middle mountains	Gravitational, mudflows, avalanches
			Chyvchyny (5.0.2)	Vaulted lump Alpine and middle mountains	Gravitational, mudflows, avalanches
	Volcanic Carpathians (6)		Makovytsia–Syniak Massif (6.0.1)	Volcanic-erosional	Erosion, debris flows
			Velykyi Dil Massif (6.0.2)	Volcanic-erosional	Erosion, debris-flows
			Tupyi-Oash Massif (6.0.3)	Volcanic-erosional	Erosion, debris-flows

		Berezne–Lipchans'ka (Turyans'ka) Valley (6.0.4)	Volcanic-erosional	Erosion, landslides
		Transcarpathian Hills (6.0.5)	Pliocene-Pleistocene denudational-accumulational	Erosion, landslides
		Irshava Valley (6.0.6)	Intermountains denudational-alluvial valley	Bank erosion, floods
Trans-carpathian Plain (7)	Chop-Mukachevo plain (7.1)	Tysa Alluvial Plain (7.1.1)	Alluvial valley	Bank erosion. Floods
		Berehove Hills (7.1.2)	Denudational-accumulational plain with volcanic remains	Sheet and gully erosion
	Upper-Tysa (Solotvyno) valley (7.2)		Denudational-accumulational low mountains with terrace complex and salt hill remnants	Erosion, salt karstification, landslides

The dividing line between the *forest* and subalpine *belt*s runs at about 1,400–1,500 m elevation and separates different systems of geomorphic processes and assemblages of landforms. The summit surfaces (at 1,400–2,000 m) and their slopes are in the *subalpine belt* (the Polonyna-Chornohora Carpathians, Lump, Dividing (Internal) Horhany Mountains and in the Marmarosh Massif), where processes of slow debris movements, gravitational, nival processes, and avalanches dominate. Here debris of different fractions (1–300 cm diameter) moves slowly downslope at 2–4 mm year^{-1} average rate. Stone streams are 1–6 m wide and, tens to hundreds of meters long.

In the strongly dissected middle to high mountains mass movements are accompanied by sheet wash and gully erosion. Most intensive gravitational and erosion processes are observed on steep slopes (with inclinations over 12°) and on precipices. Between elevations from 900–1,000 m to 1,400–1,500 m, on the wide, steep (5–12°) to very steep (12–25°), strongly dissected slopes with a discontinuous vegetation cover, slow movements of water saturated soils with various amounts of debris are observed.

In terraced and non-terraced *river valleys* (at elevations from 500–600 m to 900–1,000 m) a variety of geomorphic processes are active, but intensive river channel erosion and accumulation as well as mudflow accumulation are predominant.

Neotectonic movements also influence geomorphic processes. Measurements at research stations testify that the denudation rate in the Drohobych Highland (Precarpathians) is 0.35 mm year^{-1}. If other processes (landslides, mudflows, gully erosion etc) are also taken into account, total denudation amounts to 1.5–3 mm year^{-1}. The rate of total denudation in the Ukrainian Carpathians calculated by I. Hofstein (1995), based on measured parameters of sediment delivery from Carpathian river basins, is 2.3 mm year^{-1}. According to V. Somov and I. Rakhimova (1983) tectonic uplift in the Ukrainian Carpathians and Precarpathians is 1.5–2.5 mm year^{-1}. Thus, the comparison of denudation and tectonic movements points to a self-regulating dynamic equilibrium.

8.3 Climate and Human Impact

The main *climatic characteristics* that control geomorphic processes are the following:

1. Precipitation is 1,400–1,700 mm year^{-1} in the mountains and 700–1,000 mm year^{-1} in the Precarpathian and Transcarpathian plains.
2. Frequent heavy rainfall events over extensive areas constitute ca 40% of the monthly amounts.
3. During rainfalls above 70 mm day^{-1}, which make up 48% of all heavy rains, the infiltration capacity of brown forest soils is exceeded and intensive erosion, floods, mudflows, landslides, damaging agricultural fields, roads, bridges, dykes and dams, ensue. Heavy rains of 200–300 mm day^{-1} intensity can continue for even 3 days.

4. Rainfall intensity can reach up to 7 mm min^{-1} for 10–30 min on slopes of 20–50°. The surface runoff coefficient from catchments of 100–200 km^2 area ranges from 2.5 to 3.1 m^3 s^{-1} km^{-2} (Kovalchuk 1997), and from catchments of 300–500 km^2 it is 1–2 m^3 s^{-1} km^{-2} in the mountains and 0.5–0.2 m^3 s^{-1} km^{-2} in the plains.
5. Strong winds with speed over 25 m s^{-1} as well as squalls and hurricanes with 40 m s^{-1} are observed almost every year and cause tree uprooting, snowdrift and avalanches.
6. Strong snowfalls (up to 40 cm deep snow per event) are also quite common (with 60–80% probability). Maximal snow depth is 3–4 m. Fast thawing causes extreme winter and spring floods.
7. Total yearly runoff equals 18.3 km^3, while the average specific runoff 0.008–0.038 m^3 s^{-1} km^{-2}, with maximum values observed in the Transcarpathians.

After rainfalls of at least 20 mm day^{-1} intensity floods are formed, and over 100 mm day^{-1} extremely high floods with water levels rising 2–4 m are generated. In mountains the width of the flooded zone ranges from 20–60 (for small rivers) to 120–500 m (for medium rivers), and 600 m–2.5 km (for the medium and large rivers in the foothills of the Pre- and Transcarpathians), while in the plains it is 2.5–5 km or even more. Stream current velocity is 2–3 m s^{-1} in the foothills, 4–5 m s^{-1} in the mountains, and up to 10 m s^{-1} along small rivers.

In recent decades human impact upon the environment, including topography and geomorphic processes, became so intense that there are significant disturbances even in the forested Ukrainian Carpathians. The following *consequences of human activities* significantly influence the distribution and operation of geomorphic processes:

1. Decreasing forest cover in the Carpathians from 75–90% to 40.2% (over the past 200 years), caused by intensive and large-scale clearance for the timber industry. It is generally accepted that a forest cover area below 35% is critical from a geomorphological aspect. In the 1960s and 1970s this critical threshold was already observed in many Carpathian subregions, particularly in foothills and low mountains.
2. The 200–300 m lowering of the upper timberline due to complex reasons, cattle grazing being the principal one.
3. Destroying crooked forests that can retain snow and water on the slopes and ridges and slow down the thawing process.
4. Changes in the species structure of forests. For instance, more than 100 thousands hectares are occupied by spruce trees *(Picea abies)*, but this kind of forest cannot efficiently retain water on slopes (Hensiruk 1992). Also the changes of age structure (more than 70% of young and medium forests) are of significance. As it is known old forests can slow down runoff from a 70–100 mm rainfall event and conduct rainwater to the groundwater. Unfortunately, forests of 80 years are preserved only over 16–20% of the area.
5. Imperfect technologies and intensive forest clearance (800 000 m^3 wood per year) cause intensive erosion (removal of up to 300 t ha^{-1} soil). During the last 40 years 20 million m^3 of timber was extracted.

6. Arable land (>20% of total area) is expanding over steep slopes.
7. Heavy exploitation of mountain pastures (local name "polonyna") with a total area about 100,000 ha, deteriorating their ecological state.
8. Development of transport (wood transportation, pipelines, electricity transmission lines etc). The total length of the pipelines within the Transcarpathian region is about 2,500 km.
9. Insufficient maintenance of hydroengineering structures. In the Transcarpathian region alone there are 646 dykes and 245 km of embankments, most of them in need of restoration. There are 378 km of dykes in the Upper Dnister river basin, most of them built in the 1960s and 1970s (65% of them need reconstruction).

Geomorphic processes along the Carpathian sections of the Dnister, Prut, and Tysa Rivers greatly intensified between 1947 and 1970. With the renewal of forests some stabilization followed after 1970. But since 1990s exogenic processes have intensified again due to deforestation, causing destructive floods (in 1997, 1998, 2001, 2004, 2008, and 2010).

8.4 Water Erosion

8.4.1 Sheet Erosion

Erosional processes are widespread almost in all geomorphological regions of the Ukrainian Carpathians, but the largest areas affected by them are located in the economically most developed regions of Carpathians. The main factors of sheet erosion are slope characteristics and structural-lithological features (the occurrence of loamy sediments). An intensification of sheet erosion is also typical for the river valleys with arable land and settlements.

The findings of investigations carried out at research stations of the Precarpathian plain in 1970s (Boliukh et al. 1976) and in 1980s in the Irshava Valley of the Volcanic Carpathians (Khomyn and Bilaniuk 2010) allow determination of *erosion rates under various crops*: for grassland: 0.003 mm year^{-1}; for wheat and rye: 0.03 mm year^{-1}; for weeds: 0.4 mm year^{-1} and for summer crops 1.7 mm year^{-1}. Taking into account the harvest areas of crops, the overall average intensity of surface erosion in the Precarpathians is calculated to be 0.35 mm year^{-1}. According to I. Kovalchuk (1997), the average rate of soil removal in Precarpathians ranges from 8.7 to 45.0 t ha^{-1} year^{-1}, i.e. a soil layer of 0.86–3.2 mm year^{-1} (4–16 times above the acceptable maximum value). The rate of chemical denudation was determined for spring thawing as 13.9 kg ha^{-1}. The ratio between solid material removal and chemical denudation for 1 km^2 is 6–1.

The *erosion of banks* of ravines, streams and meandering rivers is widely observed in the Ukrainian Carpathians. The highest rates are recorded during extreme floods and flash floods in flysch (argillites and siltstones) areas and alluvial and deluvial Quartenary sediments, typically at meanders of different size and shape along the rivers of the Beskids (Stryvihor, Dnister, Stryi), which have formed not later than Pliocene–Pleistocene. Most intensive erosion processes in the Lump Carpathians are observed usually downstream 'lumps' (ridges). Heavily eroded first, second, and third terraces are found in river valleys of the Beskids Mountains. In the Horhany Mountains major erosion can be observed not only within the transversal main river valleys, but also along the longitudinal tributaries (e.g., the Limnytsia) in synclinal valleys of unconsolidated sediments. In the Polonyna–Chornohora Carpathians and the Marmarosh Massif the distribution and intensity of erosion are well correlated with vertical belts: channel erosion and bank undercutting are common at lower elevations (500–900 m). The destruction of the lower terraces in narrow V-shaped valleys is observed along all the major rivers (the Tysa, Teresva, Tereblia, Rika, Latorytsa and Uzh).

8.4.2 Gully and Ravine Erosion

Gullies and ravines are the most widespread form of erosion in the Precarpathian and Transcarpathian regions. Bank, slope, top and bottom gullies are all present. *Slope ravines* (fixed and active) are spread on the slopes of the Maidan and Sloboda-Runhuns'ka low mountains, and on the Stryvihor–Dnister, Dnister–Bystrytsia Pidbuz'ka, Bystrytsia Solotvyns'ka–Bystrytsia Nadvirnians'ka, Liuchka Pistynka–Rybnytsia, and Prut–Siret interfluves. Ravines develop in loams and clays, often covered by thick layers (5 m and more) of eluvial-deluvial sediments. Their depths vary between 2 and 6 m. *Bank ravines* are much deeper (10–12 m and more) on the interfluve between the Prut and Bystrytsia Nadvirnians'ka Rivers. Within the Outer Precarpathians (near the Dnister River valley) gully networks develop in thick layers of Badenian and Sarmatian clays.

In the Transcarpathians gully and ravine erosion is observed in foothills as well as in the Solotvyno (Upper Tysa) valley. In contrast to the Precarpathians, here ravines develop in alluvial sediments of tens of meters thickness as usual. The densest gully network (0.4–0.6 km km^{-2}) is observed on the interfluves between the Tysa and Uzh Rivers in the Transcarpathian foothills and in the Solotvyno valley.

In the Lump and Dividing Verkhovyna Carpathians most of the gullies are located in the Menilitic and Krosno sediment complexes. Bank ravines are related to the margins of medium and high terraces with relatively thick layers of alluvial sediments. In the Beskids both fixed and actively developing ravines are located between the Vyrva–Stryvihor, Stryvihor–Dnister Rivers and in the Dnister and Trukhaniv valleys.

Gully and ravine erosion in the Horhany section of the Lump Carpathians is mostly located in the outer low mountain ranges: in the Vyhoda, Pniv-Nadvirna, and Deliatyn valleys, on interfluves between the rivers Bystrytsia Solotvyns'ka–Maniavka–Bystrytsia Nadvirnians'ka, on the right side of the Prut river valley.

Somewhere the slope ravines (fixed and active) can reach significant *dimensions*, particularly on interfluves, like the Bystrytsia Solotvyns'ka–Maniavka near the villages Krychka and Maniavka in clays with up to 1.5 m thickness. Gully dimensions vary between quite wide limits: for length from 70 to 150 m, for width from 5 to 15 m, and for depth from 3 to 12 m.

In the Dividing Verkhovyna Carpathians gully and ravine erosion is common in the San river valley, between the rivers Dnister and Yablun'ka, near the village Vovche on the slopes of the Boryn'ka catchment. Slope ravines are developed in the basins of the Vicha, Rika, Studenyi, Holiatunka and Pryslip Rivers. Near the village Nyzhni Vorota gullies incise into the floors of old wide, already fixed, ravines (balkas) (Slyvka 2001).

8.5 Mass Movements

8.5.1 Debris Flows

Debris flows in the Ukrainian Carpathians fall into two locally identified types: *"water-and-stone flows"* and *"debris floods."* The first type is characterized by high sediment loads (to 30% of the volume), fine debris <10%, thalweg inclination >0.10, and bulk density of 1.15–1.55 g cm^{-3} (Rud'ko and Kravchuk 2002). This type of debris flow is rather rare, one event happens in 25–50 years. Debris floods are more frequent and contain 10–20% sediment load.

The most favorable regions for debris flows in the Ukrainian Carpathians are the Lump and the Polonyna-Chornohora Carpathians and the Marmarosh Massif. Especially intensive flows are characteristic for the boundaries between geomorphological regions: for instance, between the Lump and Dividing Verkhovyna Carpathians (Krosno Zone), between the Dividing Verkhovyna and Polonyna-Chornohora Carpathians, and between the Chornohora and Marmarosh Massif. In the Lump Carpathians they occur in almost all catchments. In the Beskids and Pokuttia–Bukovyna Carpathians debris flows are often associated with landslides and gullies. Numerous flows are observed in the Stryi and Opir catchments. Mud-and-stone flows periodically take place in the catchments of the Cheremosh and Siret Rivers as well as along the Pistynka River headwaters. Most of the debris flows occur in Horhany Mountains: in the Prut river basin along the streams on the Mahura slopes and along numerous tributaries of the Zhenets', Zhonka and Prut Rivers, induced by intensive sheet erosion and landslides. Debris flows associated with gravitational processes were described from several catchments of the Bystrytsia Solotvyns'ka basin. Large amounts of stone-loamy debris come into the main riverbed and induce debris floods as at the Limnytsia headwaters and tributaries. In the Polonyna-Chornohora Carpathians debris flows of mixed type predominate and their calculated bulk density ranges from 1,120–1,190 kg m^{-3} (along the Lems'kyi River in the Bila Tysa catchment) to 1,800–1,900 kg m^{-3} (in the Teresva and Apshytsia basins, where mean slope inclination is 0.12–0.25) (Rud'ko and Kravchuk 2002).

Debris flows in the Ukrainian Carpathians originate from the following types of source area: sliding, deluvial-debris, rockfall, mudstream, alluvial, and snow-slide centers. According to investigations at the Prutets' Chemehyvs'kyi research station, debris accumulation takes 5–7 years. During this time 10–15 thousands m^3 of the sediments accumulate in riverbeds of 3–5 km length. *Sliding centers* develop along the contact area (Krosno Zone) between the Lump and Polonyna-Chornohora Carpathians (inner flysch anticlinorium). *Deluvial-debris centers* dominate on the slopes of the Lump and Dividing Horhany, Svydivets' massifs and in the middle mountains of Pokuttia and Bukovyna. *Rockfall centers* of coarse debris and limited extension are related to high and steep slopes composed of massive sandstones, limestones or crystalline rocks. *Mudstream centers* are located in areas with shallow groundwater table and also of limited dimensions and low (0.5–0.7 m) sediment thickness. *Alluvial centers* are observed in the Mlyns'kyi catchment (Cheremosh river basin) with a total of 60 thousands m^3 of alluvium (along the 6.5 km section of the Mlyns'kyi stream, 0.94 m^3 of alluvium per 1 m of river length). *Snow-slide centers* mostly occur in the Chornohora, Marmarosh, and Svydovets' massifs. This type of mudflow is a secondary in Ukrainian Carpathians. The smallest and shortest flow centers are of sliding type, while the longest belong to the alluvial type.

The floods in 1998 and 1999 testified that critical thresholds for mudflows are 100–200 mm day^{-1} total rainfall amount and 6–12 mm h^{-1} intensity. The natural conditions are reinforced by the complete deforestation of steep slopes, log transport destroying vegetation cover and the generated soil erosion. Field investigations in the Krosno Zone allowed one to distinguish between *debris flows along thalwegs* and *blockslides* over soils due to supersaturation or man-made undercutting of footslopes. The first type is prevalent in the catchments of the Bradulovets', Sukhar, Volovets', and Ozerianka Rivers (slope angles of 40–50°, usually deforested). Debris flows along thalwegs occur at slope angles ranging from 35–50° in headwaters to 25–30° in transit zones. The second type of debris flow is characteristic for the Teresva basin on clayey flysch rocks. Dammed debris flows result from the impounding of watercourses by landslides (as in the Mokrianka and Brusturianka catchments). Recently 24 active debris flows have been mapped in Prut River basin, 15 in the Bilyi Cheremosh, and 11 in the Bystrytsia Nadvirnians'ka basins. Source basins are usually of circular form, the percentage of sediments amounts to 10–35% or even 70%, flow rate is 3–4 m s^{-1}, duration of the event is 1.5–2 h, and flow depth amounts to 2.5–3 m (Holoyad et al. 1995).

8.5.2 Landslides

Landslides are concentrated in the Precarpathians (especially the Pokuttia and Bukovyna regions), Dividing Verkhovyna Carpathians and along boundaries between the Dividing Verkhovyna with the Lump and Polonyna-Chornohora Carpathians and between the Lump Carpathians and Precarpathians. In the Pokuttia and Bukovyna sections over 60–70% of the area hazardous landslide occur. Landslides on the right side of the Prut catchment are usually of block type (Rud'ko et al. 1997). Around Chernivtsi city about 20 ha are occupied by slides. Disasters have been observed here from 1962 to present. In the foothill zone of the Pokuttia-Bukovyna most active *landslides of stream and block type* are found between the rivers Pistynka–Rybnytsia–Cheremosh as well as between the Siret and Malyi Siret. Stream landslides reactivate unexpectedly during spring thawing. Movements occur over viscous clays as slip planes. In the headwaters of Komarivtsi stream (right-bank tributary of the Siret) that landslides were stable between 1991 and 1996 was observed (Rud'ko et al. 1997) and explained by poor water saturation of Badenian clays.

In the Horhany (central) Precarpathians two zones of landslides can be defined. In the foothills landslides affect terraces margins, ravine banks, autochtonous slopes of clayey flysch, and clay molasses. On the right-side of the Dnister River basin, landslides develop in the thick layers of loess-like alluvial loams over Badenian and Sarmatian clays and Cretaceous rocks. Major landslides are also observed on the both sides of the Prut catchment. Near the mouth of the Loyovets' Stream a landslide scarp is more than 300 m long and 100 m deep. There are extensive active landslides on the right side of the Prut basin; one endangers a protected yew woodland. Between the Prut and Bystrytsia Nadvirnians'ka Rivers landslides affect ravine walls of 10–12 m height and are 25 m deep, 10–30 m wide, with failure scarps 5–20 m high. In the headwaters of all streams circular landslide headscarps 300 m across are widespread. Their walls and bottoms are often cut by small gullies with many small slips on their sides. There are also numerous man-induced landslides in the region.

The landslides in the Dnister river valley usually occur on old fixed and active ravine slopes. The scarps are 6–10 m high and 50–100 m wide. In the Pokuttia-Bukovyna section of the Lump Carpathians a continuous series of landslides follow the boundary with the Vorokhta-Putyla low mountains (the Krosno Zone). Extensive landslide tongues are mostly stabilized and grassed. In thin-bedded flysch shallow landslides can be seen along the Cheremosh River and Sukhyi Stream. The landslides developed in Eocene shales are deeper with steep scarps and large tongues and troughs (in the Rushoru and Richka valleys and on the southern slopes of the Lel'kiv and Hromova Mountains).

In the Lump Horhany stabilized landslides are widespread mostly in the contact belt with the Precarpathians. Large old fossil landslide bodies are located between the rivers Bystrytsia Nadvirnians'ka and Solotvyns'ka. On the slopes of the Mala and Velyka Hyha a fossil landslide with failure front up to 25 m high can be seen. The combination of the relict and recent landslides is typical for this area. The landslides are 15–17 m long with a failure front of up to 10 m height. Landslides in clays of the Maniavka catchment show clear step forms with many small mounds. In the headwaters of small rivers active circuses are numerous. Unstable slopes on 2 m thick deluvial loam near Krychka village show complicated landslide features.

In the western Dividing Verkhovyna large areas with fossil landslide slopes and locally active forms are characteristic (Rud'ko et al. 1997). There are two active stream-type landslides here: Tykhyi and Bystryi. Tykhyi has an area of 718,730 m^2 and 10–15 m thickness. The paradoxically slow-moving Bystryi ("quick") landslide is monitored since 1985. In the Polonyna near the Tereblia water reservoir fossil landslides have an estimated total volume between 100,000 and 200,000 m³ and affect weathered argillites in fixed ravines. Movements are intensified by economic activities. In the western Polonyna Ridge older stabilized landslides in thick debris show rates of 5–10 cm per month.

In the Transcarpathian region landslides are very unevenly distributed. They are concentrated in the Solotvyno depression (or Molass Carpathians) with characteristic inversion morphostructure, on slopes above the Tereblia and Teresva river valleys and their tributaries. Until 1998 16 landslides had been mapped here but after the floods of 1998, 1999, and 2001 33 additional landslides activated. In the Teresva River basin the total number of registered landslides is 285 (38% of the total in the Transcarpathian region). Slope stability along the Teresva River is reduced by road constructions on the river terraces and by riverbank erosion. After the major floods block landslides of step-like topography reactivated along the Mokrianka, Yalivets', and Brusturanka Rivers. Maximum landslide density reaches 1.9 landslides per km² at Ternovo, an area with susceptible rocks and high seismicity. Landslides are also common in the Tereblia basin with densities locally amounting to 0.7–1.2 units per km². The overall area affected by the landslides is 130.15 km² in the Teresva River basin, 41.44 km² in the Uzh River basin, 32.66 km² in the Rika River basin, and 30.63 km² in the Latorytsia River basin.

The landslides reactivated after flood events fall into two groups: (1) those formed on Neogene horizontal clay layers with intercalated sandstones and limestones; (2) those formed in clay-flysch rocks. Generally landslides of lamellar and structural-lamellar types are predominant (their overall volume is 1.5–3 million m³) and most hazardous for roads.

In the Transcarpathian foothills minor landslides with clear seasonal activity form in weathered volcanites (as around Uzhgorod city). Clearcutting on slopes, wood transportation, and construction significantly contribute to landslide generation, particularly in narrow river valleys between high walls with steep (30–50°) slopes.

8.5.3 Other Gravitational Processes

Other gravitational processes mostly affect the steep slopes of the Polonyna-Chornohora Lump Carpathians, Marmarosh Massif, and the Volcanic and Dividing Verkhovyna Carpathians. In the Precarpathians they are widespread on the left half of the Bystrytsia Nadvirnians'ka catchment on very steep slopes (35–40°) in argillite and sandstone debris of 3–15 cm diameter but larger (25 cm) blocks of sandstones also occur. Debris is accumulated on floodplains or reach the riverbed.

In the Lump Carpathians gravitational processes are active on mountain ridges of Paleocene and Eocene sandstones as well as on upper Cretaceous sandstone-argillite-siltstone sequences. On the gentler southwestern slopes of the Lump Carpathians barren summit *felsenmeers* gradually change into *block streams* and accumulation features at footslopes. On the steeper northeastern slopes rockfalls and felsenmeers often fixed by mountain pines and by firs and birches at lower elevations. In the Lump Horhany within the upper belt of the northeastern slopes *rockfalls* are widespread in the crooked forest zone of the Hrofa–Moloda–Popadia, Mahura, Koza, and Hora Massifs (Kravchuk 2005). The height of steep slopes (>45°) can locally reach 100 m. Most of them are not very active, some of them were active more than 100 years ago. The Dovbushanka rockfall on 27 December 1897 to the northeast of Dovbushanka mountain (1,755 m) was 500 m long and created a 150 m high scarp and a stone stream of 200–250 m width moving down the Fedotsyl River valley.

Numerous minor *rock* and *debris falls* in the Prut and Bystrytsia Nadvirnians'ka River valleys and its tributaries threaten railways and roads. Further rockfalls are known from the Bystrytsia Solotvyns'ka, Limnytsia, and Svicha River basins. Rock and debris falls are often associated with landslips, as in the Opir and Stryi River valleys. Most falls involve fine (0.03–0.15 m) material and are related to lower-rhythmical flysch complexes. Those of sandstone debris have grain sizes of 0.3–0.5 m.

8.6 Fluvial Erosion

8.6.1 River Network

The rivers of the Ukrainian Carpathians belong to the Dnister, Tysa, Siret, and Prut systems. The Dnister River and its mountain tributaries flow in northeastern direction, the Tysa with tributaries to the southwest, and the Prut and Siret in northeastern and southeastern directions. There are about 460 rivers in the Ukrainian Carpathians (40% of them belongs to the Dnister system) but only eight rivers are longer than 100 km (four in the Dnister basin, three in the Tysa basin, and the Prut River). The upper reaches of Carpathian rivers are mostly narrow, deep, with steep slopes, limited discharges, and high current velocity. At 700–1,400 m elevation channel *gradient*s are between 50 and 100 m km^{-1} with incision depth exceeding 700 m. Riverbeds contain large boulders. In mountain forelands river gradient decreases to 10–20 m km^{-1} with incision depths of 50–150 m and deposition increases. Valley cross-sections are trapezoid with braided channels and *current velocities* from 0.7 to 5.0 m s^{-1} (locally for small streams 7–10 m s^{-1}) during flood events. Reduced stream velocity involves intensive accumulation and island formation. Many islands are stabilized by bushes. The alternation of *canyon* (across mountain ridges, with waterfalls) and *broad valley* types (between mountain ridges) are characteristic.

Steep slopes, high precipitation (above 1,000 mm), low evaporation and groundwater levels result in high *drainage density* (2.0–2.8 km km^{-2} for the Upper Cheremosh; 2.0–2.3 km km^{-2} for the Upper Tysa; and 1.9–2.7 km km^{-2} for the Upper Svicha and Limnytsia Rivers). In the Transcarpathians density is 1.8 km km^{-2} and in the Precarpathians 0.9–1.3 km km^{-2}.

Average specific *sediment delivery* in the Ukrainian Carpathians ranges from also original data are possible from Resources of the surface waters of USSR.

Table 8.2 Sediment and water discharges of rivers in the Ukrainian Carpathians (Calculated by I. Kovalchuk from hydrometeorological data)

Period number	Period of measurements, years	Average specific sediment delivery modulus, t km^{-2} y^{-1}	Sediment discharge change coefficient, Mn/M1	Average runoff, mm	Runoff change coefficient, Hn/H1	Ratio between sediment and runoff change coefficients, Hn/M1
Tysa River (at Rakhiv)						
I	1947–1962	58.0	1.00	702.0	1.00	12.10
II	1963–1970	180.9	3.12	778.9	1.11	4.31
III	1971–1975	117.0	2.02	659.8	0.94	5.64
IV	1976–1980	146.0	2.52	803.4	1.14	5.50
V	1981–1985	58.8	1.01	754.0	1.07	12.82
VI	1986–1988	64.7	1.12	690.3	0.98	10.67
Dnister (at Sambir)						
I	1946–1962	62.3	1.00	249.6	1.00	4.01
II	1963–1970	255.8	4.11	452.6	0.81	1.77
III	1971–1975	236.8	3.80	458.6	1.84	1.94
IV	1976–1980	206.0	3.31	593.2	2.38	2.88
V	1981–1985	268.0	4.30	420.2	1.68	1.57
VI	1986–1988	92.3	1.48	360.0	1.44	3.90
Stryi (at Verkhnie Syniovydne)						
I	1951–1962	73.3	1.00	520.3	1.00	7.10
II	1963–1970	134.6	1.84	591.0	1.14	4.39
III	1971–1975	352.0	4.80	530.2	1.02	1.51
IV	1976–1980	320.0	4.37	655.2	1.26	2.05
V	1981–1985	228.8	3.11	608.6	1.17	2.67
VI	1986–1988	116.0	1.58	487.7	0.94	4.20
Prut (at Chernivtsi)						
I	1947–1962	111.5	1.00	283.5	1.00	2.54
II	1963–1970	292.1	2.62	325.8	1.15	1.12
III	1971–1975	388.0	3.48	351.0	1.24	0.90
IV	1976–1980	290.0	2.60	427.6	1.51	1.47
V	1981–1985	256.0	2.12	349.8	1.23	1.48
VI	1986–1988	121.7	1.09	250.3	0.88	2.06

Main hydrological parameters for 1975–1980 and whole observation period. Vol. 1 Western Ukraine and Moldova (1985). Hydrometeoizdat, Leningrad (in Russian): 60 tons year^{-1} km^{-2} in the Dnister and Tysa catchments to 400–600 ton year^{-1} km^{-2} in the Prut and Rybnyk River basins (Kovalchuk and Mykhnovych 2004) (Table 8.2), but its maximum is 4,400 ton year^{-1} km^{-2} along the Rybnyk. Its dynamics (documented for the period 1950–2000) is a good indicator of the rate of recent geomorphic processes. Change has been calculated for the Dnister (at Strilky): 300%, the Holovchanka (at Tuchlia): 50%, for the Bystrytsia (at Ozymyna): 300%, Opir (at Skole): 200%, Stryi (at Verkhnie Syniovydne): 80%, and Dnister (at Halych): 50%.

The investigations of changes of *river systems* are based on the analyses of large scale (1:100,000) topographic maps from the nineteenth to twentieth centuries, which show drainage pattern and density as well as forests, settlements, roads, and other features. It was found that in the river basins of the northeastern Ukrainian Carpathians, first- (72–80%) and second-order (14–18%) streams dominate both for number and total length (41–52% and 17–26%, resp.). Third-order rivers make up 4–5% in number and 12–17% in length and fourth- to sixth-order rivers have only a 5–7% joint share. Between 1855 and 2000 major changes took place in river system structure: a drop occurred in the number of streams due to silting and drying, erosion, and accumulation in catchments, while elsewhere the number of streams rose due to intensified groundwater drainage, water runoff, and drainage of agricultural fields. The dimensions of such changes were studied in the Svicha catchment. The number and length of first- and second-order streams both decreased by 3.2% and 3.6%, respectively.

8.6.2 Water and Sediment Discharge

Fluctuations in water and sediment discharges between 1948 and 2004 for the mountain rivers are high (for the Stryi River: from 16 t km^{-2} year^{-1} to 880 t km^{-2} year^{-1}). Sediment discharge in the period 1968–1970 was four times higher than the long-term average. Its maxima do not correlate with precipitation and water discharge (Table 8.2). This fact points to the great influence of other factors like intensive deforestation in the mountains.

The main reasons of alterations in sediment transport are *land use changes* and *gravel extraction* from the channel and the floodplain. Along with increasing precipitation in the period 1960–1975, human activities like deforestation in the late 1950s and early 1960s, and the cultivation of slopes using heavy machinery were the main drivers of the transformation. Vertical channel deformations closely follow the changes in sediment transport. For example, an increase of sediment load is observed for the Dnister River between Strilky and Sambir. River erosion intensified due to gravel extraction upstream Sambir and also near Stryi on the Stryi River.

The analysis of long-term data sets (for 1970–2005) of sediment load, precipitation, and deforestation compared with findings of field research on erosion processes in deforested areas shows correlation between precipitation (mm year^{-1}) and sediment delivery (tons km^{-2} year^{-1}) (correlation coefficient: 0.65), and no correlation between deforested area (ha) and sediment load (t km^{-2} year^{-1}) (correlation coefficient: 0.15). It is assumed that *deforestation impact* on sediment load occurs *with some time delay*. This assumption can be verified by a multiple regression analysis taking sediment delivery as a dependent variable and deforested areas, precipitation, and water discharge as independent variables. The regression analysis indicates that 1 ha annual growth in deforestation increases sediment delivery by 0.54 ton km^{-2} year^{-1}. With the present intensity of deforestation, maximum sediment delivery is expected to occur in 5 years after

Table 8.3 Rate of vertical deformations of river beds in the upper Dnister catchment (by A. Mykhnovych)

River	Locality	Period, years	Deformation, cm	Average rate, mm year^{-1}
Dnister	Strilky	29	−40	−13.79
Dnister	Sambir	32	+64	20.00
Dnister	Rozdil	29	−8	−2.76
Dnister	Zhuravno	29	−10	−3.45
Stryvihor	Khyriv	7	−1	−1.43
Stryvihor	Luky	32	−200	−62.50
Bystrytsia	Ozymyna	32	+20	6.25
Tys'menytsia	Drohobych	10	+20	20.00
Stryi	Matkiv	10	−1	−1.00
Stryi	Zavadivka	10	−5	−5.00
Stryi	Yasenytsia	10	−5,5	−5.50
Stryi	V. Syniovydne	32	−100	−31.25
Zavadka	Rykiv	7	−10	−14.29
Yablun'ka	Turka	32	−32	−10.00
Rybnyk	Maidan	7	−10	−14.29
Opir	Skole	10	−7	−7.00
Slavs'ka	Slavske	10	−35	−35.00
Holovchanka	Tukhlia	32	−60	−18.75
Oriava	Sviatoslav	10	−5	−5.00
Ruzhanka	Ruzhanka	8	−36	−45.00

forest clearance and minimum impact of deforestation will be observed in 7 years. A close correlation has been found between deforested areas, felling methods, and the rates of geomorphic processes by pair correlation and regression analysis for the Holovchanka catchment.

8.6.3 Channel Evolution

Vertical changes of riverbeds are studied along the mountain and foothill sections of river valleys. It was found that erosion dominates over most of the Ukrainian Carpathians. The highest rate of incision is observed for the Stryvihor (at Luky, 65.5 mm year^{-1}), Ruzhanka (at Ruzhanka, 45.0 mm year^{-1}), Slavs'ka (at Slavs'ka 35.0 mm year^{-1}), and Stryi Rivers (at Verkhnie Syniovydne, 31.3 mm year^{-1}) (Table 8.3). Channel accumulation has been recorded for the Dnister (at Sambir, 20.0 mm year^{-1}), Tysmenytsia (at Drohobych, 20.0 mm year^{-1}), Shchyrka (at Shchyrets, 7.8 mm year^{-1}), and Bystrytsia (Ozymyna, 6.3 mm year^{-1}).

Vertical riverbed deformations are caused by both dredging and increasing water discharge as confirmed for different periods and cross-sections in other river systems of the Ukrainian Carpathians (Holoyad et al. 1995; Kovalchuk 1997). Riverbed changes also influence other geomorphic processes (bank erosion, debris flows, and landslides).

8.6.4 River Terraces and Valley Types

Along the main river valleys of the Precarpathians five to seven *terraces* can be identified. The river valleys of the northwestern (Pre-Beskid) and central (Pre-Horhany) Precarpathians were formed along zones of tectonic weakness subsequently buried under Paleogene-Miocene sediments. Terraced river valleys mostly of Pleistocene age occupy almost 35–37% of the Precarpathians. The fragments of *seventh* (Pliocene) terrace can be seen on the right side of the Limnytsia catchment (Krasna mountain) with 160–180 m relative heights and 590 m elevation; between the rivers Svicha and Limnytsia (Zalissia mountain, 485 m) with gravels on top (relative height: 140 m) and also between the rivers Prut and Bystrytsia nadvirnians'ka. The sixth (Loyeva surface) and fifth terraces were formed in the Early Pleistocene. The *sixth terrace* is found on the five interfluves between the Dnister and Cheremosh Rivers. Its gravels of 5–10 m thickness overlie Miocene Molass sediments. Relative terrace heights here vary from 60–80 m (along the upper Dnister and Svicha) to 110–130 m (in the Limnytsia, Bystrytsia and Prut catchments). The *fifth terrace* is only conserved in fragmentary form among others in the Stryvihor catchment, between the rivers Stryvihor – Dnister, Dnister Stryi, Stryi – Svicha. The greatest thickness of alluvium was discovered in the Prut valley near Deliatyn (15 m) and in the river mouths of Svicha and Limnytsia (up to 20 m). The Middle Pleistocene *fourth* terrace occupies the largest areas in the Dnister and Stryi valleys, between Svicha and Sukil, headwaters of Vorona, on the left bank of the Prut and as fragments in the Limnytsia valley. Its relative height is between 25 and 45 m and alluvium thickness is 20 m (5–8 m of gravels covered by 8–12 m of loess loam). The maximum thickness of alluvium is 30 m (near Halych city). The Late Pleistocene *third* and *second* terraces are best preserved in the Precarpathians (in the Upper Dnister, Stryi-Zhydachiv, Kalush and Bystrytsia valleys, along the left bank of the Prut, in the Cheremosh and Siret river valleys). The relative height of the second terrace varies from 5–8 m (in the northwestern Precarpathians) to 10–15 m (in the Cheremosh valley), while the third terrace is located at 15–25 m height. Alluvium thickness reaches 20–25 m in the Stryi-Zhydachiv valley. The *first terrace* and the Holocene floodplain are present in all river valleys. The height of first terrace is 2.5–5 m. The widest terraces (4–5 km) are located in the Upper Dnister, Kalush and Bystrytsia valleys. *Floodplains* show a higher (1–2 m) and a lower (0.5–1 m) level. Their width is 200–300 m, but it can be 1 km or more. Alluvium thickness varies from 3–5 m (in the foothills) to 10–15 m (at river mouths).

In the Beskids wide *transversal valleys* dominate. In the mountains the Stryvihor, Dnister, and Stryi rivers run in winding valleys forming incised meanders. In the headwaters (Dividing Carpathians) the river valleys inherited relict longitudinal valleys of the Paleo-San River. P. Tsys' (1968) defined two types of the valleys: *epigenetic* valleys, typical for the Upper-Dnister Beskids (Stryvihor, Dnister, Stryi, Bystrytsia-Pidbuz'ka) and *erosional-tectonic valleys*, typical for the Skole Beskids (Opir, Oriava, Rybnyk, Sukil').

The rivers of the Lump Horhany Mountains (the Mizunka, Svicha, Limnytsia, Bystrytsia Solotvyns'ka, Bystrytsia Nadvirnuans'ka and Prut) are V-profile, locally canyon-type, deeply incised transversal valleys with steep slopes and usually longitudinal tributaries, which flow into the main rivers at right angles (orthogonal river system).

In contrast to the Beskids and Horhany, the rivers of the Pokuttia-Bukovyna Carpathians have thick alluvium on all terraces. In the Pokuttia the Pistynka, Rybnytsia and Cheremosh flow in deep erosional, V-shaped valleys with many terraces, while the river valleys of Bukovyna (the Siret and Malyi Siret) are canyon-like with fewer terraces.

There are many relict *longitudinal valleys* in the Dividing Verkhovyna Carpathians. Some of their fragments are incorporated into the transversal valleys (Slyvka 2001). Longer river sections run along synclines and cross anticlinal ridges in shorter canyons. The rivers of the Dividing (Inner) Horhany follow deeply incised transversal valleys with steep slopes (locally canyons).

The river valleys of the southwestern macroslopes (Tysa catchment) have a more complicated origin. Their development was impacted by Upper Pliocene volcanic activity of the Vyhorlat–Huta ridge. There are seven terraces of 2–3 km width here at relative heights of 1.5–3 m, 7–10 m, 15–20 m, 30–40 m, 60–70 m, 90–100 m, and 200 m.

8.6.5 Flood Hazard

The Ukrainian Carpathians is a region affected by very intensive *floods* induced by human factors (changes of vegetation cover, deforestation, cultivation and grazing of slopes, changes in land drainage, and channel conditions) combined with natural forces (precipitation conditions, soils properties, etc.). In addition, the flood situation is aggravated by the garbage, debris, and wood accumulation in the channels, at bridges and communication lines crossing river beds (Kovalchuk 1997; Szlávik 2002). The analysis of the monitoring data and literature sources indicate that during the past 100–120 years the highest floods in the Ukrainian Carpathians were observed in 1911, 1927, 1941, 1947, 1955, 1957, 1969, 1970, 1974, 1978, 1979, 1980, 1984, 1989, 1992, 1993, 1995, 1997, 1998, 2001, 2002, 2004, 2008, 2010 (Romashchenko and Savchuk 2002). During extreme flood events water level increases up to 10 m, inundated zones amount to tens of km^2. The settlements most often flooded are along the Upper Dnister, Tysa, and Teresva Rivers.

In the Ukrainian Carpathians floods occur almost in any season of the year and last from hours to 4–8 days. The water discharge curve shows several peaks (from 2 to 12 per event). Highest water levels and discharges are observed in summer and autumn (May to October). Water discharge during flood is 5–40-fold higher than baseflow. The most rapid water level rise is observed in the zones between mountains and foothills. Heavy rainfalls (120–250 mm per day) generate 2.5–3.1 m^3 s^{-1} km^{-2}

Table 8.4 Extreme flood events in the Transcarpathians (After Lymans'ka 2001)

	Type of flood		
	Channel building	Channel-destruction	Disastrous channel destruction
Years	1974, 1978, 1979, 1980	1992, 1993, 1995	1947, 1957, 1970, 1998, 2001
Average precipitation, mm	30–50	40–60	70–100
Maximum precipitation, mm	100–130	100–160	>160
Average rise in water level, m	2–4	2–3	2–3
Maximum rise in water level, m	6–7	8.3–8.5	4–7.5

runoff (Kovalchuk 1997) and, in turn, severe erosion and sediment transport in rivers, dam breaches, flooding agricultural land, settlements, and communication infrastructure.

Traditionally flood control in the Ukrainian Carpathians involves dykes, drainage canals, sluices, pumping stations, etc. Most of the dykes (built between 1950 and 1970) are imperfect: stand too close to riverbanks, and are technically deficient and deformed by recent geomorphologic processes. Along about 70% of the dykes there is a high risk of overtopping during 100-year floods and 20–30% during 25-year floods. Therefore, it is urgent to restore the existing flood-protection system in the Ukrainian Carpathians.

Most considerable environmental-geomorphological effects of floods are:

1. Rapid rises of water level (Dnister, Latorytsia, Uzh, Tysa), fields, communications, buildings, engineering constructions flooded and destroyed;
2. Large areas flooded for tens days in the premountain river segments (Dnister, Stryl, Uzh, Latorytsia, Tysa) after breaking and overtopping of dykes;
3. Extreme intensification of landslides in 5–20% of the region (more than 550 landslides over 16.2 km^2 area activated after flood in the Transcarpathians alone);
4. Intensification of mudflows (about 300 mudflows of 10–20 thousand m^3 activated by floods);
5. Deterioration of ecological conditions in agricultural and urban areas.

The major floods with their causes and effects are summarized according to the classification of Obodovs'kyi (2000) in Table 8.4.

A disastrous flood happened in the Transcarpathians on 4–9 March 2001. The main driver was extremely high amount and intensity of rainfall (90–296 mm) between 3 and 5 March 2001 all over the region. In addition, high air temperature (up to +14°C) caused intensive snowmelt (an additional 20–40 mm). Water level rose 4.3–8.55 m on the Tysa River (near Chop town even 9.5 m) and tributaries (Romashchenko and Savchuk 2002). Two hundred and fifty five settlements and 33,569 buildings were flooded, 1925 buildings and 52.7 km of roads were destroyed, 338 landslides, 81 debris flows and 102 sites with bank erosion were identified (Lymans'ka 2001). Most of the landslides occurred in the low mountains of the Molass Carpathians.

Flood hazard research focuses on geomorphic processes in recently deforested areas in the Beskids at four study sites in the Krasny sandstone catchment of northwestern exposition, 523.6–770.5 m elevations and 35–45° inclination. The findings for the years 2004–2006 showed intensive gully erosion after deforestation. Average gully length was 33–51 m, maximum length over 100 m, and gully width 1.52–1.70 m. In the summer (16 July–15 August 2004) thalwegs incised 3–5 cm at maximum daily rainfall on 31 July 2004, 90 mm in Sviatoslav, and 160 mm in Skole town. Sheet erosion amounted to 0.6 cm and in autumn accumulation of 0.5–7.8 cm followed. With the new vegetation period accumulation reached 5.3–9.2 cm, but another rainfall of 56 mm caused debris flows nearby and built fans of 43–164 m^3.

8.7 Karst

There are three types of karst in the Ukrainian Carpathians: carbonate, sulfate and evaporite (salt) karsts. Jurassic *carbonate karst* is found in the Chyvchyna Massif. In the Precarpathians, in the San moraine-outwash-alluvial plain both carbonate and sulfate karsts are observed. *Sulfate karst* is most widespread in the Precarpathians. The karst processes are intensified due to the impact of mining. The studies by the Geological Survey on karst development near the Yaziv sulfate mine between 1971 and 1991 found that groundwater levels dropped by 86 m within a depression area of 210 km^2. *Salt karst* is widespread in the Pre- and Transcarpathians. In the Precarpathians the karst belt of salt layers and lenses extends along a northwest to southeast axis. Its overall width is up to 40 km and maximum salt thickness is 600 m (Rud'ko and Kravchuk 2002). Karst processes are accelerated by aggressive waters. Locally the depression extends at 6–10 m year^{-1} rate.

8.8 Nival and Glacial Landforms

In the Ukrainian Carpathians *avalanches* occur on slopes of 20–45° at elevations of 300–2,000 m. Avalanche paths are 100–500 m long, the longest is more than 3 km. The core area is from 1–5 to 40–50 ha. Avalanches usually occur once a year in the catchments of the Tysa, Teresva, Shopurka, Tereblia, Latorytsia, Borzhava, and Rika Rivers, once in 2–3 years on the Verkhovyna Ridge and in the Uzh, Latorytsia, and Borzhava headwaters (Maslova et al. 1999). The volumes of moved snow vary from hundreds to tens of thousands m^3, maxima can reach 0.5–1 million m^3.

Avalanches most often develop in winter and spring (February to March). The most endangered regions are the Chornohora, Rakhiv massif, Svydovets', Krasna, Borzhava ridges, Polonyna Runa, and Lautians'ka Holytsia. For example, in

February 1999 snow amounts fourfold exceeded the usual value (300–500 thousands m^3) and caused a number of avalanches in the Transcarpathians (Maslova et al. 1999). The avalanche started on the Polonyna Krasna was 3 km long and built a fan of 20 m height.

Landform remnants of Pleistocene glacials can be seen on the summits of the ridges Chornohora, Svydivets', Marmarosh, and Horhany. Relict glacial landforms in the Chornohora are glacial *troughs, karren,* and *kettles* in three areas: in the headwaters of the Zarosliats'kyi Prut River; in the valleys of the Dantsers'kyi Prut, Hadzhyna, and Kizi Rivers and in the headwaters of the Dzembronia and Pohorilets'. *Felsenmeers* are also relict forms of Pleistocene glacials and most widespread in the Lump Horhany (on the highest lump-ridges Skole, Parashka, Zelemyanka) and also found in the Dividing (Internal) Horhany, Chornohora, Svydivets', and Marmarosh. Above the forest belt (mostly on northern slopes) they are stabilized by dwarf pines.

8.9 Conclusions

The predominant geomorphic processes active at present in the Ukrainian Carpathians are weathering, sheet, gully, river-bank, and channel erosion, landslides, debris, and mudflows, avalanches, landslips, karst, man-induced processes, and hydro-geomorphological processes, riverine, and flash floods. Among secondary geomorphic processes wind action, chemical and biogenic denudation, and accumulation can be cited. According to their distribution and the volume of material moved, the relative significance of each process group can be estimated (Table 8.5).

In the mountains the main drivers of landform evolution are erosion, landslides, and debris flows. Human activity has increased the intensity of natural geomorphic processes hundredfold.

Table 8.5 Recent geomorphic processes in Ukrainian Carpathians

Type of the process	Percentage (%) by distribution area	Type of the process	Percentage (%) by volume of the material moved
Sheet erosion	38	Landslides	41
Landslides	23	Sheet erosion	30
River-bed erosion and accumulation	12	River-bed erosion and accumulation	11
Mudflows	9	Mudflows	7
Avalanches	6	Man-induced	6
Man-induced	6	Landslips	2
Karst	3	Avalanches	1
Landslips	2	Karst	1
Biogenic	1	Biogenic	1
Total	*100*	*Total*	*100*

References

Bogacki M, Kovalchuk I, Mykhnovych A (2000) The dynamics of the river network structure in the Dnister basin as reaction on the anthropogenic changes of nature conditions. Miscellanea Geographica, Warszawa 9:11–18

Boliukh O, Kanash A, Kit M, Kravchuk Y (1976) Stationary investigations of the sheet erosion in the Precarpathians. Higher School, Lviv, 138 p (in Ukrainian)

Budz M, Kovalchuk I (2008) Geologic-geomorphological classification of mud-flows. Bull Lviv Univ Ser Geogr 35:28–33 (in Ukrainian)

Dubis L, Kovalchuk I, Mykhnovych A (2006) Extreme geomorphic processes in the Eastern Carpathians: spectrum, causes, development, activization and intensity. Studia Geomorphologica Carpatho-Balcanica 40:93–106

Habchak N (2010) Relief-forming processes and their dynamics in the Rika River basin of the Transcarpathians. Phys Geogr Geomorphol, Kyiv 58:282–288 (in Ukrainian)

Hensiruk S (1992) Regional nature use. Svit, Lviv, 336 p (in Ukrainian)

Hofstein I (1995) Geomorphological studies of Ukrainian Carpathians. Naukova Dumka, Kyiv, 84 p (in Ukrainian)

Holoyad B, Slyvka R, Panevnyk V (1995) Erosion-denudation processes in Ukrainian Carpathians. Ivano-Frankivs'k, 114 p (in Ukrainian)

Khomyn Y, Bilaniuk V (2010) Results of the stationary investigations of the sheet erosion in the Borzhava River catchment. Phys Geogr Geomorphol Kyiv 58:154–164 (in Ukrainian)

Kovalchuk I (1997) Regional ecological-geomorphologic analysis. Institute of Ukraine Studies, Lviv, 440 p (in Ukrainian)

Kovalchuk I, Kravchuk Y (2001) 50 years of the Department of Geomorphology at the Lviv National University. Bull Lviv Univ Ser Geogr 28:3–15 (in Ukrainian)

Kovalchuk I, Mykhnovych A (2004) Recent morphodynamical processes in the forested landscapes of Ukrainian Carpathians. Sci Bull For Eng Tech Technol Environ State For Univ Lviv 14(3):273–285 (in Ukrainian)

Kovalchuk I, Liashchuk B, Pakulia M (1978) Comparison of the longitudal profiles of the Precarpathian rivers (on the examples of the Bystrytsia Nadvirnians'ka and Bystrytsia Solotvyns'ka Rivers). Bull Lviv Univ Ser Geogr 11:95–99

Kovalchuk I, Mykhnovych A, Pylypovych O (2000) River systems structure and hydrological regime transformations in the Dnister River basin. Investigations of the Upper Dnister River basin. Ivan Franko National University of Lviv, Lviv, pp 34–43 (in Ukrainian)

Kovalchuk I, Mykhnovych A, Quast J, Steidl J, Ehlert V (2008) Current problems of water management in the Upper Dnister basin. In: Roth M, Nobis R, Stetsiuk V, Kruhlov I (eds) Transformation processes in the Western Ukraine: concepts for sustainable land use. Weissensee Verlag, Berlin, pp 125–135

Kowalczuk I, Mychnowicz A (2009) Transformacja struktury systemów rzecznych w Karpatach Ukraińskich (Transformation of the structure of fluvial systems in the Ukrainian Carpathians). Prace i Studia Geograficzne Warszawa 41:115–122 (in Polish)

Kowalczuk I, Pylypowycz O, Mychnowycz A (2008) Zmiany intensywności denudacji w Beskidach Skoliwśkich (Karpaty Ukraińskie) (Changes in the intensity of denudation in the Skole Beskids, Ukrainian Carpathians). In: Międzynarodowe sympozjum "Antropopresja w górach i na przedpolu. Zapis zmian w formach terenu i osadach" (Głuchołazy, 24–27 czerwca 2008), pp 36–45 (in Polish)

Kravchuk Y (1999) Geomorphology of the Precarpathians. Merkator, Lviv, 188 p (in Ukrainian)

Kravchuk Y (2001) Studies of the Ukrainian Carpathians by the Lviv geomorphologists. Bull Lviv Univ Ser Geogr 28:16–19 (in Ukrainian)

Kravchuk Y (2003) Lviv geography during last 120 years: history, personals, scientific schools. In: Proceedings of the scientific conference dedicated to 120 anniversary of geography at the Lviv University, Lviv, pp 3–16 (in Ukrainian)

Kravchuk Y (2005) Geomorphology of the Lump Carpathians. University Publishers, Lviv, 232 p (in Ukrainian)

Krawczuk J (1999) Tradycje naukove polskich i ukraińskich badań geograficznych Karpat Wschodnich (The tradition of Polish-Ukrainian geographical scientific research in the Eastern Carpathians). Geografia na przelomie wieków – jedność w różnorodności. Warszawa, pp 46–51 (in Polish)

Lymans'ka I (2001) Extremely high floods on the Transcarpathian rivers, their types and consequences. Phys Geogr Geomorphol Kyiv 41:144–149 (in Ukrainian)

Maslova T, Hryshchenko V, Susidko M (1999) Snowy winters in the Transcarpathians and the avalanches development intensity. Scientific Papers of the Ukrainian Scientific Hydrometeorological Institute, Kyiv, pp 144–149 (in Ukrainian)

Mykhnovych A (1998) The Upper Dnister river systems structure and its transformation under the impact of natural and man-made factors. Bull Lviv Univ Ser Geogr 21:161–167 (in Ukrainian)

Obodovs'kyi O (2000) River beds stability evaluation and floods classification of the mountain rivers. In: Ukraine and the global processes: geographic dimension, vol 2. Luts'k, Kyiv, pp 205–209 (in Ukrainian)

Obodovs'kyi O (2001) Hydrologic-ecological evaluation of river bed processes (on the example of the rivers of Ukraine). Nika-Centre, Kyiv, 274 p (in Ukrainian)

Oliferov A (2007) Mudflows in the Crimea and Carpathians. Tavria, Simferopil, 176 p (in Ukrainian)

Pol W (1851) Rzut oka na polnocne stoki Karpat, Publ. Tow. Przyjaciol Oswiaty, Krakow (in Polish)

Problems... (2008) Problems of the geomorphology and paleogeography of the Ukrainian Carpathians and adjoining areas Scientific papers, University Publishers, Lviv, 320 p (in Ukrainian)

Professor... (2004) Petro Tsys'. Prepared by Prof. I. Kovalchuk. Lviv, Ivan Franko National University Publishing Centre, Lviv, 433 p (in Ukrainian)

Relief... (2010) Relief of Ukraine. Vakhrushev B, Kovalchuk I, Komliev O, Kravchuk Y, Paliyenko E, Rud'ko G, Stetsiuk V. Ed. by Stetsiuk V. Slovo Publishers, Kyiv, 689 p

Romashchenko M, Savchuk D (2002) Water disasters. Carpathian floods. Statistics, causes, regulations. Agrarian Science, Kyiv, 304 p (in Ukrainian)

Romer E (1909) Proba morfometrycznej analizy grzbietów Karpat Wschodnich (Experimental morphometrical analysis of ridges in the Eastern Carpathians). Rocznik Lwów 34:5–6 (in Polish)

Rud'ko G, Kravchuk Y (2002) Engineering-geomorphologic analysis of the Ukrainian Carpathians. University Publishers, Lviv, 172 p (in Ukrainian)

Rud'ko G, Yakovliev Y, Ragozin O (1997) Monitoring of the areas of geological processes and natural-technogenous disasters risks calculations. Znannia, Kyiv, 80 p (in Ukrainian)

Shushniak V (2004) History of geomorphologic research and actual geomorphologic problems of the Carpathians. In: Problems of the geomorphology and paleogeography of the Ukrainian Carpathians and adjoining areas. Scientific papers, University Publishers, Lviv, pp 163–176 (in Ukrainian)

Skvarchevs'ka L (1962) Geomorphologic features of the river valleys in Turka district (Lviv region). Bull Lviv Univ Ser Geogr 1:34–39 (in Ukrainian)

Slyvka R (1994) Mudflows in the Ukrainian Carpathians and the methods of their regulation. Bull Lviv Univ Ser Geogr 19:131–155 (in Ukrainian)

Slyvka R (2001) Geomorphology of the Dividing Verkhovyna Carpathians. University Publishers, Lviv, 152 p (in Ukrainian)

Somov V, Rakhimova I (1983) Recent Earth's crust movement in the Carpatho-Balcanic region and adjoining structures. Naukova Dumka, Kyiv, 142 p (in Russian)

Szlávik L (2002) The development policy of flood control on river Tisza in Hungary. ERWG Letter: 13. Land and water management in Europe 1. 1–4

Tsys' P (1956) Geomorphological regions of the Ukrainian Carpathians. Geogr Bul Lviv 3:5–24 (in Ukrainian)

Tsys' P (1964) Short overview of the recent geomorphologic phenomena in the western region of Ukraine. Bull Lviv Univ Ser Geogr 2:3–10 (in Ukrainian)

Tsys' P (1968) State of the knowledge and actual problems of the Ukrainian Carpathians geomorphology. Bull Lviv Univ Ser Geogr 4:106–120 (in Ukrainian)

Tsys' P, Stadnyts'kyi D, Slyvka R, Boliukh O, Chalyk V et al (1968) Some results of the recent exogenous processes investigations in the Ukrainian Carpathians. In: Recent exogenous processes. Part 1. Academy of Sciences of USSR, Kyiv, pp 16–18 (in Russian)

Yushchenko Y (2009) River bed studies at the Chernivtsi University. In: Ukrainian historical geography and history of geography in Ukraine. Proceedings of the International Scientific Conference, Chernivtsi, pp 36–37 (in Ukrainian)

Chapter 9
Recent Landform Evolution in Hungary

Dénes Lóczy, Ádám Kertész, József Lóki, Tímea Kiss, Péter Rózsa, György Sipos, László Sütő, József Szabó, and Márton Veress

Abstract Fluvial geomorphic processes (channel and floodplain evolution) are widespread in the extensive lowlands of Hungary. Since flow regulation in the nineteenth century, river channels have shown adjustments of considerable degree. Some agricultural areas in hills and low mountain basins are seriously affected by water erosion, particularly gully development on loess. Although all sand dunes have been stabilized by now, historically wind erosion has also been a major geomorphic agent in blown-sand areas. The areas affected by mass movements and karst processes are limited but their processes still operate – partly in function of the changing climatic conditions. Applied geomorphological research focuses on ever intensifying human impact on the landscape (particularly in mining districts), which has become the primary driver of recent geomorphic evolution in Hungary, too.

D. Lóczy (✉)
Institute of Environmental Sciences, University of Pécs, Ifjúság útja 6, H-7624 Pécs, Hungary
e-mail: loczyd@gamma.ttk.pte.hu

Á. Kertész
Geographical Research Institute, Hungarian Academy of Sciences,
Budaörsi út 45, H-1012 Budapest, Hungary
e-mail: kertesza@helka.iif.hu

J. Lóki • J. Szabó
Department of Physical Geography and Geoinformatics, University of Debrecen,
Egyetem tér 1, H-4010 Debrecen, Hungary
e-mail: loki.jozsef@science.unideb.hu; szabo.jozsef@science.unideb.hu

T. Kiss • G. Sipos
Department of Physical Geography and Geoinformatics, University of Szeged,
Egyetem utca 2-6, H-6722 Szeged, Hungary
e-mail: kisstimi@gmail.com; gysipos@geo.u-szeged.hu

P. Rózsa
Department of Mineralogy and Geology, University of Debrecen,
Egyetem tér 1, H-4010 Debrecen, Hungary
e-mail: rozsa.peter@science.unideb.hu

Keywords Channel and floodplain evolution • Water erosion • Landslides • Wind erosion • Karst • Mining impact

9.1 History of Geomorphological Research in Hungary

Dénes Lóczy

Hungarian geomorphological research has developed in close association with Central European geomorphology (Lóczy and Pécsi 1989; Pécsi et al. 1993; Pécsi 1999). Its first major representative was J. Hunfalvy (1820–1888), who published a *Description of the Physical Conditions of Hungary* in three volumes between 1863 and 1865. The first systematic research into the physical geography of a region, in a true Humboldtian conception, was conducted by a populous group of scientists in the *Lake Balaton basin* under the guidance of L. Lóczy Sen. (1849–1920), the most eminent figure in Hungarian geology and geography. The findings, summarized in a monograph series of 32 volumes, include observations related to the origin and evolution of the lake basin. Lóczy's student, J. Cholnoky (1870–1950) was primarily engaged in geomorphology, particularly in the investigation of fluvial (river mechanisms, terrace formation) and aeolian action (blown-sand movements, traces of arid conditions in the Carpathian Basin), karst, and periglacial processes as well as long-term landform evolution. Cholnoky was a prolific writer, who published 53 books (mostly popularizing science at high standards) and 160 academic papers in physical geography. Gy. Prinz (1882–1973) put forward the first coherent theory (the *Tisia concept*) for the evolution of the structure of the Carpathian Basin, which had been prevalent in geosciences until the advent of plate tectonics.

In the post-war period B. Bulla (1906–1962) was an outstanding investigator of fluvial and glacial processes, loess formation, and chronological problems. His *climatogenetic* approach to *geomorphology*, similar to J. Büdel's, is manifested in his concept of rhythmical geomorphic evolution, which explains landform evolution on the basis of contemporary theories of Tertiary climatic and Pleistocene glaciation cycles. He described the geomorphology of Hungarian landscapes in his book *The Physical Geography of Hungary* (1964).

In the late 1950s and 1960s geomorphological investigations are performed in the framework of *complex landscape research* at mesoregion scale (Marosi 1979). In the monograph series *Landscapes of Hungary* the geomorphic evolution of the six

L. Sütő
Department of Geography and Tourism, College of Nyíregyháza, Sóstói út 31/B,
H-4400 Nyíregyháza, Hungary
e-mail: sutolaci@zeus.nyf.hu

M. Veress
Department of Physical Geography, University of West-Hungary, Károlyi Gáspár tér 4,
H-9701 Szombathely, Hungary
e-mail: vmarton@ttmk.nyme.hu

macroregions were presented in detail along with the assessment of utilization opportunities of landforms for various purposes. For the detailed field survey of topographic features a uniform legend for geomorphological mapping was proposed in 1963 and detailed *geomorphological maps* were drawn at 1:10,000 and 1:25,000 scales for a number of meso- and microregions. Hungarian geomorphologists had a prominent role in compiling the Geomorphological Map of the Carpathian and Balkan Region (Pécsi 1977) and the geomorphological sheet in the Atlas of Danubian Countries (Pécsi 1980). In the 1970s *applied research* was encouraged and engineering geomorphological maps for construction purposes were prepared and the areas of Hungary endangered by mass movements were surveyed. The physical environments of settlements were first studied at Pécs (F. Erdősi, Gy. Lovász).

Loess studies, led by M. Pécsi, have been central in Hungarian geomorphology. They are concerned with the definition, classification, origin, dating, and geomorphological significance of *loess deposits* (Pécsi and Richter 1996). In addition to loess-paleosol sequences, *Quaternary research* also involved the study of river terraces, travertines, and vertebrate fauna in order to solve chronological problems like identifying the Plio-Pleistocene boundary. From the analysis of finely laminated slope deposits with seasonal cycles, it was deduced that in the Pleistocene glacials *periglacial processes* (gelisolifluction, sheet wash, creep and deflation) operated over extensive surfaces and were collectively described by M. Pécsi as *derasion* (Pécsi 1967). He pointed out the significance of derasional processes in the transformation of the slopes of mountains, foothills and loess hill regions in Hungary (Mezősi 2011).

The reconstruction of long-term relief evolution has long been in the focus of geomorphological research (Pécsi and Lóczy 1986). The climatic conditions favoring the formation of *erosional surfaces* (summit planated surfaces, pediments, and glacis) were identified in the 1960s and 1970s and the polygenetic models of planation were applied to the mountains of Hungary by M. Pécsi (1970a) and his co-workers (J. Szilárd, L. Ádám, F. Schweitzer and Á. Juhász). Recently, geomorphological surfaces are reconstructed by GIS methods (e.g., Bugya and Kovács 2010).

Quantitative methods were first introduced into Hungarian geomorphology in the 1970s. *Morphometric analyses* and *modelling* of geomorphic processes based on *field and laboratory experiments* were pioneered at all the most important geomorphological schools in Hungary. Soil erosion investigations on experimental plots were launched by L. Góczán and Á. Kertész at the Transdanubian field stations of the Geographical Research Institute of the Hungarian Academy of Sciences and now continued by Z. Szalai and G. Jakab. At the University of Debrecen experiments were initiated by L. Kádár (to solve problems of river mechanism transitions, sediment transport, meandering and terrace formation) and his students, Z. Borsy (blown-sand movements, introduction of new dating techniques), Z. Pinczés (frost shattering), complemented by field surveys led by A. Kerényi (splash erosion and sheet wash), J. Szabó (mass movements) and J. Lóki (wind action). Through such investigations landscape ecological research was founded (A. Kerényi, P. Csorba). At the Eötvös Loránd University of Budapest the comparative volcano-morphological investigations of A. Székely from the 1960s to the 1980s have been further developed using remote sensing, morphometric and stratigraphic methods by his students (A. Nemerkényi, D.

Karátson). Drainage system evolution was reconstructed by Gy. Gábris applying various approaches. At the University of Szeged geoecological mapping (G. Mezősi) and fluvial geomorphology (T. Kiss) are predominant. Karst morphological research has several centers: Szeged (L. Jakucs, I. Kevei-Bárány), Budapest (L. Zámbó), Miskolc (A. Hevesi), Pécs (P.Z. Szabó, Gy. Lovász), and Szombathely (M. Veress).

The new generation of geomorphologists, who began their career after 1990, are engaged in both fundamental (novel techniques in landform dating; river, wind and soil erosion measurements; floodplain accumulation investigations; volcanic reconstructions; field surveys of anthropogenic and karst processes) and applied geomorphological research (GIS applications for landscape analyses and geomorphological hazards assessment; land reclamation and site selection for waste disposal and other infrastructural projects) – often in international cooperation. A collection of papers on the rates of active geomorphic processes have recently been published (Kiss and Mezősi 2008–2009).

9.2 Tectonic Setting

Dénes Lóczy

Hungary (93,030 km^2) is a landlocked country, located in the central Carpathian (Pannonian or Middle Danubian) Basin, which is surrounded by the Alpine, Carpathian, and Dinaric mountain ranges (Kocsis and Schweitzer 2009). The basin is a Tertiary depression, formed as a consequence of plate tectonic movements and the uplift of the encircling mountains. Naturally, the tectonic setting of the country, briefly treated here, can only be described in the context of the entire Carpatho–Balkan–Dinaric region (see Chap. 1 in this volume).

The first *plate tectonic models* for the evolution of the Carpathian Basin emerged in the 1970s and 1980s (treated in Pécsi 1999). The basement is made up of two basic components: the ALCAPA (Alpine-Carpathian-Pannonian) Mega-unit of African origin and the Tisza-Dacia Mega-unit of European origin (Haas and Péró 2004). According to a generally accepted concept, the indentation of the Adria Microplate of African origin and the subduction of the European plate margin led to the extension and eastward rotational movement ("escape") of ALCAPA and the opposite rotation of the Tisza-Dacia Mega-units during the Paleogene (Horváth and Royden 1981; Balla 1982; Kázmér and Kovács 1985; Csontos et al. 1992; Márton 1997; Fodor et al. 1999). The juxtaposition of the two mega-units completed by the Early Miocene. In the Middle Miocene *synrift stage* of the back-arc basin formation, the crustal extension resulted in andesitic and partly dacitic volcanism (Horváth 1993). The Late Miocene *post-rift stage* involved accelerated subsidence, high-rate sedimentation, and locally alkali basalt volcanism (Fodor et al. 1999). In the Pliocene north-to-south compression resulted in the termination of subsidence and even caused uplift in some parts of the basin (Horváth 1993; Horváth et al. 2006). The

Quaternary activity and geomorphological significance of the main tectonic lines and the subordinate fault patterns are strongly debated (Síkhegyi 2002).

9.3 Geomorphological Units

Dénes Lóczy

On the basis of the overall character of topography (elevation, relative relief), structure (lithology and tectonics), and geomorphic processes traditionally six geomorphological macroregions are identified (Bulla and Mendöl 1947; Pécsi and Somogyi 1969; Pécsi 1970b – Fig. 9.1):

1. The Great Hungarian or Middle Danubian Plain (Alföld or Nagyalföld);
2. The Little Hungarian or Little Danubian Plain (Kisalföld);
3. The Alpine foothills or Western Transdanubia (Alpokalja);
4. The Transdanubian Hills;
5. The Transdanubian Mountains;
6. The North-Hungarian Mountains with its intramontane basins.

With the exception of the Transdanubian Mountains, all the geomorphological units continue beyond the country's borders and their boundaries closely follow those of the highest-level landscape units (Hajdú-Moharos and Hevesi 2002).

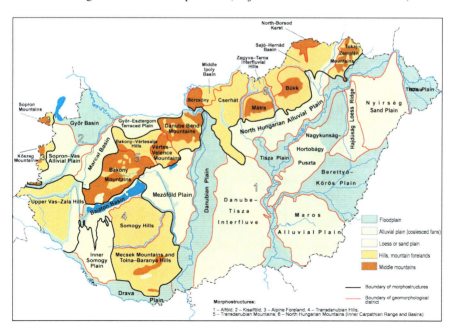

Fig. 9.1 The geomorphological divisions of Hungary (Simplified after Pécsi and Somogyi (1969); Pécsi (1970b))

The *Great Hungarian Plain* occupies two-thirds of the country's territory (62,300 km^2) and extends in all directions over the territory of neighboring countries (including the Danube Lowland, the East Croatian Plain, the Vojvodina Plain, the West-Romanian Plain, the Transcarpathian Lowland of the Ukraine, and the Eastern Slovakian Plain) with a total area of more than 100,000 km^2 and average elevation around 100 m. Along its margins there is a usually gradual transition to foothill areas. The surface was produced by Pleistocene and Holocene fluvial and aeolian accumulation and is subdivided accordingly to three types: alluvial plains, loess-mantled alluvial fans, and alluvial fans with blown-sand dunes. The protected floodplains and loess-mantled fans are presently shaped by human action (mainly arable farming), the blown-sand areas by seasonal wind action, and the active floodplains by fluvial processes.

In northwestern Hungary the *Little Plain* (5,300 km^2) encompasses the double Pleistocene alluvial fan and the right-bank terrace system of the Danube. Its subsidence and filling history resembles the evolution of the Great Hungarian Plain with the difference that most alluvial fans extended towards the center from the northwestern-northern direction. The main landscape types are cultivated floodplains and terraced plains. Today the impacts of cultivation and river deposition determine the character of the landscape.

Only a fraction of the *Alpine foreland* (750 km^2) belongs to Hungary. This is the geomorphological region with the smallest extension and the only one that – strictly speaking – falls outside of the Carpatho-Balkan-Dinaric region. Its landscapes are mostly formed on metamorphic rocks and the sediments from their erosion as well as on alluvial gravel fans. A large proportion of the vegetation cover is coniferous and deciduous forests and pastures with some arable land. Over the rock surfaces weathering and slow hillslope processes are predominant, while streams incise into the fan surfaces.

The *Transdanubian Hills* (10,200 km^2) include lowland as well as mountainous areas (Mecsek Mountains) but most of it is locally minutely dissected, loess-mantled, or sand hills of asymmetric topography and 200–300 m elevation with broad river valleys. The hill region is subdivided into independent hills and foothill areas both predominantly under crop cultivation. Sheet wash, intense rill, and local gully erosion as well as soil creep are the main geomorphic agents. Cultivation accelerates the removal of loose soil.

The *Transdanubian Mountains* (6,400 km^2) are a series of block-faulted and slightly folded-imbricated horsts of southwest to northeast general strike, separating two subbasins of the Carpathian Basin. The mountains were planated in the Mesozoic and a dense network of fractures dismembered them during the Tertiary into numerous plateaus at 300–700 m elevation (planated surfaces buried and exhumed on several occasions) separated by basins and grabens at various levels. Sparse drainage is typical with gorge-like valleys. Most of the forested, plateau-like Mesozoic and Cenozoic limestone mountains are affected by karstic processes, while surfaces of dolomite and andesite rocks show characteristic weathering phenomena and mass movements. The individual members are separated by tectonic valleys with present-day seismic activity.

Beyond the Danube Bend, the continuation of the Transdanubian Mountains of 8,000 km² area is called the *North-Hungarian Mountains*. It is a range of highly variable geological structure (member of the Inner Carpathian Range) and elevation (the only mountains reaching above 1,000 m), equally including mountains of Mesozoic-Paleogene sedimentary and Miocene-Pliocene volcanic rocks (rhyolites and andesites). The volcanic landforms are in various stages of erosional destruction. An almost continuous broad belt of pediments, glacis, and alluvial fans have formed along the southern margin of the range. The higher surfaces are forested, while foothills are used as arable land, southern slopes as gardens and vineyards. The present-day geomorphic processes also vary on a wide range: hillslope processes, mass movements, karstification, and all types of erosion by surface runoff. The members of the range are separated by broad fluvial valleys, draining to the south, towards the Tisza River.

9.4 Recent Geomorphic Processes

9.4.1 River Channel Processes

György Sipos
Tímea Kiss

The whole territory of Hungary belongs to the drainage basin of the *Danube* (total area: 817,800 km²). Arriving in the Carpathian Basin, the Danube channel slope drops to 35 cm km^{-1} (0.00035). During the Pleistocene and Holocene, the river built a double alluvial fan with anastomosing (braided and meandering) channels. Except for the Danube Bend, where the river turns from a west-east to a north-south direction, river gradient is further diminished and to the south of Budapest the Danube forms another large sandy alluvial fan, the Kiskunság (Little Cumania) fan with sparse drainage and aeolian landforms (Mezősi 2011). The discharge of the Danube is influenced by the fact that it has no major tributary on either bank between the Ipoly (Ipel') and the 749-km long Drava in Croatia.

The eastern half of Hungary is part of the catchment of the *Tisza* River (157,135 km²), the largest left-bank tributary of the Danube and the hydrological axis of the Great Hungarian Plain. The Tisza in Hungary is a typical meandering lowland river of 2 cm km^{-1} gradient (0.00002). The nineteenth-century flow regulations were even more drastic interventions into the life of the river than those on the Danube. As a consequence of numerous cut-offs river length was reduced from 1,419 to 966 km. The passage of flood waves was accelerated but the highly fluctuating river regime still causes serious problems of flood control (Szabó et al. 2008). Prolonged rainfalls or rapid snowmelt in the Slovak, Ukrainian, and Romanian Carpathians result in flood waves on both right-bank (like the Bodrog, Hernád/Hornad, Bódva and Sajó/Slaná) and left-bank tributaries (like the Szamos/Someş,

Kraszna/Crasna, Körös/Crişul, Berettyó/Barcău and Maros/Mureş). In addition to flow regulation, the greatest human interventions are the dams of Tiszalök and Kisköre on the Tisza.

In Hungary the evolution of river channels has been investigated since the mid-nineteenth century, the main period of channelization. The studies of river channel change rely on cross-section surveys made at every river kilometer (rkm) in 20–25 year time interval since the 1890s and on measurements of sediment discharge starting in the late nineteenth century. In the 1950s, river channel stabilization works gave a new impulse to hydrological mapping and analysis. Results were summarized in a series of atlases published by the VITUKI (Research Institute for Water Management) and river channel processes were described quantitatively (Szlávik and Szekeres 2003). The rates of processes like meander evolution and migration, width change, channel aggradation and incision, finally channel pattern metamorphosis are all influenced by variables of the fluvial system (water and sediment discharge, valley slope, sediment grain size, vegetation, human impact).

Before flow regulation Hungarian rivers generally flowed in anastomosing-braided-meandering channels on their gently sloping alluvial fans along mountain fronts and meandered in the flat central lowlands of the Carpathian Basin. Channelization works in the nineteenth century involved morphological changes. Comparing archive maps, Z. Károlyi (1960a) determined the changes in sinuosity and the downstream *migration of meanders*. The highest rate of migration between 1890 and 1951 was detected on the Upper Tisza (100–1,700 m, 1.6–27.9 m year^{-1}) with no bank stabilization. As pointed out by Csoma (1973) and Laczay (1982) from the analysis of hydrological maps, meanders with convex-bank point-bar formation migrate in a downstream direction. They also introduced to the Hungarian literature a meander classification scheme, which is still widely used. Somogyi (1974) investigated meander migration between 1783 and 1900 along the southern Hungarian section of the Danube River. By overlaying maps, he found that the lateral shifting and downstream migration rates of meanders were 16–50 and 20–38 m year^{-1}, respectively. Along the Hungarian upper and middle sections of the Tisza River meanders take ca 150 years to form and to be cut off (Somogyi 1978). Mike (1991) summarized the data on meander evolution on Hungarian rivers by using the VITUKI atlas series. Unfortunately, the studied parameters are not the same, and probably this is the reason why no conclusion is drawn in his study.

Meander evolution is intensive even on regulated rivers. On the Middle Tisza cross-section data indicate channel shifts of 200 m since flow regulations (rate: 1.8 m year^{-1}) and meander development *increasing channel length* by 16 km (Nagy et al. 2001). On the Lower Tisza the analyses of several cross-sections, maps, and aerial photographs (Fiala and Kiss 2005, 2006) showed that the length of a 25 rkm long study reach had increased by 6 m year^{-1} from the time of cut-offs (1860s) till the construction of revetments (1950s) (0.25 m rkm^{-1} year^{-1}), but from the 1970s this value dropped to 0.8 m year^{-1} (0.03 m rkm^{-1} year^{-1}). In the meantime the planform of some meanders has become distorted (sharpening, radius decrease). Along the Hungarian section of River Maros (Mureş) dendrogeomorphological and GIS studies (Blanka et al. 2006; Blanka and Kiss 2006; Sipos 2006) of quasi naturally developing and artificially stabilized meanders suggested that since channelization in the 1850s the apex of some unregulated meanders migrated 250–300 m (1.7–2.1 m year^{-1}) downstream. Over the past 50–60 years the formation of point bars was also intensive (10 m year^{-1}), while concave banks only retreated at a rate of 2 m year^{-1} (Blanka et al. 2006). On a 30 rkm section of the river, lacking bank stabilization the length of the center-line has grown at a rate of 2.5 m year^{-1} (0.08 m rkm^{-1} year^{-1}) since the regulations. On the River Hernád (Hornad) the length of an originally 5.7 rkm section has increased by 1.6 rkm between 1937 and 2002 (Blanka and Kiss 2006). In this case the change of meander parameters can partly be explained by tectonics. Exceptionally large floods are of increasing frequency since the 1990s (Kiss et al. 2009) and may cause bank shifting up to 16.7 m along the Hernád (as it happened in 2010), particularly along the high bluffs of overmatured meanders.

As opposed to meander evolution, *channel width* changes, partly due to natural channel development processes and partly to regulation measures, have rarely been investigated on Hungarian rivers. As consequence of the nineteenth-century artificial cut-offs, the width of regulated reaches decreased (Ihrig 1973), but after a decade the new channels began to broaden (Márton 1914; Tőry 1952; Lászlóffy 1982). Along some reaches of the Hungarian Upper Danube and channel width has generally decreased and islands coalesced with the banks parallel with riverbed aggradation (Laczay 1968). Recent studies report general width decrease along the Lower Tisza (16% since 1842, i.e., a 0.2 m year^{-1} rate) (Fiala and Kiss 2006) and Maros Rivers (Sipos 2006; Sipos et al. 2007). Channel narrowing results from bank protection and also from incision. The Maros channel broadened after regulations, but since the 1950s has narrowed by 18–20% (0.3 m year^{-1}) along sections not affected by bank stabilization (Kiss and Sipos 2003, 2004; Sipos 2006). The process is driven by the stabilizing effect of vegetation, as also observed by M. Szabó (2006) on the Szigetköz section of the Danube.

Changes in channel depth and *cross-section* shape are primarily related to incision, aggradation, or change of slope. Kvassay (1902) investigated vertical channel change along the Danube, Tisza, and their tributaries. According to his calculations, flood levels on the Lower Tisza increased by 250–300 cm, while the level of low waters decreased by around 110 cm between 1830 and 1895 (rate 1.6–1.8 cm year^{-1}). This explains cross-section alterations. Fekete (1911) compared cross-sectional surveys at three dates (1842, 1891, and 1906–1909) along the entire Tisza and found that mean depth increased by 140 cm on the Lower Tisza (2.2 cm year^{-1}), meanwhile the area of cross-sections at certain reaches changed (decreased or increased) by 100 m^2. He also calculated a riverbed aggradation of 5.1 million m^3 in 15 years (3,400 m^3 rkm^{-1} year^{-1}) for the Lower Tisza (now in Serbia). Félegyházi (1929) claims that in many cases the low stage cross-sectional area increased as the channel incised. Károlyi (1960a, b) identified up to 300 cm *incision* (a 2.6 cm year^{-1} rate in extreme cases) from low stage water level drop. Repeated channel surveys were applied by Laczay (1967) to investigate morphological change induced by the 1965 flood on the Hungarian Upper Danube. He found that during the flood certain riffles rose to the water level suitable for navigation during low stage periods. He also emphasized the role of low and medium stages in redistributing bedload. On the longer term, between 1903 and 1967, at same site 58,000 m^3 year^{-1} sediment accumulation was observed (Laczay 1968).

Dredging became necessary and resulted in 10% growth of cross-section area (a sediment loss of 1,650 m^3 rkm^{-1} year^{-1} – Csoma and Kovács 1981). From stage-discharge curves analyses Csoma (1987) identified a 30–40 cm incision for the Danube upstream of Budapest as a result of dredging. The Hungarian Lower Danube experienced scouring to 3 m depth locally since channelization.

Along a 25-km section of the Lower Tisza the consequences of various river management strategies between 1890 and 2001 were assessed by Kiss et al. (2002, 2008b). Varying with the morphological situation and the intensity of human intervention, the area of bankfull cross-sections decreased by 4–21% while depth increased by 5–45% due to building bank revetments. Novel technology (seismic measurements of river bed morphology) applied to establish the rate of alluviation on the Middle Tisza yielded no evidence of significant sedimentation and at several locations even large pools were observed (Nagy et al. 2006). In the meantime, channel capacity on the Middle Tisza at Szolnok decreased significantly (Illés et al. 2003a), since, at the same stage and water surface slope, the river channel conveyed 250–300 m^3 s^{-1} less discharge in 1998 than during the 1979 flood. By analyzing the 100-year-long data series of the Szolnok gauge station Dombrádi (2004) claims that cross-section change is no ready explanation for the recent increase of flood stages.

Riverbed morphology is associated with *sediment waves* moving through the channel. For the Hungarian Maros section, repeated channel surveys (Sipos 2006; Sipos et al. 2007), made at both similar and very different water stages, resulted in 55,000 and 89,000 m^3 sediment accumulations in two consequent years during spring floods and no variation in accumulation during low stages. From the analysis of a 15-year data series of low-water cross-sections on River Maros no channel incision could be identified (Kiss and Sipos 2007) – although the studied reach became narrower (Sipos 2006). Studied by analyzing daily cross-sectional data at gauges on the Tisza and on the Maros, the response of the riverbed to the 2000 flood event is very similar – in spite of different morphological and hydrological characteristics (Sipos and Kiss 2003; Sipos et al. 2008).

Channel pattern is a complex indicator of alluvial river behavior (Gábris et al. 2001) as it is closely related to the climate and geology of the catchment. (discharge, slope, sediment amount and type, vegetation). The issue of drivers was first raised in Hungary by Kádár (1969). Recently, Gábris et al. (2001) determined channel pattern changes over the past 20,000 years through the planimetric analyses of palaeochannels, palynology, and radiocarbon dating. Transitions between braided and meandering patterns were caused twice by climatic (water and sediment discharge) and once by tectonic (slope) change. Particularly in the Körös Basin tectonic movements could overwrite climatic changes (Gábris and Nádor 2007). The impact of neotectonic movements on channel parameters are considerable along the Tisza (Timár 2000, 2003). The reaches of high sinuosity coincide with areas of intensive subsidence, increasing local slope. At certain cross-sections the increase of flood levels is partly caused by uneven tectonic subsidence.

On the Hungarian section of the Maros River *human impact* remarkably modified natural processes (Sipos 2006; Sipos and Kiss 2006; Kiss and Sipos 2007). Due to numerous nineteenth century cut-offs the slope of the riverbed doubled and meandering-anastomosing was replaced by braiding. Where further bank stabilization did not involve channel narrowing, decrease in mid-channel island number, i.e., reduction in braided character, is observed. The recovery of a meandering pattern cannot be expected sooner than in 1,000 years.

9.4.2 Floodplain Evolution

Dénes Lóczy

Floodplain areas are the most extensive geomorphological units of the Carpathian Basin (Fig. 9.2). During the Quaternary both hydrological axes of the Carpathian Basin, the Danube (Pécsi 1959) and the Tisza (Bulla 1962), suffered major shifts in their courses that involved far-reaching consequences for the geomorphic evolution of the Great Plain. There have been several attempts at the *reconstruction of* the complete *paleo-drainage network* and, thus, the location of floodplains, for the various stages of the Quaternary (Borsy 1990; Mike 1991; Félegyházi et al. 2004). (It is interesting to note a marked extremity of opinion: the creationist or 'intelligent design' thought, popular internationally, under a mystic varnish also emerged concerning drainage evolution – Burucs 2006) The most often applied *methods* of scientific reconstructions are palynological and sedimentological analyses of channel

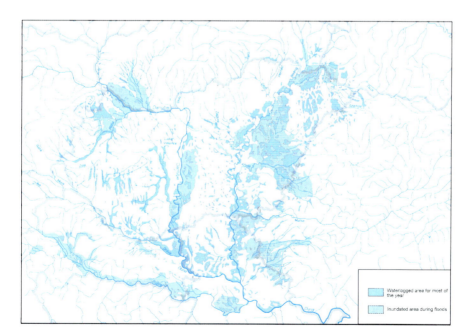

Fig. 9.2 Map of regularly inundated and seasonally waterlogged areas in Hungary before flow regulations

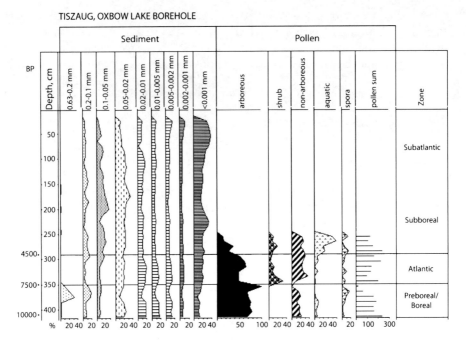

Fig. 9.3 Sedimentological and palynological profiles of an oxbow on the Tisza floodplain (By Félegyházi)

fills for the northeastern Great Hungarian Plain (Fig. 9.3 – Borsy and Félegyházi 1983; Félegyházi 2001); estimation of one-time water discharge from radioactive isotope dating and also palynological investigations for the Middle Tisza region (Gábris et al. 2001); application of remote sensing data, field surveys using GPS and dendrochronology (Sipos and Kiss 2001); GIS interpetation of archive data and channel profiling by radar (Sipos and Kiss 2003).

Channelized watercourses, flood-control dykes, and drainage ditches attest to the largest scale of human transformation of floodplains. *Channel alterations* over historical time have been recorded on the map sheets of the Research Institute for Water Management (VITUKI) Hydrographical Atlas series since the 1970s. For instance, along the Bodrog River (VITUKI 1974), the impacts of damming at Tiszalök on the channel were analyzed on the basis of maps from 1979 to 1986. The results did not show significant rates of alluviation or horizontal channel shift (Szeibert and Zellei 2003). Surveying continued along the Hungarian-Slovak joint section in 1990 and 70 cross-sections were completed. Recently, investigations resumed in order to support the hydrological modeling necessary to assess the changes that occurred during the 1998 and 1999 floods. However, the geomorphological interpretation of the data collected by hydrologists and flood control experts is usually lacking.

The monitoring of *hydrological conditions* is indispensable for the study of floodplain evolution. Water discharge measurements of scientific precision began along the Danube in the 1950s. From 1999 on discharge at extreme water levels was repeatedly recorded along all the tributaries of the Tisza. At flood stages experts of the

Eötvös József College at Baja investigated *sediment transport* along the Middle and Lower Tisza (Zellei and Sziebert 2003) and the Bodrog (Szeibert and Zellei 2003). During the March 2001 flood wave the concentration of suspended load was found to be as high as 1,372 mg L^{-1} at Kisköre and 1,195 mg L^{-1} in the Szolnok section (Szlávik and Szekeres 2003). The difference proves deposition in the Lake Tisza reservoir. It is a remarkable observation that sediment transport in suspension – at least along this river reach – attains its maximum before bankfull stage and, therefore, most of the material transported is deposited in point bars and – in spite of previous expectations – relatively small proportions serve the accretion of natural levees and more remote floodplain sections (Szlávik et al. 1996; Szlávik and Szekeres 2003; Pálfai 2003). The impact of *individual floods* on floodplains resulted in both accumulation and erosion in the Szatmár plain (Borsy 1972); along the Körös rivers (Rakonczai and Sárközi-Lőrinczi 1984) and the Maros and Middle Tisza (Oroszi et al. 2006). Recently, the hydrological, geomorphological and soil conditions of flash floods are intensively studied using GIS modeling (e.g., Czigány et al. 2010).

Human impact has been observed on all the rivers of Hungary (Somogyi 1978, 2000; Csoma 1968). Channel dredging, gravel extraction, confluence replacement, and, along the Danube, even flow diversion (Laczay 1989) brought about major channel and floodplain changes (Lóczy 2007). Since channelization in the nineteenth century sedimentation in study areas on the active Tisza floodplain amounts to 2–3 m for natural levees (Nagy et al. 2001; Schweitzer 2001), the average rate of sedimentation is ca. 3.8 mm year^{-1} (Gábris et al. 2002), for a flood year 18.9 mm year^{-1} (Sándor and Kiss 2006, 2008 – Fig. 9.4), while along the Maros it was 6.3 mm year^{-1} in backswamps and 2.3 mm year^{-1} on inactive natural levees (Oroszi and Kiss 2004; Oroszi et al. 2006) and 1.5–1.8 m on the Körös floodplains since regulations (Babák 2006). New techniques also serve this kind of research, like magnetic susceptibility analysis and x-ray measurements (Sándor and Kiss 2006). Mapping projects show that over the flood-free alluvial terrain large-scale farming gradually obliterates the traces of fluvial processes from the surface (Kis and Lóczy 1985; Balogh and Lóczy 1989; Lóczy and Gyenizse 2011). Recently available digital terrain models, however, present a still highly variable microtopography of floodplains, also useful for practical purposes like the identification of the capacity of designed emergency reservoirs along the Tisza River (Illés et al. 2003a, b; Gábris et al. 2004).

9.4.3 Soil Erosion by Water

Ádám Kertész

Soil erosion is a major environmental problem in agricultural landscapes. In Hungary about 35% of agricultural land is affected (8.5% severely, 13.6% moderately and 13.2% slightly eroded – Stefanovits and Várallyay 1992) as first shown on a soil erosion map at 1:75,000 scale (Stefanovits and Duck 1964), established according to the percentage of soil eroded from an intact soil profile (100%). For large-scale farming 50 million m^3 of annual soil removal from hillslopes in Hungary was estimated (Erődi

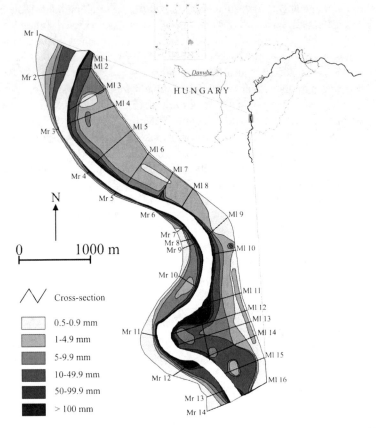

Fig. 9.4 Channel aggradation on the Lower Tisza River at Mindszent caused by the 2006 spring flood (After Sándor and Kiss (2008))

et al. 1965). Erosion-generating extreme (>30 mm day^{-1}) rainfall recurs 4–12 times a year during the growing season (Stefanovits and Várallyay 1992).

Water erosion mostly involves *sheet wash*, *rill*, and *gully erosion* on agricultural land left barren after crops are harvested (Kertész 2008). Limited infiltration due to surface compaction, sealing, crusting and subsurface pan formation intensify sheet and rill erosion on large arable fields (Kertész 1993; Kertész et al. 1995a, b, 2000, 2001, 2002). Gully erosion can reach development rates of 0.5 m per rainfall event and may supply half of sediment delivery from a catchment. Soils formed on loose deposits, such as loess or loess-like sediments in two-thirds of the area, are generally highly erodible. Even lowlands covered by deep loess mantles are prone to gully formation. Large gullies cut into riverbank bluffs (e.g., along the Danube). Loess hill regions with high relative relief (e.g., the Somogy, Tolna, and Szekszárd Hills), particularly if deforested and cultivated, also have dense gully and ravine networks (Jakab et al. 2010). Forest roads make even forested mountains prone to gully erosion.

Soil erosion research, which began in the 1950s (Mattyasovszky 1953, 1956) with the application of the Universal Soil Loss Equation (USLE), was continued with soil erosion mapping projects in the 1960s and 1970s within the framework of landscape studies (Marosi 1979). Fieldwork of landscape ecological approach was combined with laboratory experiments at the University of Debrecen, focused on rainsplash erosion (Kerényi 1991). Complex soil erosion mapping of test areas, primarily the vineyards of the Tokaj-Hegyalja region (Kerényi 1984, 1991; Boros 1996, 2003; Marosi et al. 1992), the Lake Balaton catchment (Dezsény 1984; Kertész et al. 1995a, b; Kertész and Richter 1997), and in Velence Hills (Mezősi et al. 2002) provided data on erosion rates for soils on weathered igneous rocks.

Soil erosion hazard is usually associated with slope inclination categories. In addition to large-scale arable fields of 5–12% slope, most endangered are deforested and cultivated slopes in the 17–25% and 12–17% classes, representing 3.4% and 4.2% of the country, respectively. The local rate of soil erosion is estimated at *plot, hillslope*, or *small catchment* scales (Kertész and Góczán 1988; Kertész 1993; Kertész et al. 1995a, b, 2000, 2001, 2002; Kerényi 1984, 1985, 1991; Krisztián 1998; Marosi et al. 1992; Jakab et al. 2010) with the help of rainfall simulation (Csepinszky and Jakab 1999; Centeri 2002a), pioneered in Hungary by B. Kazó (1966). Among the various factors influencing soil erosion, rainfall erosivity, soil erodibility and vegetation (Centeri 2002a, b), and land cover and tillage (Pinczés et al. 1978; Kertész 2008) were investigated.

The types and rates of hillslope erosion are studied at *test areas*, at first near the villages of Szomód and Bakonynána in northern Transdanubia (Góczán et al. 1973), then at Pilismarót in the Danube Bend (Kertész and Góczán 1988). Runoff, soil loss, redeposition on slope, and nutrient, fertilizer and pesticide losses (Farsang et al. 2006) were measured. Erosion rates under traditional and minimum tillage are studied in the Zala Hills and the efficiency of geotextiles is investigated at Abaújszántó (Northeast-Hungary).

Recently erosion models like EUROSEM (Barta 2001), EPIC (Mezősi and Richter 1991; Huszár 1999) and WEPP are applied (László and Rajkai 2003). Satellite images are also interpreted for estimating soil erosion rates on the catchment of Lake Velence (Verő-Wojtaszek and Balázsik 2008). The Soil Conservation Information and Monitoring System (TIM) operates 18 measurement sites for practical purposes (Nováky 2001). Splash, sheet, rill, and gully erosion are equally intense in the dissected hill regions of Hungary (in the Tolna Hills studied by Balogh and Schweitzer 1996 and in Tokaj-Hegyalja by Kerényi 2006). Detailed investigations of microrill, rill, and gully erosion have begun recently (Jakab et al. 2005; Jakab 2006; Kerényi 2006; Marton-Erdős 2006). The rate of gully erosion is studied using ^{137}C isotope (Jakab et al. 2005).

The rate of soil loss is characterized by the following categories: The average rate of soil formation (2 tons ha^{-1} year^{-1} – Stefanovits 1977) is regarded the upper threshold of sustainable farming and ca 80% of Hungary's area falls within this category. The next limit is 11 tons ha^{-1} year^{-1}, which involves the highest still allowable nutrient loss, based on US experience (14% of Hungary). In areas (6% of

Hungary) with soil loss higher than that arable farming is only possible observing strict regulations (Kertész and Centeri 2006).

9.4.4 Mass Movements

József Szabó

The first geological, stratigraphical, and geomorphological study of mass movements in Hungary was the description of the 1877 landslide at Döröcske in the Somogy loess hills (Inkey 1877). In gentle valley slope segments the remnants of former landslides were recognized. In his stratigraphic descriptions he mentions more than half a dozen collapses, landslides and the resulting heaps, dislocated slices, and rolling surfaces. Mass movements can be decisive in the geomorphic evolution of a region and can significantly influence human activities (Cholnoky 1926).

The disastrous nature of rapid movements along rivers and lakeshores has been long known (at Lake Balaton – Bernáth 1881 and in Buda, the hilly part of Budapest – Schafarzik 1882). When the industrial town Sztálinváros (now: Dunaújváros) was built on the *loess bluff* of the *Danube* major constructions and settlements were found to be threatened by landslides (Domján 1952; Kézdi 1952). Along the right bank of the almost 200 km long Danube section south of Budapest, steep and 20–50 m high bluffs rise (Lóczy et al. 2007). The stratigraphic analyses of Pannonian (Upper Miocene) sediments and loess sequences in the bluffs provided valuable information on landslide generation and pointed out relationships between movements and the water levels of the Danube (Kézdi 1970; Horváth and Scheuer 1976; Pécsi et al. 1979, 1987; Scheuer 1979; Pécsi and Scheuer 1979; Fodor et al. 1983; Fodor and Kleb 1986). The Dunaföldvár landslide served as a textbook example of rotational ("sliced") slides (Pécsi 1971a, b, 1979). Based on the truncated groundplans of Roman structures built right on the Danube bluff, a rate of bluff retreat of 5–15 m per 100 years over the last 2,000 years can be estimated (Lóczy et al. 1989). In the unusually wet year of 2010 loess bluff collapses became more frequent at several locations (Kulcs, Dunaszekcső-Újvári et al. 2009). Undercut high bluffs are also located along the Hernád/Hornad (Szabó 1995, 1996b, 1997) and Rába/Raab rivers.

Movements in *unconsolidated Tertiary sediments* are common in the Outer Somogy Hills (Szilárd 1967) and appear in even larger numbers in the Tolna Hills (Ádám 1969), where they are decisive agents of geomorphic evolution. In the 1950s Gy. Peja made observations in the northern foreland of Bükk Mountains of Tertiary sediments and set up a classification system based on age and landform (identifying "branch," "sister," twinned, and other types) as well as revealing the landslide-inducing impact of brown coal mining (Peja 1956, 1975). In Northern Hungarian landslide-modeled hill regions were studied by J. Szabó (the Sajó-Bódva Interfluve – Szabó 1971; the Cserehát Hills – Szabó 1978, 1985a).

In 1972 a large-scale *mass movement inventory* project was launched. It was coordinated by the Central Geological Office (KFH) and involving broad cooperation in earth sciences. By 1980 as many as 987 localities with mass movements had

9 Recent Landform Evolution in Hungary

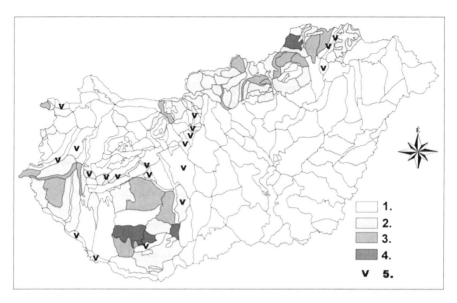

Fig. 9.5 Mass movement hazard in the microregions of Hungary. 1, negligible; 2, slight; 3, moderate; 4, serious; 5, partial areas with extremely high risk (Data from the KFH national inventory processed by Szabó et al. (2008))

been surveyed (Fodor and Kleb 1986; Pécsi and Juhász 1974; Pécsi et al. 1976), but only a fraction of the collected data have been evaluated (Szabó 1996a, b). The inventory has provided ample evidence for *landslide hazard of human origin* and for the significance of recent or presently active processes. In the North Hungarian Mountains man-induced movements amount to ca 29% of total (Fodor 1985). Their share reaches 50% in 14 landscape units (mostly in industrial or mining regions), while 15% of movements were inventoried as of mixed origin and 29% still active at the time of the survey (Fig. 9.5). In remote regions still unknown (mostly fossil) landslides were assumed (Szabó 1996a) and subsequently further areas of surface instability have been described from northern (Ádám and Schweitzer 1985) and southern Transdanubia (Juhász 1972; Lovász 1985; Fábián et al. 2006). The spatial distribution of rapid mass movement hazard can be established with high reliability today (Table 9.1).

In *mountains of volcanic origin* landslide features, only sporadically described before (Láng 1955, 1967; Székely 1989), are found to be decisive elements of the landscape in certain topographical positions – along mountain margins, in fossil caldera rims and valleys (Szabó 1991, 1996b). In addition to fossil landslides on weathered tuffs interbedded between hard lava beds or on unconsolidated bedrocks under volcanics, recent mass movements also occur in considerable number. In the heavily affected North Hungarian Mountains numerous movements are associated with roadcuts, mining, or irresponsible settlement development – as in the case of Hollóháza, North-Hungary (Zelenka et al. 1999; Szabó 2004).

Table 9.1 Description and typology of landslide processes and features in the various landscape types of Hungary (By Szabó)

Material	Location	Slip plane material	Slip plane position	Morphological position	Type	Activity
Hills						
Oligocene clays, marls, loess or travertine mantle	Around Budapest	Clay or marl	At variable depth	Terrace margins, quarry walls	Slumps, stratum slips	Man-induced (quarrying, construction)
Lower to Middle Miocene schlier (locally volcanic cover)	Heves-Borsod Hills, Baranya Hegyhát	Clayey schlier (locally sand)	Deep-seated but usually above base level	Valley slopes, tectonic lines	Stratum slips with huge heaps	Low and reducing activity (man-induced)
Pannonian clays, sands – deep loess mantle	Tolna, Zselic, Outer Somogy Hills	Pannonian beds, loam beds in loess	Deep-seated, close to base level	Marginal tectonic lines, valley slopes	Stratum slips with huge heaps	Low activity, many fossil features
Pannonian clays, sands – loam mantle	Vas Hegyhát	In or below loam	Close to surface, above base level	Valley slopes	Carpet-like slips (creeps)	Moderate activity, recent
Pannonian clays, sands – loess-free	Cserehát Hills, Sajó-Bódva Interfluve	Pannonian beds	Close to surface, above base level	Valley slopes	Stratum slips with heaps, carpet-like slips	Considerable activity, recent
River bluffs						
Pannonian clays, sands with thick loess mantle	Along the Danube	Pannonian beds, loam beds in loess	Deep-seated, often close to base level	Bluff margins	Sliced slides	Mostly recent, locally active
Pannonian clays, sands variable mantle	Lake Balaton	Pannonian beds	Deep-seated, generally close to base level	Bluff margins	Sliced slides, stratum slips with heaps	Considerable activity, mostly recent
Pannonian clays, sands with Plio-Pleistocene sand cover	Rába river	Pannonian beds	Deep-seated, close to base level	Bluff margins	Stratum slips with huge heaps	Recent

9 Recent Landform Evolution in Hungary

Pannonian clays, sands with thin or no loess mantle	Lake Fertő, Hernád river	Pannonian beds	Deep-seated, variable	Bluff margins	Stratum slips with heaps (frequent falls)	Recent, exceptionally active, maximum activity in Hungary
Mountains of volcanic origin						
Badenian and Sarmatan volcanites	Visegrád, Mátra Mountains, less typically: Börzsöny, Cserhát, Zemplén Mountains	Unconsolidated Oligo-Miocene sediments	Deep-seated, generally close to base point	Mountain margins, tectonic lines, caldera rims, slopes of deep internal valleys	Mountain slides, slides of large blocks	Mostly fossil, rarely recent, exceptionally active
Pliocene basalts	Tapolca Basin, Medves Plateau	Weathered tuff beds	Deep-seated, generally close to base point	Mountain margins	Mountain slides, slides of large blocks	Mostly fossil, rarely recent, exceptionally active

The new *mass movement map* of Hungary (Szabó et al. 2008 – Fig. 9.3) describes the risk of these processes by microregions. Among all movements 20% are considered completely inactive (Fodor and Kleb 1986). Even in microregions with the lowest risk mass movements used to be important in landform evolution (e.g., Danube Bend, Börzsöny and Mátra Mountains of andesite or basalt regions). Also in regions of low relative relief serious risk of mass movements is limited to narrow belts, almost negligible in size (e.g., high bluffs along the Danube and Hernad rivers). Such zones are marked by a V symbol (Fig. 9.3).

M. Pécsi's idea that in *landslide classification* the slip plane should be the fundamental criterion (Pécsi 1975) was elaborated into a complete system by J. Szabó (1983, 1985b). The system covers the origin of the slip plane (preformed or syngenetic), its position relative to the base of the slope, its angle with the slope surface, and the consistency of slope material, and thus achieves a natural classification.

For the purposes of prediction the dates of landslides are compared with *precipitation* amounts and the alternation of dry and wet spells (Juhász and Schweitzer 1989; Juhász 1999). The results from the Hernád valley and from hill regions (Szabó 2003) also support the claim that in Hungary mostly winter half-years of extraordinarily high precipitation generate major movements. Although according to the most probable climate change scenario for the country a trend towards Mediterranean conditions is expected, the predicted more arid climate does not reduce landslide hazard since precipitation – and particularly rainfall – will be higher in the winter half-year.

Recently, *new methods* have been introduced into landslide research. Stratigraphical analyses (Juhász 1972) are complemented by the palynological and radiocarbon dating of Holocene movements (Szabó 1997; Szabó and Félegyházi 1997). They indicate that – particularly in mountains of volcanic origin – the transitional warming period from the Late Pleistocene to Preboreal (the thawing of permafrost) significantly increased landslide activity. Most recent applied research is concerned with the complex monitoring of landslide evolution (e.g., Újvári et al. 2009), environmental impact assessment of engineering structures built in landslide-prone areas (Kleb and Schweitzer 2001) and the analysis of the relationship between landslide disasters and social responsibility (Szabó 2004).

9.4.5 Aeolian Processes

József Lóki

In Hungary, there are three major *blown-sand regions*, located on Pleistocene alluvial fans: Inner Somogy southwest of Lake Balaton (Marosi 1970; Lóki 1981), the Kiskunság (Little Cumania) on the Danube-Tisza Interfluve (Bulla 1951), and the Nyírség (Borsy 1961) in the northeastern corner of the country (Fig. 9.6).

There are considerable variations in their physical environments and, consequently, in their landforms. In the most humid of the sand regions, the lower-lying *Inner Somogy*, the depth of alluvial sand accumulation and the maximum height of sand landforms is 5–8 m. Numerous stabilized rounded blowouts of less than 40 m

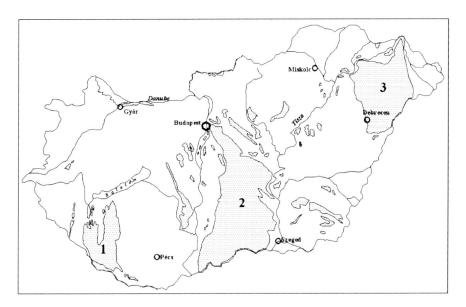

Fig. 9.6 Blown-sand regions in Hungary. 1, Inner Somogy; 2, Kiskunság; 3, Nyírség

diameter, formed on alluvial fans with shallow groundwater table, are most typical. Residual ridges between wind furrows, protected by vegetation from erosion, are also relatively common. Their maximum height does not reach 10 m. (The investigation of sand regions and wind erosion started with papers on wind furrow/blowout dune assemblages and separating residual ridges – Cholnoky 1902 – and on river bank dunes along the Tisza River – Cholnoky 1910). Extensive sand sheets also occur (Marosi 1970). In the *Kiskunság* even more extensive flat sand sheets and narrow stripes of sand accumulation are characteristic. Wind furrows are deepened into blown sands redeposited in several stages of the Last Glaciaton. Blowout dunes of longitudinal type (accumulations of sand arranged in winding ridges of up to 1,500 m length and occasionally 10 m high – Kádár 1935) are common alternating with deflation flats (locally with saline ponds). In the northeastern *Nyírség* the greatest variety of landforms occurs: markedly elongated wind furrows and blowout dunes (up to 18 m height) (Borsy 1961), while in the deep sand accumulation of the southern Nyírség, asymmetric dunes and marginal dunes are predominant. The most widespread landforms are parabolic dunes, developed around blowouts of oval shape and regarded the counterparts of desert barchans in sand regions of the temperate belt (Kádár 1938). Regular parabolic dunes are rather rare, they are mostly asymmetric with poorly developed western horns. Their shape is equally controlled by wind, topographic and vegetation conditions (Lóki 2004). The abundant recharge of sand promotes the formation of regular parabolic dunes. In the case of reduced recharge and wind blowing out sand sideways from oval blowouts asymmetric dunes with longer horns develop. The largest landforms of blown-sand regions are irregular closed deflational flats of locally more than 2 km diameter (mostly in the northern Nyírség).

Sand dune surfaces are artificially kept vegetation-free by the staff of the Kiskunság National Park in the environs of Fülöpháza in order to present sand movement to visitors. Elsewhere dunes are afforested by Robinias and Austrian pines as well as by the plantation of orchards (apricot) and vineyards.

Sand mobilization occurred in several *phases* in Hungary (Gábris 2003; Kiss in Mezősi 2011). Marosi (1967) claimed that alluvial fan building in the Carpathian Basin first ended in the Inner Somogy sand region, where aeolian action was the most enduring. In the driest spells of the Würm glaciation wind was the predominant agent of geomorphic evolution. The age of landforms is confirmed by periglacial features, ice wedge cast filled by sands with 'kovárvány' bands (Marosi 1966, 1967, 1970). In the Nyírség ^{14}C dating and on the northern Danube-Tisza Interfluve malacological and stratigraphical dating from fluvial deposits rich in shells revealed that blown sand was mobilized in the Upper Pleniglacial and again in the Dryas (Borsy et al. 1985; Lóki et al. 1993; Lóki 2004). Sand movements induced by human activities have been assumed for the Preboreal, Boreal, and Atlantic phases (Kádár 1956; Marosi 1967; Borsy 1977, 1991; Gábris 2003; Lóki 2004; Nyári and Kiss 2005; Ujházy et al. 2003). Subboreal wind erosion was also confirmed (Kiss et al. 2008a, b). Archaeological and OSL datings (Gábris 2003; Ujházy et al. 2003; Nyári and Kiss 2005, Kiss et al. 2006; Nyári et al. 2006, 2007) support the view that sand was blown on several occasions in the early Subatlantic Phase, in the Iron Age and during the Migrations. Bronze Age movements were caused by overgrazing (Lóki and Schweitzer 2001; Gábris 2003; Nyári and Kiss 2005). Historical sand movements in the eighteenth and nineteenth centuries are related to deforestation and arable farming (Marosi 1967; Borsy 1977, 1991).

In the mid-twentieth century, large-scale farming and improper agrotechnology led to increased wind damage on fields (Bodolay 1966). At *present, wind erosion* is limited to dry surfaces uncovered by vegetation, i.e., it takes place primarily on ploughed surfaces in spring or autumn as well as on surfaces without a snow cover in winter (Kiss et al. 2008a, b). At wind velocities of 5.5–6.0 m s^{-1}, sand motion reaches surprisingly high intensity (Borsy 1991). Deflation also affects dried marshes as well as heavier soils, which become dusty during tillage (Lóki 2004).

Wind erosion hazard affects 16% of Hungary (Stefanovits and Duck 1964). Not only sand and peat surfaces but even more humous and fertile soils are also affected (Stefanovits and Várallyay 1992). In soil conservation, environment-friendly solutions, optimal irrigation, vegetation cover, and tillage methods are sought (Lóki 2004). The erosion on berms of alkali flats in grasslands is also remarkable (Rakonczai and Kovács 2006). On the basis of measurements (e.g., on the Danube–Tisza Interfluve – Szatmári 1997) and experiments (Lóki and Négyesi 2003) the soils are referred to grades of wind erosion hazard (Fig. 9.7):

– Insignificant hazard (0): on silt-clay and clay soils, where >10.5 m s^{-1} wind speeds are necessary to generate sand motion (30.2% of the area of Hungary);
– Moderate hazard (1): on loam and silty loam soils with critical wind speeds between 8.6 and 10.5 m s^{-1} (43.3%);

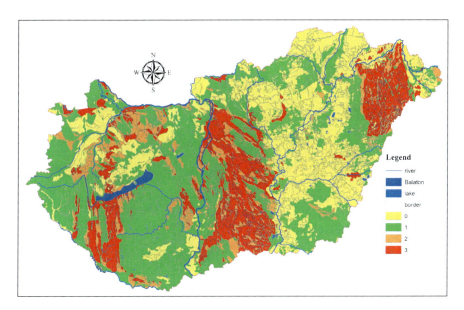

Fig. 9.7 Wind erosion hazard in Hungary (by J. Lóki). For explanation see the text

– Medium hazard (2): on sandy loam soils with thresholds of 6.5–8.5 m s^{-1} wind speeds (9.4%) and
– Serious hazard (3): on sand and loamy sand soils as well as peat and muck surfaces from which more than threefold more soil is removed by wind than in the "insignificant hazard" category. Critical wind speed here is below 6.5 m s^{-1}.

Global warming involves aridification and increasing weather extremities in Hungary. It is reflected in the alternation of spells of intensive rainfall with periods of severe drought, when wind erosion is expected to intensify.

9.4.6 Karstification

Márton Veress

Karstic rocks appear on 1.5% of Hungary's area: in the Bükk and Aggtelek Mountains in North-Hungary, in some members of the Transdanubian Mountains (Bakony, Vértes, Gerecse, Pilis and Buda Mountains), in the Mecsek Mountains and Villány Hills in Southern Transdanubia, and in the Balf Hills (Fig. 9.8). Significant parts of the Transdanubian Mountains and Mecsek Mountains are covered karst, both autogenic and allogenic, and there are also patches in the Bükk and Aggtelek Mountains.

The origin of karst features like dolines, ponors, and poljes were first explained by J. Cholnoky (1926) and the significance of climatic and biogenic factors in

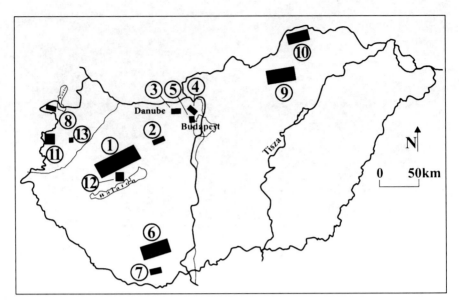

Fig. 9.8 Karst areas in Hungary (1–10) and the localities mentioned in the text (11–13). 1, Bakony; 2, Vértes; 3, Gerecse; 4, Pilis; 5, Buda Mountains; 6, Mecsek Mountains; 7, Villány Hills; 8, Balf Hills; 9, Bükk; 10, Aggtelek Mountains; 11, Kőszeg Mountains; 12, Kál Basin; 13, Bük thermal bath

karstification was underlined by L. Jakucs (1977, 1980). Based on decades of field measurements in the Aggtelek Mountains, L. Zámbó (1985) set up a hydrogeochemical model for the corrosional behavior of clay doline fills with different rates of solution in zones identified in cross-section. D. Balázs (1991) conducted laboratory experiments on karst corrosion and classified tropical karst processes and landforms observed on all continents.

Recent karst development involves primary or secondary processes. *Primary processes* are solution, mechanical erosion (in caves) and precipitation. *Solution* is the most diverse: it can be caused by *rainwater* or *thermal water* and it can take place *subsoil* and *under the karstwater table* (due to mixing corrosion or turbulent flow). In erosion caves, however, solution is only a secondary karst process. *Non-karstic processes* induced by primary karstification are collapses (in caves, on covered karsts), sediment transport, reworkings (in caves or on covered karst by pluvial erosion), accumulations and piping in sedimentary cover (in depressions of covered karst).

Among surface features, *dolines* occur in high density on the Bükk Plateau and in the Mecsek Mountains, where two generations of them are distinguished (Szabó 1968). On the planated surface of the Balf Hills, near Lake Fertő/Neusiedlersee, there are shallow *uvalas* of a few 100 m diameter, developed on the porous Leitha limestone. The uvalas are still growing today as the roofs of minor cavities in doline walls cave in (Prodán and Veress 2006). Locally, there are canyon-like *epigenetic valleys* in limestone. *Ponors* (swallow holes) developed at junctions are connected to erosional caves, mainly in the Bükk and Aggtelek Mountains (Baradla Cave – Jakucs 1977). *Karren* are less typical in Hungary. Grike karren and kamenitzas are produced by subsoil solution. Veress et al. (1998) claim that notches develop on any rock with calcareous minerals (e.g., in Kőszeg Mountains). Kamenitzas are also found on sandstones with amorphous silica (in the Kál Basin, Balaton Uplands).

Secondary karst landforms include *"transformed" ponors* (swallow dolines) in valley junctions. With the stream incising into the rock, the junction is retreating and a new ponor develops by solution and the former ponor becomes a doline (Hevesi 1980). Such forms predominate in the epigenetic valleys of the Bükk and Aggtelek Mountains. Today some of the epigenetic valleys are *blind valleys* and develop further by *corrosion* (e.g., in the Aggtelek Mountains). *Collapse dolines* do not often occur in the Hungarian karsts. Small-sized features due to collapse, however, are developing at present as the thin roofs of cavities close to the surface cave in. Such dolines are a few meters in diameter (in the Bakony Mountains – Veress 2000; in the Balf Hills – Prodán and Veress 2006).

Infiltrating waters mainly cause solution in the epizone of uncovered karst. The rock mass is fragmented by the solution of the water which percolates along crevices (Veress and Péntek 1996). As the debris is gradually dissolving, the surface is subsiding. *Solution dolines* develop in sites where the process takes place more rapidly. *Pits* develop on the limestone floor of covered karst where the sedimentary cover is thin. The void created at the pits is translated to the limestone floor and *covered karst depressions* develop (Veress 2000). *Karst gorges* occur on the covered karst of the Bakony Mountains (Kerteskő Gorge, Ördögárok). The gorge floors are above the karstwater table and function like ponors: the throughflowing waters partly seep into the karst. Solution by seepage contributes to their present incision (Veress 1980, 2000). Broad covered karst depressions develop in the sedimentary cover (like on the floor of Tábla Valley, Bakony Mountains, Veress 2006, 2007 – Fig. 9.9).

In the epiphreatic and phreatic zones, *cavities* of various size develop by mixing corrosion and turbulent flow under the karst water table, while *effluent caves* develop along the karst margin. The mode of cave development varies with to the position of the karstwater table. In Hungary there are almost 3,000 caves (cavities longer than 2 m): the longest is the Baradla Cave (25 km, Aggtelek Karst, with the highest chamber of 60 m) and the deepest is the István-lápa Shaft (250 m, Bükk Mountains).

Calcareous *precipitations* may develop from water flow on the surface or in caves (calcareous sinter), from dripping water in the epiphreatic zone (dripstones) and in the stagnant karst water zone. In Hungary major *travertine mounds* and *rimstone bars* are the Veil Waterfall (in the Szalajka Valley, Bükk Mountains), the Szinva Waterfall (in Szinva Valley, Bükk Mountains), at the Bolyamér Spring, at the Kecskekút Spring, at the Jósva Spring and along the bank of the Jósva stream (Aggtelek Mountains, Sásdi 2005), and the rimstone bars of the Melegmány Valley (Mecsek Mountains). Human-induced precipitation from the thermal (57°C) water of Bük spa (Vas County) with high Ca^{2+} and Na^+ ion concentrations (Veress 1998) is observed on the walls and pipes of the basins of the public bath.

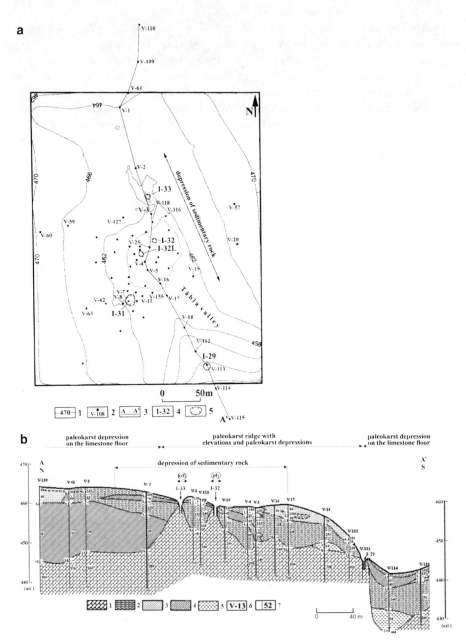

Fig. 9.9 A depression in the sedimentary cover on the floor of the Tábla Valley (Bakony Mountains – Veress 2006). (**a**) Vertical Electric Sounding (VES) profile. 1, contour line; 2, site of VES measurement; 3, location of profile; 4, mark of the covered karst feature; 5, covered karst feature. (**b**) Geological and morphological profile. 1, limestone; 2, limestone debris; 3–4, loess with debris; 5, clay; 6, VES number; 7, geoelectrical resistance (ohms)

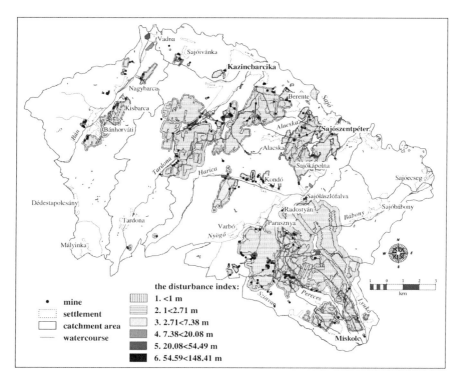

Fig. 9.10 Landscape transformation in the Borsod Coal Mining Area (Sütő 2000). The disturbance index shows how thick a layer was removed from the surface

9.4.7 Human Impact of Mining

László Sütő
Péter Rózsa

Human activity has become an immanent part of landscape evolution (Szabó et al. 2010). In 2005 there were 1217 listed mining sites in Northern Hungary alone (Farkas 2006). Major landscape transformation has been caused by the 250-year old *brown coal mining* (Fig. 9.10). During the "heroic age" of the coal mining – from the mid-eighteenth to the mid-twentieth century – manual chamber excavation with wood securing was applied. In the Borsod Coal Mining Area, near Arló village (south of Ózd town), a lake was formed by landslides as early as 1863, i.e., 11 years after the beginning of large-scale forest clearance for mining (Peja 1956; Leél-Őssy 1973). After World War II self-securing mechanized frontal excavation technology was introduced in most mines. It was more productive, but much larger connected cavities resulted.

Undermining is another great problem, for example, in Lyukóbánya and Pereces (near Miskolc (Sütő et al. 2004)), where the total length of the mine galleries is almost 50 km (equal to that of the underground tunnels in Budapest).

Similarly, in the Salgótarján–Medves Hill area (northern part of the Nógrád Coal Mining Area) 268 of 543 depressions are attributed to coal mining. Prolonged *subsidence* is typical in the foothills of the North-Hungarian, Transdanubian, and Mecsek Mountains, where mines are closed now. In the vicinity of the huge open lignite pits in the Mátra and Bükk forelands ca 0.2 m ground subsidence has been recorded until the early 1980s (Fodor and Kleb 1986) and subsidence areas are also mapped in the Transdanubian Mountains (near Dorog, Tokod, Oroszlány, and Tatabánya – Juhász 1974). Above the northern mining field at Nagyegyháza, 99.9% of the total subsidence happened 8 years after mines were closed. For another mine (Mány I/a) 99.9% of subsidence is estimated to happen in 12 years after closing (Ládai 2002).

To eliminate water inrush risk, from 1965 to 1988, 300 and 600 m^3 min^{-1} of water was pumped out of coal and the *bauxite* mines, respectively. The discharge of many streams decreased, springs dried out, the water level of the Hévíz Lake and Tapolca Cavern Lake (both are well-known tourism attractions) dropped. This led to heated public discussion, which contributed to the decline of bauxite mining. Now the opencast bauxite mine near Gánt (southern Vértes Mountains) displays 28 m high walls with tropical cone karst (Dudich 2002).

According to F. Erdősi (1987), in the environs of Mecsek Mountains, 135 millions m^3 of earth was moved by hard coal mining activities since World War II. The biggest, almost 200 m deep opencast pits are now being reclaimed at Vasas and in Karolina Valley in Pécs. The largest subsidence in Mecsek is the Pécs-Somogy trough of 13.5 km^2 area and 27 m depth – sevenfold larger than the undermined area. In the Juhász Hill, near Nagymányok, *sinkholes* formed (Erdősi 1987). In some cases the surface may rise locally after mine closure. Near Pécs and Komló rises of up to 12 cm resulted from the hydration of near-surface clays and replenishment of confined water reservoirs (Somosvári 2002).

Among *accumulation* landforms, from 1950 to 1980, 248 *spoil heaps* of ca 200 million m^3 total volume were built in five mining counties of Hungary (Baranya, Borsod-Abaúj-Zemplén, Heves, Komárom-Esztergom, Nógrád) (Egerer and Namesánszky 1990). For instance, the total volume of the terraced twin tips of Béke and István Shafts near Pécs is 11.2 millions m^3, their maximum height is 67 m, and they cover a triangular area, 750 m long and 550 m wide (Erdősi 1987). Spoil heaps are affected by physical and chemical weathering, burning, deflation, intensive mass movements, and gully erosion. One of the largest in the Borsod Coal Mining Area is more than 100 m high and accommodates 1.2 million m^3 spoil (Sütő 2000, 2007 – Fig. 9.11). Slides cause 8 m high ruptures on its slopes (Homoki et al. 2000).

In Hungary *ore mining* (copper, iron, lead, zinc, uranium, and bauxite) induced locally significant landscape changes (at Recsk, Rudabánya, Gyöngyösoroszi, Telkibánya, Kővágószőlős, Gánt, Kincsesbánya, Nyírád). At Kővágószőlős (Mecsek Mountains), to obtain 20,300 tons of uranium, 46 million tons of rock had to be moved. The subsidence area is almost 42 km^2 and the groundwater regime is also altered. Nine 17-m high percolation prisms are under reclamation.

Stone quarrying is practiced around almost every settlement in the mountains of Hungary. In Mátra Mountains 57 major quarries of 240 ha are listed, and less than 10 of them are reclaimed (Dávid 2000). On Medves Hill (its area is a mere 32 km^2)

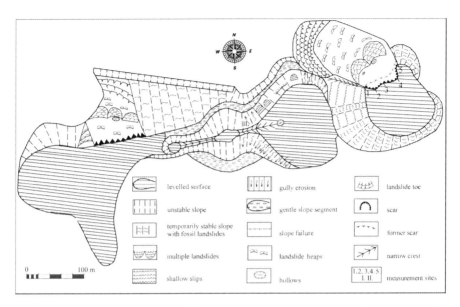

Fig. 9.11 Sketch of a spoil heap (in the Ádám Valley near Kazincbarcika) showing the geomorphic processes operating on them (Sütő 2000)

basalt quarries and their wastes cover 27 and 24 ha, respectively (Karancsi 2002; Rózsa and Kozák 1995). Quarry walls develop like natural cliffs: debris fans, gorges, block slopes, and, in pits, ponds can form. The lubrication of bedding planes by water may result in sagging (Karancsi 2002). Ponds in the abandoned quarries are often spectacular tourist attractions (like the "Tarn" in a former millstone quarry of Megyer Hill near Sárospatak town). Some quarries expose geological features and are sources of scientific information (e.g., Kálvária Hill of Tata, Ság Hill near Celldömölk, Hegyestű near Lake Balaton, Nagy Hill of Tokaj) (Rózsa and Kozák 1995; Pap 2008), presented by nature trails. Some quarries have almost demolished entire hills. (Esztramos Hill in North-Hungary was reduced from 380 to 340 m; the basalt hill near Zalahaláp was totally removed and 7 million m³ of limestone was quarried from the Bél-kő within a century (Hevesi 2002)).

Following land reclamation, part of the abandoned open pits serve as recreation or exhibition areas (Pap 2008). Elsewhere rehabilitation takes place spontaneously (Szebényi 2002; Gasztonyi 2002; Kalmár and Kuti 2005). Opportunities for the *complex rehabilitation* and use of spoil heaps were investigated, among others, by Z. Karancsi (2002) for the Medves Hill area; by Hahn and Siska–Szilasi (2004), E. Homoki and co-workers (2000) and L. Sütő (2000, 2007) for the Borsod Coal Mining Area; by L. Dávid (2000) for Mátra Mountains and by P. Csorba (1986) for the Tokaj Mountains. The Tokaj Nagy Hill of 20 km² area is one of the best preserved volcanic cones in Hungary, is built up of locally several hundred meters thick pyroxene dacite lava flows, exposed in several quarries. A volcanological nature trail and a tourist resort next to the pond in the center of the quarry pit are proposed (Rózsa and Kozák 1995).

9.5 Conclusions

Dénes Lóczy

In the present-day Hungary of lowland nature fluvial geomorphic processes (channel and floodplain evolution) are of particular significance and intensively studied. Along some rivers significant adjustments of channel geometry have been observed since flow regulation in the nineteenth century. Almost all the area of hills and low mountains (if not forested) is affected by water erosion, particularly gully development in loess regions. The topography of the three major blown-sand regions and some other sand-covered alluvial fans and terraces is mostly due to aeolian erosion and deposition. All sand dunes, however, have been stabilized. Although the areas affected by mass movements and karst processes are limited in extension, they offer a variety of landforms for geomorphologists – some still actively evolving today. The modeling of such processes and the precise measurement of the rate of landform evolution is a central task of geomorphological research. Applied studies are directed at the estimation of the significance of ever intensifying human impact on the landscape, which has become the primary driver of recent geomorphic evolution in Hungary, too.

References

Ádám L (1969) A Tolnai-dombság kialakulása és felszínalaktana (Origin and geomorphology of the Tolna Hills). Akadémiai Kiadó, Budapest, 186 p (in Hungarian)

Ádám L, Schweitzer F (1985) Magyarázó Dunaalmás-Neszmély-Dunaszentmiklós közötti terület felszínmozgásos térképéhez (Memoir to the mass movements map of the Dunaalmás–Neszmély– Dunaszentmiklós), vol 33, Elmélet – Módszer – Gyakorlat. Geographical Research Institute, HAS, Budapest, pp 108–169, (in Hungarian)

Babák K (2006) A Hármas-Körös hullámterének feltöltődése a folyószabályozások óta (Alluviation on the active floodplain of the Triple Körös river since channelization). Földrajzi Értesítő 55(3–4):393–399 (in Hungarian)

Balázs D (1991) A karsztos mélyedések globális rendszerezése. Dolinák – dolinaegyüttesek (Global systemization of karst depressions: dolines, doline assemblages). Karszt és Barlang 1–2:35–44 (in Hungarian)

Balla Z (1982) Development of the Pannonian Basin basement through the Cretaceous-Cenozoic collision: a new synthesis. Tectonophysics 88:61–102

Balogh J, Lóczy D (1989) Mapping ecological units on a Danubian flood-plain. In: Pécsi M (ed) Geomorphological and geoecological essays. Geographical Research Institute, Hungarian Academy of Sciences, Budapest, pp 73–85

Balogh J, Schweitzer F (1996) Az erózió térképezése és számítása egy Tolna-dombsági mintaterületen (Mapping and estimating soil loss in a test area of Tolna Hills). In: Schweitzer F, Tiner T (eds) Nagyberuházások és veszélyes hulladékok telephely-kiválasztásának földrajzi feltételrendszere (Geographical conditions for site selection of major investments and toxic waste disposal). Geographical Research Institute, HAS, Budapest, pp 149–165 (in Hungarian)

Barta K (2001) A EUROSEM talajeróziós modell tesztelése hazai mintaterületen (Testing EUROSEM soil erosion model in Hungarian test areas). In: Papers of the First Hungarian Geographical Conference, Szeged, 2001. CD, 9 p (in Hungarian)

Bernáth J (1881) Egy Balaton-parti földsüllyedésről (On a subsidence at Lake Balaton). Földtani Közlöny 7:137–140 (in Hungarian)

Blanka V, Kiss T (2006) Kanyarulatfejlődés vizsgálata a Maros alsó szakaszán (Meander evolution along the Lower Maros River). Hidrológiai Közlöny 86(4):19–23 (in Hungarian)

Blanka V, Sipos Gy, Kiss T (2006) Kanyarulatképződés tér- és időbeli változása a Maros magyarországi szakaszán (Spatial and temporal changes in meander formation along the Hungarian Maros River). CD. In: Third Hungarian geographical conference, Budapest (in Hungarian)

Bodolay I-né (1966) A talajművelés szerepe a szélerózió fellépésében (Tillage and wind erosion). Agrokémia és Talajtan 15:183–198 (in Hungarian)

Boros L (1996) Tokaj-Hegyalja szőlő-és borgazdaságának földrajzi alapjai és jellemzői (Geography and characteristics of vini- and viticulture in Tokaj-Hegyalja). vol 3, Észak-és Kelet-Magyarországi Földrajzi Évkönyv. Szabolcs–Szatmár–Bereg County Pedagogical Institute, Nyíregyháza. 322 p (in Hungarian)

Boros L (2003) Talajeróziós vizsgálatok Tokaj-Hegyalján (Soil erosion studies in Tokaj-Hegyalja). In: Frisnyák S, Gál A (eds) Szerencs és a Zempléni-hegység (Szerencs and the Zemplén Mountains). Nyíregyháza College, Nyíregyháza, pp 31–39 (in Hungarian)

Borsy Z (1961) A Nyírség természeti földrajza (Physical geography of the Nyírség). Akadémiai Kiadó, Budapest, 227 p (in Hungarian)

Borsy Z (1972) Üledék- és morfológiai vizsgálatok a Szatmári-síkságon az 1970. évi árvíz után (Sedimentological and geomorphological investigations in the Szatmár Plain after the 1970 flood). Földrajzi Közlemények 20(1–2):38–42 (in Hungarian)

Borsy Z (1977) A Duna-Tisza köze homokformái és a homokmozgás szakaszai (Sand landforms on the Danube–Tisza Interfluve and phases of sand movements). Alföldi tanulmányok 1:43–53 (in Hungarian)

Borsy Z (1990) Evolution of the alluvial fans of the Alföld. In: Rachocki AH, Church M (eds) Alluvial fans: a field approach. Wiley, London, pp 229–246

Borsy Z (1991) Blown sand territories in Hungary. Zeitschrift für Geomorphologie Supplement-Band 90:1–14

Borsy Z, Félegyházi E (1983) Evolution of the network of watercourses in the end of the Pleistocene to our days. Quat Stud Poland 4:115–125

Borsy Z, Csongor É, Lóki J, Szabó I (1985) Recent results in the radiocarbon dating of wind-blown sand movements in Tisza-Bodrog Interfluve. Acta Geogr Debrecina 22:5–16

Bugya T, Kovács IP (2010) Identification of geomorphological surfaces by GIS and statistical methods in Hungarian test areas. In: Lóczy D, Tóth J, Trócsányi A (eds) Progress in geography in the European capital of culture 2010. Imedias, Kozármisleny, pp 249–260

Bulla B (1951) A Kiskunság kialakulása és felszíni formái (Development and landforms of the Kiskunság). Földrajzi Könyv- és Térképtár Értesítő 10–12:101–116

Bulla B (1962) Magyarország természeti földrajza (Physical geography of Hungary). Tankönyvkiadó, Budapest, 423 p

Bulla B, Mendöl T (1947) A Kárpát-medence földrajza (Geography of the Carpathian Basin). New edition. Lucidus Kiadó, Budapest (1999), 420 p (in Hungarian)

Burucs Z (2006) A Kárpát-medence vízrajzának geometriai összefüggései. Véletlen vagy inkább teremtett a vizeink rendszere (Geometry of drainage in the Carpathian Basin: incidental or created water system?). In: Third geographical conference, Budapest, 2006. CD, 10 p (in Hungarian)

Centeri Cs (2002a) Importance of local soil erodibility measurements in soil loss prediction. Acta Agron Hung 50(1):43–51

Centeri Cs (2002b) The role of vegetation cover in soil erosion on the Tihany Peninsula. Acta Bot Hung 44:285–295

Cholnoky J (1902) A futóhomok mozgásának törvényei (Laws of blown-sand movement). Földtani Közlöny 32:6–38 (in Hungarian)

Cholnoky J (1910) Az Alföld felszíne (Surface of the Great Pain). Földrajzi Közlemények 38(4):413–436 (in Hungarian)

Cholnoky J (1926) A földfelszín formáinak ismerete (morfológia) (Earth surface landforms: morphology). Egyetemi nyomda, Budapest, 295 p (in Hungarian)

Csepinszky B, Jakab G (1999) Pannon R-02 esőszimulátor a talajerózió vizsgálatára (Rainfall simulator Pannon R-02 for soil erosion studies). 41st Georgikon Days, Keszthely, pp 294–298 (in Hungarian)
Csoma J (1968) A felső-dunai mellékágrendszerek mederváltozása (Channel changes along the distributary branch system of the Upper Danube). Földrajzi Értesítő 17(3):309–323 (in Hungarian)
Csoma J (1973) A korszerű folyószabályozás alapelvei és módszerei (Principles and methods of modern river regulation). VITUKI, Budapest (in Hungarian)
Csoma J (1987) A nagymarosi vízlépcső alatti Duna-meder vizsgálata (Investigation of the Danube channel downstream the Nagymaros Barrage). Vízügyi Közlemények 69(2):286–296 (in Hungarian)
Csoma J, Kovács D (1981) A Duna Rajka–Gönyü közötti szakaszán végzett szabályozási munkák hatásának értékelése (Assessment of the impacts of regulation measures along the Rajka–Gönyü reach of the Danube River). Vízügyi Közlemények 63(2):267–294 (in Hungarian)
Csontos L, Nagymarosi A, Horváth F, Kováč M (1992) Tertiary evolution of the Intra-Carpathian area: a model. Tectonophysics 208:221–241 (in Hungarian)
Csorba P (1986) Geoökológiai vizsgálatok a bodrogkeresztúri riolittufa meddőhányókon (Geoecological investigations on the rhyolite spoil heaps of Bodrogkeresztúr). Földrajzi Értesítő 35(1–2):57–78
Czigány Sz, Pirkhoffer E, Geresdi I (2010) Impact of extreme rainfall and soil moisture on flash flood generation. Időjárás 114(1–2):11–29
Dávid L (2000) A kőbányászat mint felszínalakító tevékenység tájvédelmi, tájrendezési és területfejlesztési vonatkozásai Mátra-hegységi példák alapján (Landscape protection, landscape management and regional development aspects of quarrying as a geomorphic activity, based on examples from the Mátra Mountains). Manuscript Ph.D. thesis. University of Debrecen, Debrecen, 160 p (in Hungarian)
Dezsény Z (1984) A lehetséges erózió térképezése és az erózióveszély vizsgálata a Balaton-vízgyűjtő területén (Mapping potential erosion and study of erosion hazard in the Lake Balaton catchment). Vízügyi Közlemények 66:311–324 (in Hungarian)
Dombrádi E (2004) Vízhozam- és vízállás-idősorok analízise a folyómeder állapotváltozásainak kimutatására (Analysis of water discharge and water level series to survey changes in the conditions of river channels). Hidrológiai Közlöny 84(1):57–60 (in Hungarian)
Domján J (1952) Közép-dunai magaspartok csúszásai (Landslides of bluffs along the middle Danube section). Hidrológiai Közlöny 32:416–423 (in Hungarian)
Dudich E (2002) Gánt (Vértes hegység). GEO 2002. Sopron 2002. augusztus 21-25. Kirándulásvezető (Field Guide). Magyarhoni Földtani Társulat, Budapest (in Hungarian)
Egerer F, Namesánszky K (1990) Magyarország meddőhányóinak katasztere (Inventory of spoil heaps in Hungary). Manuscript. University of Miskolc, Miskolc, 71 p + 6 maps (in Hungarian)
Erdősi F (1987) A társadalom hatása a felszínre, a vizekre és az éghajlatra a Mecsek tágabb környezetében (Impact of society on the surface, water and climate in the broader environs of the Mecsek Mountains). Akadémiai Kiadó, Budapest, 228 p (in Hungarian)
Erődi R, Horváth V, Kamarás M, Kiss A, Szekrényi B (1965) Talajvédő gazdálkodás hegy- és dombvidéken (Soil-conserving cultivation in hills and mountains). Mezőgazdasági Kiadó, Budapest, 240 p (in Hungarian)
Fábián SzÁ, Kovács J, Babák K, Lampért K, Lóczy D, Nagy A, Schweitzer F, Varga G (2006) Geomorphic hazards in the Carpathian foreland, Tolna Hills. Studia Geomorphologica Carpatho-Balcanica 40:107–118
Farkas J (2006) A Magyar Geológiai Szolgálat 2005. évi működési jelentése (Annual report of the Hungarian Geological Survey). Hungarian Geological Survey, Budapest, 38 p (in Hungarian)
Farsang A, Kitka G, Barta K (2006) A talajerózió szerepe a talaj foszfor-háztartásában (The role of soil erosion in the phosphorus budget of soils). In: Kiss A, Mezősi G, Sümeghy Z (eds) Táj, környezet és társadalom (Landscape, environment and society). University of Szeged, Szeged, pp 179–189 (in Hungarian)

Fekete Zs (1911) A Tisza folyó medrének közép-keresztszelvényei (Cross-sections of the Tisza River). Vízügyi Közlemények 4–6:141–148 (in Hungarian)

Félegyházi P (1929) A Tisza folyó jellegzetes szakaszainak és az egész Tiszának átlagos szelvényadataiban a szabályozás kezdete óta 1922. évig beállott változások és azok összehasonlítása (Changes in cross-sections of the Tisza River and along its characteristic sections from regulation to 1922). Vízügyi Közlemények 9:93–102 (in Hungarian)

Félegyházi E (2001) A Berettyó-Kálló vidék és az Érmellék medertípusainak elemzése (Analysis of channel types in the Berettyó-Kálló and Érmellék region). In: Papers of Hungarian geographical conference 2001, Szeged. CD, 7 p (in Hungarian)

Félegyházi E, Szabó J, Szántó Zs, Tóth Cs (2004) Adalékok az Északkelet-Alföld pleisztocén végi, holocén felszínfejlődéséhez újabb vizsgálatok alapján (Contributions to the Late Pleistocene-Holocene geomorphic evolution of the Northeast-Alföld based on recent research). In: Current results in Hungarian geography: papers of the second Hungarian geographical conference, Szeged. CD, 8 p (in Hungarian)

Fiala K, Kiss T (2005, 2006) A középvízi meder változásai az 1890-es évektől az Alsó-Tiszán (Changes in medium flow channel of the Lower Tisza River since the 1890s) I-II. Hidrológiai Közlöny 85(3): 60–68 and 86(5): 13–17 (in Hungarian)

Fodor P (1985) Észak-Magyarország nyugati részének felszínmozgásai (Mass movements in western North-Hungary). Mérnökgeológiai Szemle 34:31–44 (in Hungarian)

Fodor T, Kleb B (1986) Magyarország mérnökgeológiai áttekintése (Engineering geological overview of Hungary). MÁFI, Budapest, 199 p (in Hungarian)

Fodor P, Horváth Zs, Scheuer Gy, Schweitzer F (1983) A Rácalmás-kulcsi magaspartok mérnökgeológiai térképezése (Engineering geomorphological mapping of the Rácalmás-Kulcs bluffs). Földtani Közlöny 113:313–332 (in Hungarian)

Fodor L, Csontos L, Bada G, Győrfi I, Benkovics L (1999) Tertiary tectonic evolution of the Pannonian Basin system and neighbouring orogens: a new synthesis of paleostress data. In: Durand B, Jolivet L, Horváth F, Séranne M (eds) The Mediterranean Basins: tertiary extension within the Alpine Orogen. Geology Society, London, Special Publication 156, pp 295–334

Gábris Gy (2003) A földtörténet utolsó 30 000 évének szakaszai és a futóhomok mozgásának főbb periódusai Magyarországon (Division of the last 30,000 years of earth history and phases of blown-sand movement in Hungary). Földrajzi Közlemények 127(1–4):1–14 (in Hungarian)

Gábris Gy, Nádor A (2007) Long-term fluvial archives in Hungary: response of the Danube and Tisza rivers to tectonic movements and climatic changes during the Quaternary: a review and new synthesis. Quat Sci Rev 26:2758–2782

Gábris Gy, Félegyházi E, Nagy B, Ruszkiczay Zs (2001) A Középső-Tisza vidékének negyedidőszak végi folyóvízi felszínfejlődése (Late Quaternary fluvial evolution of the Middle Tisza region). In: Papers of the Hungarian geographical conference 2001, Szeged. CD, 10 p (in Hungarian)

Gábris Gy, Telbisz T, Nagy B, Bellardinelli E (2002) A tiszai hullámtér feltöltődésének kérdése és az üledékképződés geomorfológiai alapjai (The alluviation of the Tisza active floodplain and geomorphology). Vízügyi Közlemények 84(3):305–322 (in Hungarian)

Gábris Gy, Timár G, Somhegyi A, Nagy I (2004) Árvízi tározás vagy ártéri gazdálkodás a Tisza mentén (Flood storage or flood economy along the Tisza). In: Current results in Hungarian geography: papers of the 2nd Hungarian geographical conference), Szeged, 2004. CD, 18 p (in Hungarian)

Gasztonyi É (2002) A recski mélyszinti érckutatás külszíni környezeti hatásai (Environmental impacts of deep-level ore exploration at Recsk). Földtani kutatás 39(2):35–39 (in Hungarian)

Góczán L, Schőner I, Tarnai P (1973) Új típusú berendezés a geomorfodinamikai folyamatok analíziséhez, talaj és környezetvédelmi kontrolljához (New equipment for analyzing and controlling geomorphic processes). Földrajzi Értesítő 22(4):479–482 (in Hungarian)

Haas J, Péró Cs (2004) Mesozoic evolution of the Tisza Mega-unit. Int J Earth Sci 93:297–313

Hahn Gy, Siska–Szilasi B (2004) Bányameddők és bányatavak rekultivációja Magyarországon (Reclamation of spoil heaps and mine ponds in Hungary). In: Dövényi Z, Schweitzer F (eds) Táj és környezet: Tiszteletkötet a 75. éves Marosi Sándornak. MTA FKI, Budapest, pp 213–225 (in Hungarian)

Hajdú-Moharos J, Hevesi A (2002) A kárpáti-pannon térség tájtagolódása (Landscape divisions of the Carpatho-Pannonian region). In: Karátson D (ed) Magyarország földje – kitekintéssel a Kárpát-medence egészére (The land of Hungary – with outlook to the entire Carpathian Basin), 2nd edn. Magyar Könyvklub, Budapest, pp 294–306 (in Hungarian)

Hevesi A (1980) Adatok a Bükk-hegység negyedidőszaki ősföldrajzi képéhez (On the Quaternary paleogeography of the Bükk Mountains). Földrajzi Közlemények 110(3–4):540–550

Hevesi A (2002) Napjaink földtudományainak kibontakozása (Earth sciences today). In: Karátson D (ed) Magyarország földje – kitekintéssel a Kárpát-medence egészére (The land of Hungary – with outlook to the entire Carpathian Basin), 2nd edn. Magyar Könyvklub, Budapest, pp 17–18 (in Hungarian)

Homoki E, Juhász Cs, Baros Z, Sütő L (2000) Anthropogenic geomorphological research on waste heaps in the East-Borsod coal basin (NE Hungary). In: Rzętały M (ed) Z badań nad wpływen antropopresji na środowisko. Studenckie Koło Naukowe Geografów UŚ, Sosnowiec, pp 24–31

Horváth F (1993) Towards a mechanical model for the formation of the Pannonian basin. Tectonophysics 226:333–357

Horváth F, Royden L (1981) Mechanism for the formation of the intra-Carpathian basins: a review. Earth Evol Sci 1:307–316

Horváth Zs, Scheuer Gy (1976) A dunaföldvári partrogyás mérnökgeológiai vizsgálata (Engineering geological analysis of the bank collapse at Dunaföldvár). Földtani Közlöny 106(4):425–440 (in Hungarian)

Horváth F, Bada G, Szafián P, Tari G, Ádám A, Cloeting S (2006) Formation and deformation of the Pannonian Basin: constraints from observational data. In: Gee DG, Stephenson R (eds) European lithosphere dynamics. Geology Society Memoir 32. Geology Society, London, pp 191–206

Huszár T (1999) Talajerózió-becslés az EPIC-EROTÓP módszerrel (Soil erosion estimation by the EPIC–erotope method). Földrajzi Értesítő 48(1–2):189–198 (in Hungarian)

Ihrig D (ed) (1973) A magyar vízszabályozás története (History of river channelization in Hungary). VÍZDOK, Budapest, 398 p (in Hungarian)

Illés L, Kerti A, Bodnár G (2003a) A beregi öblözet lokalizációs terve (Localization plan for the Bereg Bight). Vízügyi Közlemények Special Issue 4: 231–275 (in Hungarian)

Illés L, Konyecsny K, Kovács S, Szlávik L (2003b) Az 1998. novemberi árhullám hidrológiája (Hydrology of the November 1998 flood wave). Vízügyi Közlemények Special Issue I: 47–77 (in Hungarian)

Inkey B (1877) Földcsuszamlás Somogy megyében (Landslide in Somogy county). Földtani Közlöny 7:125–128 (in Hungarian)

Jakab G (2006) A vonalas erózió megjelenésének formái és mérésének lehetőségei (Linear types of erosion and measurement opportunities). Tájökológiai Lapok 4(1):17–33 (in Hungarian)

Jakab G, Kertész Á, Papp S (2005) Az árkos erózió vizsgálata a Tetves-patak vízgyűjtőjén (Gully erosion studies in the catchment of the Tetves stream). Földrajzi Értesítő 54(1–2):149–165 (in Hungarian)

Jakab G, Kertész Á, Szalai Z (2010) Scale dependence of gully investigations. Hung Geogr Bull 59(3):319–330

Jakucs L (1977) Morphogenetics of Karst Regions. Akadémiai Kiadó, Budapest, 284 p

Jakucs L (1980) A karszt biológiai produktum! (Karst is a biogene product). Földrajzi Közlemények 104(3–4):330–344 (in Hungarian)

Juhász Á (1972) Sásd környékének csuszamlásos tömegmozgásos jelenségei (Landslides around Sásd). Földrajzi Értesítő 21(3–4):471–474 (in Hungarian)

Juhász Á (1974) Anthropogene Einwirkungen und Geoprozesse in der Umgebung von Komló. Földrajzi Értesítő 23(2):223–224

Juhász Á (1999) A klimatikus hatások szerepe a magaspartok fejlődésében (Climatic controls of high bluff evolution). Földtani Kutatás 36(3):14–20 (in Hungarian)

Juhász Á, Schweitzer F (1989) A Balatonkenese és Balatonvilágos közötti magaspartok felszínmozgásos formatípusai (Mass movement features along the bluffs between Balatonkenese and Balatonvilágos). Földrajzi Értesítő 38(3–4):305–318 (in Hungarian)

Kádár L (1935) Futóhomok tanulmányok a Duna-Tisza közén (Blown sand studies on the Danube–Tisza Interfluve). Földrajzi Közlemények 63(1):4–15 (in Hungarian)
Kádár L (1938) A széllyukakról (On blowouts). Földrajzi Közlemények 66(1):117–121 (in Hungarian)
Kádár L (1956) A magyarországi futóhomok-kutatás eredményei és vitás kérdései (Achievements and debated issues of blown-sand research in Hungary). Földrajzi Közlemények 80:143–163 (in Hungarian)
Kádár L (1969) Specific types of fluvial landforms related to the different manners of load transport. Acta Geogr Debr 8–9:115–178
Kalmár J, Kuti L (2005) Meddőhányók és zagytározók természetes rehabilitációjának földtani körülményei (Geological conditions for the natural reclamation of spoil heaps and sludge reservoirs). Földtani kutatás 42(2):4–11 (in Hungarian)
Karancsi Z (2002) Természetes és antropogén eredetű környezetváltozás a Medves térség területén (Environmental transformation of natural and anthropogenic origin in the Medves region). Manuscript Ph.D. thesis. University of Szeged, Szeged, 117 p (in Hungarian)
Károlyi Z (1960a) A Tisza mederváltozásai – különös tekintettel az árvízvédelemre (Channel changes of the Tisza River with special regard to flood control). VITUKI, Budapest, 102 p (in Hungarian)
Károlyi Z (1960b) Zátonyvándorlás és gázlóalakulás – különös tekintettel a magyar Felső-Dunára (Bar shifting and ford formation along the Hungarian Upper Danube). Hidrológiai Közlöny 40(5):349–358 (in Hungarian)
Kázmér M, Kovács S (1985) Permian-Paleogene paleogeography along the eastern part of the Insubric-Periadriatic lineament system: evidence for continental escape of the Bakony-Drauzug unit. Acta Geol Hung 28:71–84
Kazó B (1966) A talajok vízgazdálkodási tulajdonságainak meghatározása mesterséges esőztető készülékkel (Identifying soil water budget properties by rainfall simulation). Agrokémia és Talajtan 15(2):329–352 (in Hungarian)
Kerényi A (1984) A talajerózió vizsgálatának laboratóriumi kísérleti módszere (Laboratory experiment method for soil erosion studies). Földrajzi Értesítő 33(3):266–276 (in Hungarian)
Kerényi A (1985) Surface evolution and soil erosion as reflected by measured data. In: Pécsi M (ed) Environmental and dynamic geomorphology, vol 17, Studies in geography in Hungary. Akadémiai Kiadó, Budapest, pp 79–84
Kerényi A (1991) Talajerózió. Térképezés, laboratóriumi és szabadföldi kísérletek (Soil erosion: mapping and field and laboratory experiments). Akadémiai Kiadó, Budapest, 219 p (in Hungarian)
Kerényi A (2006) Az areális és lineáris erózió mennyiségi értékelése bodrogkeresztúri mérések alapján (Quantitative evaluation of sheet and gully erosion based on measurements at Bodrogkeresztúr). In: Csorba P (ed) Egy szakmai életút eredményei és színhelyei (Results and scenes of a life). University of Debrecen, Debrecen, 67–77 (in Hungarian)
Kertész Á (1993) Application of GIS methods in soil erosion modelling. Comput Environ Urban Syst 17:233–238
Kertész A (2008) Water and wind erosion in Hungary. In: Kertész A, Kovács Z (eds) Dimensions and trends in Hungarian geography, vol 33, Studies in geography in Hungary. Geographical Research Institute, HAS, Budapest, pp 47–54
Kertész Á, Góczán L (1988) Some results of soil erosion monitoring at a large-scale farming experimental station in Hungary. Catena Suppl 12:175–184
Kertész Á, Richter G (1997) The Balaton project. ESSC Newsl 1997:2–3
Kertész Á, Centeri Cs (2006) Hungary. In: Boardman J, Poesen J (eds) Soil erosion in Europe. Wiley, Chichester, pp 139–153
Kertész Á, Huszár T, Lóczy D (1995a) Land use changes in Lake Balaton catchment. In: Simmons IG, Mannion AM (eds) The changing nature of the people–environment relationship: evidence from a variety of archives. Charles University, Prague, pp 69–78
Kertész Á, Márkus B, Richter G (1995b) Assessment of soil erosion in a small watershed covered by loess. GeoJournal 36(2–3):285–288

Kertész Á, Huszár T, Tóth A (2000) Soil erosion assessment and modelling. In: Bassa L, Csuták M, Kertész Á, Schweitzer F (eds) Physico-geographical research in Hungary. Geographical Research Institute, Hungarian Academy of Sciences, Budapest, pp 63–74

Kertész Á, Tóth A, Jakab G, Szalai Z (2001) Soil erosion measurements in the Tetves Catchment, Hungary. In: Helming K (ed) Multidisciplinary approaches to soil conservation strategies.In: Proceedings, international symposium, ESSC, DBG, ZALF. Müncheberg, 11–13 May 2001, pp 47–52

Kertész, Á, Csepinszky B, Jakab G (2002) The role of surface sealing and crusting in soil erosion. In: Technology and method of soil and water conservation Vol. III. Proceedings, 12th international soil conservation organization conference, Beijing, China. Tsinghua University Press, Beijing, 26–31 May 2002, pp 29–34

Kézdi A (1952) A Balaton északkeleti peremén bekövetkező mozgások vizsgálata (Investigation of movements on the northeastern shore of Lake Balaton). Hidrológiai Közlöny 32:403–408 (in Hungarian)

Kézdi Á (1970) A dunaújvárosi partrogyás (Bank collapse at Dunaújváros). Mélyépítés-tudományi Szemle 20:281–297 (in Hungarian)

Kis E, Lóczy D (1985) Geomorphological mapping in an alluvial plain and the assessment of environmental quality. In: Pécsi M (ed) Environmental and dynamic geomorphology. Case studies in Hungary, vol 17, Studies in geography in Hungary. Akadémiai Kiadó, Budapest, pp 181–192

Kiss T, Sipos Gy, Fiala K (2002) Recens üledékfelhalmozódás sebességének vizsgálata az Alsó-Tiszán (Investigation of the rate of recent alluviation along the Lower Tisza). Vízügyi Közlemények 84(3):456–472 (in Hungarian)

Kiss T, Sipos Gy (2003) Investigation of channel dynamics on the lowland section of River Maros. Acta Geogr Szeged 39:41–49

Kiss T, Sipos Gy (2004) A Maros medermintázatának megváltozása a szabályozások hatására (Changes in the channel pattern of the Maros river after channelization). In: Füleky Gy (ed) Víz a tájban (Water in the landscape). Környezetkímélő Agrokémiáért Alapítvány, Gödöllő, pp 183–189 (in Hungarian)

Kiss T, Nyári D, Sipos Gy (2006) Homokmozgások vizsgálata a történelmi időkben Csengele területén (Blown sand movement in historical times at Csengele). In: Táj, környezet és társadalom (Landscape, environment and society). University of Szeged, Szeged, pp 373–383 (in Hungarian)

Kiss T, Sipos Gy (2007) Braid-scale geometry changes in a sand-bedded river: significance of low stages. Geomorphology 84(3):209–221

Kiss T, Mezősi G (eds) (2008–2009) Recens geomorfológiai folyamatok sebessége Magyarországon (The rate of recent geomorphic processes in Hungary). Szegedi Egyetemi Kiadó, Szeged, 204 p (in Hungarian)

Kiss T, Nyári D, Sipos Gy (2008a) Történelmi idők eolikus tevékenységének vizsgálata: a Nyírség és a Duna-Tisza köze összehasonlító elemzése (Study of aeolian action in historical times: a comparison between the Nyírség and the Danube–Tisza Interfluve). In: Geographia generalis et specialis. University of Debrecen, Debrecen, pp 99–106 (in Hungarian)

Kiss T, Fiala K, Sipos Gy (2008b) Altered meander parameters due to river regulation works, Lower Tisza, Hungary. Geomorphology 98(1–2):96–110

Kiss T, Blanka V, Sipos Gy (2009) Morphometric change due to altered hydrological conditions in relation to human impact, River Hernád, Hungary. Zeitschrift für Geomorphologie 53(Supplement 2):197–213

Kleb B, Schweitzer F (2001) A Duna csuszamlásveszélyes magaspartjainak település környezeti hatásvizsgálata (Complex environmental impact statement of Danubian bluffs with mass movement hazard to settlements). In: Földtudományok és a földi folyamatok kockázati tényezői (Hazards in earth sciences and terrestrial processes). Hungarian Academy of Sciences, Budapest, pp 169–194 (in Hungarian)

Kocsis K, Schweitzer F (2009) Hungary in maps. Geographical Research Institute, Hungarian Academy of Sciences, Budapest, 211 p

Krisztián J (1998) Talajvédelem (Soil conservation). GATE Faculty of Agriculture, Gyöngyös (in Hungarian)
Kvassay J (1902) A szabályozások hatása a folyók vízjárására Magyarországon (Impact of channelization on river water regime in Hungary). Vízügyi Közlemények 15(1):8–27 (in Hungarian)
Laczay I (1967) Az 1965. évi árvíz tetőző vízszintjei a felsődunai hullámtérben. Az árvíz hatása a mederalakulásra (Maximum water levels of the 1965 flood on the active floodplain of the Upper Danube and its impacts on channel evolution). Vízügyi Közlemények 49(1):119–127 (in Hungarian)
Laczay I (1968) A cikolaszigeti mellékágrendszer mederváltozásának vizsgálata (Studies on the changes of the distributary channels of the Danube River at Cikolasziget). Vízügyi Közlemények 50(2):245–255 (in Hungarian)
Laczay I (1982) A folyószabályozás tervezésének morfológiai alapjai (Morphological foundations of planning river channelization). Vízügyi Közlemények 64(2):235–256 (in Hungarian)
Laczay I (1989) Ipari kotrások hatása a Komárom-Nagymaros közötti Duna-szakasz mederviszonyaira (Impact of industrial dredging on channel conditions along the Danube reach between Komárom and Nagymaros). Vízügyi Közlemények 70(3):547–567 (in Hungarian)
Ládai JT (2002) A külszíni süllyedés időbeni lefolyásának vizsgálata (Study of the temporal course of ground subsidence). Bányászati és Kohászati Lapok 136:22–26 (in Hungarian). http://www.ombkenet.hu/bkl/banyaszat/2003/bklbanyaszat2003_1.pdf
Láng S (1955) A Mátra és a Börzsöny természeti földrajza (Physical geography of Mátra and Börzsöny Mountains), vol 1, Földrajzi monográfiák. Akadémiai Kiadó, Budapest, 512 p (in Hungarian)
Láng S (1967) A Cserhát természeti földrajza (Physical geography of Cserhát Mountains), vol 7, Földrajzi monográfiák. Akadémiai Kiadó, Budapest, 376 p (in Hungarian)
László P, Rajkai K (2003) A talajerózió modellezése (Modelling soil erosion). Agrokémia és Talajtan 52(3–4):427–442 (in Hungarian)
Lászlóffy W (1982) A Tisza – vízi munkálatok és vízgazdálkodás a tiszai vízrendszerben (The Tisza: water management in the water system of the Tisza). Akadémiai Kiadó, Budapest, 610 p (in Hungarian)
Leél-Őssy S (1973) Természeti antropogén folyamatok és formáik vizsgálata Ózd és Arló környékén (Study of natural and anthropogenic processes and landforms in the Ózd and Arló area). Földrajzi Értesítő 22(2–3):195–213 (in Hungarian)
Lóczy D (2007) The changing geomorphology of Danubian floodplains in Hungary. Hrvatski Geografski Glasnik 69(2):5–20
Lóczy D, Gyenizse P (2011) Micromorphology influenced by tillage on a Danubian floodplain in Hungary. Zeitschrift für Geomorphology 55(Supplementary Issue 1):67–76
Lóczy D, Pécsi M (1989) Geomorphology in Hungary. Trans Jpn Geomorphol Union 10-B:103–107
Lóczy D, Balogh J, Ringer Á (1989) Landslide hazard induced by river undercutting along the Danube. Supplementi di Geografia física e quaternaria dinamica 2:5–11
Lóczy D, Fábián SzÁ, Schweitzer F (2007) River action and landslides in Hungary. In: Basu SR, De SK (eds) Issues in geomorphology and environment. acb publications, Kolkata, pp 1–15
Lóki J (1981) Belső-Somogy futóhomok területeinek kialakulása és formái (Development and features of blown- sand areas in the Inner Somogy). Acta Geogr Debr 1979–1980:81–111 (in Hungarian)
Lóki J (2004) Wind erosion surface formation in Hungary during the Holocene. In: Formy I Osady Eoliczene, Stowarziszenie Geomorfologów Polskich, Poznań, pp 14–21
Lóki J, Négyesi G (2003) A talajfelszíni kéreg képződése és hatása a széleróziora (Soil crusting and wind erosion). Acta Geogr Debr 36:55–64 (in Hungarian)
Lóki J, Schweitzer F (2001) Fiatal futóhomokmozgások kormeghatározási kérdései a Duna–Tisza közi régészeti feltárások tükrében (Dating recent blown-sand movements in the light of archaeological excavations on the Duna–Tisza Interfluve). Acta Geogr Debr 35:175–183 (in Hungarian)
Lóki J, Hertelendi E, Borsy Z (1993) New dating of blown sand movement in the Nyírség. Acta Geogr Debr 1994:67–76 (in Hungarian)

Lovász Gy (1985) Csuszamlásos folyamatok Orfű térségében (Landslides around Orfü). In: Ádám L, Pécsi M (eds) Mérnökgeomorfológiai térképezés (Engineering geomorphological mapping). Geographical Research Institute, HAS, Budapest, pp 95–107 (in Hungarian)

Marosi S (1966) Kovárványrétegek és periglaciális jelenségek összefüggésének kérdései a belsősomogyi futóhomokban (Relationships between 'kovárvány' bands and periglacial phenomena in the blown sand of Inner Somogy). Földrajzi Értesítő 15(1):27–40 (in Hungarian)

Marosi S (1967) Megjegyzések a magyarországi futóhomok területek genetikájához és morfológiájához (Remarks on the origin and morphology of blown sand areas in Hungary). Földrajzi Közlemények 91(3–4):231–255 (in Hungarian)

Marosi S (1970) Belső-Somogy kialakulása és felszínalaktana (Development and geomorphology of the Inner Somogy region). Akadémiai Kiadó, Budapest, 169 p (in Hungarian)

Marosi S (ed) (1979) Applied geographical research in the Geographical Research Institute of the Hungarian Academy of Sciences, vol 21, Geographical abstracts from Hungary. Geographical Research Institute, HAS, Budapest, 174 p

Marosi S, Juhász Á, Kertész Á, Kovács Z (1992) An estimation and mapping method for erosion hazard in the catchment of Lake Balaton. In: New perspectives in Hungarian geography, vol 27, Studies in geography in Hungary. Akadémiai Kiadó, Budapest, pp 9–20

Márton Gy (1914) A Maros alföldi szakasza és fattyúmedrei (The lowland section of the Maros River and its anastomosing channels). Földrajzi Közlemények 52(3):282–301 (in Hungarian)

Márton E (1997) Paleomagnetic aspects of plate tectonics in the Carpatho-Pannonian region. Mineralium Deposita 32(5):441–445. doi:10.1007/s001260050112

Marton-Erdős K (2006) A Bodrogkeresztúri-katlan talajeróziójának formái és területi elterjedése (Types and distribution of soil erosion in the Bodrogkeresztúr half-basin). In: Csorba P (ed) Egy szakmai életút eredményei és színhelyei (Results and scenes of a life). University of Debrecen, Debrecen, pp 39–54 (in Hungarian)

Mattyasovszky J (1953) Észak-dunántúli talajok eróziós viszonyai (Erosion of soils in northern Transdanubia). Agrokémia és Talajtan 2:333–340 (in Hungarian)

Mattyasovszky J (1956) A talajtípus, az alapkőzet és a lejtőviszonyok hatása a talajeróziós folyamatok kialakulására (Impact of soil type, parent material and slope conditions on soil erosion). Földrajzi Közlemények 4(4):355–364 (in Hungarian)

Mezősi G (2011) Magyarország természetföldrajza (Physical geography of Hungary). Akadémiai Kiadó, Budapest, 393 p

Mezősi G, Richter G (1991) Az EPIC (Erosion-productivity impact calculator) modell tesztelése (Testing the EPIC model). Agrokémia és Talajtan 40:461–468 (in Hungarian)

Mezősi G, Albrecht V, Bódis K (2002) A környezeti veszélyek térképezése a Velencei-tó északi részén (Mapping environmental hazards north of Lake Velence). In: Abonyi-Palotás J, Becsei J, Kovács Cs (eds) A magyar társadalomföldrajzi kutatás gondolatvilága (Concepts in Hungarian human geographical research). Ipszilon, Szeged, 95–106

Mike K (1991) Magyarország ősvízrajza és felszíni vizeinek története (Paleohydrography of Hungary and the history of its surface waters). Aqua Kiadó, Budapest, 698 p (in Hungarian)

Nagy I, Schweitzer F, Alföldi L (2001) A hullámtéri hordalék-lerakódás (Accumulation on the active floodplain). Vízügyi Közlemények 83(4):539–564 (in Hungarian)

Nagy ÁT, Tóth T, Sztanó O (2006) Új, kombinált módszerek a Közép-Tisza jelenkori mederképződményeinek jellemzésére (New combined methods to describe recent channel features for the Middle Tisza River). Földtani Közlöny 136(1):121–138 (in Hungarian)

Nováky B (2001) A TIM eróziós mérőpontok észleléseinek elemzése és értékelése (Assessment of TIM measurement points). In: Papers of the first Hungarian geographical conference, Univesity of Szeged, Szeged, 2001. CD, 10 p (in Hungarian)

Nyári D, Kiss T (2005) Homokmozgások vizsgálata a Duna–Tisza közén (Sand movements on the Danube–Tisza Interfluve). Földrajzi Közlemények 129(3–4):133–147 (in Hungarian)

Nyári D, Kiss T, Sipos Gy (2006) Történeti időkben bekövetkezett futóhomok mozgások datálása lumineszcenciás módszerrel a Duna–Tisza közén (Dating historical blown-sand movements by luminescence methods on the Danube–Tisza Interfluve). In: Papers of the third Hungarian geographical conference. Geographical Research Institute, HAS, Budapest, 2006 (in Hungarian)

Nyári D, Kiss T, Sipos Gy (2007) Investigation of Holocene blown-sand movement based on archaeological findings and OSL dating, Danube–Tisza Interfluve, Hungary. Journal of Maps, Kingston University, Kingston-upon-Thames. 2007. pp 47–56 http://journalofmaps.com/student/07_05_Nyari.pdf

Oroszi VGy, Kiss T (2004) Folyószabályozás hatására felgyorsult hullámtér feltöltődés vizsgálata a Maros magyarországi szakaszán (Alluviation of active floodplain accelerated by river channelization along the Hungarian Maros river). In: Current results in Hungarian geography: papers of the second Hungarian Geographical Conference, Szeged, 2004. CD. 20 p (in Hungarian)

Oroszi VGy, Kiss T, Bottlik A (2006) A 2005. évi tavaszi áradás üledékfelhalmozó hatása a Maros hullámterén (Alluviation by the 2005 spring flood on the active floodplain of the Maros). In: Papers of the third geographical conference, Geographical Research Institute, HAS, Budapest, 2006. CD, 10 p (in Hungarian)

Pálfai I (2003) Oxbow-lakes in Hungary. Ministry of Environmental Protection and Water Management, Budapest, 125 p

Pap S (2008) Felhagyott külszíni bányák mint földtani bemutatóhelyek (Abandoned opencast mines as sites of geological education). Természet Világa 139(5):212 (in Hungarian)

Pécsi M (1959) A magyarországi Duna-völgy kialakulása és felszínalaktana (Development and geomorphology of the Hungarian Danube valley), vol 3, Földrajzi Monográfiák. Akadémiai Kiadó, Budapest, 345 p (in Hungarian)

Pécsi M (1967) Relationship between slope geomorphology and Quaternary slope sedimentation. Acta Geol 11(1–3):307–321

Pécsi M (1970a) Surfaces of planation in the Hungarian mountains and their relevance to pedimentation. In: Pécsi M (ed) Problems of relief planation, vol 8, Studies in geography in Hungary. Akadémiai Kiadó, Budapest, pp 29–40

Pécsi M (1970b) Geomorphological regions of Hungary, vol 6, Studies in geography in Hungary. Akadémiai Kiadó, Budapest, 45 p

Pécsi M (1970c) Geomorphological regions of Hungary. Akadémiai Kiadó, Budapest, 45 p

Pécsi M (1971a) A földcsuszamlások főbb típusai (Main types of landslides). Földrajzi Közlemények 95(1–2):125–143 (in Hungarian)

Pécsi M (1971b) Az 1970. évi dunaföldvári földcsuszamlás (The Dunaföldvár landslide in 1970). Földrajzi Értesítő 20(3–4):233–238 (in Hungarian)

Pécsi M (1975) Geomorphology. Mass movement of the Earth's surface. UNESCO International Postgraduate Course on the Principles and Methods of Engineering Geology. MÁFI, Budapest, 241 p

Pécsi M (1977) Geomorphological map of the Carpathian and Balkan Regions (1:1,000,000). Studia Geomorphologica Carpatho-Balcanica 11:3–31

Pécsi M (1980) Erläuterung zur geomorphologischen Karte des "Atlases der Donauländer". Österreichische Osthefte 22(2):141–168

Pécsi M (ed) (1999) Landform evolution studies in Hungary, vol 30, Studies in geography in Hungary. Akadémiai Kiadó, Budapest, 216 p

Pécsi M, Juhász Á (1974) Kataster der Rutschungsgebiete in Ungarn und ihre kartographische Darstellung. Földrajzi Értesítő 23(3–4):192–202

Pécsi M, Lóczy D (eds) (1986) Physical geography and geomorphology in Hungary. Geographical Research Institute, Hungarian Academy of Sciences, Budapest, 122 p

Pécsi M, Richter G (1996) Löss: Herkunft – Gliederung – Landschaften, Zeitschrift für Geomorphologie, Supplement-Band 98. Gebrüder Borntraeger, Berlin, 391 p

Pécsi M, Scheuer Gy (1979) Engineering geological problems of the Dunaújváros loess bluff. Acta Geol 22(1–4):345–353

Pécsi M, Somogyi S (1969) Subdivision and classification of the physiographic landscapes and geomorphological regions of Hungary. In: Sárfalvi B (ed) Research problems in Hungarian applied geography. Akadémiai Kiadó, Budapest, pp 7–27

Pécsi M, Juhász Á, Schweitzer F (1976) A magyarországi felszínmozgásos területek térképezése (Mapping mass movements in Hungary). Földrajzi Értesítő 25(3–4):223–238 (in Hungarian)

Pécsi M, Scheuer Gy, Schweitzer F (1979) Engineering geological and geomorphological investigation of landslides in the loess bluff along the Danube in the Great Hungarian Plain. Acta Geol Hung 22:327–343

Pécsi M, Lóczy D, Marosi S, Somogyi S, Gábris Gy, Mezősi G, Szabó J (1993) An history of geomorphology in Hungary. In: Walker HJ, Grabau H (eds) The evolution of geomorphology. Wiley, Chichester, pp 189–199

Peja Gy (1956) Suvadástípusok a Bükk északi (harmadkori) előterében (Types of slumps in the northern Tertiary foreland of the Bükk Mountains). Földrajzi Közlemények 80(3):217–240 (in Hungarian)

Peja Gy (1975) Geomorfológiai megfigyelések az Északi-középhegység laza kőzetű tömegmozgásos lejtőin (Geomorphological observations on unconsolidated slopes of North Hungarian Mountains with mass movements). Földrajzi Értesítő 24(1–2):123–140 (in Hungarian)

Pinczés Z, Kerényi A, Marton-Erdős K (1978) A talajtakaró pusztulása a Bodrogkeresztúri-félmedencében (Soil erosion in the Bodrogkeresztúr half-basin). Földrajzi Közlemények 142(3–4):210–236 (in Hungarian)

Prodán T, Veress M (2006) Adalékok a Balfi-tönk felszíni karsztszerű képződményeinek morfológiájához és kialakulásához (On the morphology and origin of surface karstic formations on the Balf Hills). Karszt és Barlang 2006/1–2:41–48 (in Hungarian)

Rakonczai J, Kovács F (2006) A padkás erózió folyamata és mérése az Alföldön (Berm erosion and its measurement in the Great Plain). Agrokémia és Talajtan 55(2):329–346 (in Hungarian)

Rakonczai J, Sárközi-Lőrinczi M (1984) Az 1980. évi árvíz talajtani hatásai a Kettős-Körös mentén (Impact of the 1980 flood on soils along the Körös rivers). Alföldi Tanulmányok 8:31–41 (in Hungarian)

Rózsa P, Kozák M (1995) Protection of volcanological natural monuments on the Miocene paleovolcano of Tokaj-Nagyhegy (NE Hungary). Geol Soc Greece Spec Publ 4(3):1069–1073

Sándor A, Kiss T (2006) A hulláméri üledék felhalmozódás mértékének vizsgálata a Közép- és az Alsó-Tiszán (Investigation of the rate of active floodplain alluviation along the Middle and Lower Tisza). Hidrológiai Közlöny 86(2):58–62 (in Hungarian)

Sándor A, Kiss T (2008) Floodplain aggradation caused by the high magnitude flood of 2006 in the Lower Tisza Region, Hungary. J Environ Geogr 1(1–2):31–39

Sásdi L (2005) Az Aggtelek-Budabányai-hegység édesvízi mészkő előfordulásai (Travertines in the Aggtelek–Rudabánya Mountains). Karsztfejlődés, Szombathely 10:137–151 (in Hungarian)

Schafarzik F (1882) Földcsuszamlás Buda határában (Landslide at Buda). Földtani Értesítő 3:123 (in Hungarian)

Scheuer Gy (1979) A dunai magaspartok mérnökgeológiai vizsgálata (Engineering geological investigation of Danubian bluffs). Földtani Közlöny 109:230–254 (in Hungarian)

Schweitzer F (2001) A magyarországi folyószabályozások geomorfológiai vonatkozásai (Geomorphological aspects of river channelization in Hungary). Földrajzi Értesítő 50(1–4):63–72 (in Hungarian)

Síkhegyi F (2002) Active structural evolution of the western and central parts of the Pannonian basin: a geomorphologial approach. EGU Stephan Mueller Special Publication Series 3:203–216. http://www.cosis.net/members/journals/df/abstract.php?a_id=2698

Sipos Gy (2006) A meder dinamikájának vizsgálata a Maros magyarországi szakaszán (Channel dynamics along the Hungarian Maros section). Manuscript Ph.D. thesis. University of Szeged, Szeged (in Hungarian)

Sipos Gy, Kiss T (2001) Egy szigetrendszer morfodinamikájának vizsgálata a Maros apátfalvi szakaszán (Dynamics of an island system along the Apátfalva Maros reach). In: Papers of the first Hungarian geographical conference, Szeged, 2001. CD, 9 p (in Hungarian)

Sipos Gy, Kiss T (2003) Szigetképződés és -fejlődés a Maros határszakaszán (Island formation and evolution along the border section of the Maros River). Vízügyi Közlemények 85(3):477–498 (in Hungarian)

Sipos Gy, Kiss T (2006) A medertágulatok szerepe a síksági folyók morfológiai stabilitásában a Maros példáján (Role of channel embayments in the stability of lowland rivers: example of the

Maros). In: Third Hungarian geographical conference, Geographical Research Institute, HAS, Budapest, 2006. CD, 10 p (in Hungarian)
Sipos Gy, Kiss T, Fiala K (2007) Morphological alterations due to channelization along the Lower Tisza and Maros Rivers. Geographica Física e Dinamica Quaternaria 30:239–247
Sipos Gy, Fiala K, Kiss T (2008) Changes of cross-sectional morphology and channel capacity during an extreme flood event, lower Tisza and Maros rivers, Hungary. J Environ Geogr 1(1–2):41–51
Somogyi S (1974) Meder- és ártérfejlődés a Duna sárközi szakaszán az 1782–1950 közötti térképfelvételek tükrében (Channel and floodplain evolution along the Sárköz section of the Danube as reflected by map surveys between 1982 and 1950). Földrajzi Értesítő 23(1):27–36 (in Hungarian)
Somogyi S (1978) Regulated rivers in Hungary. Geogr Pol 41:39–53
Somogyi S (ed) (2000) A XIX. századi folyószabályozások és ármentesítések földrajzi és ökológiai hatásai (Geographical and ecological impacts of river channelization and flood control measures). Geographical Research Institute, HAS, Budapest, 302 p (in Hungarian)
Somosvári Zs (2002) Felszínemelkedés jelenségek okai (Causes of surface rising). A sorozat Bányászat 62:35–46, A Miskolci Egyetem Közleményei
Stefanovits P (1977) Talajvédelem – környezetvédelem (Soil conservation – environmental protection). Mezőgazdasági Kiadó, Budapest (in Hungarian)
Stefanovits P, Duck T (1964) Talajpusztulás Magyarországon (Soil erosion in Hungary). OMMI, Budapest (in Hungarian)
Stefanovits P, Várallyay Gy (1992) State and management of soil erosion in Hungary. In: Proceedings of the soil erosion and remediation workshop, US–Central and Eastern European Agro-Environmental Program, Budapest, April 27–May 1, 1992, pp 79–95
Sütő L (2000) Mining agency in the east Borsod Basin, North-East Hungary. In: Jankowski AT, Pirozhnik II (eds) Nature use in the different conditions of human impact. Studenckie Koło Naukowe Geografów UŚ, Minsk/Sosnowiec, pp 116–123
Sütő L (2007) A szénbányászat geomorfológiára és területhasználatra gyakorolt hatásainak vizsgálata a Kelet-borsodi-szénmedencében (Impact of coal mining on the landscape and land use in the eastern Borsod Basin). Manuscript Ph.D. thesis. University of Debrecen, Debrecen. 177 p + 14 maps (in Hungarian)
Sütő L, Homoki E, Szabó J (2004) The role of the impacts of coal mining in the geomorphological evolution of the catchment area of the mine at Lyukóbánya (NE-Hungary). In: Lóki J, Szabó J (eds) Anthropogenic aspects of landscape transformations, vol 3. Kossuth University Press, Debrecen, pp 81–92
Szabó PZ (1968) A magyarországi karsztosodás fejlödéstörténeti vázlata (Evolutionary sketch of karst formation in Hungary). In: Studies of the Transdanubian Scientific Institute, HAS, 1967–1968. Akadémiai Kiadó, Budapest, pp 13–25
Szabó J (1971) Geomorphology of the region between the Rivers Sajó and Bódva. Acta Geogr Debr 1969–70:179–196
Szabó J (1978) A Cserehát felszínfejlődésének fő vonásai (Geomorphic evolution of Cserehát Hills). Földrajzi Közlemények 102(3–4):246–268 (in Hungarian)
Szabó J (1983) Thoughts on the general characterization of landslide processes with special respect to the problems of classification). Acta Geogr Debr 1981:83–114
Szabó J (1985a) Csuszamlásvizsgálatok a Csereháton (Landslide studies in Cserehát Hills). Földrajzi Értesítő 34(3–4):409–429 (in Hungarian)
Szabó J (1985b) Landslide typology in hilly regions of Northern Hungary. In: Pécsi M (ed) Environmental and dynamic geomorphology. Akadémiai Kiadó, Budapest, pp 171–180
Szabó J (1991) Landslide processes and forms in the Hungarian mountains of volcanic origin. In: Kertész Á, Kovács Z (eds) New perspectives in Hungarian geography. Geographical Research Institute, HAS, Budapest, pp 63–75
Szabó J (1995) Stellenwert der Rutschungsprozesse bei der morphologischen Entwicklung der Hochuferstrecken von Flüssen – dargelegt am Beispiel des Hernad-Tales in Ungarn. Mitteilungen der Österreichischen Geographischen Gesellschaft, Wien 137:141–160
Szabó J (1996a) Results and problems of cadastral survey of slides in Hungary. In: Chacon J, Irigaray C, Fernandez T (eds) Landslides. A.A. Balkema, Rotterdam/Brookfield, pp 71–78

Szabó J (1996b) Csuszamlásos folyamatok szerepe a magyarországi tájak geomorfológiai fejlődésében (Landslides in the geomorphic evolution of landscapes of Hungary). Kossuth Egyetemi Kiadó, Debrecen, 223 p (in Hungarian)

Szabó J (1997) Magaspartok csuszamlásos lejtőfejlődése a Hernád-völgyben (Slope evolution of high bluffs of the Hernád valley through landslides). Földrajzi Közlemények 121(1–2):17–46 (in Hungarian)

Szabó J (2003) The relationship between landslide activity and weather: examples from Hungary. Nat Hazards Earth Syst Sci (European Geosciences Union) 3:43–52

Szabó J (2004) A hollóházai földcsuszamlások (1999) az időjárás és a társadalmi felelősség tükrében (The landslides of 1999 at Hollóháza with view of weather and social responsibility). In: Tar K (ed) Földtudományi tanulmányok (Studies in earth sciences). Tiszteletkötet Dr. Justyák János 75. születésnapjára. Debreceni Egyetem, pp 173–180 (in Hungarian)

Szabó M (2006) A vegetáció foltmintázata és a szukcesszió lehetséges útjai a Szigetközben a lipóti övzátony példáján (Vegetation pattern and possible succession paths in the Szigetköz on the example of the point bar at Lipót). Third Hungarian geographical conference, Geographical Research Institute, HAS, Budapest, 2006. CD. (in Hungarian)

Szabó J, Félegyházi E (1997) Problems of landslide chronology in the Mátra Mountains in Hungary. Eiszeitalter und Gegenwart 47:120–128

Szabó J, Lóki J, Tóth Cs, Szabó G (2008) Natural hazards in Hungary. In: Kertész A, Kovács Z (eds) Dimensions and trends in Hungarian geography, vol 33, Studies in geography in Hungary. Geographical Research Institute, HAS, Budapest, pp 55–68

Szabó J, Dávid L, Lóczy D (2010) Anthropogenic geomorphology: a guide to man-made landforms. Springer, Dordrecht/Heidelberg, 298 p

Szatmári J (1997) Wind erosion risk on the southern Great Hungarian Plain. Acta Geogr Szeged 36:121–135

Szebényi G (2002) Ércbányászati eredetű környezetföldtani tényezők és veszélyforrások Recsk–Parádfürdő térségében (Environmental geological factors and hazards of ore mining origin in the Recsk–Parádfürdő). Földtani kutatás 39(2):28–35 (in Hungarian)

Szeibert J, Zellei L (2003) A Bodrog országhatár-torkolat közötti szakaszának árvízi vízszállítása (Flood discharge along the Bodrog between the national border and the confluence). Vízügyi Közlemények Special Issue IV:145–172 (in Hungarian)

Székely A (1989) Geomorphology of the Intra-Carpathian volcanic range. In: Pécsi M (ed) Geomorphological and geoecological essays, vol 25, Studies in geography in Hungary. Akadémiai Kiadó, Budapest, pp 49–60

Szilárd J (1967) Külső-Somogy kialakulása és felszínalaktana (Development and geomorphology of Outer Somogy), vol 7, Földrajzi tanulmányok. Akadémiai Kiadó, Budapest, 150 p (in Hungarian)

Szlávik L, Szekeres J (2003) Az árvízi vízhozammérések kiértékelésének eredményei és tapasztalatai (1998–2001) (Evaluation of the results and experiences of discharge measurements during floods, 1998–2001). Vízügyi Közlemények Special Issue IV:45–58 (in Hungarian)

Szlávik L, Kiss A, Galbáts Z (1996) Az 1995 decemberi Körös-völgyi árvíz és a szükségtározások hidrológiai elemzése és értékelése (Hydrological analysis and evaluation of the 1995 flood in the Körös valley and of emergency storage). Vízügyi Közlemények 78(1):69–104 (in Hungarian)

Timár G (2000) Földtani folyamatok hatása a Tisza alföldi szakaszának medermorfológiájára (Impact of geological processes on the channel morphology of the lowland Tisza River section). Manuscript Ph.D. thesis. ELTE, Budapest (in Hungarian)

Timár G (2003) Controls on channel sinuosity changes: a case study of the Tisza River, the Great Hungarian Plain. Quat Sci Rev 22:2199–2220

Tőry K (1952) A Duna és szabályozása (Flow regulation of the Danube). Akadémiai Kiadó, Budapest, 454 p (in Hungarian)

Ujházy K, Gábris Gy, Frechen M (2003) Ages of periods of sand movement in Hungary determined through luminescence measurements. Quat Int 111:91–100

Újvári G, Mentes Gy, Bányai L, Kraft J, Gyimóthy A, Kovács J (2009) Evolution of a bank failure along the River Danube at Dunaszekcső, Hungary. Geomorphology 109:197–209. doi:10.1016/j.geomorph.2009.03.002

Veress M (1980) A Csesznek környéki völgyoldalak barlang-torzóinak vizsgálata (Investigation of cave torsos on valley sides around Csesznek). Karszt és Barlang 1980/2:65–70 (in Hungarian)

Veress M (1998) A Vas megyei Bükfürdő édesvízi mészkőkiválás formáinak morfogenetikai csoportosítása (Morphogenetic classification of travertine precipitations at Bük Spa, Vas County). Hidrológiai Közlöny 4:214–222 (in Hungarian)

Veress M (2000) Covered karst evolution in the Northern Bakony Mountains, W-Hungary. Bakony Natural Sciences Museum, Zirc, 167 p

Veress M (2006) Adatok a Tési-fennsík két részletének fedett karsztosodásához (On covered karst formation in two terrains of the Tés Plateau). Karsztfejlődés, Szombathely 11:171–184 (in Hungarian)

Veress M (2007) Adalékok az Eleven-Förtési töbörcsoport (Bakony-hegység) karsztosodásához (On the karstification of the Eleven-Förtés doline group, Bakony Mountains). Karsztfejlődés, Szombathely 12:171–192 (in Hungarian)

Veress M, Péntek K (1996) Theoretical model of surface karstic processes. Zeitschrift für Geomorphologie 40(4):461–476

Veress M, Szabó L, Zentai Z (1998) Mésztartalomhoz köthető felszínfejlődés a Kőszegi-hegységben (Geomorphic evolution in the Kőszeg Mountains controlled by carbonate content). Földrajzi Értesítő 47(4):495–514 (in Hungarian)

Verő-Wojtaszek M, Balázsik V (2008) A talajerózió követése űrfelvételek alkalmazásával a Tetvespatak példáján (Monitoring soil erosion using satellite images: example of the Tetves stream). Agrokémia és Talajtan 57(1):21–36 (in Hungarian)

VITUKI (1974) Bodrog, vol 20, Vízrajzi Atlasz sorozat. VITUKI, Budapest (in Hungarian)

Zámbó L (1985) The role of clay deposits in the geomorphic evolution of dolines. In: Pécsi M (ed) Environmental and dynamic geomorphology, vol 17, Studies in geography in Hungary. Akadémiai Kiadó, Budapest, pp 97–108

Zelenka T, Trauer N, Szabó J (1999) A hollóházai földmozgások földtani okai (Geological explanation of movements at Hollóháza). Földtani Kutatás 36:27–33 (in Hungarian)

Zellei L, Sziebert J (2003) Árvízi áramlásmérések tapasztalatai a Tiszán (Experience from current measurements during floods on the Tisza). Vízügyi Közlemények Special Issue IV:133–144 (in Hungarian)

Chapter 10
Recent Landform Evolution in the Romanian Carpathians and Pericarpathian Regions

Dan Bălteanu, Marta Jurchescu, Virgil Surdeanu, Ion Ionita, Cristian Goran, Petru Urdea, Maria Rădoane, Nicolae Rădoane, and Mihaela Sima

Abstract In the Romanian Carpathians, developed on crystalline and volcanic rocks, the main geomorphological processes are rockfalls, debris flows, and topples. In the eastern part of the Eastern Carpathians, built up of Cretaceous and Paleogene flysch, landslides and mudflows are of major significance. High and middle mountain karst features and cave systems are also widespread. In the alpine area of the Southern and Eastern Carpathians, avalanches are common on the steep slopes of

---

D. Bălteanu • M. Jurchescu (✉) • M. Sima
Institute of Geography, Romanian Academy, D. Racovița Str. 12,
023993 Bucharest, Romania
e-mail: igar@geoinst.ro; marta_jurchescu@yahoo.com; simamik@yahoo.com

V. Surdeanu
Faculty of Geography, Babeș-Bolyai University of Cluj-Napoca, Clinicilor Str. 5-7, 400006 Cluj-Napoca, Romania
e-mail: surdeanu_v@yahoo.com

I. Ionita
Department of Geography, "Alexandru Ioan Cuza" University of Iasi,
Carol I Blvd., 20 A, 700505 Iasi, Romania
e-mail: ion.ionita72@yahoo.com

C. Goran
"Emil Racoviță" Institute of Speleology, Romanian Academy,
Calea 13 Septembrie Str. 13, 050711 Bucharest, Romania
e-mail: cristian.goran@gmail.com

P. Urdea
Department of Geography, Faculty of Chemistry, Biology and Geography,
West University of Timișoara, Pestalozzi Blvd. 16, 300115 Timișoara, Romania
e-mail: urdea@cbg.uvt.ro

M. Rădoane • N. Rădoane
Faculty of History and Geography, "Ștefan cel Mare" University of Suceava,
Universitatii Str., 13, 722029 Suceava, Romania
e-mail: radoane@usv.ro; nicolrad@yahoo.com

glacial cirques and valleys. Landslides also develop on high quarry slopes, waste dumps and tailing dams characteristic of the mining sites of the Apuseni Mountains. High discharges along the Carpathian rivers cause intense erosion and the undercutting of slopes, favoring landslides and flooding. Although in fluvial erosion channel incision is predominant (for half of all river sections studied), riverbed aggradation is also observed locally. On the agricultural lands of the Subcarpathians and in the Transylvanian Depression slopes are degraded by sheet and gully erosion, landslides, and mudflows. On the Moldavian Plateau soil erosion, gullying, and landslides are major exogenous geomorphic processes. The country-wide spatial distribution of these geomorphological hazards has been evaluated by several authors (e.g., Geografia României I. 1983; Bălteanu 1997).

Keywords Landslides • Soil erosion and gullying • Karst • Periglacial processes • Fluvial processes • Mining impact

10.1 Major Geomorphological Units

Dan Bălteanu

The Romanian section of the *Carpathian Mountains* occupies 66,303 km² (27.8% of the country's territory) and stretches along 910 km between the Tisza Valley and the Danubian Gorges, with an extension, the Apuseni Mountains, up to the Someş Valley (Romania. Space, Society, Environment 2006). The Carpathian arch is bordered by a hill and tableland region (the Subcarpathians, the Banat and Crişana Hills, the Getic Piedmont, and the Moldavian Plateau) and encircles the Transylvanian Depression. Average altitude is 1,136 m, the highest summit is Moldoveanu Peak in the Făgăraş Massif (2,544 m). This range has a very complex morphology and structure, being also very much fragmented. The Carpathians in Romania can be divided into the following distinct units (Romania. Space, Society, Environment 2006).

10.1.1 The Eastern Carpathians

The Eastern Carpathians (33,584 km²), between Romania's northern border and the Prahova Valley, are structured in three distinct longitudinal units: a *crystalline* unit in the central part, with the highest peaks in the Rodna Mountains (Pietrosu 2,303 and Ineu 2,279 m); a *sedimentary* unit in the east, built up from Cretaceous and Paleogene Flysch; and a *Neogene volcanic* unit in the west, at the contact with the Transylvanian Depression. In the southeast there are the Curvature Carpathian and Subcarpathian areas, where the Vrancea Seismic Region of high seismicity potential is located (three over 7 M Richter scale earthquakes/century on the average).

10.1.2 The Southern Carpathians

The Southern Carpathians (15,000 km²) extend from east to west between the Prahova Valley and the Timiş-Cerna Corridor. They represent the highest and most compact section of the Romanian Carpathians, with heights above 2,500 m, large denudational levels, and a characteristic alpine relief dotted with numerous glacial cirques and valleys. At altitudes above 2,000 m, extended alpine and subalpine meadows are found. These mountains are made up of crystalline schists and Mesozoic sedimentary deposits of distinct elevations. The main tectogenetic phase is dated to the Upper Cretaceous.

10.1.3 The Banat and Apuseni Mountains

The Banat and Apuseni Mountains (17,714 km²), spread out between the Danube and the Someş rivers, include two distinct subunits: the Banat Mountains in the south up to the Mureş Corridor, and the Apuseni Mountains (highest peak Curcubata, 1,847 m) north of the Corridor. They have a complex horst and graben structure, being built of crystalline rocks, limestones, and volcanic rocks with important ferrous and non-ferrous ore deposits.

10.1.4 The Transylvanian Depression

The Transylvanian Depression (25,029 km²) lies between the three units of the Romanian Carpathians. It is a tableland with heights of 400–800 m and has a tectonic origin. The basement is Carpathian with a post-tectonic mantle of Upper Cretaceous to Lower Miocene age. The depression is filled by clastic rocks, marine tuffs, a salt formation, marls, sands, and clays (Miocene to Pliocene), with monoclinal and dome-like structures (Romania. Space, Society, Environment 2006).

The *Pericarpathian region* includes the following units. The *Mehedinţi Plateau* (785 km²) is situated in the southwest of Romania between the Danube and the Motru rivers, at heights of 500–600 m. It is actually a lower compartment of the Southern Carpathians and consists of crystalline schists and limestones. The *Subcarpathians* (16,409 km², 400–900 m altitude), which border the Carpathians on the east and south along 550 km length, are built up of folded-faulted Neogene molasses and are affected by neotectonic uplift (Zugrăvescu et al. 1998). Human pressure in the area is particularly severe and has contributed to an intense remodeling of stream channels and slopes. The *Getic Piedmont* (12,940 km²) is located to the south of the Southern Carpathians at altitudes of 200–700 m and consists of gravels and sands with intercalations of marls and clays. It was formed in the Romanian-

Quaternary interval and is basically a relict piedmont fragmented by large, consequent allochthonous valleys. Piedmont catchment basins are affected by gully erosion and landslides, their intensity decreasing from north to south. The *Moldavian Plateau* (23,085 km²) lies in eastern Romania and is developed on a platform basement covered by sedimentary formations deposited in several cycles. The *Banat* and *Crişana Hills* and *Plain* (28,640 km²) are situated in the western part of the Banat and Apuseni Mountains, are discontinuous and have a predominant piedmont character. The plain lies to the west and it was formed during several stages after the recession of the Pannonian Lake and the accumulation of the fluvio-lacustrine and lacustrine sediments.

10.2 History of Geomorphological Research

Dan Bălteanu and Marta Jurchescu

The research of geomorphological processes in the Romanian Carpathians has evolved differently for the various types of processes but more common features appeared in the phase of a more rapid development since 1990, corresponding to relatively suddenly opened access to international literature.

Landslide investigations have a long tradition in Romania. Landslide-related studies started especially in the 1920s (Mihăilescu 1926). In the following decades, numerous articles and books addressed this subject, in a more descriptive manner, either with the aim of classifying, presenting some local cases, or zoning landslides across geomorphic units or all over the country (e.g., Mihăilescu 1939b; Tufescu 1964, 1966; Morariu and Gârbacea 1968b; Ielenicz 1970). Regionally, the Curvature Carpathians and Subcarpathians, being among the most complex units in terms of lithological and structural conditions and part of the Vrancea Seismic Region with the most active subcrustal earthquake activity in Europe are intensely modeled by a wide range of landslide processes and benefitted from a long history of observations (e.g., Mihăilescu 1939a; Tufescu 1959; Posea and Ielenicz 1970). The relatively recent development could be summarized by a two-level analysis: regional assessments and site-specific studies. *Regional assessments* focus on study-areas ranging from small-size catchments to larger geomorphic units. In an earlier stage, direct qualitative methods and some of the first quantitative ones were employed, involving geomorphological classification and large-scale mapping (1:5,000, 1:10,000) (Bălteanu 1975, 1983; Ielenicz 1984). This was done repeatedly in the course of time, allowing to differentiate between annual, seasonal, and monthly changes, with the aim of assessing the morphodynamic trends of slopes and of elaborating morphodynamic maps (Bălteanu 1975). From 1980 to 2000, the research focus was placed on elaborating the methodology of general geomorphological mapping at regional scales of 1:25,000 and 1:200,000 and on synthesizing available landslide information over wider areas (e.g., Gârbacea 1992; Irimuş 1998; Surdeanu 1996,

1998; Bălteanu 1997; Dinu and Cioacă 2000). These regional studies were the precursors of the first landslide susceptibility and hazard maps produced after 2000. Direct susceptibility and hazard maps, based on expert judgment, have been elaborated on test areas at large scales. Indirect susceptibility assessments were based first on heuristic methods associated with the use of GIS techniques (Mihai 2005), and subsequently, with the constant construction or improvement of some landslide databases (e.g., Şandric and Chiţu 2009), through statistical analyses of empirical data (Şandric 2005, 2008; Micu 2008; Bălteanu and Micu 2009; Micu and Bălteanu 2009; Chiţu et al. 2009; Mihai et al. 2009; Constantin et al. 2011; Chiţu 2010). In the last few years, quantitative research was also extended to other processes, like debris flows or rockfalls (Ilinca 2010; Pop et al. 2010; Surdeanu et al. 2010). Recently, by analyzing either the variability of rainfall as triggering factor or the historical frequency of landslides, it has been possible to make primary estimations on the temporal probabilities of landslide occurrence and produce landslide hazard maps (Dragotă et al. 2008; Micu 2008; Şandric 2008; Bălteanu and Micu 2009; Chiţu 2010). Qualitative, expert-based hazard maps have also been drawn over wider areas (Micu et al. 2010). In some cases it has been proven that validated susceptibility and hazard assessments are in agreement with the detailed morphodynamic mapping made in the past (Bălteanu and Micu 2009).

Investigations at the *local scale* included repeated mapping projects, measurements, estimations of movement rates, collection of soil samples and determination of geotechnical properties (Bălteanu and Teodoreanu 1983; Bălteanu 1983, 1986), topographical, inclinometrical measurements (Surdeanu 1998), the analysis of internal landslide structures by geophysical techniques (Andra and Mafteiu 2008; Urdea et al. 2008b; Chiţu 2010) and eventually the application of deterministic methods (Micu 2008; Constantin et al. 2010; Chiţu 2010).

The landslide susceptibility map of Romania (Bălteanu et al. 2010), based on a semi-quantitative method, offers a general view on landslide occurrence in the Romanian Carpathians.

The study of *soil erosion* and *gullying processes* has focused on two major directions: (1) the long-term monitoring on some experimental plots, conducted at several research stations located in different environmental conditions, and (2) the inventorying of gullies, as well as repeated surveying of some representative gullies over longer periods of time. Experimental results on runoff plots enabled the formulation of empirical methods of *soil erosion prediction* (Moţoc 1963; Moţoc et al. 1975; Traci 1979; Dârja et al. 2002; Moţoc and Mircea 2002; Ionita et al. 2006), later tested in various areas (e.g., Patriche et al. 2006). The second field of investigation (e.g., Moţoc et al. 1979a; Mihai and Neguţ 1981; Ichim et al. 1990; Rădoane and Rădoane 1992; Ionita 1998, 2000a, 2006; Rădoane 2002) covered both types of *gullies*, continuous and discontinuous (Ionita 2003), on which measurements of head regression and area and volume growth have been performed over time. Among the techniques used, the one involving ^{137}Cs isotope provided information on temporal variations in sediment deposition and soil erosion rates (e.g., Ionita and Margineanu 2000; Ionita et al. 2000). Furthermore, based on repeated surveys, it became possible to apply statistical and deterministic models to gully evolution (e.g., Rădoane et al. 1995, 1997; Ionita 1998, 2003, 2006; Mircea 2002; Ionita et al. 2006). Models capable to predict the initiation of future gully processes within a catchment have also been developed (Moțoc 2000 cited in Mircea 2002; Moțoc and Mircea 2005).

Moreover, total erosion and *sediment delivery* in small catchments were assessed either in an empirical or deterministic manner (Moțoc et al. 1979b; Ichim and Rădoane 1984; Ionita 1999, 2008; Mircea 2006).

The earliest maps or descriptions of *karst forms (caves)* on the present territory of Romania are as old as the late seventeenth up to the nineteenth centuries. Nevertheless, a systematic scientific research of karst forms and processes began with the establishment of the Institute of Speleology by E.G. Racoviță in the city of Cluj in 1920. Reorganization of the old Institute led to the elaboration of monographical studies devoted to different karst areas (Bleahu and Rusu 1965; Orghidan et al. 1965). Starting with 1965, simultaneously with the specialization of researchers in different areas of interest, the speleological activity was enriched by amateur contributions to cave exploration and survey (*sports speleology*). The unprecedented progress in discoveries occurred in close collaboration with researchers who constantly verified and standardized new data (e.g., Bleahu and Povară 1976; Goran 1980, 1982; Lascu and Sârbu 1987).

The more recent activity of the Institute of Speleology has been divided among several fields of theoretical and applied karstology and has developed both in Bucharest (e.g., Constantin 1992; Constantin et al. 2001) and in Cluj (e.g., Onac 2002; Racoviță et al. 2002; Perșoiu et al. 2011), using modern equipment and investigation methods (absolute datings, paleoclimate reconstructions, chemical and drainage analysis of karst aquifers, etc.). Besides, an important element is the continuous updating of the Romanian cave inventory.

The first observations on some specific *periglacial elements*, though not defined as such, date back to the end of the nineteenth and the beginning of the twentieth centuries (de Martonne 1900, 1907; etc.). Corresponding to an international trend, a sudden concern for both actual and Pleistocene periglacial issues occurred only after 1955. An abundance of works followed, including some syntheses on the whole territory of the country (e.g., Niculescu and Nedelcu 1961; Niculescu 1965; Naum 1970; Schreiber 1974; Mihăilescu and Morariu 1957; Morariu et al. 1960; Ichim 1980). Until 1990, however, only the identification of specific phenomena, alongside the description and explanation of their occurrence and age, was aimed at. After that year, specific methods and techniques started being applied in the alpine areas of the Romanian Carpathians for various purposes. Rock glacier investigations, BTS-measurements and geophysical tomographies were performed and a solar radiation model was applied to investigate the presence of permafrost in the Carpathians (Urdea 1991, 1993, 1998a, 2000; Urdea et al. 2001–2002; 2008a). Using geophysical techniques the inner configuration of some periglacial deposits could also be analyzed (Urdea et al. 2008a, c). Dendrogeomorphological methods served to date periglacial landforms or reconstruct their evolution (Urdea 1998b). Local measurements of *active processes* included: the study of frost weathering in rocks by means of thermal infrared images; the monitoring of the movements produced by frost heave and frost thrusting, as well as by pipkrake, needle ice, and frost sorting processes, through the use of elevationmeters and cryometers (Urdea et al. 2004). Other periglacial phenomena, such as solifluction, the movement of ploughing blocks, talus and rock creep, or nivation-related processes were also monitored in order to estimate movement rates or the occurrence frequency of events (Urdea et al. 2004; Voiculescu 2009; Voiculescu and Popescu 2011).

The study of *fluvial* geomorphic *processes* took on a pronounced quantitative character when, starting with the 1960s, the necessary hydrometrical measurements increased in number and quality and mainly focused on the rivers draining the outer flanks of the Eastern Carpathians, and, to a lesser extent, on rivers crossing the Southern or the Banat and Apuseni Mountains (including the Danube Gorge). The aim was to identify the behavior of riverbed systems in relation to natural and especially to human controlling factors: man-made reservoirs (an earlier overview in Ichim and Rădoane 1986), river channelization (Hâncu 1976), river straightening, embankments, in-stream sand mining, etc. The major issues were the vertical and planform mobility of river channels, followed both in river *longitudinal profiles* and in *cross-sections*, statistically analyzed by employing databases on numerous cross-sections over short and long periods of time, in order to detect evolution trends (e.g., Diaconu et al. 1962; Ichim and Rădoane 1980, 1981; Bătucă 1978; Ichim et al. 1979; Bondar et al. 1980; Rădoane et al. 1991; Feier and Rădoane 2008; Perșoiu 2008); the contribution of slopes to riverbeds and *sediment budget estimations* by indirect methods (e.g., Gașpar and Untaru 1979; Ichim et al. 1998; Rădoane and Rădoane 2003b; Dumitriu 2007; Feier 2007; Burdulea-Popa 2007); the past and future evolution of drainage basins in terms of the degree of *concavity* of the longitudinal profiles modeled mathematically (Rădoane et al. 2003); geometrical, physical and petrographic analyses of current stream *channel deposits* (Ichim et al. 1996, 1998) in the light of two distinct laws: downstream fining and channel material bimodality (Rădoane and Rădoane 2003a; Rădoane et al. 2008a), using specific field sampling and laboratory processing techniques.

10.3 Recent Landform Evolution in the Carpathian and Pericarpathian Regions

10.3.1 Landslides

Dan Bălteanu, Virgil Surdeanu and Marta Jurchescu

Landslides are among the most widespread geomorphological processes in the hilly regions built of Neogene molasse deposits, as well as in the mountainous regions of Romania developed on Cretaceous and Paleogene flysch. Primarily due to the presence of these sedimentary rocks consolidated to various degrees and to *tectonic influences*, landslides are most common in the Subcarpathian Region and in the Eastern Carpathians – particularly in the Curvature area of high seismicity and active neotectonic activity. The uneven incision of the drainage network has generated variable *relative relief* ranging from 50–350 m for smaller streams to 350–700 m for major rivers (like the Moldova, Bistriţa, Trotuş, Buzău, and Prahova).

Annual average *precipitation*, ranging from 600 to 1,000 mm, falls within 85–125 days, the snow pack lasts for 100–180 days. The pluviometric regime alternates between wet and dry periods, triggering and maintaining the masses of earth in a dynamic state for a long period of time. In the last 130 years, the four intervals of pluviometric excess (1912–1913; 1939–1942; 1970–1972 and 2004–2005) led to a recrudescence of landslide processes in the flysch area (Surdeanu 1996).

As two-thirds of the flysch mountains are below 1,000 m altitude, large areas are accessible to human settlement and the development of *economic activities*. As a result, the equilibrium of their slopes has been upset. Although around 40% of Romania's forests are concentrated in the Eastern Carpathians, yet *deforestation*, gradually expanding from mountain foot to top during the Middle Ages and at even faster rates in the nineteenth, twentieth and twenty-first centuries, mostly in the mountain basins of the Buzău, Trotuş, Bistriţa, Moldova rivers and in the Curvature area, has contributed to the reactivation of old slides. In many cases, the valleys of first to third order rivers, developed on marls and clays, are filled with landslide colluvia (the basins of the Buzău, Putna, Trotuş, Bistriţa, etc.), and have a specific slide valley morphology.

Mining works, oil drilling, ever more densely *built-up areas* and *communication* routes (at densities of 6.4 km km^{-2} in the oil fields), as well as the hydrotechnical structures raised in the Buzău, Bistriţa, Argeş, Olt, and Someş valleys have challenged the stability of slopes and extended landslide-prone areas, particularly on the slopes of reservoirs in the Buzău and Bistriţa valleys and in the Trotuş Mountains oil fields. The mining areas of the Apuseni Mountains and the Eastern Carpathians present a special situation: some areas are affected by landslides related to waste dumps and tailings dams. Unstable waste dumps in the mining zones of Baia Mare, Ostra – Tarniţa and Călimani (the Eastern Carpathians) and Certej – Săcărâmb (the Apuseni Mountains) pollute rivers over long distances. In the late twentieth and the early twenty-first centuries, *salt extraction* in the inner and sub-montane depressions (of Maramureş at Coştui, and of Târgu Ocna, respectively), as well as in the Carpathian Foreland (Ocnele Mari) led to the collapse of galleries, triggering large-scale slidingcollapsing processes.

The geomorphological mapping projects, field surveys and laboratory tests undertaken in the Carpathian area (Mihăilescu 1939a; Badea 1957; Donisa 1968; Barbu 1976; Posea and Ielenicz 1976; Ichim 1979; Untaru 1979; Surdeanu 1979, 1998; Bălteanu 1983; Ielenicz 1984; Micu 2008; Mureşan 2008, etc.) have revealed some *regularities* in the great diversity of mass movements:

- Most landslides would affect the surface deposits lying at the base of slopes of 35–62% clays and, as a rule, slides with a 1:20 to 1:50 length-to-width ratio prevail.

- In the mountain region, landslides occurring at 700–800 m maximum elevation have the greatest impact on landform evolution (given that human pressure, such as raw materials extraction, is also at its peak – up to 1,100 m in the Tarcău Mountains).
- Nearly 75% of the active landslides recorded in the mountains occur on deforested slopes. There are instances, however, when big landslides also develop on forested slopes (in the Ceahlău and the Buzău Mountains).
- Shallow slides last for a couple of weeks. Deep-seated slides may extend over years, scores of years with episodes of rapid and slower rates of material removal.

Having analyzed over 500 active landslides in the Eastern Carpathian Flysch zone north of the Trotuș River over the past four decades (Surdeanu 1979, 1987, 1996, 1998), the following features have been distinguished:

- *Translational shallow slides* with a 0.5–1.0 m circular scarp and a short dynamic phase (days), 55 m long and 22 m wide on the average;
- *Rotational slides* with a linear scarp, on the average 70 m long and 31 m wide;
- *Slumps* with a micromorphology of monticles and waves, and a 2–30 m high circular or linear scarp; on the average 95 m long and 25 m wide;
- *Valley slides*, developed in the upper part of drainage basins, with a 2–3 m high scarp, on the average 180 long and 90 m wide;
- The volume of material entailed is of the order of hundreds and thousands (sometimes even over one million) cubic meters;
- The landslide-induced *denudation rate* in the flysch mountains was estimated at 2–40 mm year^{-1} for slide-affected areas (Surdeanu 1998).

A special case is represented by the Curvature Carpathians, where tectonic uplift and the seismicity specific of the Vrancea area favor the occurrence of deep-seated landslides (Bălteanu 1983; Ielenicz 1984) (Fig. 10.1).

In the Subcarpathians, with a dominantly argillaceous Neogene molasse substrate and a high content of montmorillonite and illite (quick clays), landslides have a greater share in the modeling of the slopes. Shallow translational slides and moderate to very deep-seated rotational slides, alongside mudflows, are common.

Recent assessments (Micu and Bălteanu 2009; Bălteanu et al. 2010) have shown that the most affected areas lie in the Curvature Subcarpathians (Fig. 10.2). Besides rainfall, shocks induced by strong earthquakes (magnitude over 7 M on the Richter scale and return period of 35–40 years), localized in the Vrancea Seismic Region, play a major role in activating deep-seated slides, rockfalls and debris flows. In some areas denudation rates were estimated at 0.5–10 mm year^{-1} corresponding to years of high precipitation with a return period of 5–7 years (Bălteanu 1983). The slopes of this unit, mostly covered by landslide deposits, range from highly stable to unstable with an annual frequency in landslide reactivations. These affect primarily the lower part of slopes associated with the land use changes of the post-communist period and with the higher frequency of torrential rainfalls.

In the Getic Subcarpathians and the Getic Piedmont, the spatial distribution of slides shows a correlation to small catchments and gully erosion. In addition,

Fig. 10.1 Deep-seated landslide on the righthand slope of the Siriu Reservoir, Buzău Mountains (Curvature Carpathians) (*left*) and its scarp (*right*) (Photos: Dan Bălteanu)

Fig. 10.2 Active mudflow at Malu Alb, Pătârlagele (Curvature Subcarpathians), in 2010 (Photo: Laurențiu Niculescu)

degradation caused by coal and salt mining over large areas led to an increase of landslide susceptibility (Bălteanu et al. 2010) (Figs. 10.3 and 10.4).

A great diversity of landslides occurs on vast stretches of land in the Transylvanian Depression. In the Transylvanian Plain, mainly built of clayey rocks, shallow and

Fig. 10.3 Deep-seated slump at the margin of a coal quarry, Berbeşti (Getic Subcarpathians) (Photo: Marta Jurchescu)

Fig. 10.4 Waste dump affected by a recent deep-seated landslide, Mateeşti (Getic Subcarpathians) (Photo: Marta Jurchescu)

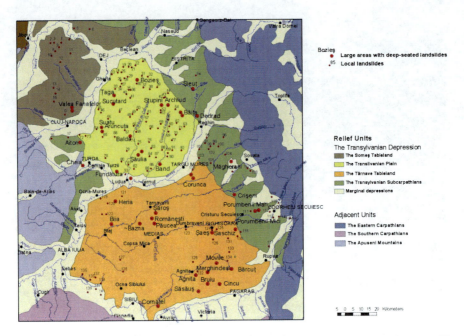

Fig. 10.5 Deep-seated landslides ("Glimee") in the Transylvanian Depression (based on over 50 scientific papers, books and PhD theses published between 1950 and 2002) (Bălteanu and Jurchescu 2008)

medium-deep slides prevail. In the central and southern part of the Depression we find large-sized, deep-seated landslides of *lateral spreading* type, produced on a substrate of marl-clayey complexes with intercalations of Sarmatian sandstone and sand. Spores and pollen analyses dated them to the humid Boreal stage (Morariu et al. 1964; Morariu and Gârbacea 1966, 1968a, b; Gârbacea and Grecu 1983; Grecu 1992; Gârbacea 1992, 1996; Irimuş 1996, 1998) (Fig. 10.5).

Landslides are also common for the homoclinal relief of the Moldavian Plateau, particularly on cuesta fronts. Pujina and Ionita (1996), investigating a time span of 170 years, found that periods of landslide reactivation outnumbered those with first time failures (Bălteanu et al. 2010).

10.3.2 Soil Erosion and Gullying

Ion Ionita

One of the most severely eroded *agricultural areas* in Romania is the Moldavian Plateau (27,000 km²). Clayey-sandy Miocene-Pliocene layers with a gentle north-northwest to south-southeast dip of 7–8 m km^{-1} outcrop from the sedimentary substrate (Jeanrenaud 1971). The climate is temperate continental (mean annual temperature: 8.0–9.8°C; average annual precipitation: 460 mm at lower elevations in the south and 670 mm in the central and northwestern area rising to 587 m). Natural vegetation cover was drastically changed by human action, particularly over the past two centuries. Mollisols and argiluvisols (forest soils) used for crop production are the most common (arable land 58%, pastures and meadows 16% and forests 13%).

Currently soil erosion and gullying are assessed from long-term *monitoring* of experimental plots and repeated field surveys of *gullies*. The effect of soil cover on runoff and soil losses was studied on runoff plots over the period 1970–1999 at the Central Research Station for Soil Erosion Control Perieni-Bârlad. Substantial field databases have resulted from almost 20 years of monitoring representative gullies located in the southern part of the Moldavian Plateau near the city of Bârlad, using aerial photographs of the 1960 and the 1970 flights, classical leveling work and repeated surveys through a particular close stakes grid after 1980.

The Perieni runoff plots were established on the left valley side of Tarina catchment with 12% slope and slightly eroded mollisol. Generally, data collected here over a 30-year period on soil and water losses indicate the following (Ionita 2000a; Ionita et al. 2006):

- Mean annual precipitation is 504.3 mm, and the precipitation that causes runoff and erosion falls as rain during the growing season from May to October;
- About 26% (133.5 mm) of the annual precipitation induced runoff/erosion on continuous fallow and 18.5% (93.5 mm) for maize;
- Runoff ranges from 36.5 mm under continuous fallow with the peak of 12.0 mm in July and 17.7 mm under maize with the peak of 6.5 mm in June;
- Average soil loss is 33.1 t ha^{-1} year^{-1} for continuous fallow with the peak of 12.8 t ha^{-1} in July and 7.7 t ha^{-1} year^{-1} for maize with the peak of 3.7 t ha^{-1} in June.

It has to be remarked that on heavily eroded forest soils the value of the soil loss is doubled.

According to Moțoc and co-workers (1998) data collected from the continuous fallow plot and processed using a 3-year moving average revealed that over the period 1970–1999 there were three soil erosion peaks, in 1975, 1988 and 1999.

Radoane and co-workers (1995) identified two areas with a higher gully density: the first in the north, where mostly small discontinuous gullies have developed on clays and the second in the south, around the city of Bârlad, where, on loamy-sandy layers, valley-bottom continuous gullies prevail. Results on discontinuous gullies have indicated that during a variable period of 6–18 years the gully head retreated 0.92 m year^{-1} on average with a range from 0.42 to 1.83 m year^{-1}. The mean growth of gully area was 17.0 m^2 year^{-1} and varied between 3.2 and 34.3 m^2 year^{-1}. Both values indicate a slow erosion rate (Ionita 2000b, 2003, 2006; Ionita et al. 2006). Moreover, the annual regime of gullying shows great fluctuations and 60% of total *gully growth* took place in only 5 years (1980, 1981, 1988, 1991 and 1996).

Conventional measurements on sedimentation using check-iron plates along the floor of discontinuous gullies over the period 1987–1997, indicate a higher rate of aggradation in the upper half of the gully floor. This finding supports the development of a short steeper reach within the gully floor as a critical location for gullying renewal. Similar values were obtained from the ^{137}Cs depth profile. Furthermore, it was possible to date gullies at 23–48 years and to claim that discontinuous gullies deliver most of the sediment needed for their own aggradation. The evolution pulses reflect a dynamic balance between two simultaneous processes, erosion and sedimentation, within a single system. As for continuous gullies, linear gully head retreat, areal gully growth and erosion rates were established for three periods (1961–1970, 1971–1980 and 1981–1990). Results indicate that gully erosion has decreased since 1960 (Ionita 2000b, 2006; Ionita et al. 2006). Average gully head retreat ranged from 19.8 m year^{-1} in the 1960s and 12.6 m year^{-1} in the 1970s to 5.0 m year^{-1} during the 1980s. This decline is due to the rainfall distribution, and the increased influence of soil conservation. The mean gully head retreat of 12.5 m year^{1} over the 30-year period (1961–1990) was accompanied by a mean gully area growth of 366.8 m^2 year^{-1} and a mean erosion rate of 4,168 t year^{-1}. The continuous gullies also developed in pulses. Gullying in the 1981–1996 period concentrated mostly on the 4 months from mid-March to mid-July in an area with mean annual precipitation around 500 mm. Another main finding of this 16-year stationary monitoring was that 57% of the total gullying occurred during the cold season, especially in March due to freeze-thaw cycles, with the remainder during the warm season. Of the total gully growth, 66% results from only 4 years (1981, 1988, 1991 and 1996) when precipitation was higher.

Field measurements performed in small catchments during flash streamflows allowed the identification of two types of sediment delivery scenarios, synchronous and asynchronous (Ionita 1999, 2000b, 2008). Even for very rare events, the synchronous scenario, mostly associated with quick thawing, shows very high sediment concentration, exceeding 300.0 g L^{-1} at the basin outlet and low values, up to 40.0 g L^{-1}, upstream of gullies in the upper basin. Gullying is the major source of sediment. The asynchronous scenario commonly occurs and is characterized by higher water discharges and fluctuating sediment concentration (Piest et al. 1975). Total erosion in the Moldavian Plateau of Eastern Romania averages 15–30 t ha^{-1} year^{-1} (Moțoc 1983).

Since 1991, by implementing the new *Land Reform* (Acts nos 18/1991 and 01/2000), the previous area under conservation practices was gradually converted to the traditional downhill farming system. Under these circumstances the rate of soil erosion and sedimentation doubled (Ionita et al. 2000).

10.3.3 Karst

Cristian Goran

The carbonate karst of the Carpathians occupies 3,700 km^2, representing 5.6% of the mountainous area and 82% of the total karst area of Romania. Karst areas occur in all Carpathian units, being the most extensive in the central and northern parts of

Fig. 10.6 The distribution of karst regions in the Romanian Carpathians (Modified after Bleahu and Rusu 1965)

the Eastern Carpathians, in the west of the Southern Carpathians and in the Banat and Apuseni Mountains (Fig. 10.6). The karst terrains are distributed as follows: 16% in the Eastern Carpathians, 27% in the Southern Carpathians, 26% in the Banat Mountains and 31% in the Apuseni Mountains (Bleahu and Rusu 1965). Karst-prone rocks mainly outcrop in the Mesozoic mountain massifs in contact with uplifted blocks of crystalline schists. The widest limestone outcrops, relatively unitary, are located in the Reşiţa–Moldova Nouă Syncline from Banat (over 600 km^2), in the Bihor Massif and in the Pădurea Craiului Mountains. Karst structures were identified up to a maximum elevation of more than 2,400 m (in Negoiu, Făgăraş Mountains) and down to a minimum elevation, reached under the sea level (in the Danube Gorges, Almăj Mountains).

The distribution of the karst units is correlated with the morphology and the structure of limestone areas (Fig. 10.7). Following the stages of tectonic uplift, the Mesozoic carbonate platforms were divided into *morphotectonic types*: *plateaus* (in the mountains with elevated crystalline basement), *limestone bars* (along the margins of mountains borders or along the tectonic corridors) and *isolated massifs* (in the area of high ridges, by the outcropping of crystalline limestones and around tectonic depressions, by the fracturing of the limestone cliffs). These morphotectonic types were further diversified during the Quaternary morphohydrographical

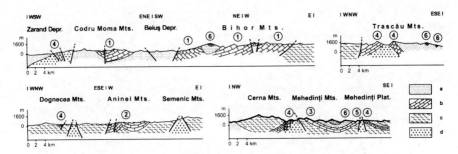

Fig. 10.7 The relationship between the mountain area massiveness and the distribution of the karst types in the Apuseni Mountains, Banat Mountains and the west of the Southern Carpathians (Modified after Goran 1983). 1, unitary elevated plateaus; 2 dissected elevated plateaus; 3, unitary bars; 4, fragmented bars; 5, leveled bars; 6, isolated massifs; a, non-carbonate sedimentary rocks; b, carbonate rocks; c, crystalline schists; d, magmatic intrusions

and karst evolution, producing the following genetic-evolutive types of karst massifs (Goran 1983):

1. Unitary and elevated karst plateaus (Pădurea Craiului Mountains, Bihor, Vașcău);
2. Elevated and hydrographically divided karst plateaus, represented by the karst of the Reșița-Moldova Nouă Synclinorium (Banat Mountains);
3. Subsided karst plateaus, located around the mountainous area (Romanian Plain, Western Plain and Hills);
4. Unitary limestone bars (Hăghimaș, Piatra Craiului, Buila-Vânturarița, Mehedinți and Scărița-Belioara);
5. Fragmented limestone bars (Trascău, Cerna Valley, Cernădia-Cerna Oltețului area and Casimcea Plateau);
6. Leveled limestone bars (Mehedinți Plateau and Moneasa area);
7. Isolated limestone massifs (located in the majority of the mountainous units); according to their morphology, they are further divided in *isolated ridge massifs* (Maramureș, Rarău and Giumalău Mountains), *isolated slope massifs* (Postăvaru, Trascău) and *isolated valley massifs* (Perșani, Ciucaș, Bucegi, Leaota).

The present-day karst inherits some of the Mio-Pliocene paleokarst features and structures, but was mainly formed due to the Quaternary uplift and fragmentation of the Carpathians. Related also to the Quaternary evolution is the partial covering of the karst terrains with detrital deposits or soil, epikarst areas being limited to mountain ridges or isolated massifs. From this point of view, the karst of Romania can be considered a transitional, moderately developed karst, which evolves now in pluviokarst regime (elevated or isolated authigenic karst units) and in fluviokarst regime (allogenic, hydrographically dissected karst units).

From the point of view of the karst systems recharge and of the intensity of the dissolution processes, a high-mountain karst (developed at elevations above

1,500–1,700 m), with water recharge from snow and rain, mainly as authigenic (unitary) karst, with well-represented epikarst, and karst areas of medium and low mountains, hydrographically divided, with lower relief energy, binary karst functioning and covered by forests, meadows and limited arable land, can be identified. *High-mountain karst* presents the following genetic types: limestone ridges (Piatra Craiului, Buila–Vânturarița, Oslea, Scărița–Belioara), unitary and elevated plateaus (Hăghimaș, Retezat, and Bihor–Vlădeasa) or isolated ridge massifs (mountains such as Maramureș, Rodna, Bistrița, Postăvaru, Bucegi, Făgăraș, Parâng, and Bihor). On this karst landscape, there are extensive areas occupied by karren and large sinkholes, while the endokarst consists mainly of potholes, genetically related to underground drainage networks of a few kilometers in length. Among the high mountains karst units of the massifs rise higher than 2,000 m and provide evidence for the presence of Quaternary glaciers (Rodna, Făgăraș, Parâng, and Retezat). There are also nivokarst structures, represented by karren, chimneys, dissolution arches, and sinkholes continued with shafts in the Bihor, Retezat, Piatra Craiului, and Bucegi Mountains. Another feature of the high-mountain karst is the presence of potholes and caves sheltering perennial ice deposits, also affected by the glaciokarst or nivokarst processes (Scărișoara Glacier Cave, Borțig Pothole, Glacier from Zgurăști, Zăpodie Cave from the Bihor Mountains, the potholes from Stănuleți, Retezat, Soarbele, and Albele Mountains). The landscape of the *middle and low mountains* is marked by planation and recent debris deposits. The limestones outcrop in isolated peaks and ridges or on surfaces where the clay or soil cover has been eroded. This landscape pertains to the Pericarpathian erosion levels (Râul Șes and Gornovița), consisting of unitary and elevated karst plateaus (Pădurea Craiului Mountains, Vașcău and Dumbrăvița plateaus from the Codru-Moma Mountains, Vf. lui Stan–Domogled Ridge from the Mehedinți Mountains, the west of the Șureanu Mountains), elevated and fragmented plateaus (Banat and Vâlcan Mountains), dissected limestone bars (Trascău Mountains, "Ciucevele" and "Râmnuțele" from the Cerna Valley, Galbenul–Olteț–Cerna Vâlceană area), leveled limestone bars (Mehedinți Plateau) and many isolated massifs. On the karst surface, sinkhole fields or valleys with sinkholes are frequently developed, while along the massif margins, at the inlets, along the lithological contacts, blind valleys and fluviokarst depressions are formed. The endokarst is represented by outlet and multilevel caves, located on the valley slopes.

Over 12,000 *caves* have been recorded in the Romanian Carpathians (cavities more than 5 m long), this figure accounting for 96% of the caves of Romania (Goran 1982). The distribution of the caves on mountain units is depending on the size of the karst area and on the intensity of the karst processes. Therefore, from the Eastern Carpathians, 810 caves have been registered, from the Southern Carpathians, 5,697 caves, from the Banat Mountains, 1,553, and from the Apuseni Mountains – 3,960 caves. An exceptional concentration of caves is found in the Bihor Mountains (1,299), Retezat Unit (more than 2,800), Anina Mountains (1,041) and Pădurea Craiului Mountains (800). The Vântului Cave (Pădurea Craiului Mountains) is the longest (more than 50 km long) cave, while the V5 Pothole (Bihor Mountains) is the deepest (−641 m).

10.3.4 Periglacial Processes

Petru Urdea

Romanian geoscientists agree that the landscape of Romania finally took shape in the Quaternary period. In this complex process, periglaciation played a decisive part – especially in the mountains. The relict elements have to be distinguished from present-day periglacial processes. Obviously, *in lowlands and hills* periglacial elements have a *relict* character, in contrast to *uplands* (alpine regions), where some of the *recent periglacial elements* have reached their climax.

The *climatic conditions* in the periglacial zone of the Romanian Carpathians are exemplified by a mean annual air temperature of 3°C at Cozia (1,577 m), 1.0°C at Vlădeasa, 0.2°C at Bâlea Lake (2,038 m), −0.5°C at Țarcu (2,180 m) and −2.5°C at Omu (2,505 m), where the absolute minimum temperature is −38°C. Mean annual precipitation is 844.2 mm at Cozia, 1151.3 mm at Vlădeasa, 1,246 mm at Bâlea Lake, 1,180 mm at Țarcu and 1,280 mm at Omu. The continentality index according to Gams (CIG) is over 50°. Snow depth is highly variable, between 50 and 370 cm, according to the wind action. About 60–75% of precipitation falls as snow, and the snow cover in the region lasts 150–210 days of the year. The 3°C mean annual isotherm, i.e., the lower limit of periglacial environment, runs at ca 1,700 m elevation. In the periglacial region there are three zones, the *solifluction zone* between the 3°C and 0°C isotherm, the *zone of complex periglacial processes* between 0°C and −3 (−2)°C isotherm, and the *cryoplanation zone* of intense mechanical weathering between the −3 (−2)°C and −6°C isotherm. From the Peltier diagram the morphoclimatic systems are periglacial system with physical dominance for Omu and Bâlea Lake and boreal system for Țarcu (Urdea and Sîrbovan 1995).

BTS measurements and the low water temperatures (<2°C) of rock glaciers outlets prove the existence of *permafrost* in rock glaciers and scree deposits. It was an interesting and amazing discovery that permafrost is also present at low elevations, at 1,100 m at Detunata Goală (Apuseni Mountains) (Urdea 2000), proved by the solar radiation model (Urdea et al. 2001–2002). Recently, the existence of permafrost was documented by geophysical investigations, especially for rock glaciers (Urdea et al. 2008a).

Periglacial deposits, important paleoclimatic indicators, are formed by freeze-thaw action (blockfields, talus cones, scree, stone streams, rock rivers), by solifluction and aeolian deposition (loess, nivo-eolian deposits). For scree slopes at Bâlea Lake and Văiuga, geoelectrical DC tomography (Urdea et al. 2008a), based on the Wenner–Schlumberger array layout configuration, shows the presence of distinct layers specific to stratified slope deposits (Sass 2006), in fact *"éboulis ordonée"* (Urdea 1995). The resulted models for the solifluction lobe Paltina, in dipole-dipole configuration (suitable for vertical structures), and at an equal distance of 1 m between the electrodes, permit a differentiation of distinct layers of 40–50 cm and undulating solifluctional layers (Fig. 10.8). In the case of the *fossil patterned ground* Paltina–Piscul Negru, the 2D electrical resistivity tomography model in Wenner

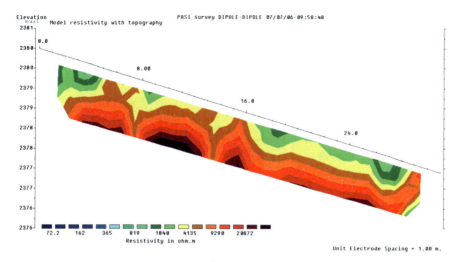

Fig. 10.8 Electrical resistivity tomography profile (inversion model) on the solifluction lobe Paltina (Făgăraş Mountains) (Urdea et al. 2008d)

configuration, presents distinct layers with a special undulating and pocket design formed by frost heaving and frost sorting. For *rock glaciers* (Ana and Pietrele in the Retezat Mountains, Capra Tunnel and Văiuga in the Făgăraş Mountains) electrical tomography reveals typical structures. In the bottom part high resistivity (>700 kΩ m) points to rock bodies rich in ice (Urdea et al. 2008a).

Although the alpine area of the Romanian Carpathians belongs to the periglacial and boreal morphoclimatic altitudinal zones, present-day geomorphological processes are very complex (Urdea et al. 2004). *Frost weathering* is regarded as a particularly effective geomorphic agent, a combination of frost shattering and frost wedging. The occurrence of freshly split blocks, gravelly and sandy regolith in the granitoid massifs (Retezat, Parâng, Vârfu Pietrii, and others) with tors indicates that the process of grussification and rock weathering, and the production of coarse loose debris, monitored in the Lolaia Mountain area (Retezat Mountains) and in the Transfăgărăşan area, are still active and continuous under present-day climatic conditions. The information obtained from thermal images (a thermoinfrared camera Fluke Ti20) confirms surface temperature variations between minerals in granitoid rocks under short-term (diurnal) temperature fluctuations controlled by color and crystal size. Surface temperatures variations between minerals (quartz, feldspar, mica, amphibole) cause differential thermal expansion, strain and disintegration and, with the contribution of nivo-eolian processes, cavernous weathering or the formation of *honeycomb* microforms (Fig. 10.9). The predominantly upward directed frost heaving and predominantly lateral frost thrusting, induced by ice segregation in the ground, produce *pipkrakes* or needle ice, which can heave stones as large as cobbles and frost sorting, very active on the surfaces of the solifluction terraces. Sorted patterned ground is characteristic of the strandflats of glacial lakes (like Ana and Valea Rea, Retezat Mountains) (Fig. 10.10).

Fig. 10.9 Honeycombs on Lolaia Mountain (1,750 m, Retezat Mountains)

Fig. 10.10 Frost sorting on the Ana Lake strandflat (at 1,975 m in the Retezat Mountains)

The monitoring of two areas on the Muntele Mic Mountain, i.e., a field of periglacial earth hummocks at the 1,765 m and an area with flat surface at 1,774 m elevation, using elevationmeters BAC and Danilin cryometers, shows the values between 30 and 72 mm for earth hummocks and between 8 and 35 mm for flat ground (Urdea et al. 2004). The differential downslope displacement of colluvial deposits and rocks through *gelifluction processes* and frost creep produce a range of landforms, like gelifluction lobes, gelifluction sheets, gelifluction benches, and plowing blocks. The movement of the plowing blocks was monitored on Muntele Mic and Parângul Mic Mountains (Table 10.1).

Frost creep is controlled by the number of freeze-thaw cycles, slope angle and ground moisture content. The talus cones and scree slopes affected by frost creep have a distinct aspect. Creep is also important for rock glaciers (Fig. 10.11). The rock debris of scree cones on Lolaia Mountain and "stone banked lobes" or "rocky lobes" (e.g., on Gemănarea Mountain in the Parâng Mountains) show differentiated rates of movement for the different parts, ranging from 1.22 to 3.78 cm year^{-1} (Urdea et al. 2004).

Nivation, embracing all processes associated with enduring snow patches, transport debris by snow creep and slopewash by melting snow. Monitored in the Muntele Mic area, it has been found still active in the present-day morphodynamics of the Carpathians periglacial belt (Urdea et al. 2004). The combination of nivation and other periglacial processes is responsible for the development of erosional features, such as nivation hollows, benches, niches (Fig. 10.12), cryoplanation terraces and others. In the Romanian Carpathians, *avalanches* affect steep slopes with a frequency of 2–20 events per year.

10.3.5 Fluvial Processes

Maria Rădoane and Nicolae Rădoane

In Romania there are over 250 reservoirs, with ca 500 km branches and supplies and 16,000 km of river embankments. Thus, fluvial processes have been a major concern of the researchers in the last 30 years (Diaconu et al. 1962; Grumăzescu 1975; Hâncu 1976; Panin 1976; Pascu 1999; Bondar et al. 1980; Armencea et al. 1980; Ichim et al. 1989; Ichim and Rădoane 1990; Rădoane et al. 1991, 2003, 2008a, b, c; Amăriucăi 2000; Rădoane 2004; Rădoane and Rădoane 2005; Dumitriu 2007; Burdulea-Popa 2007; Canciu 2008; Feier and Rădoane 2008; Perşoiu 2008).

Quantitative research of fluvial geomorphology has focused mainly on the main rivers in the eastern part of Romania, respectively the Prut and Siret rivers with their major tributaries and drainage basins of over 70,000 km^2.

The geomorphological analysis of longitudinal profiles (Rădoane et al. 2003) involved mathematical models and coefficients of variation to select a *concavity index* (Fig. 10.13). The index values tend to increase from north to south in the Eastern Carpathians. The explanation for this situation has called for a review of

Table 10.1 Results of monitoring ploughing blocks in the Parângu Mic area

#	Altitude (m above sea level)	Block size					Block movement (mm)			Block azimuth (°)		
		Slope (°)	Length (m)	Width (m)	Height (m)	Relative volume (m³)	2000–2001	2001–2003	Annual	2000	2001	2003
1	2,035	26	1.45	0.54	0.115	0.09	11.4	25.6	12.3	195	195	199
2	2,037	18	2.26	0.69	0.25	0.38	9.9	14.1	8	202	198	202
3	2,050	16	1.83	1	0.25	0.45	12.6	22.4	11.6	228	220	226
4	2,045	15	2	0.91	0.16	0.29	6.2	8.8	5	213	212	217

Fig. 10.11 Talus rock glaciers at Știrbu (Retezat Mountains) affected by frost creep

Fig. 10.12 Nivation niches on Țarcu Mountain at 2,020 m above sea level

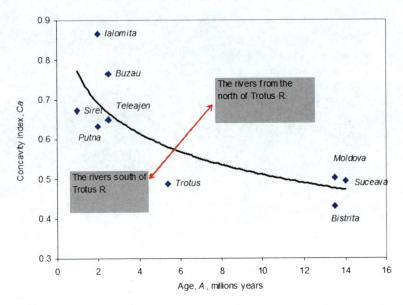

Fig. 10.13 Illustration of the relationship between the shape and the age of longitudinal profiles for the rivers in the eastern and southeastern part of Romania (Rădoane et al. 2003)

ideas on the stages of evolution of the drainage network in the region. The relationship between river age and longitudinal profile shape shows that the geomorphologic evolution of the river has not made a footprint in a decisive way on the shape of the longitudinal profile as, for instance, the Davisian erosional cycle concept suggests. The rivers from the north of Trotuș river (ages of 13–14 million years on the same course) have longitudinal profiles apparently less evolved (reduced concavity, increased slope) (Fig. 10.13). In contrast, the rivers south of Trotuș (Putna, Buzău, Prahova, Ialomița), whose courses have undergone major changes, interruptions, tectonic uplift, subsidence in the approximately 2.5 million years of evolution, are characterized by highly concave longitudinal profiles. However, in accordance with the classical Davisian conceptual and modern models (Snow and Slingerland 1987), the latter profiles should have a much higher concavity coefficient. The linear–exponential equilibrium expresses a balance between erosion and accumulation, therefore, it is a characteristic *profile of transport*, with a high slope, which the rivers north of Trotuș have preserved, with some variations, for 14 million years. In accordance to Hack's dynamic equilibrium theory, a form of relief preserves those characteristics that ensure a state of equilibrium in the exchange of mass and energy with other forms of relief (Hack 1960). This applies to the shape of the longitudinal profiles of rivers in the Eastern Carpathians.

The geomorphic analysis of longitudinal profiles is linked to the processes of downstream fining and *channel material* bimodality. These processes were studied for the six major Carpathian rivers (Rădoane et al. 2008a). The investigations on the bed material variability of the Siret Basin rivers were mainly focused on verifying the

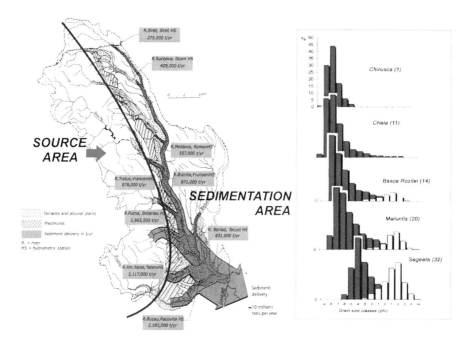

Fig. 10.14 Map of suspended sediment transport in the Siret drainage basin. The transport of fine sediment represented by arrows is superimposed on coarse sediment transport, identified by the extension of piedmonts and alluvial terraces. The central line divides the two main areas of the sediment system: source area and sedimentation area (*to the left*). Bimodality is seen at the intersection of both distributions on the example of the Buzău River. *Numbers* indicate sampling sites from source to mouth (*to the right*) (Rădoane et al. 2008a)

exponential model of reduction in the *sediment size* along the river, according to Sternberg's law, which shows that the river bed particles reduce their dimension proportionally with the mechanical work made against friction along the river. Depending on the length of the river, the median diameter (D_{50}) is reduced overall exponentially, but on important lengths of the rivers this exponential decrease is acutely disturbed. From this point of view as well, the Eastern Carpathian rivers record many deviations from the conceptual model. The Trotuș and Siret Rivers even show an increase in the material's dimension along most of their lengths. The only rivers that nearly relate to the exponential model on their entire length are the Suceava and the Moldova. The main cause for which the Sternberg model does not fit to the other four rivers lies in the contribution of the tributaries with a massive sediment input in the rivers in question, a lot greater than their ability to modify (Fig. 10.14, left).

The bedload has a distinct *bimodality*, the two peaks in the grain size distribution curve being separated by a small gravel fraction of 1–8 mm diameter. This bimodality of fluvial deposits may be explained by the different origin of the bedload. For the Carpathian tributaries of the Siret River, coarse gravel joins a unimodal distribution presenting a right skewness with enhanced downstream fining. The source of the coarse material is the river channel itself. A second distribution with a sandy mode is, in general, skewed to the left. The source of the second peak is the amount of sand that reached the riverbed from erosion on hillslopes. The tails of the histograms skewed to the right (for the gravel) and skewed to the left (for the sands) intersect. The intersection of the two modes occurs in the area of fractions from the 0.5–8 mm range. This explains the penury of particles between 0.5 and 8 mm. For the rivers where the sources of fine sediment are low, the 0.5–8 mm fractions are more frequent than the factions under 1 mm (Fig. 10.14, right). For the Siret River itself, bed sediment bimodality is greatly enhanced due to the fact that the second mode represents more than 25% of the full sample. As opposed to its tributaries, the source of the first mode, of gravel, is allochthonous to the Siret River, generated by the massive input of coarse sediment from Carpathian tributaries, while the second mode, of the sands, is local.

The riverbed material is subject to *vertical mobility* in the longitudinal profile and in cross section. Change in the bed elevation of alluvial rivers, in a positive or a negative way as related to a reference point, is a direct response to a sediment-supply deficit or surplus. Data from 63 cross sections of the Siret River basin were analyzed, in particular those from the right side of the river. From monitoring bed elevation in the cross sections for a period of over 70 years, degradation (100–120 cm) was found in almost half of the cases and aggradation (80–100 cm) in less than 30% of cases. Stability of the riverbed, the vertical oscillation of the riverbed below the value of 50 cm, characterizes a little over 20% of the cases.

The most abrupt *human intervention* into the river systems is the construction of dams and reservoirs. There are ca 250 reservoirs in Romania, mainly on the Bistrița, Siret, and Prut Rivers. On the Prut River, for instance, a dam was built at Stânca–Costești (Fig. 10.15), resulting in capturing almost all (over 95%) the sediment load in the upper drainage basin. Consequently, immediately downstream the dam, river is entirely devoid of *suspended load*. Along the next 500 km downstream the river attempts to compensate for the sediment load lost, but only achieves to raise its suspended sediment transport to 63%, as measured at the river mouth. The amount of water discharge, however, has not been affected by dam construction, only the regime was modified through human regulation.

Current measurements at seven gauging stations on aggradation-degradation processes along the Prut River are available for the period 1975–2005. Naturally, upstream of Stânca Reservoir, the riverbed shows slight aggradation, probably as a response to sediment storage at the end of the reservoir. Immediately downstream of the dam, degradation is the dominant process as a direct effect of a drastic reduction of sediment load. *Incision*, however, is not linear, but there are also areas where the riverbed is slightly aggrading. Overall, the effect of the dam is transmitted downstream the Prut riverbed over a distance of 400 km, with an incision rate of over 4 m^3 $year^{-1}$. Only towards the confluence with the Danube the rate drops below 0.5 m^3 $year^{-1}$ (Fig. 10.15).

In conclusion, the fluvial processes in Romania follow the *tendencies* observed for European rivers under prolonged human impact (Petts et al. 1989). At the beginning of the nineteenth century, the process of aggradation was dominant, while in the twentieth century the complexity of anthropogenic interventions resulted in a

10 Recent Landform Evolution in the Romanian Carpathians... 275

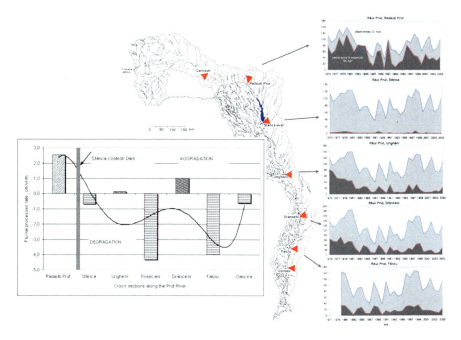

Fig. 10.15 Rate of fluvial processes (aggradation or degradation) along the Prut River

deepening and narrowing of riverbeds. In Romania, however, there is a certain delay in channel response. Although the incision is dominant (for over 50% of all sections studied), riverbed aggradation is still present.

10.3.6 Mining Activities and Environmental Impacts

Mihaela Sima

The Romanian Carpathians are rich in mineral deposits, some having been exploited since ancient and even pre-historical times (e.g., gold in the Apuseni Mountains) (Fig. 10.16). Although mining is practiced on a fairly small scale, its *environmental impact* is extremely severe: contamination of waters with heavy metals from the exploitation and processing of non-ferrous ores; acid drainage from coal mines and metallurgical plants; suspended load mainly from coal mines; radioactive ores; air pollution related to flotation, burning and processing plants (sulfur and nitrogen oxides, carbon-dioxide and methane); *topographic changes* (waste dumps, tailings ponds, and underground galleries) affecting the environmental and degrading lands, soils, flora and fauna and, most importantly, human health.

After 1990, Romanian mining industry experienced a major restructuring: the majority of mines, still economically efficient, were gradually closed down, an action

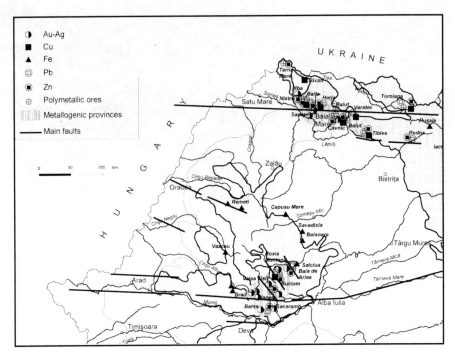

Fig. 10.16 Distribution of ore deposits in north-western Romania (Şerban et al. 2004)

that continued after 2000 and particularly in 2006–2007. Most mining sites are under *environmental rehabilitation*, but, because of money shortage, during the past few years rehabilitation programs were either not launched, or implemented only in certain areas. In 2001 technological *accidents* in the tailing dams of Maramureş Country had a significant cross-border impact. The loud international response led to fundamental changes in European legislation (e.g., the Seveso II Directive on the control of major accidents involving dangerous substances covers also mining activities). In the course of the accidents several rivers (the Lăpuş/Someş, Novăţ/Vişeu, the Tisza and the Danube) on the territory of neighboring countries were polluted. One of the main results obtained by studying the situation in those regions is that the pollution found at all the major observation points of Maramureş and Satu Mare counties and from the Apuseni Mountains, was caused by waste spills from active mines being either untreated or their waste water treatment plants were out of operation (Fig. 10.17). The contamination, especially of surface and groundwaters, with pollutants of mining origin extends only up to 5 km from the observation point, affecting a corridor some 1 km wide along the river channel. In general, between these observation points, both river and groundwater lie within EU quality standard limits, although the metal concentrations found in the river and floodplain sediments are somewhat higher due to historical pollution (Brewer et al. 2002; Macklin et al. 2003; Bird et al. 2003, 2008; Şerban et al. 2004).

Fig. 10.17 River affected by acid mine drainage in the Southern Apuseni Mountains, Romania

Under the current technical and technological conditions mining is no longer economically efficient, and many pits are already exhausted. Unfortunately, mines that have been closed down do not benefit from ecological rehabilitation programs or from adequate conservation measures either and continue to yield high quantities of pollutants, thereby being detrimental to the environment. However, there are companies that have maintained or even plan to extend their activity (e.g., at Roşia Montană). Non-ferrous metal ores are no longer extracted or processed in Romania for lack of state subventions to update obsolete technology. Unless countermeasures are taken, environmental degradation is expected to go on for scores and hundreds of years with a serious long-term impact.

References

Amăriucăi M (2000) Şesul Moldovei extracarpatice dintre Păltinoasa şi Roman. Studiu geomorphologic şi hidrologic (The Extra-Carpathian floodplain of Moldova River between Păltinoasa and Roman. Geomorphological and Hydrologic Study). Edit. Carson, Iaşi, 180 p (in Romanian)

Andra A, Mafteiu M (2008) The landslides from Tigveni–Momaia. In: Bălteanu D (ed) IAG Regional conference on geomorphology "Landslides, floods and global environmental change in mountain regions". Field Guidebook, University Publishing House, Bucharest, pp 59–64

Armencea Gh, Marinescu Gh, Stoicescu H, Lup I (1980) Aspecte ale prognozei procesului de coborâre al albiei râurilor aval de baraje (On the river bed incision prognosis downstream of dams). Hidrotehnica 25(5):101–103 (in Romanian)

Badea L (1957) Observații asupra unor alunecări din bazinul Buzăului (Some remarks on the landslides of the Buzău drainage basin). Probleme de Geografie 4:388–392 (in Romanian)

Bălteanu D (1975) Un eșantion de hartă morfodinamică din Subcarpații Buzăului (A morphodynamic map in the Buzău Subcarpathians). In: Lucrările Colocviului Național de Geomorfologie Aplicată și Cartografiere Geomorfologică, Iași, pp 349–354 (in Romanian)

Bălteanu D (1983) Experimentul de teren în geomorfologie (Field experiment in geomorphology). Edit. Academiei Române, București, 156 p (in Romanian)

Bălteanu D (1986) The importance of mass movement in the Romanian Subcarpathians. Z Geomorphol 58(Supplement-Band):173–190

Bălteanu D (1997) Geomorphological hazards of Romania. In: Embleton C, Embleton-Hamann Ch (eds) Geomorphological hazards of Europe. Elsevier, Amsterdam, pp 409–427

Bălteanu D, Jurchescu M (2008) Deep-seated landslides (Glimee) in the Transylvanian Depression. Map in: Bălteanu D (ed) IAG Regional conference on geomorphology "Landslides, floods and global environmental change in mountain regions". Field Guidebook, University Publishing House, Bucharest

Bălteanu D, Micu M (2009) Landslide investigation: from morphodynamic mapping to hazard assessment. A case study in the Romanian Subcarpathians, Muscel catchment. In: Malet J-Ph, Remaitre A, Bogaard T (eds) Landslide Processes. From geomorphologic mapping to dynamic modelling. CERG Editions, Strasbourg, pp 235–241

Bălteanu D, Teodoreanu V (1983) The mass movement from Malu Alb. In: Excursion guidebook symposium "The role of geomorphological field experiments in land and water management", Bucharest, pp 77–80

Bălteanu D, Chendeș V, Sima M, Enciu P (2010) A country-wide spatial assessment of landslide susceptibility in Romania. Geomorphology 124(3–4):102–112

Barbu N (1976) Obcinele Bucovinei. Edit. Stiint. si Enciclop, București, 316 p (in Romanian)

Bătucă D (1978) Aspecte ale morfologiei generale a albiilor râurilor din bazinul hidrografic Mureș superior, (On the general morphology of the riverbeds in the Upper Mureș drainage basin). Hidrotehnica 23(6):121–124 (in Romanian)

Bird G, Brewer P, Macklin M, Balteanu D, Driga B, Serban M, Zaharia S (2003) The solid state portioning of contaminant metals and As in river channel sediments of the mining affected Tisa drainage basin, northwestern Romania and eastern Hungary. Appl Geochem 18:1583–1595

Bird G, Brewer P, Macklin M, Balteanu D, Serban M, Driga B, Zaharia S (2008) River system recovery following the Novat-Rosu tailings dam failure, Maramureș County, Romania. Appl Geochem 23(12):3498–3518

Bleahu M, Povară I (1976) Catalogul peșterilor din România (The catalogue of the caves of Romania). Edit. C.N.E.F.S, București, 53 p

Bleahu M, Rusu T (1965) Carstul din România (Karst in Romania). Lucrările Institutului de Speologie "Emil Racoviță", București 4:59–73 (in Romanian)

Bondar C, State I, Dediu R, Supuran I, Vașlaban G, Nicolau G (1980) Date asupra patului albiei Dunării în regim amenajat pe sectorul cuprins între Baziaș și Ceatal Izmail (Data on the Danube managed riverbed between Baziaș and Ceatal Izmail). Studii și Cercetări de Hidrologie 48:145–168 (in Romanian)

Brewer P, Macklin M, Balteanu D, Coulthard T, Driga B, Howard A, Bird G, Zaharia S, Serban M (2002) The January and march tailings dam failures in Maramures county, Romania, and their transboundary impacts on the river systems. In: Proceedings of advanced research workshop "Approaches to handling environmental problems in the mining and metallurgical regions of NIS counties", Mariupol, 5–7 Sept 2002, pp 56–64

Burdulea-Popa A (2007) Geomorfologia albiei râului Siret (The geomorphology of the Siret River Channel). Manuscript PhD thesis. Al. I. Cuza University of Iași (in Romanian)

Canciu C (2008) Valea Dunării între Brăila și Pătlăgeanca – studiu geomorfologic (The Danube Valley between Brăila and Pătlăgeanca – Geomorphological Study). Manuscript PhD thesis, University of Bucharest (in Romanian)

Chițu Z (2010) Predicția spațio-temporală a hazardului la alunecări de teren utilizând tehnici S.I.G. Studiu de caz arealul subcarpatic dintre Valea Prahovei și Valea Ialomiței (Spatial and temporal prediction of landslide hazard using GIS. Case-study: the Subcarpathian area between the Prahova and Ialomița valleys). Manuscript PhD thesis, University of Bucharest, 295 p (in Romanian)

Chițu Z, Șandric I, Mihai B, Săvulescu I (2009) Evaluation of landslide susceptibility using multivariate statistical methods: a case-study in the Prahova Subcarpathians, Romania. In: Malet JPh, Remaitre A, Bogaard T (eds) Landslide Processes. From geomorphologic mapping to dynamic modelling. CERG Editions, Strasbourg, pp 265–270

Constantin S (1992) The intra-Aptian karstification phase and the paleokarst associated deposits in the southern sector of Locva Mountains (South-West Romania). Theor Appl Karstol 5:83–92

Constantin S, Lauritzen S-E, Stiuca E, Petculescu A (2001) Karst evolution in the Danube Gorge from U-series dating of a bear skull and calcite speleothems form Pestera de la Gura Ponicovei (Romania). Theor Appl Karstol 13–14:39–50

Constantin M, Trandafir AC, Jurchescu MC, Ciupitu D (2010) Morphology and environmental impact of the Colți-Aluniș landslide (Curvature Carpathians), Romania. Environ Earth Sci 59(7):1569–1578

Constantin M, Bednarik M, Jurchescu MC, Vlaicu M (2011) Landslide susceptibility assessment using the bivariate statistical analysis and the index of entropy in the Sibiciu basin (Romania). Environ Earth Sci 63(2):397–406. doi:10.1007/s12665-010-0724-y

Dârja M, Budiu V, Tripon D, Păcurar I, Neag V (2002) Eroziunea hidrică și impactul asupra mediului (Erosion by water and its impact on the environment). Edit. Risoprint, Cluj-Napoca, 100 p (in Romanian)

de Martonne E (1900) Contribution a l'etude de la période glaciaire dans les Karpates Méridionales. Bull Soc Géol de la France 28(3):275–319

de Martonne E (1907) Recherches sur l'évolution morphologique des Alpes de Transylvanie (Karpates Méridionales). Revue de géographie annuelle I (1906–1907), 286 p

Diaconu C, Ciobanu S, Avădanei A, Motea I, Stănescu S (1962) Despre stabilitatea albiilor râurilor României (On the stability of river channels in Romania during the last 30–40 years). Studii de Hidrologie 3:53–66 (in Romanian)

Dinu M, Cioacă A (2000) Rolul hazardelor naturale în evoluția localităților din România (On the role of natural hazards in the evolution of settlements in Romania). Analele Universității Spiru Haret, București 3:43–52 (in Romanian)

Donisa I (1968) Geomorfologia văii Bistriței (Geomorphology of the Bistrița Valley). Edit. Academiei Române, București, 285 p (in Romanian)

Dragotă C, Micu M, Micu D (2008) The relevance of pluvial regime for landslide genesis and evolution. Case study: Muscel basin (Buzău Subcarpathians, Romania). In: Present environment & sustainable development. 2. Edit. Universității "Al. I. Cuza", Iași, pp 242–257

Dumitriu D (2007) Sistemul aluviunilor din bazinul râului Trotuș (Alluvial system in the Trotuș River basin). Edit. Universității, Suceava, 259 p (in Romanian)

Feier I (2007) Evoluția istorică a migrării albiei râului Someșu Cald, (Historical evolution of the Someșu Cald riverbed mobility). Manuscript PhD thesis, "A. I. Cuza" University, Iași (in Romanian)

Feier I, Rădoane M (2008) Dinamica in plan orizontal a albiei minore a raului Somesu Mic inainte de lucrarile hidrotehnice majore (1870–1968) (Channel planform dynamics of the Someșu Mic River before the major human modifications, 1870–1968). Analele Universității Suceava 16:13–26 (in Romanian)

Gârbacea V (1992) Harta glimeelor din Câmpia Transilvaniei (Map of the "glimee"-type landslides distribution in the Transylvanian Plain). Studia Universitatis Babeș-Bolyai, Geographia 37(1–2):21–24 (in Romanian)

Gârbacea V (1996) Remarques sur le relief de "glimee" en Roumanie. Geografia Fisica e Dinamica Quaternaria 19:219–221

Gârbacea V, Grecu F (1983) Relieful de glimee din Podișul Transilvaniei și potențialul lor economic (The glimee landforms of the Transylvanian Plateau and their economic potential). Memoriile Secțiilor Științifice ale Academiei R.S.R. 4(2):305–312 (in Romanian)

Gaşpar R, Untaru E (1979) Contribuţii la studiul transportului de aluviuni în bazinele torenţiale parţial împădurite (Contributions to the study of sediment transport in partially forested torrential catchments). Buletinul Informativ ASAS 8:87–95 (in Romanian)

Goran C (1980) Catalogul peşterilor din România – 1979 (The Romanian Caves Register – 1979). Buletinul Informativ CCSS 4:172–179 (in Romanian)

Goran C (1982) Catalogul sistematic al peşterilor din România (A systematic catalogue of caves in Romania). Edit. Sport-Turism, Bucureşti, 496 p (in Romanian)

Goran C (1983) Les types de relief karstique de Roumanie. Trav Inst Spéol "Emile Racovitza", Bucureşti 22:91–102

Grecu F (1992) Bazinul Hârtibaciului – Elemente de morfohidrografie (The Hârtibaciu Basin – Elements of Morphohydrography). Edit. Academiei Române, Bucureşti, 160 p (in Romanian)

Grumăzescu C (1975) Depresiunea Haţegului. Studiu geomorfologic (The Haţeg Depression. Geomorphological study). Edit. Academiei Române, Bucureşti, 148 p (in Romanian)

Hack JT (1960) Interpretation of erosional topography in humid temperate regions. Am J Sci 258-A(67):219–230

Hâncu S (1976) Regularizarea albiilor râurilor (Channelization of Riverbeds). Edit. Ceres, Bucuresti, 144 p (in Romanian)

Ichim I (1979) Muntii Stanisoarei. Studiu geomorfologic (The Stânişoara Mountains. Geomorphic Study). Ed. Acad, Bucuresti, 121 p (in Romanian)

Ichim I (1980) Probleme ale cercetării periglaciarului în România (Issues on the periglacial research in Romania). Studii şi Cercetări Geol Geofiz Geogr–Geografie 27(1):127–135 (in Romanian)

Ichim I, Rădoanc M (1980) On the anthropic influence time in morphogenesis with a special regard on the problem of channel river dynamics. Revue Roumaine Géol Géophys Géogr, série Géographie 24:35–40

Ichim I, Rădoane M (1981) Contribuţii la studiul dinamicii albiilor de râu în perioade de timp scurt şi de timp îndelungat (Contributions to the study of riverbed dynamics during short and long time periods). Hidrotehnica 25(5):135–138 (in Romanian)

Ichim I, Rădoane M (1984) Cercetări privind sursele de aluviuni şi energia potenţială de eroziune, cu exemplificări din regiunea Vrancei (Research on the sediment sources and the potential energy of erosion, with examples from the Vrancea region). Hidrotehnica 29(6):183–187 (in Romanian)

Ichim I, Rădoane M (1986) Efectele barajelor în dinamica reliefului. Abordare geomorfologică (The effects of dams on relief dynamics. A geomorphological approach). Edit. Academiei R. S. România, Bucureşti, 157 p (in Romanian)

Ichim I, Rădoane M (1990) Channel sediment variability along a river: a case study of the Siret River, Romania. Earth Surf Process Landf 15(3):211–226

Ichim I, Rădoane M, Rădoane N, Surdeanu V, Amăriucăi M (1979) Problems of meander geomorphology with particular emphasis on the channel of the Bârlad River. Revue Roumaine Géol Géophys Géogr, série Géographie 23:35–47

Ichim I, Bătucă D, Rădoane M, Duma D (1989) Morfologia şi dinamica albiilor de râu (River channel morphology and dynamics). Edit. Tehnică, Bucureşti, 407 p (in Romanian)

Ichim I, Mihaiu G, Surdeanu V, Rădoane M, Rădoane N (1990) Gully erosion in agricultural lands in Romania. In: Boardman J, Foster IDL, Dearing JA (eds) Soil erosion on agricultural land. Wiley, Chichester, pp 55–68

Ichim I, Rădoane M, Rădoane N, Miclăuş C, Grasu C (1996) Sediment budget of the Putna drainage basin (Vrancea). Revue Roumaine Géol Géophys Géogr, série Géographie 40:125–132

Ichim I, Rădoane M, Rădoane N, Grasu C, Miclăuş C (1998) Dinamica sedimentelor. Aplicaţii la râul Putna – Vrancea (Sediments dynamics. Applications to the Putna River, Vrancea). Edit. Tehnică, Bucuresti, 192 p (in Romanian with English summary)

Ielenicz M (1970) Zonele cu alunecări de teren din ţara noastră (Landslides in our country). Terra 2(1):31–40 (in Romanian)

Ielenicz M (1984) Munţii Ciucaş-Buzău. Studiu geomorfologic (The Ciucaş-Buzău Mountains. Geomorphic Study). Edit. Academiei Române, Bucureşti, 146 p (in Romanian)

Ilinca V (2010) Valea Lotrului. Studiu de geomorfologie aplicată, (The Lotru Valley. Applied geomorphological study). Manuscript PhD thesis, University of Bucharest, 217 p (in Romanian)

Ionita I (1998) Studiul geomorfologic al degradarilor de teren din bazinul mijlociu al Barladului. (Geomorphological study of the land degradation in the middle catchment of Barlad river). Manuscript PhD thesis, University "Alexandru Ioan Cuza", Iași, 287 p (in Romanian)

Ionita I (1999) Sediment delivery scenarios for small watersheds. In: Proceedings of the symposium "Vegetation, land use and erosion processes". Institute of Geography, Bucharest, pp 66–73

Ionita I (2000a) Geomorfologie Aplicata. Procese de degradare a terenurilor deluroase (Applied Geomorphology. Processes of hilly terrain degradation). Edit. Univ. "Al.I. Cuza", Iași, 247 p (in Romanian)

Ionita I (2000b) Formarea si evolutia ravenelor din Podisul Barladului. (Forming and evolution of gullies in the Bârlad Plateau). Edit. Corson, Iași, 169 p (in Romanian)

Ionita I (2003) Hydraulic efficiency of the discontinuous gullies. In: Poesen J, Valentin C (eds) Gully erosion and global change. Catena 50(2–4):369–379

Ionita I (2006) Gully development in the Moldavian Plateau of Romania. In: Special Issue Helming K, Rubio JL, Boardman J (eds) Soil erosion research in Europe. Catena 68(2–3):133–140

Ionita I (2008) Sediment movement from small catchments within the Moldavian Plateau of Eastern Romania. In: Schmidt J, Cochrane T, Phillips Ch, Elliot S, Davies T, Basher L (eds) Sediment dynamics in changing environments. IAHS Publication 325, IAHS Press, Wallingford, pp 316–320

Ionita I, Margineanu RM (2000) Application of ^{137}Cs for measuring soil erosion/deposition rates in Romania. Acta Geologica Hispanica 35(3–4):311–319

Ionita I, Margineanu R, Hurjui C (2000) Assessment of the reservoir sedimentation rates from ^{137}Cs measurements in the Moldavian Plateau. In: Queralt I, Zapata F, Agudo G (eds) Assessment of soil erosion and sedimentation through the use of the ^{137}Cs and related techniques. Acta Geologica Hispanica Special 35(3–4):357–367

Ionita I, Rădoane M, Mircea S (2006) Chapter 1.13 "Romania". In: Boardman J, Poesen J (eds) Soil erosion in Europe. Wiley, Amsterdam–London–New York, pp 155–166

Irimuş I-A (1996) La corréllation des glissements de terrain avec les types de dômes péripheriques dans le bassin de Transylvanie (Roumanie). Geografia Fisica e Dinamica Quaternaria 19:245–248

Irimuş I-A (1998) Relieful pe domuri și cute diapire în Depresiunea Transilvaniei (Landforms Developed on Domes and Diapir Folds). Edit. Presa Universitară Clujeană, Cluj, 300 p (in Romanian)

Jeanrenaud, P (1971) Geologia Moldovei centrale dintre Siret și Prut (Geology of the Central Moldavia between the Siret and Prut rivers). Manuscript abstract of the PhD thesis, Al. I. Cuza University, Iași (in Romanian)

Lascu C, Sârbu Ș (1987) Peșteri scufundate (Underwater caves). Edit. Acad, București, 255 p (in Romanian)

Macklin M, Brewer P, Bălteanu D, Colthard T, Driga B, Howard A, Zaharia S (2003) The long-term fate and environmental significance of contaminant metals released by the January and March 2000 mining tailings dam failures in Maramureș County, upper Tisa Basin, Romania. Appl Geochem 18:241–257

Micu M (2008) Evaluarea hazardului legat de alunecari de teren in Subcarpatii dintre Buzau si Teleajen (Landslide Hazard Assessment in the Buzău – Teleajen Subcarpathians). Manuscript PhD thesis, Institute of Geography, Bucharest, 242 p (in Romanian)

Micu M, Bălteanu D (2009) Landslide hazard assessment in the Bend Carpathians and Subcarpathians, Romania. Z Geomorphol 53(Supplement 3):49–64

Micu M, Sima M, Bălteanu D, Micu D, Dragotă C, Chendeș V (2010) A multi-hazard assessment in the Curvature Carpathians of Romania. In: Malet J-Ph, Glade T, Casagli N (eds) Mountain risks: bringing science to society. CERG Editions, Strasbourg, pp 11–18

Mihai B (2005) Munții Timișului (Carpații Curburii). Potențialul geomorfologic și amenajarea spațiului montan (The Timiș Mountains (Curvature Carpathians): Geomorphic Potential and Mountain Landscape Planning). Edit. Universității din, București

Mihai Gh, Neguț N (1981) Observații preliminare privind evoluția ravenelor formate pe alternanțe de orizonturi permeabile și impermeabile (Preliminary observations on the evolution of gullies formed on alternating permeable and impermeable horizons). Studii și Cercetări de Geografie 28:117–126 (in Romanian)

Mihai B, Șandric I, Săvulescu I, Chițu Z (2009) Detailed mapping of landslide susceptibility for urban planning purposes in Carpathian and Subcarpathian towns of Romania. In: Gartner G, Ortag F (eds) Cartography in central and eastern Europe, Lecture notes in geoinformation and cartography. Springer, Heidelberg/Berlin, pp 417–429

Mihăilescu V (1926) Despre frane sau forme de teren rezultate din acțiunea de dărâmare a agenților externi (o propunere) (On landslides and landforms issued from the demolition action of external agents. A suggestion). Buletinul Societății Române de Geografie 45:101–110 (in Romanian)

Mihăilescu V (1939a) Porniturile de teren din regiunea Nehoiaș, (Landslides in the Nehoiaș region). Buletinul Societății Române de Geografie 58:191–193 (in Romanian)

Mihăilescu V (1939b) Porniturile de teren și clasificarea lor (Landslides and their classification). Revista Geografică Română 2(2–3):106–113 (in Romanian)

Mihăilescu V, Morariu T (1957) Considerații generale asupra periglaciarului și stadiul cercetărilor actuale în România (General aspects on the periglacial and overview of current research in Romania). Studii și Cercetări Geol série Géographie, Acad. Rom., Filiala Cluj 8(1–2):21–44 (in Romanian)

Mircea S (2002) Formarea, evoluția și strategia de amenajare a ravenelor (Formation, evolution and management strategy of gullies). Edit. Bren, București, 209 p (in Romanian)

Mircea S (2006) Contribuții la cunoașterea evoluției formațiunilor în adâncime în bazinele hidrografice, Rezumatul tezei de doctorat, (Contributions to the study of the evolution of gully formations in drainage basins), Manuscript abstract of the PhD thesis. University of Agricultural Sciences and Veterinary Medicine, București, 85 p (in Romanian)

Morariu T, Gârbacea V (1966) Quelques observations au sujet des processus de versant de la Depression Transylvanique. Revue Roumaine Géol Géophys Géogr, série Géographie 10(2):147–165

Morariu T, Gârbacea V (1968a) Studii asupra proceselor de versant din Depresiunea Transilvaniei (Studies on slope processes in the Transylvanian Depression). Studia Universitatis Babeș-Bolyai, Geol Geogr, Cluj 1:81–90 (in Romanian)

Morariu T, Gârbacea V (1968b) Déplacements massifs de terrain de type glimee en Roumanie. Revue Roumaine de Géol Géophys Géogr, serie Géographie 12(1–2):13–18

Morariu T, Mihăilescu V, Dragomirescu SȘ, Posea Gr (1960) Le stade actuel de recherches sur le périglaciaire de la R.P. Roumaine. In: Recueil d'études géographiques concernant le territoire de la R.P. Roumaine. Edit. Academiei, București, pp 45–53

Morariu T, Diaconeasa B, Gârbacea V (1964) Age of land-slidings in the Transylvanian Tableland. Revue Roumaine de Géol Géophys Géogr, série Géographie 8:149–157

Moțoc M (1963) Eroziunea solului pe terenurile agricole și combaterea ei (Soil erosion on agricultural lands and its control). Edit. Agrosilvică, București, 318 p (in Romanian)

Moțoc M (1983) Ritmul mediu de degradare erozionala a solului în R. S. Romania (Mean rate of soil degradation by erosion in Romania). Buletinul Informativ al ASAS, 13. București, (in Romanian)

Moțoc M, Mircea S (2002) Evaluarea factorilor care determină riscul eroziunii hidrice în suprafață (Evaluation of the factors determining the soil erosion risk). Edit. Bren, Bucuresti, 60 p (in Romanian)

Moțoc M, Mircea S (2005) Unele probleme privind formarea viiturilor și eroziunea în bazine hidrografice mici (Some problems regarding flash-flood formation and erosion in small catchments). Edit. Cartea Universitară, București, 104 p (in Romanian)

Moțoc M, Munteanu St, Băloiu V, Stănescu P, Mihai Gh (1975) Eroziunea solului și metodele de combatere (Soil erosion and the control methods). Edit. Ceres, București, 301 p (in Romanian)

Moțoc M, Taloescu I, Neguț N (1979a) Estimarea ritmului de dezvoltare a ravenelor (Assessment of the development rate of gullies). Buletinul Informativ ASAS, 8:77–86 (in Romanian)

Moțoc M., Stănescu, P., Taloescu, I., (1979b) Metode de estimare a eroziunii totale și a eroziunii efuente pe bazine hidrografice mici (Methods for assessing total erosion and sediment delivery within small catchments). Buletinul I.C.P.A., București, 38 p (in Romanian)

Moțoc M, Ionita I, Nistor D (1998) Erosion and climatic risk at the wheat and maize crops in the Moldavian Plateau. Rom J Hydrol Water Resour 5(1–2):1–38

Mureșan A (2008) Geomorfodinamica vailor de pe versantul vestic al Muntilor Maramureșului. (Geomorphodynamics of the valleys on the western slope of the Maramureș Mountains). Manuscript PhD thesis, Babeș–Bolyai University, Cluj Napoca (in Romanian)

Naum T (1970) Complexul de modelare nivo-glaciar din masivul Călimanului (Carpații Orientali), (The nivo-glacial modeling complex in the Călimani Massif, Eastern Carpathians). Anal Univ Buc–Geogr 19:67–75 (in Romanian)

Niculescu Gh (1965) Munții Godeanu. Studiu geomorfologic (The Godeanu Mountains. Geomorphological study). Edit. Academiei, București, 339 p (in Romanian)

Niculescu Gh, Nedelcu E (1961) Contribuții la studiul microreliefului crio-nival din zona înaltă a munților Retezat, Godeanu–Țarcu și Făgăraș–Iezer (Contribution to the study of crio-nival microrelief in the high zone of the Retezat, Godeanu–Țarcu and Făgăraș–Iezer Mountains). Probleme de Geografie 8:87–121 (in Romanian)

Onac BP (2002) Caves formed within upper Cretaceous skarns at Băița, Bihor county, Romania: mineral deposition and speleogenesis. Can Mineral 40:1693–1703

Orghidan T, Pușcariu V, Bleahu M, Decu V, Rusu T, Bunescu A (1965) Harta regiunilor carstice din România (Map of the karst regions in Romania). Lucrările Institutului de Speologie "Emil Racoviță" 4:75–104, (in Romanian)

Panin N (1976) Some aspects of fluvial and marin processes in the Danube Delta. Anuarul Institutului de Geologie și Geofizică 50:149–165

Pascu M (1999) Cercetări privind influența regularizării radicale a albiilor de râuri asupra stabilității unor construcții aferente și a mediului înconjurător – cu referire la bazinul hidrografic al râului Prahova (Research on the influence of radical Riverbeds Channelization on the Stability of Nearby Buildings and Environment – Case Study: Prahova Drainage Basin). Manuscript PhD thesis, "Gh. Asachi" Technical University, Iași (in Romanian)

Patriche CV, Căpățână V, Stoica DL (2006) Aspects regarding soil erosion spatial modeling using the USLE/RUSLE within GIS. Geographia Technica 2:87–97 (in Romanian)

Perșoiu I (2008) Time and space adjustments of Someșu Mic River. Geophysical research abstracts 10, EGU General Assembly, Vienna. EGU2008-A-00826

Perșoiu A, Onac BP, Wynn JG, Bojar A-V, Holmgren K (2011) Stable isotopes behavior during cave ice formation by water freezing in Scarisoara Ice Cave, Romania. J Geophys Res, Atmos 116 (D02111):8 p, doi: 10.1029/2010JD014477

Petts GE, Möller H, Roux AL (eds) (1989) Historical changes of large alluvial rivers in western Europe. Wiley, Chichester/London, pp 323–352

Piest RF, Bradford JM, Wyatt MG (1975) Soil erosion and sediment transport from gullies. ASCE J Hydraul Div 101(1):65–80

Pop O, Surdeanu V, Irimuș I-A, Guitton M (2010) Distribution spatiale des coulées de debris contemporaines dans le Massif du Căliman (Roumanie). Studia Universitatis Babeș-Bolyai, Geographia, Cluj-Napoca 55(1):33–44

Posea Gr, Ielenicz M (1970) Alunecările de teren de pe Valea Buzăului (sectorul montan) (Landslides along the Buzău Valley (mountain sector)). Analele Universității București, Geografie: 59–66 (in Romanian)

Posea G, Ielenicz M (1976) Types de glissements dans les Carpathes de la courbe (Bassin du Buzău). Revue Roumaine Géol Géophys Géogr, série Géographie 20:63–72

Pujina D, Ionita, I (1996) Present-day variability and intensity of the sliding processes in the Barlad Tableland. In: Proceedings of international conference on disasters and mitigation Madras, India, pp 4–35

Racoviță Gh, Moldovan OT, Onac BD (2002) Monografia Carstului din Munții Pădurea Craiului (The Karst of Padurea Craiului Mountains. Monographic Study). Institutul de Speologie "Emil Racoviță", Cluj-Napoca, 263 p (in Romanian with summary in English)

Rădoane N (2002) Geomorfologia bazinelor hidrografice mici (Geomorphology of small catchments). Edit. Universității Suceava, Suceava, 255 p (in Romanian)

Rădoane M (2004) Dinamica reliefului în zona lacului Izvoru Muntelui (The Relief Dynamics in the Izvoru Muntelui Reservoir Area). Edit. Universității Suceava, Suceava, 218 p (in omanian)

Rădoane M, Rădoane N (1992) Areal distribution of gullies by the grid square method. Case study: Siret and Prut interfluve. Revue Roumaine Géol Géophys Géogr, série Géographie 36:95–98

Rădoane M, Rădoane N (2003a) Morfologia albiei râului Bârlad și variabilitatea depozitelor actuale (The Bârlad riverbed morphology and variability of current deposits). Revista de Geomorfologie 4–5:85–97

Rădoane N, Rădoane M (2003b) Cercetări geomorfologice pentru evaluarea rolului albiei râului Olteț ca sursă de aluviuni (Geomorphic research for the assessment of the Olteț riverbed role as a sediment source). Analele Universității "Ștefan cel Mare", Suceava, Geografie 10(2001):27–35 (in Romanian)

Rădoane M, Rădoane N (2005) Dams, sediment sources and reservoir silting in Romania. Geomorphology 71:217–226

Rădoane M, Ichim I, Pandi G (1991) Tendințe actuale în dinamica patului albiilor de râu din Carpații Orientali (Present-day trends in the river channels changes in the Eastern Carpathians). Studii și Cercetări Geol Geofiz Geogr–Geografie 38:21–31 (in Romanian)

Rădoane M, Rădoane N, Ichim I (1995) Gully distribution and development in Moldavia, Romania. Catena 24:127–146

Rădoane M, Rădoane N, Ichim I (1997) Analiza multivariată a geomorfologiei ravenelor din Podișul Moldovei (Multivariate analysis of the geomorphology of gullies in the Moldavian Plateau). Analele Universității "Ștefan cel Mare", Suceava, pp 19–32 (in Romanian)

Rădoane M, Rădoanc N, Dumitriu D (2003) Geomorphological evolution of river longitudinal profiles. Geomorphology 50:293–306

Rădoane M, Rădoane N, Dumitriu D, Miclăuș C (2008a) Downstream variation in bed sediment size along the East Carpathians Rivers: evidence of the role of sediment sources. Earth Surf Process Landforms 33(5):674–694

Rădoane M, Rădoane N, Cristea I, Oprea-Gancevici D (2008b) Evaluarea modificărilor contemporane ale albiei râului Prut pe granița românească (Assessment of the contemporary changes of the Prut riverbed on the Romanian border). Revista de Geomorfologie 10:57–71 (in Romanian)

Rădoane M, Feier I, Rădoane N, Cristea I, Burdulea A (2008c) Fluvial deposits and environmental history of some large Romanian rivers. Geophysical research abstracts 10, EGU General Assembly, Vienna. EGU 2008 1MO3P-0399

Șandric I (2005) Aplicații ale teoriei probabilităților condiționate în geomorfologie (Application of the conditional probability theory in geomorphology). Analele Universității București 54:83–97 (in Romanian)

Șandric I (2008) Sistem informațional geografic temporal pentru analiza hazardelor naturale. O abordare bayesiană cu propagare a erorilor (Temporal geographic information system for the analysis of natural hazards. A Bayesian approach with error propagation). Manuscript PhD thesis, University of Bucharest, 243 p (in Romanian)

Șandric I, Chițu Z (2009) Landslide inventory for the administrative area of Breaza, Curvature Subcarpathians, Romania. J Maps 2009/7:75–86. doi:10.4113/jom.2009.1051

Sass O (2006) Determination of the internal structure of alpine talus deposits using different geophysical methods (Lechtaler Alps, Austria). Geomorphology 80:45–58

Schreiber WE (1974) Das Periglazialrelief des Harghita-Gebirges. Revue Roumaine Géol Géophys Géog, série Géographie 18(2):179–187

Șerban M, Macklin M, Brewer P, Bălteanu D, Bird G (2004) The impact of metal mining activities on the upper Tisa River Basin, Romania, and transboundary river pollution. Studia Geomorphologica Carpatho-Balcanica 38:97–111

Snow RS, Slingerland RL (1987) Mathematical modelling of graded river profiles. Geology 95:15–33

Surdeanu V (1979) Recherches experimentales de terrain sur les glissements. Studia Geomorphologica Carpatho-Balcanica 15:49–64

Surdeanu V (1987), Studiul alunecărilor de teren din valea mijlocie a Bistriței (zona munților flișului) (The study of landslides from the middle Buzău Valley, area of the flysch mountains). Manuscript PhD thesis, Iași, 192 p (in Romanian)

Surdeanu V (1996) La repartition des glissements de terrain dans le Carpates Orientales (zone du flysch). Geografia Fisica e Dinamica Quaternaria 19(2):265–271

Surdeanu V (1998) Geografia terenurilor degradate. I Alunecari de teren (Geography of Degraded Lands. I. Landslides). Edit. Presa Universitară Clujeana, Cluj-Napoca, 274 p (in Romanian)

Surdeanu V, Pop O, Chiaburu M, Dulgheru M, Anghel T (2010) La dendrogéomorphologie appliquee a l'étude des processus geomorphologiques des zones minieres dans le Massif du Calimani (Carpates Orientales, Roumanie). In: Surdeanu V, Stoffel M, Pop O (eds) Dendrogéomorphologie et dendroclimatologie – méthodes de reconstitution des milieux géomorphologiques et climatiques des régions montagneuses. Presa Universitară Clujeană, Cluj-Napoca, pp 107–124

Traci C (1979) Aspecte privind rolul culturilor forestiere în combaterea proceselor de eroziune și în ameliorarea solului (Aspects regarding forest cultivations in mitigating erosion processes and improving soil). Buletinul Informativ ASAS 8:125–130

Tufescu V (1959) Torenți de noroi în Vrancea (Mud-torrents in Vrancea). Comunicările Academiei RPR 1:67–72 (in Romanian)

Tufescu V (1964) Typologie des glissements de Roumanie. Revue Roum Géol Géophys Géogr 7:140–147

Tufescu V (1966) Modelarea naturală a reliefului și eroziunea accelerată (Natural relief modelling and accelerated erosion). Edit. Acad. R.S.R, București, pp 155–256 (in Romanian)

Untaru E (1979) Contribuții la prevenirea alunecărilor de teren din bazinele hidrografice ale Milcovului și Câlnăului prin culturi forestiere de protecție (Contributions to Landslides Prevention within the Catchments of the Milcov and Câlnău Rivers through Protective Reforestations). Manuscript PhD thesis, ASAS, Bucharest (in Romanian)

Urdea P (1991) Rock glaciers and other periglacial phenomena in the Southern Carpathians. Analele Universității Oradea, Geografie 13–26

Urdea P (1993) Permafrost and periglacial forms in the Romanian Carpathians. In: Proceedings of sixth international conference on Permafrost, Beijing, 5–9 July 1993, South China University of Technology Press 1, pp 631–637

Urdea P (1995) Quelques considérations concernant des formations de pente dans les Carpates Méridionales. Permafr Periglac Process 6:195–206

Urdea P (1998a) Rock glaciers and permafrost reconstruction in the Southern Carpathians Mountains, Romania. Permafrost. In: Seventh international conference proceedings, Yellowknife, Canada, University Laval, Collection Nordicana 57, pp 1063–1069

Urdea P (1998b) Considerații dendrogeomorfologice preliminare asupra unor forme periglaciare din Munții Retezat (Preliminary dendrogeomorphic considerations on some periglacial forms in the Retezat Mountains). Analele Universității Craiova, Geografie, Serie nouă 1:41–45 (in Romanian)

Urdea P (2000) Un permafrost de altitudine joasă la Detunata Goală (Munții Apuseni). (A low elevation permafrost at Detunata Goală, Apuseni Mountains). Revista de Geomorfologie 2:173–178 (in Romanian)

Urdea P, Sîrbovan C (1995) Some considerations concerning morphoclimatic conditions of the Romanian Carpathians. Acta Climatologica Szegediensis 28–29:23–40

Urdea P, Török-Oance M, Ardelean M, Vuia F (2001–2002) Aplicații ale S.I.G. în investigarea permafrostului sporadic de la Detunata Goală (Munții Apuseni). (GIS-applications in investigating sporadic permafrost at Detunata Goală, Apuseni Mountains). Analele Universității Vest Timișoara, Geografie 11–12:7–16 (in Romanian)

Urdea P, Vuia F, Ardelean M, Voiculescu M, Torok-Oance M (2004) Investigations of some present-day geomorphological processes in the alpine area of the Southern Carpathians (Transylvanian Alps). Geomorphologia Slovaca 4(1):5–11

Urdea P, Ardelean F, Onaca A, Ardelean M, Török-Oance M (2008a) Application of DC resistivity tomography in the alpine area of Southern Carpathians (Romania). In: Kane DL, Hinkel K (eds) Ninth International conference on Permafrost, Institute of Northern Engineering, University of Alaska, Fairbanks, pp 323–333

Urdea P, Ardelean F, Onaca A, Ardelean M (2008b) Deep-seated landslides (glimee) in the Saschiz and Șoard-Seciunei area. Geophysical investigations. In: Bălteanu D (ed) IAG Regional conference on geomorphology "Landslides, floods and global environmental change in mountain regions". Field Guide. University Publishing House, București, pp 32–33

Urdea P, Török-Oance M, Ardelean F, Onaca A, Ardelean M, Voiculescu M (2008c) The Făgăraș Mountains: Bâlea-Capra area. In: Bălteanu D (ed), IAG Regional conference on geomorphology "Landslides, floods and global environmental change in mountain regions". Field Guide. University Publishing House, București, pp 42–48

Urdea P, Ardelean M, Onaca A, Ardelean F (2008d) An outlook on periglacial of the Romanian Carpathians. Analele Universității Vest Timișoara, Geografie 18:5–28

Voiculescu M (2009) Snow avalanche hazards in the Făgăraș massif (Southern Carpathians) – Romanian Carpathians. Management and perspectives. Nat Hazards 51:459–475

Voiculescu M, Popescu F (2011) Management of Snow Avalanche Risk in the Ski Areas of the Southern Carpathians – Romanian Carpathians. Case Study: The Bâlea (Făgăraș Massif) and Sinaia (Bucegi Mountains) Ski Areas. In: Zhelezov G (ed) Sustainable development in mountain regions, southern Europe. Springer, Heidelberg/Berlin, pp 103–122

Zugrăvescu D, Polonic G, Horomnea M, Dragomir V (1998) Recent vertical crustal movements on the Romanian territory, major tectonic compartments and their relative dynamics. Revue Roumaine Géol Géophys Géogr, série Géographie 42:3–14 (in Romanian)

(1983) Geografia României I. Geografia Fizică (Geography of Romania, I. Physical Geography). Edit. Academiei, București, pp 171–194 (in Romanian)

(2006) Romania. Space, society, environment. Publishing House of the Romanian Academy, Bucharest, pp 60–81

Chapter 11
Recent Landform Evolution in Slovenia

Blaž Komac, Mauro Hrvatin, Drago Perko, Karel Natek, Andrej Mihevc, Mitja Prelovšek, Matija Zorn, and Uroš Stepišnik

Abstract Slovenia is the *locus classicus* of karst phenomena and karst geomorphology is the most intensively practiced branch. According to their general morphology, hydrology and evolution, Alpine karst, isolated karst, and Dinaric karst are distinguished. New measurement techniques revealed changes in the intensity of corrosion, for example, a reduction in intensity in many caves at the Pleistocene/Holocene boundary. The study of unroofed caves provided information on the dynamics of karst systems. Fluvial geomorphic processes are also intensive. Landslide hazard is the greatest in the alpine region, where surfaces susceptible to landslides cover one fifth of the area. In spite of intensive postglacial erosion, some traces of Pleistocene glacial and periglacial processes are still found in the Slovenian Alps and Dinaric Mountains.

Keywords Karst • Fluvial erosion • Fluvial accumulation • Mass movements • Glacial and periglacial processes

---

B. Komac (✉) • M. Hrvatin • D. Perko • M. Zorn
Anton Melik Geographical Institute, Scientific Research Centre of the Slovenian Academy of Sciences and Arts, Novi trg 2, SI–1000 Ljubljana, Slovenia
e-mail: blaz.komac@zrc-sazu.si; mauro@zrc-sazu.si; drago@zrc-sazu.si; matija.zorn@zrc-sazu.si

K. Natek • U. Stepišnik
Department of Geography, Faculty of Arts, University of Ljubljana, Aškerčeva cesta 2, SI–1000 Ljubljana, Slovenia
e-mail: karel.natek@guest.arnes.si; uros.stepisnik@gmail.com

A. Mihevc • M. Prelovšek
Karst Research Institute, Scientific Research Centre of the Slovenian Academy of Sciences and Arts, Titov trg 2, SI–6230 Postojna, Slovenia
e-mail: mihevc@zrc-sazu.si; mitja.prelovsek@zrc-sazu.si

11.1 Introduction

Blaž Komac

Slovenia's relief is characterized by extraordinary diversity. The main reason for this situation is that Slovenia lies at the contact of four major relief units: the Alps, the Dinaric Mountains, the Adriatic Sea Basin, and the Pannonian Basin. Because of diverse relief, landform is often the most important factor in distinguishing between regions and is an important element of geographical typology and regionalization in Slovenia (Perko 2001, 2007). Consequently, Slovenian geographers have developed several relief-based landform classifications.

11.2 Relief Types

Mauro Hrvatin
Drago Perko

The oldest *landform typology* was developed by A. Melik (1935), who distinguished 14 landform units: 5 types of Alpine Mountains, 2 types of Dinaric Plateaus, 1 type of Dinaric Basins and Lowlands, 2 types of Mediterranean Hills, 2 types of Pannonian Hills, and 2 types of Plains. After a long pause, K. Natek (1993a) prepared a new landform classification, distinguishing between eight landform types: Plains, Low Hills, Low Mountains, High Mountains, Low Karst, High Karst, Low Fluviokarst, and High Fluviokarst.

The first *computerized typology* was developed by D. Perko (1992, 2001), who identified surface *roughness* using a relief coefficient based on surface height and slope variability. Later he used a similar method to define eight landform units using a new relief coefficient on the basis of height and aspect variability (Perko 2007, 2009): Plains, Rough Plains, Low Hills, Rough Low Hills, High Hills, Rough High Hills, Mountains, and Large Valleys.

Another classification was developed by M. Gabrovec and M. Hrvatin (1998) for the *Geografski atlas Slovenije (Geographical Atlas of Slovenia)*. They divided Slovenia into six morphological units (Hrvatin and Perko 2009): Plains, Hills, Low Mountains, High Mountains (above 1,700 m), Low Plateaus, High plateaus (above 700 m Fig 11.1); as well as six genetic types: Destructive Fluvial-denudational, Accumulative Fluvial-denudational, Glacial, Limestone Karst, Dolomite Karst, and Coastal (Fig. 11.2).

The genetic relief types for Slovenia were first presented on the 1:500,000 scale Geomorphological Map of Yugoslavia in 1992 (Andonovski et al. 1992).

According to the 25-meter DEM average slope inclination in Slovenia is 14°, and the average altitude is 557 m (two thirds of the world's average). Its highest point is the peak of Mount Triglav (2,864 m), and its lowest is the coastline (0 m). The altitude belt from 0 to 200 m, which includes the Pannonian and Mediterranean plains, altogether encompasses less than one tenth of Slovenia's surface; the belt from 200 to

11 Recent Landform Evolution in Slovenia

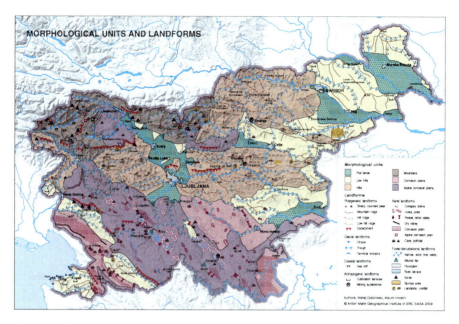

Fig. 11.1 The main morphological units and major landforms in Slovenia

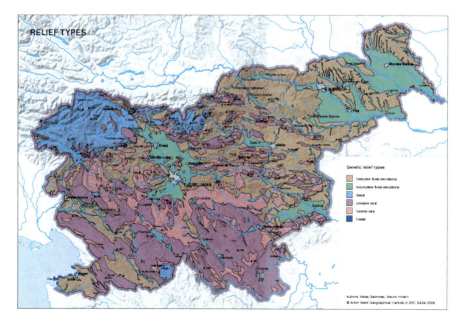

Fig. 11.2 Fluvial, glacial, karst, and coastal relief types in Slovenia

400 m that includes mainly the Pannonian and Mediterranean low hill areas and the Alpine plains encompasses almost one third; the belt from 400 to 800 m that includes the majority of the Alpine and Dinaric hills encompasses almost two fifths; the belt from 800 to 1,200 m that includes mainly high Dinaric plateaus and the highest Alpine hills one eighth; and the belt above 1,200 m only 6%. Due to the influence of altitude, various *threshold limits* are identified. The snow line in Slovenia is around 2,700 m, and the tree line lies between 1,600 and 1,700 m in the Julian Alps, between 1,700 and 1,800 m in the Kamnik-Savinja Alps, between 1,800 and 1,900 m in the Karavanke Mountains and only a little above 1,500 m in the Snežnik Mountain. The average limit of human settlement runs about 500 m below the tree line. The highest farmsteads are found at 1,300 m in the eastern Karavanke mountains. The maximum elevation for corn cultivation is 800 m and for vineyards 500 to 600 m (Perko 2001).

11.3 History of Geomorphological Research in Slovenia

Karel Natek

The earliest period of geomorphology in Slovenia until the beginning of the twentieth century is defined by the absence of domestic scientific institutions and researchers, a distinct dominance of German (Austrian) geomorphology, and a lack of contact with the development of geomorphology in English-speaking countries.

In the same period, the unique karst surface of Trieste's hinterland and the northwestern Dinaric Mountains encouraged general interest in the unusual *karst phenomena*. One of the first researchers was the Slovenian nobleman J. Vajkard Valvasor (J. Weickhardt Valvasor), author of *Die Ehre des Herzogthums Crain* (The Glory of the Duchy of Carniola, 1689) and a scientific treatise on the functioning of the intermittent Cerknica Lake. The latter won him the membership of the Royal Society in London. In the eighteenth century the Habsburg rulers encouraged a number of researchers, such as F. A. Steinberg, J. Nagl, and T. Gruber, to explain unusual karst phenomena in the province of Carniola.

In the nineteenth century, economic interests came into the foreground of karst research, related primarily to the water supply of Trieste and the Southern Austrian Railway, leading to the exploration of the Škocjan Caves after 1838 and the prevention of floods on karst poljes in which the Czech engineer W. Putick played an important role. In the late nineteenth century, the exploration of karst caves flourished although some of today's tourist caves had long been open for tourists, including Vilenica Cave (since the eighteenth century), the inner parts of Postojna Cave (discovered in 1818), and Križna jama Cave (discovered in 1838).

Interest in "non-karst" geomorphology only began in the second half of the nineteenth century, linked to the first geological mapping of Austrian provinces. The presence of geomorphology was insignificant in the first stage of mapping (M. V. Lipold, the first Slovenian geologist). Only well after 1890, along with structural geology, researchers (F. Kossmat, F. Teller, J. Dreger) began to address the basic issues of landform evolution. The results of these studies were published in German, but at the same time Slovenian geographical and geomorphological terminology began to develop to meet the needs of the emerging Slovenian secondary education system (Jesenko 1874). The final achievements of the "pre-scientific" period of Slovenian geomorphology are the historical-geographical descriptions of Slovenia's landscapes published between 1892 and 1926. Among these descriptions, Seidl's work on the Kamnik–Savinja Alps (1907–1908) stands out for its geomorphological content.

The foundation of the *University of Ljubljana* in 1919 and its *Department of Geography* represents a milestone in the development of Slovenian geomorphology. The first comprehensive work on landform evolution was published by J. Rus (1924) but it was A. Melik's regional-geographical monograph that had the greatest influence on the development of geomorphology in Slovenia (Melik 1935). Applying Davis' theory on the cyclical development of relief and the denudational chronology theory, A. Melik, the geologist I. Rakovec and their colleagues attempted to explain geomorphic evolution in the Tertiary and Quaternary from evidence of the remnants of previous summit levels. This work continued in the 1960s with detailed studies of summit levels across Slovenia. These studies, however, failed to explain the differences in the intensity of tectonic uplift from the early Pliocene on and also failed to provide any useful results because they did not sufficiently consider the climate changes at the end of the Pleistocene (Winkler v. Hermaden 1957).

In the period between the two World Wars, studies focused on the development of the *river network*, the formation of epigenetic valleys and Pleistocene *glaciations*. Penck and Brückner's work on the glaciation of the Alps (1909) was the starting point for the latter. By World War II, studies had been done on the extent of the largest glaciers and fluvio-glacial terraces and they received a new impetus in the 1950s and 1960s with the introduction of the concept of climatic geomorphology (Šifrer 1961, 1971, 1983, 1998). Further work in this field included the studies of Quaternary sediments and their exploitation in Slovenia. Outside specific areas of Pleistocene glaciers and fluvioglacial deposits, relatively little attention was devoted to studying periglacial processes in the Pleistocene (Šifrer 1998; Komac 2003; Zupan Hajna 2007; Natek 2007).

Geomorphological studies of non-karst surfaces only moved away from the traditional Davisian school with the climatic-geomorphological approach in the 1960s and 1970s, but for a variety of reasons they did not reorient toward the emerging quantitative geomorphology. Frequent floods and particularly the severe consequences of torrential floods were the spur for the long-term geographical research program *Studies of flood areas in Slovenia* that contributed substantially to our knowledge of the Quaternary development of fluvial surfaces, and the geomorphologic effects of *floods* (Komac et al. 2008). For the most part, quantitative studies of fluvial dynamics were performed in the framework of hydrology, and over the last decades geographers have undertaken detailed studies primarily of the direct geomorphic and broader geographical effects of torrential floods (Natek 1993b; Zorn et al. 2006; Komac et al. 2008).

Already in the 1960s, the effects of summer thunderstorms aroused interest in *landslides*, one of the leading agents of the recent reshaping of the surface on tectonically active, poorly conglomerated and impermeable Tertiary sediments (Gams and Natek 1981; Natek 1989; Zorn and Komac 2004). In recent years, special attention has been devoted to major movements of slope masses in the alpine world, prompted by an earthquake in the Bovec region on April 12, 1998 (Natek et al. 2003; Zorn et al. 2006; Komac and Zorn 2007; Zorn and Komac 2008a). Great progress has also been achieved in landslide modeling and prediction (Zorn and Komac 2008a).

Quantitative studies of *slope processes* do not have a long tradition in Slovenian geomorphology, but recent long-term studies of these processes in the Dragonja River basin have provided the first detailed insight into their intensity (Zorn 2009a, b). Until recently mostly general assessments of karst and mountain *geomorphosites* have been elaborated in Slovenia. Comprehensive geomorphosite assessment studies have been carried out only recently (Erhartič 2010).

11.4 Recent Geomorphological Processes and Landforms

11.4.1 Karst

Andrej Mihevc

Karst is a type of landscape with special surface, underground and hydrological features, and phenomena. Karst in Slovenia developed only on limestone and dolomite and covers 8,800 km^2 (43%) of the country. According to their general morphological and hydrological conditions and its evolutionary history, Alpine karst, isolated karst, and Dinaric karst are distinguished (Fig. 11.3). *Alpine karst* is characterized by large vertical gradients and a mixture of fluvial and karst elements in the landscape resulting in deep fluvial valleys between limestone mountains and karst plateaus. A deep vadose zone and deep cave systems typically developed here. At altitudes above 1,200 m, the karst relief was modified by glacial or periglacial processes. Slovenia's alpine karst belongs to the Southern Calcareous Alps. *Isolated karst* developed in small limestone areas surrounded by impermeable rocks that controlled the evolution of the encircled patches of karst. The majority of karst areas in Slovenia belong to the *Dinaric karst* type, most prominent in terms of size, geomorphology, hydrology, and underground features. The dominant landforms are extensive level surfaces at various elevations, closed depressions, dolines, uvalas, poljes, and conical hills. Allogenic surface rivers flowing from non-carbonate regions either sink at the edge of the karst forming blind valleys or larger contact karst features or cross the karst in canyons. The interaction between surface and underground features has also resulted in collapse dolines, cave entrances, and unroofed caves exposed by denudation.

Descriptions of karst geomorphology in Slovenia, mostly of the Dinaric karst, can be traced back to the seventeenth century. In the nineteenth century, attention focused mainly on caves and large sinking rivers or karst springs. Due to its distinctive

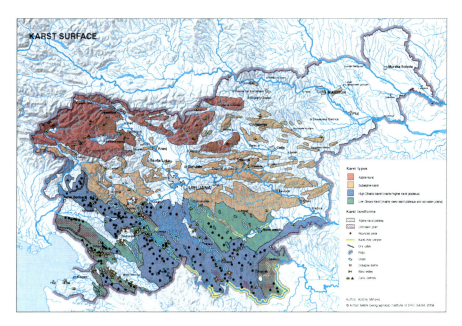

Fig. 11.3 Dinaric, Alpine, and Subalpine karst surfaces cover almost one half of Slovenia's territory

landforms and the considerable amount of research done here, this region became a *locus classicus* in the nineteenth century, and "Kras," the name of the karst plateau, entered the international *scientific terminology* as "karst" (Gams 1974, 2003).

Researchers who played an important role in geomorphology include the speleologists, geographers and geologists Schmidl (1854), Kraus (1894), Martel (1894) and Grund (1914). Their research work was amalgamated by J. Cvijić (1893), who put the research emphasis on geomorphology. He also made the first synthetic overview of geomorphology of the Dinaric karst. Cvijić and other geomorphologists also transferred Davis' *geomorphological cycle* model to karst. This approach, however, failed to contribute to a better understanding of the development of karst. In describing individual forms, they assigned the forms to a stage in the geomorphological cycle. Thus, for example, collapse dolines supposedly only occurred in the mature and senile stages of the cycle. Thinking in the framework of the model led to the conclusion that some forms evolve into other forms, for example, dolines to uvalas and then to poljes – although in reality these are forms of completely different origin. Fluvial elements such as dry and blind valleys were explained by a pre-karst development phase.

The cyclical model of karst evolution was later modified on a *climatic geomorphological* basis (Roglič 1957; Radinja 1972, 1985). Karstologists saw two distinct types of features in the relief that they attempted to explain by the influence of climate on morphological processes: during the warmer periods of the Pliocene, the dominant morphogenetic factor was *dissolution* that accelerated the formation of dolines, poljes, level surfaces, and conical hills and the formation of fine sediments. They also believed that later colder climate periods within the Pleistocene subjected karst to pronounced *mechanical processes* that created coarser clastic sediments that blocked underground drainage and partly renewed fluvial elements on karst. Therefore, smooth hillslopes, collapses in caves, fluvial deposits in contact karst areas, sediments in karst poljes, and karst canyons were explained by cold climatic intervals in the Pleistocene (Melik 1955; Gospodarič 1985; Šušteršič and Šušteršič 2003).

A less schematic approach to karst geomorphology in recent decades was enabled by the development of new *measurement* methods and the general progress in earth sciences along with those in surface geomorphology established firmer connections with speleological karst research. The largest caves and cave systems at the edge of karst poljes, caves carved by sinking streams, and caves, remnants of former systems, were studied (Gospodarič 1970; Mihevc 2001). The dating of flowstones (Gospodarič 1981; Zupan Hajna 1991; Mihevc 2001) pointed primarily to younger events in the caves related to climate change.

In the 1950s, I. Gams abandoned Cvijić's theory and devoted his studies to the *corrosional origin* of karst, and thus contributed significantly to the understanding of the intensity and significance of recent corrosion by measuring water hardness at springs and the loss of weight in limestone plates (Gams 1957, 1963, 1965, 1969, 1973, 1974, 1981, 1985, 1993, 1995a, 1999, 2000, 2003). J. Kunaver (1978, 1982, 1983) studied corrosion and its effects in the high mountain karst of the Julian Alps. The *intensity of corrosion* or dissolution, which provides information on the rate of the most important geomorphic process in the underground, has been studied to a lesser degree. Intensity in the past can be determined with the help of dating, and recent intensity by *in situ* measurements. The first modest measurements were conducted by Gams (1962), followed by A. Mihevc with measurements in the Škocjanske jame (Mihevc 2001), Ponikve v Odolini (1993) and Križna jama caves (1997). The improved methodology for measuring limestone plates allowed an actual insight into the spatial and time distribution of the intensity of corrosion. The results of measurements in permanently or periodically flooded phreatic and epiphreatic zones surprisingly indicate that in some caves (Škocjanske jame, Postojnska jama, Planinska jama, Križna jama, Tkalca jama) flowstone deposition prevails over corrosion – even alongside the course of larger rivers (Reka, Pivka). Today these caves are no longer expanding because the modest corrosion in periods of very high water is outpaced by the somewhat larger *deposition of flowstones* at the lower high, medium and low water levels. This is a very significant finding because it indicates that corrosion was once more intense. The reduction in the intensity of corrosion in many caves occurred at the transition from the Pleistocene to the Holocene. This is linked to warming, vegetation changes, and consequently the presence of more CO_2 in the soil and the water, which became harder immediately below the surface. Upon entering larger ventilated cavities, the water oversaturated with CO_2 begins to degas due to lower partial pressure of CO_2 and flowstones deposit. This phenomenon, which until recently was identified only for the growth of speleothems in the vadose zone, has much wider, regional geomorphic significance since it influences all areas

Fig. 11.4 A dell ("dolec"), a small and shallow dry valley, is a characteristic form of dolomite karst relief

linked with karst; for example, the geomorphic development of contact karst and areas directly downstream from karst springs (Prelovšek 2009).

Much attention was also devoted to the study of individual *landforms*. Dells (Fig. 11.4) on dolomite were described by Gams (1968), Gabrovec (1995), and Komac (2003, 2006). The construction of highways crossing karst areas revived interest in studying subsoil karst forms (Gams 1971; Slabe 1997, 1999; Knez and Slabe 2006). Some recent studies treat karst from the theoretical viewpoint as the result of the previously more or less excluded effect of corrosion (Šušteršič 1996; Dreybrodt et al. 1997, 2005; Gabrovšek and Dreybrodt 2000, 2001; Gabrovšek 2000, 2007; Dreybrodt and Gabrovšek 2002).

The morphology, climate effects and land use of *dolines* and the sediments in them were studied by Gams (2000) and their mathematical properties by Šušteršič (1985). The proportion of dolines on the surface and various genetic types of dolines on the Divača karst were described by Mihevc (2001, 2007a). The structural conditions for the formation of large collapse dolines were studied by Šušteršič (1973), Čar (2001), and Šebela (1996). Mihevc (2001) described the formation of large *chambers in caves* and collapse dolines on the surface above caves on the Divača karst as a special speleogenetic process fostered by the oscillating level of the karst

Fig. 11.5 The formation of large chambers in caves and collapse dolines on the surface above caves on the Divača karst has been described as a special speleogenetic process fostered by the oscillating level of the karst water table

water table (Fig. 11.5). Stepišnik mapped collapse dolines and studied the processes that transform them (Stepišnik 2004, 2006, 2010).

Gams (1962, 1965, 1994, 1995a, b, 2001) described *blind valleys* and the distribution of *ponor contact karst* as a special type of fluviokarst and as forms that occur with locally intensified corrosion (Kranjc 1989; Stepišnik et al. 2007a, b), while Mihevc (1994, 2001) described the various forms of blind valleys, marginal flat surfaces or depressions and dry valleys. The forms of contact karst reflect the water conductivity of the entire karst between ponors and springs and the changes in the conductivity of karst over time that are the consequence of changes, often of tectonic origin, at the regional level (Mihevc 2007a).

P. Habič (1982, 1992) studied tectonic landforms and divided the Dinaric karst in Slovenia into morphological and hydrographical units, described their morphological characteristics along with the distribution of karst elevations, closed depressions and plains. However, he did not evaluate the dynamics of tectonic movements or determine a time frame for them.

Initially geological data were used to determine the *age of karst*. The starting point was the date of the end of marine sedimentation, which defined the beginning of the last karst phase of geomorphic evolution. Later karstologists attempted to date landforms using past climate conditions, which presumably foster forms or sediments characteristic of warmer or colder periods. Structural geology provides dating of individual *tectonic phases*. The northward movement of the Adriatic microplate caused contraction deformations in the late Tertiary. The contraction ended about six million years ago and was followed by a rotation accompanied by uplift, folding and the formation of strike-slip basins (Vrabec and Fodor 2006). These changes are all reflected in larger karst units.

The study of *unroofed caves* opened to the surface by denudation introduced a completely new approach to studying the age of karst and landform evolution on karst (Mihevc 1996, 1999, 2007a, b) and made it possible to explain the development of the three-dimensional karst geomorphic systems in time. This approach requires profound knowledge on landforms and on locations of suitable unroofed caves filled with sediment as well as on relict and active caves. The close connection between surface and underground karst processes has become increasingly evident in studies of collapse dolines (Šušteršič et al. 2003; Stepišnik 2006; Zupan Hajna et al. 2008a).

The use of various methods for the absolute and relative *dating of cave sediments* (U/Th method, paleomagnetism, fossils) further substantiated findings about the "rapid" corrosion-related lowering of the karst surface and the old age of karst caves or the sediments deposited in them (Zupan Hajna 1993; Mihevc 2002; Zupan Hajna et al. 2008b). Using the denudation rate (between 20 and 80 m Ma^{-1} on karst) the sedimentation column and the assumed depth of the removed rock determined by the sediments remaining in the caves, the age of individual caves on the Kras Plateau was determined to be between two and five million years (Mihevc 2001). This was confirmed later by paleontological and paleomagnetic dating, which provided even more detailed results (Bosák et al. 1998, 1999, 2004; Zupan Hajna et al. 2008a). There are no remnants of the pre-karst phase on the karst surface where these caves are found. The remains of cave animals found attached to the walls of unroofed caves made it possible to reconstruct the development of a three-dimensional segment of karst over the last four million years in the Podgorski karst (Bosák et al. 2004; Mihevc 2007a) and the southwestern part of the Kras Plateau. During this time the surface was denuded by about 200 m, and the level of the karst water table and underground rivers in the caves dropped around 350 m. The lowering of the water table underground is reflected in the sequence of forms of ponor contact karst. All these phenomena were caused by regional tectonic uplift, which made possible the precise dating through geomorphological analyses (Mihevc and Zupan Hajna 1996; Mihevc 2007b; Zupan Hajna 2004; Zupan Hajna et al. 2008a).

11.4.2 Fluvial Geomorphology

Matija Zorn
Blaž Komac

Although fluvial processes dominate in Slovenia (Gabrovec and Hrvatin 1998), geomorphologists primarily studied the glacial and limestone karst relief of Slovenia in the past. As a result, numerous geomorphic processes have been poorly studied in Slovenia. There is increasing evidence that fluvial geomorphic processes are currently very intensive. *High magnitude/low frequency events* contributed more to relief development as it has been accepted until now. It is claimed that

Fig. 11.6 The hilly region along the Drava River

surface development was strongly influenced not only by the climate change at the end of the Pleistocene (Šifrer 1983). A combination of high precipitation and large quantities of mobile slope material on steep Alpine and pre-Alpine slopes resulted in several large prehistoric and historic rockfalls and debris flows (Šifrer and Kunaver 1978; Komac and Zorn 2007; Zorn and Komac 2008b).

Hydrogeomorphic processes greatly influence relief development. Slovenia has more than 26,000 km of rivers and streams (Fig. 11.6); approximately 1,700 km of these are distinctly torrential and transport over two million cubic meters of sediment annually, mostly during short-duration storms (Komac et al. 2008). Among the worst disasters were the floods in the Gradaščica and Sora river basins west of Ljubljana (1926, 2007 and 2010), in the Savinja River basin, and the town of Celje (1954, 1990, 1998, 2007), in the Haloze hills on the western edge of the Pannonian Plain south of Ptuj (1989). In addition, local flash floods occur almost annually (Komac et al. 2008).

The average annual total *sediment yield* of Slovenian rivers is estimated at 3.9–5.7 million cubic meters (Zorn and Komac 2005). About half to three-fifths of the eroded material remain on slopes, in tali and fans, while about a quarter is deposited on footslopes. The data on material deposited in river basins indicate that streams and rivers carry ca 15.2 t ha^{-1} year^{-1} of material in the Soča River basin, ca 6.3 t ha^{-1} year^{-1} in the Sava River basin, ca 5.6 t ha^{-1} year^{-1} in the Drava River basin and ca 2.6 t ha^{-1} year^{-1} in the Kolpa River basin. In the low coastal hills, streams and rivers transport ca 6.4 t ha^{-1} year^{-1} of material (Komac and Zorn 2005). Almost 70% of the annual suspended load in the streams and rivers of Slovenia only moves during a few high-water events (Ulaga 2002).

In addition to being carried by water, mobile particles are also moved to lower positions by *hillslope processes*: mainly by flowing (soil creep, solifluction, and debris flow), sliding (landslides) and falling (rockfall). On average *landslide* hazard is the greatest in the alpine region, where surfaces susceptible to landslides cover a fifth of the area. In some regions landslides are significant geomorphic processes (Zorn and Komac 2009). The Dinaric region, predominantly built of limestone, is the least threatened (with 7% susceptible to landslides) (Zorn and Komac 2008a).

We tend to attribute the responsibility for most major *rockfalls* to *earthquakes*. The 1998 earthquake (5.6 M; EMS intensity VII–VIII) caused significant geomorphic changes at over one hundred sites of the Upper Soča valley in northwestern Slovenia. Two of the 52 recorded rockfalls were of a larger scale (on Krn Mountain and Osojnica Mountain in the Julian Alps) with a total volume of over a million m^3 of material moved (Natek et al. 2003). The quantity of material displaced exceeded the average annual sediment production in the region (Zorn 2002). In approximately the same area 44 rockfalls were recorded in the case of the 2004 earthquake (4.9 M; EMS intensity VI–VII) (Mikoš et al. 2005). On the occasion of the Villach earthquake (6.5 M; intensity X; on January 25, 1348) about 148 million m^3 of material was released from Dobratsch Mountain in Carinthia, covering an area of 6.11 km^2. The Idrija earthquake (7.2 M; intensity X; on March 26, 1511) triggered several rockfalls and made two opposing hills collapse, while the Friuli earthquake (6.5 M; intensity X; on May 6, 1976) induced several minor movements. In some cases lakes formed behind rockfall accumulations in Alpine regions and were subsequently filled with fine-grained deposits (Komac and Zorn 2007).

Despite their rarity, *debris flows* are important geomorphic processes in alpine areas. While studying the debris flow in Log pod Mangartom in 2000, we determined that similar larger-scale phenomena often occurred in the past (Komac et al. 2008; Zorn and Komac 2008b). In the past 150,000 years there appear to be at least two generations of such flows: the first 150,000–120,000 years ago and the second at the end of the last glacial period (20,000–10,000 years ago) (Bavec et al. 2004).

Flysch *cliffs* towering up to 70 m height are the most characteristic coastal (abrasive) landforms found along a short stretch of Slovenia's Adriatic coast between Izola and Piran (Fig. 11.7). There are several estimates of the rate of the retreat (Žumer 1990), ranging from 1 cm year^{-1} to almost 0.5 m year^{-1}. Measurements of rockwall retreat of steep bare (Eocene) flysch slopes in the inland flysch badlands (Ogrin 1992) and measurements on erosion plots (Zorn 2009b) indicate that the sediment production of flysch rocks totals around 80 kg m^{-2} year^{-1} on the average, which means that steep bare flysch rockwalls retreat at a rate of 3.5–5 cm year^{-1}.

The most intensive *soil erosion* research (Komac and Zorn 2005; Hrvatin et al. 2006) was done in flysch regions (Zorn 2009a). Soil erosion in Slovenia was mostly modeled using Gavrilović's model, the USLE, RUSLE, RUSLE2 and WATEM/SEDEM models. Measurements on erosion plots have been seldom performed so far.

Accumulational processes are found in plains, basins and on the floors of wider valleys and karst poljes. The depth of Quaternary deposits differs significantly, ranging from a few to two hundred meters. On older terraces, limestone gravels

Fig. 11.7 Coastal relief with flysch cliffs and alluvial flood plains between Izola and Piran

cemented into conglomerates that already karstified in some places such as the Udin boršt terrace near Kranj. Recent tali are found in Alpine areas whereas fossil tali are also characteristic of the Dinaric regions (Šifrer 1983).

Accumulational landforms mostly reflect the intensity of geomorphic processes inland. The last sedimentation phase began when the level of the Adriatic Sea started to rise. The sedimentation rate since the postglacial Holocene period (Preboreal) has been 2.9 mm year^{-1} in the Sečovlje salt pans. Currently, average sedimentation at the mouths of the rivers on the Slovenian coast is around 1 mm year^{-1}. Some 93–95% of the material annually carried to the sea by the rivers is deposited at their mouths and only a small proportion actually reaches the deep sea.

11.4.3 Glacial Geomorphology

Uroš Stepišnik

In the Pleistocene the territory of Slovenia was influenced by glacial and periglacial processes. The most striking evidence of glacial erosion and accumulation is found in the Slovenian Alps and intramontane basins (Fig. 11.8). Investigations on glacial action were stimulated by the basic scientific work of Penck and Brückner (1909), but their findings for the Slovenian Alps were relatively general. Traces of Pleistocene glaciations were also identified in the Dinaric Mountains and to a limited extent in the subalpine mountains (Šifrer 1959, 1961; Habič 1968; Meze 1979; Bavec et al. 2004;

Fig. 11.8 Glacial and destructive fluvial-denudational relief in the Loška Koritnica Valley with Rombon Mountain (2,208 m) in the Julian Alps

Natek 2007; Gams 2008). The chronology and extent of the Pleistocene glaciations in Slovenia is still debated.

All parts of the Alps in Slovenia bear traces of multiple *glaciations*, with three major glaciers filling the present Soča, Sava (Fig. 11.9) and Savinja valleys. The Soča glacier extended more than 30 km along the Soča River valley to the present town of Tolmin. Recent studies of the Soča glacier refute former views on the large extension of late Pleistocene glaciers and associate the glacial deposits at Tolmin with the Riss or Mindel glaciations (Bavec et al. 2004). Some sediments previously regarded glacial are now considered to be of rockfall origin (Šifrer 1998). The effects of glaciation on karst areas and the karstification of the postglacial surface have been studied on the high alpine plateau of Kanin Mountain (Kunaver 1982, 2009). Late Pleistocene sediments in the Bovec basin (Kuščer et al. 1955) indicate that all sedimentation and resedimentation took place under paraglacial conditions, thus excluding the presence of glacial ice in the basin itself. On the other hand, laminated fine-grained lacustrine deposits of more than 200 m thickness indicate the existence of a large Late Pleistocene lake in the area of Bovec (Bavec and Verbič 2004; Bavec et al. 2004), connected to a large rockfall (Zorn 2002).

Ice from the eastern Julian Alps and the Karavanke Mountains formed two major lobes, the Bohinj and Dolina *glaciers*, which merged at Bled to form the Sava

Fig. 11.9 In the foreground is the Sava Plain (Savska ravan) with forest-covered conglomerate and cleared fluvio-glacial gravel terraces near Radovljica in the northwestern part of the Ljubljana Basin

Glacier. Ice from the northern Julian Alps and the Karavanke Mountains feeding the main Dolina glacier reshaped numerous smaller preglacial fluvial valleys. Ice from the eastern Julian Alps feeding the Bohinj glacier (Melik 1930) accumulated mostly in icefields on high alpine plateaus (Kunaver 1978, 1999). The effects of glacial erosion on such karst plateaus were recently reinterpreted (Šmuc and Rožič 2009) as a result of lithological structural conditions and neotectonic activity rather than glacial remodeling. The maximum extent of the Sava glacier was established by the position of *terminal moraines* in the vicinity of Radovljica. The chronology of the deposits is based on lithological and morphological correlations and on one of the first datings of sediments in Slovenia (Šifrer 1969).

The Savinja glacier was fed by ice that accumulated in the Kamnik–Savinja Alps. The extent of the glacier was determined solely by the general valley morphology since no other traces of glaciation were found due to complete or almost complete erosion (Meze 1979).

The subalpine Pohorje mountain range (highest peak: 1,543 m) was too low for the formation of an icefield but still exhibits a limited extent of glaciation. The existence of at least one *cirque glacier* was confirmed, and the lowest identified terminal moraine lies at 1,220 m (Natek 2007; Gams 2008).

At least two of the high plateaus of the Dinaric Mountain in southern Slovenia exhibit evidence of Pleistocene glaciation. The Snežnik Mountain (1,796 m) supported an icecap. The glacier reached down to approximately 900 m, where the lowest terminal moraines were recorded and where extensive karst depressions are filled with fluvioglacial debris (Šifrer 1959; Marjanac and Marjanac 2004). Similar conditions occurred on the Trnovski gozd karst plateau, where an icecap covered the highest parts of the plateau down to 1,300 m. Ice covered numerous *konta*s (closed karst depressions remodeled during Pleistocene glaciation) and fluvioglacial debris filled extensive karst depressions (Habič 1968).

Periglacial processes occurred between the equilibrium line altitude of the last glacial maximum established at 1,300–1,400 m above sea level and the tree line established at 600–800 m. They mainly resulted in intense frost weathering, increased mass movements and the rapid accumulation of fluvio-periglacial material in areas that had no direct contact with glaciated areas (Natek 2007). Extensive *fluvio-glacial* and *fluvio-periglacial fans* cover the intramontane Ljubljana basin. The proglacial fan deposits in the northern part of the Ljubljana basin were also supplied by smaller valley glaciers in the southern Karavanke Mountains. The accumulation of fluvioglacial gravel in the Ljubljana basin dammed the flow into the subsiding tectonic basin of the Ljubljana moor, causing the sedimentation of more than 100 m of fine-grained sediment and the formation of an extensive flood plain (Šifrer 1984, 1998).

Since intense postglacial erosional processes in the Slovenian Alps obliterated most of the glacial and periglacial features and the stratigraphic record of glacial events, it is rarely possible to establish a precise chronology of glaciations and the maximum extent of the Pleistocene glaciers. Apart from Šifrer's dating of the Quaternary terraces of the Ljubljana basin and some recent work in the Julian Alps, most of the research on glacigenic sediments in Slovenia was conducted decades ago; detailed studies using modern research techniques are yet to come.

11.5 Conclusion: Towards a New Geomorphological Paradigm

Andrej Mihevc
Blaž Komac

Recent breakthroughs in Slovenian geomorphology are associated with karst research. The discovery and explanation of roofless caves, which resulted in a completely new insight into the *dynamics of the karst geomorphic system* (Mihevc 1996, 2001, 2007a, b; Mihevc and Zupan Hajna 1996; Mihevc et al. 1998; Šušteršič 2000), are undoubtedly the most important recent contributions to the understanding of karst evolution. The interpretation of karst phenomena in the north-western part of the Dinaric mountains and in the Alps is based on considering the gradual and differentiated development of individual morphostructural units of karst in conditions of active tectonics (Fig. 11.10). This development has been taking place since the Oligocene when sedimentation ceased, leading to the exposure of karstic rocks or the beginning of the current phase of karstification. Since then individual structurally-conditioned segments of the karst were gradually included in the formation of karst landscape depending on the changeable hydrological conditions. These conditions were modified by altitude or hydrological barriers. Thus large karstic landforms such as karst poljes, karst plateaus, and mountains were created.

Fig. 11.10 Dinaric landscapes: in the foreground, the Dinaric karst plateau of Nanos (1,262 m) rises steeply above the Mediterranean flysch Vipava Valley (Vipavska dolina) and the low Vipava Hills (Vipavska brda); in the background, beyond the Pivka valley system (Pivško podolje) and Postojna, Dinaric plateaus and valley systems alternate

Denudation removed several hundreds of meters of rock and only in the later phases of development easily recognizable and diagnostic karst forms appeared on larger units, including dolines, blind valleys, gable valleys, dry valleys, conical hilltops, and level surfaces. The current karst surface cuts across older underground karst features, exposing roofless caves and the majority of cave entrances. Through these entrances it is possible to enter relict caves or reach still active water caves. Collapse dolines, locally predominant landforms, developed from the bottom up due to speleological processes that locally caused the ceilings of mostly active underground caves to cave in.

Also fluvial geomorphology has had a few paradigm shifts in the last decades, from the Davisian cycle to climatic geomorphology and modern quantitative approaches. The new methods of *quantitative geomorphology* rely on quantification of geomorphic processes and allow an analysis of the past and prediction of the future relief development. Relying on direct measurements in various environments, fluvial, tectonic, glacial, and hillslope geomorphology is a big step forward. The recent recognition of the *stochastic* nature of geomorphic *processes* in geomorphic studies has led to an increased emphasis on statistical analyses. Yet, the study of the *frequency and magnitude* relations of geomorphic processes still seems to be the most challenging issue. Only recently we began to better understand the role of high magnitude earthquakes as an important triggering factor for rockfalls and landslides in Alpine environment.

Geomorphic studies are not limited only to the high-energy fluvial relief of the Slovenian Alps shaped by rockfalls, landslides, and debris flows, but also to Prealpine Slovenian regions where landslides and soil erosion are common phenomena. Recent erosion and denudation measurements in coastal flysch regions pose questions of the relation between the geomorphic processes and *recent tectonic activity* with regard to Holocene evolution, including the eustatic changes at the Slovenian Adriatic coast.

It is probably too early to make any major syntheses and more detailed time determinations of the formation of karst and fluvial landforms since we lack reliable measurements on the intensity and distribution of geomorphic processes and sufficient information from the dating of sediments that are mostly preserved in past or recent contact karst areas. A better understanding of the relations between karst and non-karst regions remains to be one of the biggest challenges of Slovenian geomorphology along with the still insufficient reliable data on the magnitude and frequency of interacting geomorphic processes in a changing environment.

References

Andonovski T, Ahmetaj I, Bogdanović Ž, Bognar A, Blazek I, Bugarski D, Burić M, Bušatlija I, Davidović R, Ćurčić S, Lješevac M, Manaković D, Menković Lj, Martinović Ž, Miljković Lj., Mirković M, Natek K, Radojičić B, Tomić P, Zeremski M (1992) Geomorfološka karta Jugoslavije 1:500.000 (Geomorphological map of Yugoslavia). Jovan Cvijić Geographical Institute of the Serbian Academy of Sciences and Arts, Beograd

Bavec M, Verbič T (2004) The extent of Quaternary glaciations in Slovenia. Dev Quaternary Sci 2–1:385–388

Bavec M, Tulaczyk SM, Mahan SA, Stock GM (2004) Late Quaternary glaciation of the Upper Soča River Region (Southern Julian Alps, NW Slovenia). Sediment Geol 165(3–4):265–283

Bosak P, Mihevc A, Pruner P, Melka K, Venhodova D, Langrova A (1999) Cave fill in the Črnotiče quarry, SW Slovenia: palaeomagnetic, mineralogical and geochemical study. Acta carsol 28(2):15–39

Bosák P, Pruner P, Zupan Hajna N (1998) Palaeomagnetic research of cave sediments in SW Slovenia. Acta carsol 27(2):151–179

Bosák P, Mihevc A, Pruner P (2004) Geomorphological evolution of the Podgorski Karst, SW Slovenia: contribution of magnetostratigraphic research of the Črnotiče II site with Marifugia sp. Acta carsol 33(1):175–204

Čar J (2001) Structural bases for shaping of dolines. Acta carsol 30(2):239–256

Cvijić J (1893) Das Karstphänomen. Hölzel, Wien, 113 p

Dreybrodt W, Gabrovšek F (2002) Climate and early karstification: what can be learned by models? Acta Geol Pol 52(1):1–11

Dreybrodt W, Gabrovšek F, Siemers J (1997) Dynamics of the evolution of early karst. Theor Appl Karstology 10:9–27

Dreybrodt W, Gabrovšek F, Perne M (2005) Condensation corrosion: a theoretical approach. Acta carsol 34(2):317–348

Erhartič B (2010) Geomorphosite assessment. Acta geogr Slov 50(2):295–319

Gabrovec M (1995) Dolomite areas in Slovenia with particular consideration of relief and land use. Geogr zb 35:7–44

Gabrovec M, Hrvatin M (1998) Površje. In: Fridl J, Kladnik D, Orožen Adamič M, Perko D, Zupančič JM (eds) Geografski atlas Slovenije. Mladinska knjiga, Ljubljana, 360 p

Gabrovšek F (2000) Evolution of early karst aquifers: from simple principles to complex models. ZRC SAZU, Ljubljana, 150 p

Gabrovšek F (2007) On denudation rates in karst. Acta carsol 36(1):7–13

Gabrovšek F, Dreybrodt W (2000) Role of mixing corrosion in calcite-aggresive H_2O–CO_2–$CaCO_3$ solutions in the early evolution of karst aquifers in limestone. Water Resour Res 36(5): 1179–1188

Gabrovšek F, Dreybrodt W (2001) A model of the early evolution of karst aquifers in limestone in the dimensions of lenght and depth. J Hydrol 240:206–224

Gams I (1957) O intenzivnosti recentnega preoblikovanja in o starosti reliefa v Sloveniji (On the intensity of karstification and on the age of the relief in Slovenia). Geogr vestn 27–28:310–325 (in Slovenian)

Gams I (1962) Slepe doline v Sloveniji (Blind valleys in Slovenia). Geogr zb 7:263–306 (in Slovenian)

Gams I (1963) Meritve korozijske intenzitete v Sloveniji in njihov pomen za geomorfologijo (Measurements of corrosional intensity in Slovenia and their significance for geomorphology). Geogr vestn 34:3–20 (in Slovenian)

Gams I (1965) Types of Accelerated Corrosion. Problems of Speleological Research. In: Proceedings of the International speleological conference, Brno, 29 June–4 July1964

Gams I (1968) Geomorfološko kartiranje na primeru Rakitne in Glinic (Geomorphological mapping on the example of Rakitna and Glinice). Geogr vestn 40:69–88 (in Slovenian)

Gams I (1969) Ergebnisse der neueren Forschungen der Korrosion in Slowenien (NW-Jugoslawien). Problems of the karst denudation. Studia Geographica, Brno 5:9–20

Gams I (1971) Podtalne kraške oblike (Sub-soil karst features). Geogr vestn 43·27–45 (in Slovenian)

Gams I (1973) Die zweiphasige quartärzeitliche Flächenbildung in den Poljen und Blindtälern des nordwestlichen Dinarischen Karstes. Neue Ergebnisse der Karstforschung in den Tropen und im Mittelmeerraum. Vorträge des Frankfurter Karstsymposiums 1971. Erkundliches Wissen 32:143–149

Gams I (1974) Kras. Slovenska matica, Ljubljana, 359 p (in Slovenian)

Gams I (1981) Types of accelerated corrosion. In: Sweeting M (ed) Karst geomorphology, Benchmark Papers in Geology 59. Hutchinson Ross, Stroudsburg, pp 26–132

Gams I (1985) Mednarodne primerjalne meritve površinske korozije s pomočjo standardnih apneniških tablet (Comparative measurements of surface solution by means of standard limestone tablets). Razprave 26:361–386 (in Slovenian)

Gams I (1993) Karst denudation measurements in Slovenia and their geomorphological value. Naše jame 35(1):21–30 (in Slovenian)

Gams I (1994) Types of contact karst. Geogr fis din quart 17:37–46

Gams I (1995a) Die Rolle der beschleunigten Korrosion bei der Entstehung von Durchbruchstälern. Mitteilungen der Österreichischen Geographischen Gesellschaft 137:105–114

Gams I (1995b) Types of the contact karst. Studia Carsologica 6:98–116

Gams I (1999) Geomorphogenetics of the classical Karst – Kras. Acta carsol 27(2):181–198

Gams I (2000) Doline morphogenetic processes from global and local viewpoints. Acta carsol 29(2):123–138

Gams I (2001) Notion and forms of contact karst. Acta carsol 30:33–46

Gams I (2003) Kras v Sloveniji v prostoru in času (Karst in Slovenia in space and time). ZRC SAZU, Ljubljana, 516 p (in Slovenian)

Gams I (2008) Geomorphology of the Pohorje mountains. Acta geogr Slov 48(2):185–254

Gams I, Natek K (1981) Geomorfološka karta 1:100.000 in razvoj reliefa v Litijski kotlini (Geomorphological map 1:100,000 and the relief evolution of Litija basin). Geogr zb 21:5–67 (in Slovenian)

Gospodarič R (1970) Speleološke raziskave Cerkniškega jamskega sistema (Speleological investigations of the Cerknica cave system). Acta carsol 5:109–169 (in Slovenian)

Gospodarič R (1981) Generacije sig v klasičnem krasu Slovenije (Sinter generations on the classical karst of Slovenia). Acta carsol 9:91–110 (in Slovenian)

Gospodarič R (1985) Age and development of collapse dolines above the cave systems. Annales de la Société Géologique de Belgique 108:113–116

Grund A (1914) Der geographishe Zyklus um Karst. Zeitschrift der Gesellschaft für Erdkunde 52:621–640

Habič P (1968) Kraški svet med Idrijco in Vipavo. Prispevek k poznavanju razvoja kraškega reliefa (The Karstic region between the Idrijca and Vipava rivers: a contribution to the study of development of the Karst relief), Dela SAZU 21. Slovenska akademija znanosti in umetnosti, Ljubljana, pp 1–243 (in Slovenian)

Habič P (1982) Kraški relief in tektonika (Karst relief and tectonics). Acta carsol 10:23–43

Habič P (1992) Geomorfological classification of NW Dinaric karst. Acta carsol 20:133–164

Hrvatin M, Perko D (2009) Suitability of Hammond's method for determining landform units in Slovenia. Acta geogr Slov 49(2):343–366

Hrvatin M, Komac B, Perko D, Zorn M (2006) Slovenia. In: Boardman J, Poesen J (eds) Soil erosion in Europe. Wiley, Chichester, pp 297–310

Jesenko J (1874) Prirodoznanski zemljepis (Physical geography). Matica Slovenska, Ljubljana, 399 p (in Slovenian)

Knez M, Slabe T (2006) Dolenjska subsoil stone forests and other karst phenomena discovered during the construction of the Hrastje–Lešnica motorway section (Slovenia). Acta carsol 35(2):103–109

Komac B (2003) Dolomite relief in the Žibrše Hills. Acta geogr Slov 43(2):7–31

Komac B (2006) Dolec kot značilna oblika dolomitnega površja (Dell as a typical dolomite relief form), Založba ZRC, Ljubljana, 171 p

Komac B, Zorn M (2005) Soil erosion on agricultural land in Slovenia – measurements of rill erosion in the Besnica valley. Acta geogr Slov 45(1):53–100

Komac B, Zorn M (2007) Pobočni procesi in človek (Slope processes and a man). Založba ZRC, Ljubljana, 217 p

Komac B, Natek K, Zorn M (2008) Geografski vidiki poplav v Sloveniji (Geographical aspects of floods in Slovenia). Založba ZRC, Ljubljana, 180 p

Kranjc A (1989) Recent fluvial cave sediments, their origin and role in speleogenesis, Dela SAZU 27. Slovenska Akademija znanosti in umetnosti, Ljubljana, pp 164–166 (in Slovenian)

Kraus F (1894) Höhlenkunde. Wege und Zweck der Erforschung unterirdischer Räume. Gerold, Wien, 308 p

Kunaver J (1978) Intenzivnost zakrasevanja in njegovi učinki v zahodnih Julijskih Alpah – Kaninsko pogorje (The intensity of karst denudation in the western Julian Alps and its measurement). Geogr vestn 50:33–50

Kunaver J (1982) Geomorfološki razvoj Kaninskega pogorja s posebnim ozirom na glaciokraške pojave (Geomorphology of the Kanin Mountains with special regard to the glaciokarst). Geogr zb 22:197–346 (in Slovenian)

Kunaver J (1983) The high mountains karst in the Slovene Alps. Geogr Iugosl 5:15–23

Kunaver J (1999) Geomorfološki razvoj doline Krnice in njene zadnje poledenitve (Geomorphological development of the Krnica valley and its last glaciations). Dela 13:63–75

Kunaver J (2009) The nature of limestone pavements in the central part of the southern Kanin Plateau (Kaninski podi), western Julian Alps. In: Ginés A, Knez M, Slabe T, Dreybrodt W (eds) Karst rock features: karren sculpturing. Založba ZRC, Ljubljana, pp 299–312

Kuščer D, Grad K, Nosan A, Ogorelec B (1955) Geology of the Soča valley between Bovec and Kobarid. Geologija 17:425–476 (in Slovenian)

Marjanac L, Marjanac T (2004) Glacial history of the Croatian Adriatic and Coastal Dinarides. Quaternary glaciations. Dev Quaternary Sci 2(1):19–26

Martel EA (1894) Les Abimes. Delagrave, Paris, 580 p

Melik A (1930) Bohinjski ledenik (The Bohinj glacier). Geogr vestn 1(4):1–39 (in Slovenian)

Melik A (1935) Slovenija. Geografski opis. Splošni del (Slovenia. Geographical description. General part), 2 vols. Slovenska matica, Ljubljana, 700 p (in Slovenian)

Melik A (1955) Kraška polja Slovenije v pleistocenu (Les Poljé Karstiques de la Slovénie au pléistocène). Dela Inštituta za geografijo SAZU 3:163 p (in Slovenian with French summary)

Meze D (1966) Gornja Savinjska dolina. Nova dognanja o geomorfološkem razvoju pokrajine (La vallée supérieure de la Savinja: nouvelles constatations dans le développement géomorphologique de la région), Dela 10. Slovenska akademija znanosti in umetnosti, Ljubljana, 199 p (in Slovenian with French summary)

Meze D (1979) Razvoj reliefa Slovenije u pleistocenu (Pleistocene relief evolution in Slovenia). In: Benac A (ed) Praistorija jugoslavenskih zemalja 1. Paleolitsko i mezolitsko doba. Svjetlost, Sarajevo, pp 123–128 (in Croatian)

Mihevc A (1993) Micrometric measurements of the corrosion rate on the cave wall inscription in a swallet-cave of Odolina (Slovenia). In: Proceedings of the European Conference on Speleology (Helecine, Belgium) 2:93–96

Mihevc A (1994) Contact Karst of Brkini Hills. Acta carsol 23:100–109 (in Slovenian)

Mihevc A (1996) Brezstropa jama pri Povirju (Roofless cave at Povir). Naše jame 38:65–75 (in Slovenian)

Mihevc A (1997) Meritve hitrosti rasti sige na pregradah v Križni jami (Flowstone growth-rate measurements on the barriers of the cave Križna jama). Naše jame 39:110–115 (in Slovenian)

Mihevc A (1999) Unroofed caves, cave sediments and karst surface geomorphology – case study from Kras, W Slovenia. Naš krš 19: 3–11 (in Croatian)

Mihevc A (2001) Speleogeneza Divaškega krasa (Speleogenesis of the Divača karst). Založba ZRC, Ljubljana, 180 p (in Slovenian)

Mihevc A (2002) Postojnska jama cave system. U/Th datation of the collapse processes on Velika Gora (Point 4). In: Gabrovšek F (ed) Evolution of karst: from prekarst to cessation. Založba ZRC, Ljubljana, pp 14–15

Mihevc A (2007a) The age of karst relief in West Slovenia. Acta carsol 36(1):35–44

Mihevc A (2007b) Nove interpretacije fluvialnih sedimentov na Krasu (New interpretations of fluvial sediments from the Kras). Dela 28:15–28

Mihevc A, Zupan Hajna N (1996) Clastic sediments from dolines and caves found during the construction of the motorway near Divača, on the classical Karst. Acta carsol 25:169–191

Mihevc A, Slabe T, Šebela S (1998) Denuded caves – an inherited element in the karst morphology; the case from Kras. Acta carsol 27(1):165–174

Mikoš M, Fazarinc R, Vidrih R, Ribičič M (2005) Estimation of increasing sediment transport along watercourses in the Upper Soča river valley after recent strong earthquakes and large landslides. Geophysical Research Abstracts 7.

Natek K (1989) Vloga usadov pri geomorfološkem preoblikovanju Voglajnskega gričevja (The role of landslides in the processes of landform transformation in the Voglajna hills east of Celje, Slovenia). Geogr zb 29(1):37–77 (in Slovenian)

Natek K (1993a) Tipi površja v Sloveniji (Types of relief in Slovenia) 1. Geogr zb 40(4):26–31 (in Slovenian)

Natek K (1993b) Geoecological research into the catastrophic floods of November 1, 1990, in the Savinja River basin and its role in the mitigation of future disasters. Geogr zb 33:85–101

Natek K (2007) Periglacial landforms in the Pohorje Mountains. Dela, Slovenian Academy of Sciences and Arts 27:247–263

Natek K, Komac B, Zorn M (2003) Mass movements in the Julian Alps (Slovenia) in the aftermath of the Easter earthquake on April 12, 1998. Stud Geomorphol Carpatho-Balc 37:29–43

Penck A, Brückner E (1909) Die Alpen im Eiszeitalter 3. Die Eiszeiten in den Südalpen im Bereich der Ostabdachung der Alpen. Chr. Herm. Tauchnitz, Leipzig, 1199 p

Perko D (1992) Zveze med reliefom in gibanjem prebivalstva 1880–1981 v Sloveniji (Correlations between relief and population changes in Slovenia in the period 1880–1981). Doctoral thesis. University of Ljubljana, Faculty of Arts, Department of Geography, Ljubljana, 183 p (in Slovenian)

Perko D (2001) Analiza površja Slovenije s stometrskim digitalnim modelom reliefa (Analysis of the surface of Slovenia using the 100-meter digital elevation model). Založba ZRC, Ljubljana, 229 p (in Slovenian)

Perko D (2007) Morfometrija površja Slovenije (Morphometry of Slovenia's surface). Založba ZRC, Ljubljana, 92 p (in Slovenian)

Perko D (2009) Morfometrični kazalniki enot oblikovanosti površja v Sloveniji (Morphometric indicators of landform units in Slovenia). Geogr vestn 81(1):77–97 (in Slovenian)

Prelovšek M (2009) Present-day speleogenetic processes, factors and features in the epiphreatic zone. Doctoral thesis, University of Nova Gorica, 297 p. http://www.p-ng.si/~vanesa/doktorati/krasoslovje/3Prelovsek.pdf

Radinja D (1972) Zakrasevanje v Sloveniji v luči celotnega morfogenetskega razvoja (La karsification et l'évolution générale du relief en Slovénie). Geogr zb 13:197–243 (in Slovenian)

Radinja D (1985) Kras v luči fosilne fluvialne akumulacije (The karst in the light of fossilized deposition). Acta carsol 14–15:99–108 (in Slovenian)

Roglič J (1957) Zaravni u vapnencima (Karst plains in limestones). Geografski glasnik 19:103–134

Rus J (1924) Slovenska zemlja. Kratka analiza njene zgradnje in izoblike (Slovene lands. Short analysis of its composition and forms), Ljubljana, 48 p (in Slovenian)

Schmidl A (1854) Die Grotten und Höhlen von Adelsberg, Lueg, Planina und Laas. Kaiserliche Akademie der Wissenschaften, Wien, 316 p

Šebela S (1996) The dolines above the collapse chambers of Postojnska jama. Acta carsol 25:241–250

Seidl F (1907–1908): Kamniške ali Savinjske Alpe, njih zgradba in njih lice (Kamnik and Savinja Alps, their composition and face). Matica Slovenska, Ljubljana, 255 p (in Slovenian)

Šifrer M (1959) Obseg pleistocenske poledenitve na Notranjskem Snežniku (The extent of Pleistocene glaciation on Notranjski Snežnik). Geogr zb 5:27–83 (in Slovenian)

Šifrer M (1961) Porečje Kamniške Bistrice v pleistocenu (The drainage basin of Kamniška Bistrica during the Pleistocene). Dela SAZU 10, Ljubljana, 211 p (in Slovenian)

Šifrer M (1969) Kvartarni razvoj Dobrav na Gorenjskem (The Quaternary development of Dobrave in Upper Carniola (Gorenjska) Slovenia). Geogr zb 11:99–221 (in Slovenian)

Šifrer M (1971) Methoden und Ergebnisse der Untersuchung fluvialer Terassen in Slowenien (NW-Jugoslawien). Acta Geographica Debrecina 10:199–207

Šifrer M (1983) The main results of the geomorphological research of the Slovene Alpine area. Geogr Jugosl 5:25–30 (in Croatian)

Šifrer M (1984) Nova dognanja o geomorfološkem razvoju Ljubljanskega barja (New findings on the geomorphological development on the Ljubljansko barje). Geogr zborn 23:5–55 (in Slovenian)

Šifrer M (1998) Površje v Sloveniji (Relief of Slovenia). Geografski inštitut Antona Melika ZRC SAZU, Ljubljana (in Slovenian)

Šifrer M, Kunaver J (1978) Poglavitne značilnosti geomorfološkega razvoja Zgornjega Posočja (Main characteristics of geomorphic development of the Upper Soča valley). Zgornje Posočje: zbornik 10. zborovanja slovenskih geografov. Geografsko društvo Slovenije, Ljubljana, pp 67–81 (in Slovenian)

Slabe T (1997) Karst features discovered during motorway construction in Slovenia. Environ Geol 32(3):186–190

Slabe T (1999) Subcutaneous rock forms. Acta carsol 28(2):255–271

Šmuc A, Rožič B (2009) Tectonic geomorphology of the Triglav Lakes Valley (easternmost Southern Alps, NW Slovenia). Geomorphology 103:597–604

Stepišnik U (2004) The origin of sediments inside the collapse dolines of Postojna karst (Slovenia). Acta carsol 33:237–244

Stepišnik U (2006) Loamy sediment fills in collapse dolines near the Ljubljanica river springs, Dinaric karst, Slovenia. Cave Karst Sci 33(3):105–110

Stepišnik U (2010) Udornice v Sloveniji (Collapse dolines in Slovenia). Znanstvena založba Filozofske fakultete, Ljubljana, 118 p

Stepišnik U, Černuta L, Ferk M, Gostinčar P (2007a) Reliktni vršaji kontaktnega krasa severozahodnega dela Matarskega podolja (Relict alluvial fans of northwestern part of Matarsko podolje). Dela SAZU 28:29–42 (in Slovenian)

Stepišnik U, Ferk M, Gostinčar P, Černuta L, Peternelj K, Štembergar T, Ilič U (2007b) Alluvial fans on contact karst: an example from Matarsko podolje, Slovenia. Acta carsol 36(2):209–215

Šušteršič F (1973) K problematiki udornic in sorodnih oblik visoke Notranjske (On the problem of collapse dolines and al lied forms of high Notranjsko, South-Central Slovenia). Geogr vestn 45:71–86 (in Slovenian)

Šušteršič F (1985) Metoda morfometrije in računalniške obdelave vrtač (A method of doline morphometry and computer processing). Acta carsol 13:79–98 (in Slovenian)

Šušteršič F (1996) The pure karst model. Cave and Karst Sci 23(1):25–32

Šušteršič F (2000) The role of denuded caves within the karst surface. Mitteilungen des Verbandes der deutschen Höhlen- und Karstforscher 46(1–2):105–112

Šušteršič F, Šušteršič S (2003) Formation of the Cerkniščica and the flooding of the Cerkniško polje. Acta carsol 32(2):121–136

Šušteršič F, Šušteršič S, Stepišnik S (2003) The late Quaternary dynamics of Planinska jama, south-central Slovenia. Cave Karst Sci 30(2):89–96

Ulaga F (2002) Koncentracija suspendiranega materiala v slovenskih rekah (Concentrations of suspended sediment in Slovene rivers). Ujma 16:211–215 (in Slovenian)

Valvasor JW (1689) Die Ehre des Hertzogthums Crain: das ist, Wahre, gründliche, und recht eigendliche Belegen- und Beschaffenheit dieses Römisch-Keyserlichen herrlichen Erblandes. Laybach

Vrabec M, Fodor L (2006) Late Cenozoic tectonics of Slovenia: structural styles at the Northeastern corner of the Adriatic microplate. The Adria microplate: GPS geodesy, tectonics and hazards. NATO Science Series, IV, Earth and Environmental Sciences 61. Springer, Dordrecht, pp 151–168

Winkler v. Hermaden A (1957) Geologisches Kräftespiel und Landformung. Springer, Wien, 822 p

Zorn M (2002) Rockfalls in Slovenian Alps. Acta Geogr 42:124–160

Zorn M (2009a) Erosion processes in Slovene Istria: part 1. Acta geogr Slov 49(1):39–87

Zorn M (2009b) Erosion processes in Slovene Istria: part 2. Badlands. Acta geogr Slov 49(2):291–341

Zorn M, Komac B (2004) Recent mass movements in Slovenia. In: Orožen Adamič M (ed) Slovenia: a geographical overview. Založba ZRC, Ljubljana, pp 73–80

Zorn M, Komac B (2005) Soil erosion on agricultural land in Slovenia. Ujma 19:163–174 (in Slovenian)

Zorn M, Komac B (2008a) Zemeljski plazovi v Sloveniji (Landslides in Slovenia). Založba ZRC, Ljubljana, 159 p (in Slovenian)

Zorn M, Komac B (2008b) The debris flow in Log pod Mangartom, NW Slovenia. In: Monitoring, simulation, prevention and remediation of dense and debris flow II. WIT Press, Southampton, pp 125–133

Zorn M, Komac B (2009) The importance of landsliding in a flysch geomorphic system: the example of the Goriška brda Hills (W Slovenia). Z geomorph, suppl 53(2):57–78

Zorn M, Natek K, Komac B (2006) Mass movements and flash-floods in Slovene Alps and surrounding mountains. Stud Geomorphol Carpatho-Balc 40:127–145

Žumer J (1990) Recentni razvoj klifov na obalah Istrske Slovenije (Recent development of cliffs on the coast of the Slovenian Istria). In: Natek K (ed), Geomorfologija in geoekologija – zbornik referatov 5. znanstvenega posvetovanja geomorfologov Jugoslavije. ZRC SAZU, Ljubljana, pp 143–147 (in Slovenian)

Zupan Hajna N (1991) Flowstone datations in Slovenia. Acta carsol 20:187–204

Zupan Hajna N (1993) Survey of published flowstone datations from Slovenia. Naše jame 35(1):83–87

Zupan Hajna N (2004) The caves of the contact karst of Beka and Ocizla, in SW Slovenia. Acta carsol 33(2):91–105

Zupan Hajna N (2007) Barka depression, a denuded shaft in the area of Snežnik Mountain, Southwest Slovenia. J Caves Karst Stud 69(2):266–274

Zupan Hajna N, Mihevc A, Pruner P, Bosák P (2008a) Palaeomagnetism and magnetostratigraphy of karst sediments in Slovenia. Založba ZRC, Ljubljana, 266 p

Zupan Hajna N, Pruner P, Mihevc A, Schnabl P, Bosák P (2008b) Cave sediments from the Postojnska-Planinska cave system (Slovenia): evidence of multi-phase evolution in epiphreatic zone. Acta carsol 37(1):63–86

Chapter 12
Recent Landform Evolution in the Dinaric and Pannonian Regions of Croatia

Andrija Bognar, Sanja Faivre, Nenad Buzjak, Mladen Pahernik, and Neven Bočić

Abstract This chapter deals with main geomorphological properties of Croatia. The key elements of the geological predispositions are outlined as well as those of geodynamic evolution which, in conjunction with numerous exogenous processes, result in recent landform development. The main points of all important exogenous processes and landforms are presented: coastal, fluvial and fluvio-denudation, slope, aeolian and polygenetic. The main attention has been paid to karst and fluviokarst which cover almost 44% of the territory (the submerged karst excluded) therefore numerous recent research topics are outlined such as the development of structural geomorphological methods in karst, the research on coastal karst, glaciokarst, karst denudation measurements and cave microclimate. All this made a good basis for a detailed geomorphological regionalisation given in the chapter as well.

Keywords Dinarides • Pannonian basin • Geomorphology • Regionalisation • Karst • Croatia

---

A. Bognar • S. Faivre (✉) • N. Buzjak • N. Bočić
University of Zagreb, Faculty of Science, Department of Physical Geography,
Marulićev trg 19/II, 10 000 Zagreb, Croatia
e-mail: andrija.bognar@zg.htnet.hr; sfaivre@geog.pmf.hr; nbuzjak@geog.pmf.hr; nbocic@geog.pmf.hr

M. Pahernik
Croatian Defence Academy, Ilica 256 b, 10 000 Zagreb, Croatia
e-mail: mladen.pahernik@zg.t-com.hr

12.1 History of Geomorphological Research

The development of geomorphology as a scientific discipline in Croatia started in the second half of the nineteenth century, primarily within the framework of physical geography, although geologists and foreign scientists also contributed to relief research, which was mainly directed towards the karst area. After the first period, which had brought a number of world-famous geomorphological works, between the world wars geomorphology experienced an obvious stagnation. The exception was the valuable karst investigations by Josip Roglić. Unfortunately, structural geomorphology was neglected and structural relief forms were identified with geological structures. In geography teaching fictitious conceptions about the characteristics and development of the Earth's crust have persevered almost up to our days (Bognar 2000).

In the transitional period between the 1950s and the mid-1970s an increased interest of the society in ecology (especially in environmental protection), energy, and food problems, as well as in the necessity of the functional organization of living space has brought essential changes. By assembling geomorphologists in the framework of the state scientific project "Geomorphological mapping of the Republic of Croatia," led by academician Andrija Bognar, in 1982, systematic geomorphological investigations and morphogenetic and morphostructural mapping started.

Modern approaches and research methods are being introduced gradually. Morphostructural analysis is for the first time used in solving the problems of endogenic relief. Various quantitative methods of morphometry, morphography, and morphodynamics are applied for defining particular stages in geomorphic evolution, as well as modern methods of field work, geomorphological mapping on morphogenetic and morphostructural principles, and remote sensing. It should be pointed out that if needed, geomorphologists also contribute to geological mapping, sedimentological, chemical, and mineralogical analyses, especially those concerning loess.

The systematic education of a new generation of young scientists was an important condition for a renewal of geomorphology and physical geography in Croatia and for the successful completion of the geomorphological mapping project of the Republic of Croatia. The Croatian Geomorphological Society was established in 2003 and it is a member of the IAG (International Association of Geomorphologists). Croatian geomorphologists are also active in the INQUA (International Union of Quaternary Research). During the past 25 years research has been dedicated to exogenic processes, structural quantitative, and regional geomorphology as well as to geomorphological mapping. The application of geomorphological research findings becomes more and more widespread in practice as well. In that way the reputation of geomorphology has increased in the earth and natural sciences, as well as in ecology and socio-economic development of the Republic of Croatia.

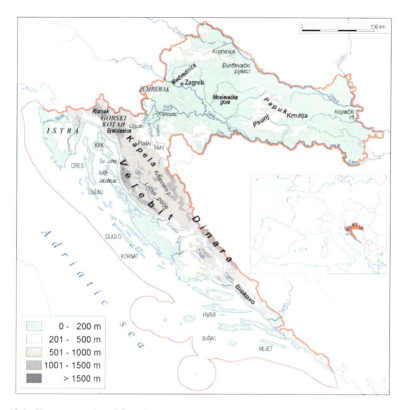

Fig. 12.1 The topography of Croatia

12.2 Geological and Geodynamic Setting

From the geodynamic point of view the Republic of Croatia (Fig. 12.1) belongs to the Late Cainozoic Alpine-Himalayan mountain belt. The Croatian territory, with a surface of 56,594 km^2 (Statistical Yearbook 2009) encompasses three main geomorphological units: the Pannonian Basin (lowlands, hills and scattered mountains), the mountain system of the Dinarides, and the Adriatic Basin (Bognar 2001).

The Dinarides of ca 700 km length represent a complex fold-and-thrust belt developed along the eastern margin of the Adriatic microplate (Dewey et al. 1973), connecting the Southern Alps with the Albanides and Hellenides. They were formed as a result of collision between the Adriatic and European plates, beginning in the late Cretaceous and reaching its peak in the Oligocene and Miocene epoch and finally resulting in the uplift of the mountain chain (Velić et al. 2006). The Dinarides can be divided into two genetically different parts: the Outer (Karst or External) Dinarides along the Adriatic Sea, composed mostly of the remnants of the Adriatic Carbonate Platform, its basement and overlying deposits, and the Inner (Internal)

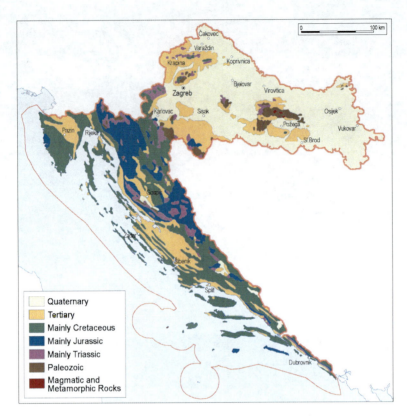

Fig. 12.2 Overview geological map of the Republic of Croatia (Croatian Geological Society, Zagreb 1990)

Dinarides, situated between the Outer Dinarides and the Pannonian Basin, composed of passive and active continental margin rocks including ophiolites (Pamić et al. 1998). The Outer Dinarides are characterized by four major sequences (Velić et al. 2006) ranging from Middle Carboniferous to Eocene/Oligocene, all within different paleogeographic settings (Fig. 12.2):

1. Middle Carboniferous – Middle Triassic
   Over more than 80 Ma the area represented an epeiric carbonate platform belonging to the NE margin of the Gondwanan part of Pangea producing mixed siliciclastic-carbonate depositions.
2. Upper Triassic – Toarcian
   45 Ma of shallow-marine carbonate deposition from the Late Triassic to Toarcian took place on a very wide carbonate platform area formed on the Adriatic basement, partially separated from Gondwana, and therefore lacking any continental influence. Consequently, the area of the future Karst Dinarides represented only a small part of this huge depositional area (Southern Tethyan Megaplatform of Vlahović et al. 2005) characterized by very similar depositional sequences.

3. Toarcian – Cretaceous

   This major sequence encompasses almost 120 Ma and it is characterized by a specific paleogeographic entity – the Adriatic Carbonate Platform (AdCP). This platform become a separate entity during the Toarcian Oceanic Anoxic Event, when former megaplatform become disintegrated into several smaller "isolated" platforms (including the Apenninic, Apulian, and Adriatic Carbonate Platforms) surrounded by deeper marine basins (including the Adriatic basin and the Bosnian and Slovenian troughs) (Vlahović et al. 2002, 2005).

4. Palaeocene – Oligocene

   The final major sequence comprises approximately 40 Ma of deposition within specific environments, mostly controlled by intense synsedimentary tectonics resulting in significant compression of the area and formation of asymmetrical flysch basins within the former platform (Velić et al. 2006).

The development and tectonics of the (Outer) Dinarides are very important for geomorphic evolution in Croatia. The whole mountainous system is named after the mountain Dinara (1,830 m). Besides a great geological, geophysical, and geomorphological diversity, as well as an intensive disruption, the Outer Dinarides represent a cohesive geotectonic entity of uniform geodynamic evolution. They also reveal similar structural properties like the predominantly Dinaric strike (NW-SE), apart from Central Dalmatia, where the Hvar strike (east to west) becomes increasingly obvious, as well as the predominance of tangential folded and faulted southwest to south verging structural units.

Three main structural units can be differentiated in the Outer Dinarides (Herak 1991; Prelogović et al. 2004): the Dinaric (2), the Adriatic (3), and the Adriatic microplate (4), (Fig. 12.3). The contact with the Inner Dinarides, Supradinaric (1) is also shown in Fig. 12.3. The Dinaric and Adriatic structures are characterized by inverse and thrust relations. They are the consequence of gradual movements of the Adriatic microplate (Adria) and of the Earth's crust deformation in the narrow zone of the European plate. The process lasted for millions of years. The Adriatic microplate is pushed by the African plate in northeastern and northern directions. The strong indentation of the small Adriatic microplate to the large European plate results in a complete closure of the areas around the present Adriatic Sea. The remaining narrowed area of the Adriatic microplate still relies on the African plate and moves towards the north-northwest. The boundaries of the structural units are defined by major faults (Fig. 12.3). On the surface they represent zones of few kilometers width. The faults mostly run parallel to steep mountain slopes or rows of islands.

The lineaments that reveal the borders of the Adriatic microplate (4) and the Adriatic (3) must be emphasized. They are the first thrust faults that marked the strong tectonically and seismically active zones. These faults are represented by three large zones: the Trieste–Učka–Lošinj (7), the Dugi otok (8), and the Vis-Lastovo-Dubrovnik (9) lineaments. The boundaries of the Adriatic and the structure of Dinaric are determined by the following inverse faults: the Trnovski-Gozd-Ilirska Bistrica-Rijeka-Vinodol (2), the Velebit (3), the Promina-Moseč (4),

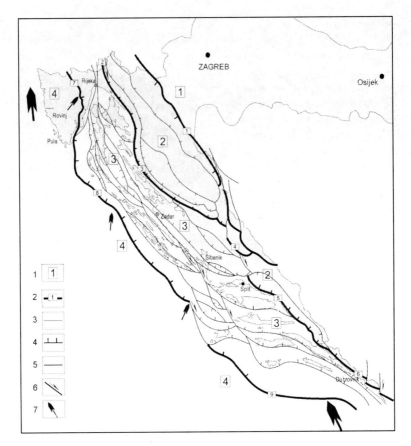

Fig. 12.3 Structural map of the Outer Dinarides (Prelogović et al. 2004). *1* Main structural units: Supradinaric (1), Dinaric (2), Adriatic (3), Adriatic microplate (4), *2* Main lineaments: [1 – Idrija–Čabar–Ogulin–Bihać fault; 2 – Fault Trnovski gozd–Ilirska Bistrica–Rijeka–Vinodol; 3 – Velebit fault; 4 – Fault Promina–Moseć; 5 – Fault Mosor–Biokovo; 6 – Fault Ploče–Dubrovnik–Bar; 7 – Fault Trieste–Učka–Lošinj; 8 – Fault of Dugi otok; 9 – Fault Vis–Lastovo–Dubrovnik], *3* Faults which delimit structural units and faults within the zones, *4* Reverse fault, *5* Normal fault, *6* Strike-slip fault, *7* Movement direction of the Adriatic microplate

the Mosor-Biokovo (5), and the Ploče-Dubrovnik-Bar (6) lineaments. The boundary between the Dinaric and the Supradinaric units (1) follows the Idrija-Čabar-Ogulin-Bihać faults (1). The faults are of variable inclination and character. Dextral and transcurrent movements are important along some sections of the faults.

Tectonic activity and the deformations of geological structures are also determined by the disposition of different rock masses, their sizes and locations within the regional structural Dinaric unit (2). The masses of this unit directly resist the movements of the Adriatic microplate. The unit of the Adriatic (3) represents in fact a large area of Earth's crust deformed by tectonic processes of action and reaction between the Dinaric and Adriatic. This is the reason why the area of Outer Dinarides is characterized by thrust structures and shallow crustal folds. The rock masses of the Adriatic microplate underlie the structures of the Adriatic (3), where they also come into contact with the rock masses of the Dinaric (2) in oblique inclined zones that are seismotectonically the most active.

The tangential movements seem to be relatively small and only locally achieve 10 km. Nevertheless, if the entire rock mass is taken into account up to the Mohorovičić discontinuity, the amplitudes are probably bigger. Apart from tangential movements, there are also horizontal movements of the neighboring tectonic units or blocks, especially along north-northwest–south-southeast faults. The amplitudes of neotectonic subsidence in the Adriatic basin vary up to 6,000 m, while the highest mountains along the coast reach 1,500 m elevation (Prelogović et al. 1982).

On the basis of the highly accurate GPS measurements of the geodynamic network CRODYN, a geodetic model of recent tectonic movements was established by Altiner (1999). The model encompasses GPS points in Croatia and Slovenia. For these points the values of heights, horizontal and vertical components of the movement velocity vector were calculated (Fig. 12.4).

Surface deformation analyses indicate zones of various deformations in the Outer Dinarides. Two main zones of extension and two zones of compression were identified. A zone with extraordinary extension, 6 mm year^{-1} per 10 km, was discovered in central Istria and southern Slovenia. This is at the same time the zone of maximum total deformation of the upper Earth's surface. A major zone of maximum compression has been discovered northeast and southeast of the area. The magnitudes of compression are somewhat smaller (3–5 mm year^{-1} per 10 km) (Altiner 1999). The geological evidence on tectonic movements of the Adriatic microplate and the high concentration of the earthquakes confirm the existence of compression zones in this area. Almost all the velocity vectors (Fig. 12.4) indicate north-northwest direction, only some of them, like the Bakar, Split, and Sveti Ivan, show northeast direction. This indicates a moderate rotation of the Adriatic microplate.

The Croatian part of the Pannonian Basin is basically a mountain and basin region today. The southwestern part of Croatia represents a contact zone of the Pannonian Basin in the west with Fore-alpine structures in the north and with the Dinarides in the south. This is mostly due to the fact that four major tectonic boundaries join here (Fig. 12.5): the Periadriatic (Insubric) Lineament (PAL), the Mid-Hungarian Lineament (MHL), the Sava Fault (SaF), and Drava Fault (DF) (Tomljenović and Csontos 2001). This Zagorje–Mid–Transdanubian Zone (ZMTZ) is composed of mixed Alpine and Dinaric lithologies (Tomljenović 2000). The precise locations of these boundaries and their kinematic history in the area are still ambiguous due to a complex Neogene deformation and a thick southeastern Pannonian Basin cover. Characteristic major tectonostratigraphic units of the ZMTZ are best preserved in the Mt. Medvednica, which can be traced along a northeast strike for about 40 km (Tomljenović and Pamić 1998).

In the north, the Dinarides are bounded by the Tisza Unit, a fragment detached from the Eurasian southern margin (Pamić et al. 2000). Paleozoic crystalline

Fig. 12.4 Horizontal and vertical components of the movement velocity vectors (Altiner 1999; Cigrovski-Detelić 1988). Legend: *1* horizontal component of the movement velocity vector, *2* vertical component of the movement velocity vector, *3* movement of the Adriatic microplate

rocks of the South Tisza, outcrop in the Moslavačka gora and the Slavonian Mountains (Psunj, Papuk and Krndija) and are today disconformably overlain by much more subordinate Mesozoic sedimentary rocks. Rocks of both age groups were penetrated by numerous oil wells (Pamić 1986) beneath a thick sedimentary fill of the South Pannonian Basin. Predominant Paleozoic rocks are represented by regionally metamorphosed sequences, migmatites and granitoides (Pamić et al. 2000). Regionally metamorphosed sequences are composed mainly of paragneisses, mica schists, and subordinate interlayers of orthoamphibolites and marbles in their medium grade parts, and greenschists, phyllites, and chloritoid schists in their lower grade parts.

Migmatites originated from the highest-grade paragneisses and mica schists of the surrounding regionally metamorphosed sequences. The petrography of the migmatites is the same as the petrography of the adjacent S-granite plutons.

S-type granites make up the cores of the Mts. Moslavačka gora and Papuk plutons. These rocks crystallized from magmas originating from partial melting of

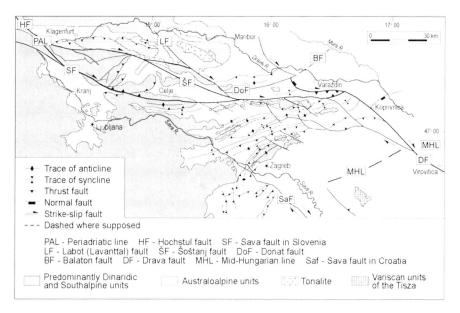

Fig. 12.5 Structural map of the contact zone between the Alps, Dinarides and Pannonian Basin (Tomljenović and Csontos 2001)

the surrounding regionally metamorphosed rocks, but also from the metasediments included in the underlying continental crust (Pamić et al. 1996). The Slavonian granitoides are of Variscan age (Pamić et al. 1988). I-type granitoids occur as small bodies only in higher-grade parts of the metamorphic sequences of the Slavonian Mountains. The heterogeneous association of igneous rocks is of oceanic crust origin, suggesting that they might have been generated along an active Paleotethyan margin, subduction zone (Pamić et al. 1996).

Field relations and petrological-geochemical data including radiometric ages provide evidence that regionally metamorphosed sequences, migmatites and granitoids represent a single geological-petrological entity of Variscan origin. The regionally metamorphosed sequences originated from a Silurian-Devonian sedimentary complex interlayered with tholeiitic basalts along the active Paleotethyan margin together with I-type granitoids of subduction-related plutonism (Pamić et al. 1996).

Mesozoic formations are of more limited extension both on the surface and in the Pannonian basement. 50–40 Ma ago, in the Eocene, the South Tisza Unit come into contact with the Dinarides. (The Pannonian Basin did not yet exist at that time.) In the Oligocene (35–25 Ma ago) along the contact of the South Tisza Unit and the just elevated Northern Dinarides major fault movements produced elongated separate basins with saline to fresh water sedimentation accompanied locally by strong volcanic activity. Those sediments principally encountered in oil wells point to the initial formation of the Pannonian Basin (Pamić 1999). In the early

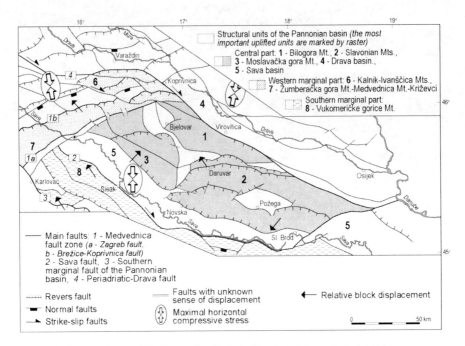

Fig. 12.6 Structural map of the Pannonian Basin in Croatia (Prelogović et al. 1998)

Miocene (19–18 Ma ago) important changes led to the formation of the Pannonian Basin (Royden et al. 1983). The changes have been initiated by the uplifting of the upper mantle to the shallow depth of 25 km. Extensional tectonics in the southeast Pannonian Basin started between the Oligocene and the Early Miocene. Three main stages in the structural development of the Croatian part of the Pannonian Basin are differentiated:

- The Oligocene and the Early Miocene characterized by extensional tectonics;
- The Early and Middle Miocene with major extensional processes and
- The Pliocene and Quaternary with prevailing transpression (Prelogović et al. 1998).

In the main extensional phase during the Early and Middle Miocene, the Sava and Drava Basins and minor sub-basins took shape. The extension-related depositional cycle ended in Pontian times and a new tectonic phase began in the Pliocene. A transpressional wrench-fault model is established for the Pliocene to recent evolution of the SW Pannonian basin, which is regarded as a neotectonic period (Fig. 12.6). Thick alluvial deposits accumulated in the major valleys of the Sava and Drava Rivers.

The recently studied mechanisms of earthquakes in Northwestern Croatia after 1908 consistently indicate predominantly compressional tectonics with reverse faulting in the central part versus strike-slip motions in the western and eastern sectors (Herak et al. 2009).

Table 12.1 Hypsometry of Croatia (Bognar 1996a)

Elevation above sea level	Landforms	%
<200 m	Lowlands	53.42%
200–499 m	Hill regions	25.61%
500–1,500 m	Low to medium-height mountains	20.82%
>1,500 m	High mountains	0.15%

12.3 Geomorphological Regions

As far as exogenic processes are concerned, Croatia belongs to the temperate fluvio-erosional zone, dominated by fluvial processes in the north and karst and fluviokarst processes in the south. A complex tectonic evolution is coupled with the influence of different exogenic processes (slope, fluvial, abrasional, glacial, periglacial, karst, fluviokarst, and aeolian processes) in the various zones of elevation (Table 12.1).

Owing to its geomorphological and geostructural position and to its form and size Croatia is divided into three main geomorphological regions: the Dinaric mountain system, the Adriatic, and the Pannonian Basin, which are further subdivided into several macro-geomorphological regions (Fig. 12.7):

1. *Panonnian Basin*

    1.1. East Croatian Plain with Upper Podravina
    1.2. Slavonian block mountains with Požega basin and Sava River valley
    1.3. Basin of NW Croatia
    1.4. Mountainous-basin region of NW Croatia

2. *Dinaric Mountain Belt*

    2.1. Mountainous Croatia
    2.2. Istrian peninsula with Kvarner coastal region and archipelago
    2.3. NW Dalmatia with archipelago
    2.4. Central Dalmatia with archipelago
    2.5. Southern Dalmatia with archipelago

3. *Relief of the Adriatic Sea Bottom*

    3.1. Adriatic shelf (part)
    3.2. Central Adriatic bank
    3.3. Southern Adriatic Basin

The further subdivisions into meso, sub, and microregions (Bognar 2001) is based on morphostructural, morphogenetical, lithological, and orographical conditions. For the identification of geomorphological regions, morpholithogenic factors were evaluated individually and integrally. Basically, every region was identified by the homogeneity and similarity of particular conditions. When delimiting mountain regions and the sea-bed of the Adriatic basin, structural-geomorphological and

Fig. 12.7 Geomorphological units in Croatia (*mega* and *macroregion*s – Bognar 2001)

morphoevolutionary conditions were decisive, and for lowland regions, the morphogenetic and lithological conditions were the most important. In certain cases it is necessary to use the criterion of spatial connections.

On the basis of the above mentioned principles seven types of geomorphological regions are distinguished in Croatia: *mountain, hill, plateau, basin, valley, lowland, insular, and submarine* regions. Three *types of mountain* macro and mesoregions were identified:

1. Remobilized block (faulted-folded and faulted-folded-imbricated) mountain ranges and massifs of Slavonia and basin-and-range regions of North-Western Croatia;
2. Mountain ranges and massifs and mountain groups of folded-faulted-thrusted structures of the Dinaric mountain system;
3. Mountain ranges, mountain ridges and mountain groups of folded-faulted-imbricated structures of the Outer Dinarides geotectonical zone (Fig. 12.8).

Different types of *hill regions* formed by derasional-erosional and derasional-corrosional processes during the Tertiary-Quaternary in clastic sediments and carbonate rocks on neotectonically elevated structures. *Plateaus* are of polygenetic

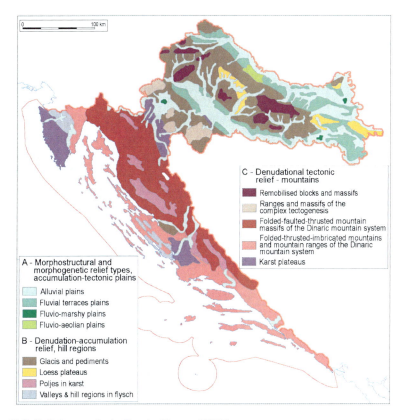

Fig. 12.8 Relief categories in Croatia (Bognar 1995b)

origin, or they are also formed in thick loess accumulations associated with neotectonically elevated block structures. The plateaus on carbonates form meso and subregions in the mountain system of Dinarides, while loess plateaus occur in the Pannonian Basin. As the expression of complex geotectonical structure and geomorphic evolution, as well as of differential geotectonical movements in the mountain system of the Dinarides and in the Pannonian Basin, smaller *basins* have been formed. They are mostly intermontane basins, which are, on the basis of their morphostructural and morphogenetic development, included into larger mesoregions. *Valleys* represent separate sub and microregions. *Lowland* geomorphological regions can be subdivided into three basic types: (1) fluvial floodplains and low-level terrace plains, (2) fluvial-marshy plains, and (3) fluvio-aeolian plains (Fig. 12.8). The *insular relief* of the northeast part of the Adriatic Sea maritime zone geotectonically belongs to the zone of the Outer Dinarides with prevailing folded-faulted-imbricated geological structure. As a rule, particular mesoregions are identified on the principle of the homogeneity of morpholithogenic conditions, and of spatial connections: the islands of the Kvarner Bay, North, Central and South Dalmatia. *Submarine types* of geomorphological regions have been delineated primarily on the principle of morphostructural homogeneity: the macroregions of the North-Adriatic Shelf, the Central Adriatic Bank, and the South Adriatic Basin.

About two-thirds of Croatian territory have been geomorphologically mapped to various scales. The Geomorphological map of the Republic of Croatia was published in the framework of the geomorphological mapping of Yugoslavia (Bognar and Blazek 1992). The new geomorphological map of Croatia at 1:500,000 scale is recently published (Bognar and Pahernik 2011). Modern geomorphologic mapping relies on a geomorphological database of 1:100,000 scale.

12.4 Karst and Fluviokarst Research

Karst and fluviokarst are particularly important relief types since they cover almost 43.7% of Croatian territory (Figs. 12.7 and 12.9; Table 12.2) (the submerged karst of the Adriatic Sea bottom excluded). It consists of two spatial units: The larger classic Dinaric karst belt (2 on Fig. 12.7) is formed of thick Mesozoic and Paleogene carbonates (Fig. 12.2). This is the area with typical karst hydrology and diverse karst phenomena. All types of corrosion features such as rillenkarren, covered karren, wall karren, meandering karren, fissures and network karren, exhumed karren, solution pans, pot-like karren, root karren, debris karren, karst tables, karren wells, karren fields, and surf karren can be found out (Perica et al. 1999–2001). The karren in central and lower parts of the Dinaric Mountains are by the rule larger than their higher-located counterparts. A good example of this can be found on the southwest flank of the Velebit Mountains.

Dolines as diagnostic karst features are particularly numerous in the Dinaric part of Croatia. In the geodatabase 349,324 dolines are recorded. The density is highest in the Gorski kotar area (242 dolines km^2). Dolines can appear individually, as is the case on the littoral slopes of the Dinarides and on different islands, but usually they form doline fields and long rows along contacts, on fractures, joints and smaller faults. Near mountain summits they appear as doline fields. The extremely high density makes some parts look like a cockpit karst. Mountain tops are also characterized by megadolines, more than 100 m deep (e.g., at the Rožanski kukovi in the Velebit Mountain, Biokovo). The elevated relief of the Velebit (but also the Risnjak and Biokovo) Mountains was glaciated during the last glacial maximum so the pre-existing karst depressions have been transformed by ice (Fig. 12.9) into glacial cirques and megadolines (Bognar et al. 1991; Bognar and Faivre 2006). The final melting of glaciers also left imprints in caves (Bočić et al. 2008). Today, during winter, dolines accumulate snow and ice on their bottoms with thermal inversions that explains the persistence of karstification throughout the year. At present, periglacial and nival processes (Perica et al. 2002) also strongly influence geomorphic evolution.

A digital database of karst depressions in Croatia has been recently set up and 161 large depressions have been distinguished (Pahernik et al. 2010). *Poljes* are typical landscape features in karst terrain with extensive flat bottoms used as arable land, with springs, ponors, estavelas, and sinking streams. The evolution of these large polygenetic depressions are determined by geological structure, karst,

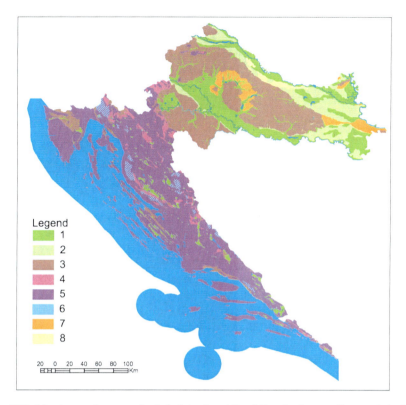

Fig. 12.9 Morphogenetic types of relief of the Republic of Croatia, Source: Geomorphological map of Croatia, 1:500,000 (Bognar and Pahernik 2011). *1* Fluvial and denudational-accumulational relief, *2* Terraces plains *3* Fluvial-denudational relief, *4* Fluviokarst relief, *5* Karst, *6* Glacial and periglacial, *7* Piping relief, *8* Aeolian relief

Table 12.2 Morphogenetic types of relief of the Republic of Croatia

Morphogenetic categories	km²	%
Accumulational (fluvial and denudational) relief	15,771	27.9
Fluvio-denudational relief	14,056	24.9
Piping relief	1,626	2.9
Fluviokarst	1,772	3.1
Karst	22,130	39.1
Glacial and periglacial relief developed on karst	846	1.5
Eolian relief	337	0.6
Total	56,538[a]	100

Source: Geomorphological map of Croatia at the scale 1: 500,000 (Bognar and Pahernik 2011)
[a]Total area obtained by digitizing the 1:500,000 map of the Republic of Croatia

fluvial, and slope processes. Their dominant morphological characteristic is the elongated Dinaric axis (northwest–southeast); they are several times longer than wide. Their surfaces are covered with deposits of mostly Quaternary and Neogene age. Despite regular inundations caused by difference between large inflow and limited ponor capacities, poljes are well-populated areas. They are surrounded by rocky hills and mountains covered with forests and grasslands important in traditional karst economy. The largest among them is Ličko polje (465 km^2). The largest stream is Lika river (78 km long – the longest sinking river in Croatia and the second longest in Europe), which flows towards NW to the Lipovo polje, where it sinks in the 2,240 m long Markov ponor of the Velebit Mountain. It flows further underground towards springs and Adriatic submarine springs between Sv. Juraj and Jablanac.

Karst plateaus are, together with karst poljes, the most extensive landforms in the Dinaric karst. They are characterized by polygenetic and polyphase origin. At a small scale, karst plateaus are large flat areas with very low relative relief. The Dinaric karst plateaus are sometimes isolated plateaus on the margins of the Dinaric karst. In the second case they can be almost regarded as a part of the polje floor, and in the third case, they appear as small tectonically dismembered and uplifted plateaus in mountainous areas. In the Croatian part of Dinaric karst there are three major karst plateaus: the Istrian, Karlovac (Una-Korana), and North Dalmatian plateaus, together ca 4,000 km^2 in area (7% of Croatian territory and 14% of total Croatian karst area). Numerous investigations explored the origin of karst plateaus (e.g., Pavičić 1908; Roglić 1951, 1957; Herak 1986; Bahun 1990; Bognar 1994, 2006). Three groups of theories were proposed: erosional, corrosional, and abrasional theories. Their genesis can be placed between the Mesozoic and the Pleistocene. Recent research is conducted in the area of Slunj plateau (Bočić 2009), on the extensive Karlovac (Una-Korana) karst plateau expanding between the Dinaric System in the southwest and in the Pannonian Basin in the northeast. The oldest rocks are Permian sandstones, and most of the area is built of the Mesozoic platform carbonate rocks, in places transgressively covered by Miocene, Pliocene, and Quaternary lacustrine and alluvial deposits. Main structures and faults stretch in the Dinaric direction. Periods of intensive exogenic processes principally occurred during and after orogenic phases interrupted by transgressions. In the post-Miocene period, the geomorphic evolution was significantly influenced by the denudation of Neogene clastic sediments and by the gradual exhumation of the carbonate bedrock. During that process, the karst relief area increased at the expense of the fluvio-denudational relief (Fig. 12.10). It resulted in development of numerous karst forms (dolines, grikes, uvalas), but also of karstified remnants of the surface paleodrainage network (dry and blind valleys). In those conditions a number of, mainly horizontal, caves developed (Bočić 2003a, b).

The Dinaric karst is rich in underground karst features (Bočić and Kuhta 2004). There are over 9000 caves and shafts. The longest caves (four of them longer than 6 km) are mostly found in the Inner karst belt, particularly between Kordun and Ogulin-Plaški depression. Caves form under vadose, epiphreatic, and shallow phreatic conditions and therefore branched multilevel caves are the most common type. The longest one is the Đulin ponor–Medvedica cave system (16.4 km long) developed by action of the Dobra sinking river. The deepest shafts occur in thick Mesozoic limestones and Paleogene Jelar breccia sediments in high karst belt of the Dinarides, especially in the Velebit Mountain (the 1421 m deep Lukina jama–Trojama system shaft and the recently discovered, 561 m deep and 20 km long, Kita Gaćešina–Draženova puhaljka system).

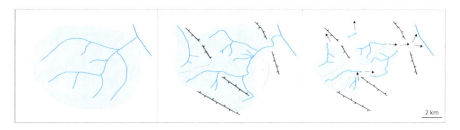

Fig. 12.10 Denudation of the clastic cap-rock with a gradual exhumation of the carbonate bedrock and disintegration of the surface paleodrainage network in the area of Slunj karst plateau (Bočić 2009)

A smaller part of Croatian karst is insular karst, scattered in the Pannonian Basin (Medvednica, Ravna gora, Papuk), developed in Mesozoic and thinner Neogene carbonates (1 on Figs. 12.7 and 12.9). It is characterized by smaller depth of karstification (mostly up to 100 m). Main surface forms are small and shallow dolines, caves, and shafts (mostly up to 50 m long and deep). The single largest cave is the 7.1 km long Veternica cave (in the Medvednica Mountain) formed by the action of former stream at the transgressive contact of Upper Triassic dolomite and dolomitic limestone with Neogene Lithotamnium limestone (Buzjak 2002).

12.4.1 Structural-Geomorphological Research in Karst Terrains

The relation between the geological structure and the landscape has always been an important subject in geomorphology, but, in the last few decades, interest in this topic has further increased. Recent papers show that for the explanation of karstification, analyses of the geometry of fissures network and of the orientation of the stress that has affected the massif are necessary. This means that the reconstruction of tectonic history is inevitable for the study of karstification.

Due to this strong relationship between karst and geological structure different karst landforms have been used in the structural geomorphological analysis. The most often analyzed forms are dolines, uvalas, karst poljes, and caves, followed by the analyses of crests and drainage networks. Various methods have been applied in the Velebit Mountains (Faivre 2000).

Recent *morphostructural analyses* are focused on dolines and on the relations between surface and underground karst features using GIS and combining geomorphological and speleological research. The significant prevalence of carbonate rocks (limestones and, to a lesser extent, dolomites) affected by tectonic movements enables the formation of typical karst landscape. Tectonic movements are crucial to provide the necessary relief (potential energy) for karstification. The application of geomorphometric techniques to karst features, particularly to dolines, began long ago and is still an important tool in karst geomorphology. The use of GIS allows ever-refined morphometric analyses that were applied in the study of Croatian karst by Mihljević (1994, 1996), Faivre and Reiffsteck (1999a), Pahernik (2000), and Faivre and Pahernik (2007).

Recent research in the Croatian karst shows that the spatial distribution of dolines is closely related to the recent deformation of the area (Faivre and Reiffsteck 1999a, b, 2002). In the karst domain, dolines can be used as a sensitive indicator of tectonic activity. The highest density of dolines is observed in the central parts of structural blocks (Faivre and Pahernik 2007), while in the main fault zones they occur rather sparsely. Analyzing their distribution according to the distance from major faults, it is found that close to the faults dolines show (near) clustered distribution, while with increasing distance, in the central parts of the blocks, where density as a rule increases, the distribution approaches random (Faivre and Pahernik 2007). This confirms that the spatial distribution of dolines seems to be of distinctive type, taking into account their position according to major faults.

The close relationship between the distribution and type of landforms and tectonic forces has been also observed studying the shape of ridge crests. For example, by the change in stress during the latest tectonically active stage, the rotation of morphostructures started, followed by gradual arc bending of the Velebit Mountains lineament. The rotation is followed by transport and stuffing of rock masses mainly towards the southeast. This is reflected in the increasing heights of mountain ridges in this very direction (Mihljević 1995). As rotations have been prograding, a differentiation between morphostructural units occurred. The consequences are arc-shaped faults and the related arc-shaped mountain ridges. The kinematic model of counterclockwise rotation of structures is recognized as the fundamental cause of arc-shaped mountain ridges, characteristic of the entire Velebit Mountains (Faivre 2007). The different tectonic regimes acting on the studied area are one of the main causes of the spatial distribution of landforms in the present-day landscape.

Another aspect of the close relationship between karst and tectonics can be demonstrated again on the example of the Velebit Mountain, which is extremely rich in caves. The mountain summits are characterized by numerous extremely deep *pits*. From the 20 deepest pits in Croatia, 12 of them are in the Velebit area, and 9 of them in the northern Velebit. The formation of pits is greatly predisposed with tectonic and hydrogeological properties resulting from prolonged karstification also influenced by past glacial processes.

12.4.2 Exogenic Processes in Karst Terrains

The measurements of the karst denudation rates, using the method of limestone tablets, have intensified since 2007. Tablets are set in the soil and in the soil/bedrock contact at numerous sites like Žumberak, Velebit Mountain, Gorski kotar, Krka river valley, Cres, and Vis islands. Karst denudation is also measured in submerged cave passages in phreatic conditions like in the karst springs of Zagorska Mrežnica, Cetina, and Slunjčica rivers. Earlier research in the Kapela and Velebit Mountains

(Pahernik 1998; Perica 1998) shows positive correlation between surface corrosion intensity and amount of precipitation but negative correlation with air temperature above the tree line. In the Bjelolasica area Pahernik (1998) measured 35–40% higher intensity for subcutaneous corrosion (up to 7.28 mm 10^{-3} year^{-1}) compared with surface corrosion (up to 6.74 mm 10^{-3} year^{-1}).

The geoecological research of cave and *doline microclimates* has become particularly intensive and shows that doline microclimate is conditioned by geographical position, altitude, and morphology as well as by local vegetation cover. According to yet unpublished findings, microclimate modifies karst and ecological processes as well as the flora in the dolines, especially in mountain areas with distinct temperature inversion. Soil temperature affects soil biochemical processes and, through CO_2 production, the rate of subcutaneous solution. Microclimate measurements, floristic, and pedogeographical investigations are carried out in four major dolines in the Risnjak, Kapela, Velebit, and Žumberak Mountains. In a 1-year long measurements in the Balinovac doline (Velebit Mountain) mean annual air temperature varies from 6.06°C at the doline bottom (at 1,434 m elevation) to 7.32°C at the top of the slope of southern exposure (1,513 m). Air temperature inversion typical of the mountain dolines and reflected in vegetation was recorded. Relative air humidity is higher on north-exposed slopes (84–86%), as expected; however, this is insignificant compared to southern slopes (80–81%). It has been established that 149 plant taxa grow in the studied part of the Balinovac doline, in three main soil types: haplic Cambisol, colluvic, rhodic molic, umbric Leptosol, calcaric and leptic, caltic Luvisol, abruptic, skeletic, clayic. Based on the indirect gradient analysis of ecological parameters, it has been determined that habitats in half-shade with arid soils poor in nutrients and humus, in which plants of the mountainous belt mostly grow, dominate the southern slopes. Shady habitats prevail on the northern slope with more humid soils, richer in nutrients and humus, where plants of the sub-alpine belt mostly grow. According to the indirect gradient analysis, on the doline bottom transitional habitats occur.

The research of *cave microclimate* (measured in caves of different climatic zones: Cfa, Cfb, and Df according to Köppen) shows that microclimate depends on the geographical position, altitude, influence of the sea, characteristics of the entrance zone (its exposure, morphology, vegetation cover), number and elevation of entrances (controls air circulation through passages), and cave hydrology. As the cave microclimate is a very important factor for cave management, microclimate has been studied in 50 caves of different karst regions in Croatia (Buzjak and Paar 2009). The measurements were performed with classical mechanical bimetal thermohygrographs and data loggers. If cave microclimate is compared to surface climate, lower air temperatures in the warm periods and higher air temperatures in the colder periods are observed. Relative humidity is high (95–100%). All cave microclimate parameters show smaller diurnal, monthly, and seasonal oscillations compared to the surface. Near the entrance of the cave the oscillations of all measured parameters are the greatest due to the surface atmosphere and hydrosphere influences. Further back from the entrance the surface effect quickly decreases and all amplitudes are reduced (Buzjak 2007). Air temperature decrease with depth depends on passage morphology and hydrology, and possibly on the geothermal effect.

The highest mean annual cave air temperatures were recorded in the Mediterranean part of Croatia (Cres, Lošinj, Krk and Rab Islands; climate type Cfa) – up to 16°C. In the caves of insular karst of continental Croatia (climate type Cfb) it is up to 10°C. The coldest cave microclimates are in the highlands and mountains of Lika and Gorski kotar area – depending on altitude ranging from 0°C to 8°C. Where mean annual temperature is close to 0°C and in descending passages that capture heavier cold air, permanent or seasonal deposits of snow are found that has fallen from the surface or ice forms by the freezing of dripping water. In these frozen zones there are very intensive periglacial processes indicated by the high rates of the mechanical destruction of fractured rocks and larger amounts of cryogenic debris.

12.5 Croatian Shoreline Evolution During the Late Holocene

The major part of the Croatian coastal area is formed in Cretaceous carbonates (Fig. 12.2) accumulated on the ancient Adriatic carbonate platform. Palaeogene limestones, calcareous breccias, and flysch are also widespread. Deposits of carbonates and flysch are for the most part irregularly covered by relatively thin Quaternary deposits, mainly of alluvial and colluvial origin but also loess, terra rossa, and other types of soils and paleosols.

The recent shape of the carbonate rocky coast is primarily a consequence of submergence of karst relief due to the relative sea-level rise. The coastal zone is part of the Outer Dinarides with recent geodynamic activity determined by a northwest motion of the Adriatic microplate and overthrusting effects along the Southern Alps. The shape of the coast strongly depends on *lithology*. According to the prevalence of limestones and dolomites erosional rocky coast predominate. Accumulational coasts are primarily linked with river valleys and torrents. Several river mouths existed as large and deep sea bays up to historical times. The rivers that originate in the flysch area are today completely buried due to the high sediment input supplied by water flows, e.g., Mirna river (Faivre et al. 2011a) and Raša river (Benac 1992). These valleys were formed by river erosion cutting into the carbonate bedrock during the Pleistocene and early Holocene, in the times when the sea level was much lower than today (Šegota 1982; Surić et al. 2005). During the last marine transgression the sea reached far into the mainland and inundated valleys. Under postglacial conditions or rapidly rising sea level, the dominant geomorphic process shifts from erosion to deposition. Finer sediments were deposited in newly flooded bays and estuaries. Decelerating sea level rise at about 6,000 BP created favorable conditions for intense sedimentation and for a shifting coastline in the inverse direction. Due to the coastal progradation, rivers like the Mirna, Rječina, Dubračina, Cetina, and Neretva belong to coasts in advancements. Rivers that pass only through karst areas, often in canyons, represents today deep sea bays (e.g., estuary of Krka River) because karst areas do not contribute significantly to the river load (Janeković et al. 1995).

At the end of numerous drowned torrent valleys beaches are formed. Occasional flows bring pebbles and cobbles that form *gravel beaches* (Faivre et al. 2011b). Fine-grained beaches are of much lesser extent. Human impact on the natural coastline is very strongly expressed today. Changes of the beach equilibrium induced by human activities are observed on numerous beaches. The beaches formed of the fan material of the drowned torrent karst valleys could be affected due to interventions along the watercourses that reduce the natural recharge of beach material. On the other hand manmade constructions along the coast also supply surplus of material that enlarges natural beaches (Rajčić et al. 2010).

The effect of *abrasion* (wave erosion) is best expressed along the external row of Croatian islands. In the areas of less resistant rocks abrasional notches and cliffs develop. The pebble beaches at the foot of the cliffs are usually not wider than 10 m (Juračić et al. 2009) due to the relatively short period of wave action on the coast, less than 6,000 years. Most of the steep slopes are structurally predisposed like scarps along the Kornat, Dugi, or Hvar islands.

Working on the 2,000-year evolution of the Croatian coast since 1999, the geomorphological and archaeological markers have been correlated systematically (Fouache et al. 2000, 2005; Faivre and Fouache 2003; Faivre et al. 2010) all along the coast. Tidal notches were used as the best indicator of sea level change.

Three main different sections can be distinguished along the Croatian coast. The northern section is characterized by widespread, well developed and well-preserved *submerged notches* between −0.45 m and −0.7 m below the present mean sea level (Fouache et al. 2000; Benac et al. 2004). The existence of notches points to the relative sea-level stabilization that, according to recent investigations, occurred around 1,500 AD (Faivre et al. 2011a). In the central section no notch has been identified until now, but archaeological remnants point to at least 1.5 m submergence during the last 2,000 years (Florido et al. 2011). The southern section has sporadically developed submerged notches and small sections with recent notches (Faivre et al. 2010). As the archaeological sea-level markers are situated below the predicted sea-level curve (Lambeck et al. 2004) on the north, central, and central-south area the influence of tectonic subsidence is discerned.

12.6 Fluvial and Fluvio-Denudational Relief

Hydrogeological structure and geographical position divide the Republic of Croatia into two parts: Continental (Pannonian) Croatia with prevailing surface outflow, and Dinaric (karst) Croatia with a specific hydromorphology. The hydrogeological characteristics have greater influence on fluvial outflow and valley network development than the hydrometereological conditions.

Croatia has the longest NE Adriatic coast, but most streams flow towards and *drainage basins* (62%) belonging to the Black Sea, and only (38%) to the Adriatic Basin (Roglić 1988). Such an *asymmetry* is conditioned by the marginal position of the Adriatic-Black-Sea watershed in Mountainous Croatia, but is also the reflection

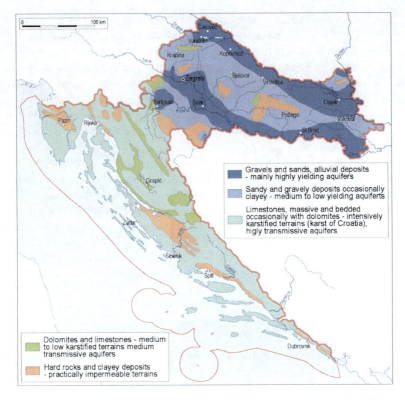

Fig. 12.11 Hydrogeological map of Croatia (Modified after Majer and Prelogović 1993)

of geographical-geological characteristics. Carbonate rocks, mostly limestones, and karst landscapes dominate the Dinarides (40.6% of Croatian territory) and control drainage (Fig. 12.11). Coastal mountain chains have the highest precipitation, but most often water disappears underground because of fractured carbonate complexes. Therefore, it is difficult to *define the watershed* between the Adriatic and Black-Sea Basins. Complex relations of underground water circulation through the karst are characteristic for almost all drainage basins (Roglić 1988). There is a marked difference between precipitation and water flowing through karst passages, the latter being often much higher. The hydrological basin is, therefore, much larger, accordingly significantly more water inflows by the underground streams through fissure systems. Sinking streams (the Lika River is the longest of them), as well as submarine springs are characteristic for karst areas.

Continental or Pannonian Croatia, especially its NW part, has the greatest concentration of surface water and the highest drainage density. The Sava River is the hydrogeographical axis draining the largest part of the area. By density and almost radial distribution of its tributaries and their valleys, it defines hydromorphology (Riđanović 1983).

The Pannonian Basin is a prevailingly lowland subsidence area dominated by *large aluvial fans*, wide *floodplains*, and fluvial *terrace* plains, which are often additionally elevated by thick deposits of *loess* and loess-like sediments (Bognar 1985a). All tributaries of the Sava River have formed composite valleys. Almost in all floodplains one can distinguish a higher and a lower level associated with enduring flooding and low water, respectively. Rivers have no marked valleys, so one can speak only about fluvial plains. Their properties are attributed to neotectonics (changes of tectonic stress) and to recent tectonics (especially on the changes of river mechanism), as well as to regulation and land reclamation works since the eighteenth century (Bognar 1985a, 1995a, 2011). As a consequence of changes of tectonic movements, river terraces are not continuous. Younger Pleistocene and Holocene terraces prevail. There are no older terraces due to subsidence. The exception is the upstream part of the Drava River (from Koprivnica towards the Slovenian border), where even four Pleistocene terraces and one old Holocene terrace have been identified. Recent subsidence created the fluvial-marshy plains of the Kopački Rit (Bognar 1985b).

In the framework of the Dinaric mountain system, valleys of complex development prevail (alternating gorges and intermountain basins), which can be explained by neotectonic movements and heterogeneous carbonate (limestones and dolomites) deposits. Springs of the allogenous streams are located in the mountain hinterland, on impermeable rocks. The streams make their way through the limestone-dolomite belt owing to their abundance of water. Formation of canyons is characteristic for the areas of pure limestone (the Čikola and Krka Rivers, partly the valleys of the Zrmanja and Cetina Rivers, etc.). Floodplains are very narrow or even absent in gorges and canyons. Valley formation is highly influenced by corrosion, consequently a great number of valleys have fluviokarst characteristics (Roglić 1988; Bognar in press). River terraces are fragmentary and, by their development, most often connected with basins and karst denudational plains. The river mouths into the Adriatic Sea are mostly of the *rias* type. Deltas are rare but an exception is the *delta* of the Neretva River, formed in several (younger Pleistocene and Holocene) development stages.

12.7 Slope Processes and Landforms

Scattered mountains, hill regions, pediments (glacis), and loess plateaus in the Pannonian Basin on the Tertiary and Quaternary clastic deposits are areas where intensive slope (sheet wash, solifluction, gullying, landslide, rockslide, and rockfall), pseudokarst and fluvial processes are active. As a rule, on sandstones and microtectonically fragmented Palaeozoic rock complexes of more inclined block mountains, gullying, slope-wash, and rockfall prevail, while on finer clastites (with higher clay content) and leached loess, sheet wash, solifluction and landslides dominate (Fig. 12.2). Landslides are the most destructive, especially those of the stratified, rotational, or step-like types. *Hill regions* are characterized by

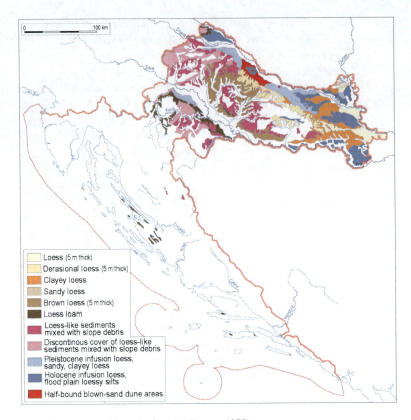

Fig. 12.12 Distribution of loess in Croatia (Bognar 1979)

greater dynamics and a relatively great impact of tectonics on their geomorphic evolution. The uplifted parts of the Pannonian Basin have been transformed by fluvial and colluvial processes, as well as by aeolian accumulation of loess, loess-like sediments and sands. *Pediments* (glacis) are gently sloping foothills directly connected to the scattered mountains in the Pannonian Basin and characterized by the alternation of parallel transversal ridges and fluvial valleys (Bognar 1980). They have been formed in marine and limnic sediments, which periclinally lean on the geologic structure of the basement. During the Pliocene, the Tertiary basement was cut in by slope and fluvial activities (parallel retreat of slopes) and was inclined (2–4°) towards lower fluvial plains. Correlative sediments of prevailingly colluvial origin from the neighboring mountains covered the entire Tertiary basement forming the pediment (glacis). During warmer and more humid interglacials, glacis were dissected into gentle hills. On the contrary, during colder and drier glaciations, a complete glacis was formed again. Consequently, glacis development was interrupted by dissection to resume again during glaciation. In the Quaternary, younger tectonic movements also had a relatively great impact on glacis formation. *Loess plateaus* represent a denudational-accumulational type of morphostructures. They have developed by pseoudokarst processes (corrosion–piping) on the loess plateaus of eastern Croatia with typical loess and by slope-fluviodenudational processes on brown noncarbonated loess, which form accumulation concave covers on more or less planated, relatively less elevated tectonic blocks of subsidence structures.

The Outer Dinarides are characterized by relatively more prominent younger tectonic movements (at 1,000–2,000 m elevation), high inclinations and relative relief, as well as by more intensive processes of mechanic weathering and regelation on dominantly carbonate base (limestones and dolomites). There rockslide (Bognar 1983, 1996b) and rockfall, gulling, sheet wash, and, in the highest parts, gelisolifluction prevail and traces of the Pleistocene glaciation (Risnjak, Velebit, Dinara, and Biokovo) are found. Dry valleys and hill regions in flysch, as well as tectonized carbonate breccias and diapiric elevations are characterized with appearance of gulling, sheet wash, sliding (block and blanket landslides, landslide streams) and rock-fall (Fig. 12.2). Pediments were formed at the foot of the most important mountains (Velebit, Dinara, Kapela, and other mountains). They represent the traces of slope processes influenced by neotectonics during the Pliocene and part of the Quaternary. These are gentle slopes deformed and fragmented by younger tectonic movements (Bognar 1992, 1994, 2006).

12.8 Polygenetic Relief

The present erosional surface on the Dinarides is found at various levels in Croatia (South-Istrian, North-Dalmatian, Zadvarje, Lika), erosional surface on Northern Velebit, and also in nearby countries (e.g., Brotnjo-Dubrava, on the mountain massifs of Čvrsnica, Bjelašnica and Treskavica in Bosnia and Herzegovina and Durmitor in Montenegro) consequently, almost from the sea-level to 1,000 m and even above 2,000 m elevation. They represent the traces of one, once larger, complete, and older planation surface. During the neotectonic stage, this surface was intensively disarranged and fragmented, and its preserved sections were transported to various hypsometric positions. Some evidence of this model can be deduced by the observation of regionally extensive bauxites, treated as correlative weathering sediments in tropical-equatorial climatic conditions. The existence of such a climatic regime has never been proposed for the Cainozoic; consequently, this model describes a unique Mesozoic planated surface (Bognar 1994, 2006).

Erosional surfaces can be observed on the majority of the mountain ranges and massifs of folded-thrust and folded-imbricated structures in the Outer Dinarides mountain zone. These are characterized by a specific step-like outline of their transversal profiles. This scenario supports their complex geomorphologic evolution marked by the alternation of uplift stages with those of relative tectonic inactivity favorable to denudation-planation processes (weathering, marine erosion, pediplanation, pedimentation, corrosion, etc.) and the relations proved by the existence of two clearly marked stepped pediments. Exceptions are the fragments of older erosional levels, i.e., the surface, which are, in view of their morphological position, connected with the mountain summits, then with the parts of the intermountaine basins bottoms, the marginal parts of the Outer Dinarides towards the Adriatic Basin, and with the Pannonian Basin (Čečura and Bognar 1989).

12.9 Studies on Loess

Loess and loess-like sediment cover approximately 20,000 km^2, 35.7% of the total area of the Republic of Croatia. The thickness of loess and loess-like deposits varies considerably from 0.5 to 60 m or more in the inland areas and between 0.5 and 60 m along the coast. The grain size distribution of loess and loess-like sediments exhibit a well-marked zonation, from the east to the west in the inland areas. The chemical properties of loess and loess-like materials in Croatia are essentially quite similar in all analyzed samples. Differences may be perceived in terms of CaCO$_3$ content, which shows some regional variety in amount. The mineralogical composition of loess and loess-like sediments is fairly uniform. Local deviations are due to differences in area of origin of the silt substance. The distribution and mineralogical analyses of loess and loess-like sediments indicate that the sources of silt should be found out in the neighboring localities and also in the Alps and Dinaric Alps. The distribution and mineralogical composition of loess and loess-like materials may be correlated with the *alluvial deposits* of the Drava, Sava and Danube Rivers and their tributaries. The silt material of aeolian origin was accumulated mostly by north and northeastern winds. Conditions for aeolian activity were optimal during the summer seasons of the glacial stage coldest peaks. The same conclusion applies to the accumulation of silt that was later transformed into deluvial or fluvial loess or into loess-like material (Bognar 1977, 1979) (Fig. 12.12).

The loess of Eastern Croatia contains mostly intercalated layers of *fossil* *steppe soils* or forest-steppe soils. A gradual change is apparent towards the west. As a result of higher humidity during the Pleistocene the western areas are noted for the development of predominantly brown soils, pseudogleyed loess and pseudogley soils. The modifying influence of relief on pedogenic processes should always be kept in mind. As a result, even within relatively short distances we may find that forest soils and brown soils may occur simultaneously in nearly profiles. This is particularly true in the case of loess profiles on the Ðakovo Plateau. Fundamental changes in the lithological properties of loess and fossil soils occur on the plateau within 15–20 km due to the influence of the Slavonian fault-block mountains. Fossil soils were also discovered in the loess profiles and in loess-like materials of fluvial origin. These are mostly marshy and hydromorphic soils, that is, intrazonal formations. Due to frequent changes in the condition of sedimentation (migration of riverbeds, etc.) the development of these hydromorphic soils was usually short-lived. Their role is of no particular significance in the chronostratigraphical subdivisions of the strata under investigation.

The loess plateaus of Eastern Croatia with domination of typical carbonate loess are marked by formation of *pseudokarst forms* (loess dolines, loess wells, loess chasms, loess pyramids, loess circuses, surducs, and loess steep sections along the banks of the Danube River and on the Island of Susak) by corrosion-piping processes. While loess dolines have developed primarily by corrosion and piping, colluvial processes, abrasion and fluvial erosion also contributed to the development of other landforms (Bognar 2003).

12.10 Aeolian Relief

Besides the typical loess of aeolian origin, predominantly accumulated in loess plateaus (and the Island of Susak in the Kvarner Bay), as well as the loess which raised the older fluvial terraces in the Croatian Pannonian area (Bognar 1980), the forms developed by deposition of aeolian sand, especially in the Drava River plain (Fig. 12.12) were also described. Sands redeposited by wind from fluvial sediments formed deflational depressions with *sand covers* and immovable dunes (garmadas – because of the proximity to underground water) during the late Pleistocene and early Holocene. While the Pleistocene dunes were formed by the northeastern and northern winds, the early Holocene ones were accumulated by northwestern winds. In the anthropogenic period, pastoral nomads cut trees in the Croatian Military Border (from the sixteenth to seventeenth century) and consequently caused a revival of the aeolian activity and development of the youngest deflation and accumulation forms (by northwest winds) in the Drava River plain. In the second half of the nineteenth century, wind action was interrupted by afforestation on aeolian sands (Đurđevački pijesci). In the Dinaric region the aeolian sand deposits (sand covers and dunes) have been discovered in the Krbavsko Polje (Laundonov Gaj), as well as in the southeastern part of the Island of Mljet, where they date from the Pleistocene and are of fluvial origin. They were redeposited by northeastern and northern winds from the paleofans in the present-day channel of Mljet (Bognar et al. 1992).

12.11 Conclusions

Recent geomorphological studies principally focus on the Karst Dinarides. The main targets were the creation of a digital geodatabase of typical karst landforms like dolines and poljes. Particular concern was also paid to different aspects related to caves. Another important research topic is the study of sea-level change with the aim to discern the tectonic activity from the isostatic-eustatic component.

Geomorphological investigations are concentrated around two main research projects: the *Geomorphological mapping of the Republic of Croatia* and the *Geomorphological and geoecological investigations of the karst area in Croatia* financed by the Ministry of Science, Education and Sport of the Republic of Croatia.

The application of the findings of geomorphological research in regional planning and ecological aspects is increasing day by day, so if this tendency continues in the future, more rapid progress is to be expected.

Acknowledgements The research presented in the paper was supported by the Ministry of Science, Education and Sport of Republic of Croatia (Geomorphological and geoecological investigation of the karst area in Croatia No. 119-1191306-1305 and Geomorphological mapping of the Republic of Croatia No. 119-0000000-1299). We would like to thank Ivica Rendulić from the Geography Department at the Zagreb University for the drawing of maps.

References

Altiner Y (1999) Analytical surface deformation theory for detection of the earth's crust movements. Springer Verlag, Berlin/Heidelberg/New York

Bahun S (1990) Stupnjevi razvoja zaravni u Dinarskom kršu (Stages of formation of the plains in the Dinaric karst). KRS Jugosl 12(6):147–158

Benac Č (1992) Recentni geomorfološki procesi i oblici u području Riječkog zaljeva (Recent geomorphological processes and landforms in the area of the Rijeka Bay). Geografski glasnik 54:1–18

Benac Č, Juračić M, Bakran-Petricioli T (2004) Submerged tidal notches in the Rijeka Bay NE Adriatic Sea: indicators of relative sea-level change and of recent tectonic movements. Mar Geol 212(1–4):21–33

Bočić N (2003a) Relation between karst and fluviokarst relief on the Slunj plateau (Croatia). Acta Carsologica 32(2):137–146

Bočić N (2003b) Basic morphogenetic characteristics of caves in the Grabovac valley (Slunj, Croatia). Geoadria 8(1):5–16

Bočić N (2009) Geomorphological characteristics of the Slunj Plateau. Unpublished dissertation, University of Zagreb, Faculty of Science, Department of Geography, Zagreb, 270 p

Bočić N, Kuhta M (2004) Neki geomorfološki aspekti speleoloških istraživanja u kršu Hrvatske (Some geomorphological aspects of speleological explorations in Croatia). 3. hrv. geogr. kongres (Proceedings), HGD, Zagreb, pp 220–233

Bočić N, Faivre S, Kovačić M (2008) Underground Karst features and Fluvioglacial sediments, example from "Ledenica u Štirovači" cave on the Velebit Mt. (Croatia). In: Zupan Hajna N, Mihevc A (eds) 16th international karstological school "Classical karst" Karst sediments, guide book & abstracts. Karst Research Institute ZRC SAZU, Postojna, pp 62–66

Bognar A (1977) Genesis of loess and loess-like sediments of fluvial origin in Baranja, JAZU, Centar za znanstveni rad, Vinkovci, Special issues IV, pp 164–188

Bognar A (1979) Distribution, properties and types of loess and loess-like sediments in Croatia. Acta Geol Acad Sci Hung Budapest 22(1–4):267–286

Bognar A (1980) Tipovi reljefa kontinentalnog dijela Hrvatske (Types of relief in the continental part of Croatia). Zbornik 30. Obljetnice Geografskog društva Hrvatske, Zagreb

Bognar A (1983) Tipovi klizišta u Hrvatskoj. Zbornik jugoslavenskog simpozija "Privredne nepogode u Jugoslaviji". SGDJ, Ljubljana

Bognar A (1985a) Basic geomorphological problems of the Drava river plain in SR Croatia. Geographical papers, Zagreb 6:99–105

Bognar A (1985b) Basic geomorphologic characteristic of tke Kopač swamp. Geografski glasnik 46:5–20

Bognar A (1992) Pediments of the South Velebit mountain range. Geografski glasnik 54:19–32

Bognar A (1994) Some basic characteristics of the evolution of the pediments in the mountain zone of the Outer Dinarides. Geografski glasnik 56:21–32

Bognar A (1995a) Regulations and his influences on the geomorphologic development of the Drava and Danube rivers in Croatia. I. In: Croatian conference of the waters, Dubrovnik, pp 449–462

Bognar A (1995b) Fizičko-geografske pretpostavke regionalnog razvoja Hrvatske. Zbornik radova I. Hrvatskog geografskog kongresa, Hrvatsko geografsko društvo, Zagreb, pp 51–65

Bognar A (1996a) Croatia – the land and natural features. GeoJournal 38(4):407–416

Bognar A (1996b) Tipovi klizišta u Republici Hrvatskoj i Republici Bosni i Hercegovini – geomorfološki i geoekološki aspekti. Acta Geogr Croat 31:27–39

Bognar A (2000) Geomorphology and its development in Croatia. In: Proceedings II. Croatian geographical congress, Zagreb

Bognar A (2001) Geomorfološka regionalizacija Hrvatske (Geomorphological regionalization of Croatia). Acta Geogr Croat 34:7–29

Bognar A (2003) Geomorphological conditions of the Susak island. In: Bognar A, Schweitzer F, Szöőr Gy (eds) Environmental reconstruction of a loess island in the Adriatic. Geographical Research Institute, Hungarian Academy of Sciences, Budapest, pp 31–44

Bognar A (2006) The Geomorphological evolution of the Dinarides. In: Adria 2006, International Geological Congress on the Adriatic area. Field trip Guide, University of Urbino, Italy/ University of Zagreb, Croatia/Geological Survey of Croatia/Ina–naftaplin Croatia, pp 23–26

Bognar A (2011) Geomorphologic characteristics of the Pazinčica valley. Investigations of the natural characteristics of the area of Rijeka. Natural History Museum Rijeka (in print)

Bognar A (in print) Geomorphology of the Neretva river delta. Hrvatski geografski glasnik

Bognar A, Blazek I (1992) Geomorphological map of the Republic of Croatia in the scale 1:500,000 as a part of the geomorphological map of Jugoslavija 1:500,000. Geokarta, Beograd

Bognar A, Faivre S (2006) Geomorphological traces of the younger Pleistocene glaciation in the central part of the Velebit mountains. Hrvatski geografski glasnik 68(2):19–30

Bognar A, Pahernik M (2011) Geomorphologic map of Republic of Croatia in the scale 1:500,000. Cro. Geom. Soc. and Department of Geography, Faculty of Science, Zagreb

Bognar A, Faivre S, Pavelić J (1991) Tragovi oledbe na Sjevernom Velebitu. Geografski glasnik 53:27–39

Bognar A, Klein V, Mesić I, Culiberg M, Bogunović M, Sarkotić-Šlat M, Horvatinčić N (1992) Quarternary sands at south-eastern part of the Mljet Island. In: Proceedings of the international symposium "geomorphology and sea", Department of Geography, Faculty of Science, University of Zagreb, Zagreb, pp 99–110

Buzjak N (2002) Speleološke pojave u Parku prirode "Žumberak–Samoborsko gorje" (Speleological features of "Žumberak–Samoborsko gorje" Nature park). Geoadria 7(1):31–49

Buzjak N (2007) Mikroklima kao komponenta geoekološkog vrednovanja spilja – primjer Spilje u Belejskoj komunadi (Belej, otok Cres) (Microclimate as a component of geoecological evaluation of caves – example of the Cave in Belejska komunada, Belej, Cres island). Geoadria 12(2):97–110

Buzjak N, Paar D (2009) Recent cave microclimate research in Croatia. In: Abstracts,17th international Karstological School "Classical Karst", Postojna, p 53

Croatian Geological Society (1990) Geological map of the Republic of Croatia. Zagreb.

Čečura Ž, Bognar A (1989) Osnovna problematika morfogeneze denudacijskih i akumulacijskih nivoa u zavali Livanjskog polja (Basic problems of the origin of denudational and accumulational surfaces of Livanjsko polje). Hrvastki geografski glasnik 51:21–28

Cigrovski-Detelić B (1988) The use of GPS measurements and geotectonic information in the analysis of the CRODYN geodynamic network. Unpublished PhD thesis, University of Zagreb, 145 p

Dewey JF, Pitman WC, Ryan WBF, Bonnin J (1973) Plate tectonics and the evolution of the Alpine system. Geol Soc Am Bull 84:265–283

Faivre S (2000) Landforms and tectonics of the Velebit mountain range (External Dinarides, Croatia). PhD thesis, Zagreb University, Faculty of Science, Department of Geography, Croatia/ Université Blaise Pascal, Clermont-Ferrand II, France, 360 p

Faivre S (2007) Analyses of the Velebit mountain ridge crests. Hrvastki geografski glasnik 69(2):21–40

Faivre S, Fouache E (2003) Some tectonic influences on the Croatian shoreline evolution in the last 2000 years. Z Geomorph NF 47(4):521–537

Faivre S, Pahernik M (2007) Structural influences on the spatial distribution of dolines, Island of Brač, Croatia. Z Geomorph N F 51(4):487–503

Faivre S, Reiffsteck Ph (1999a) Spatial distribution of dolines as an indicator of recent deformations on the Velebit mountain range. Géomorphol Relief Process Environ 1999(2):129–142

Faivre S, Reiffsteck Ph (1999b) Measuring strain and stress from sinkhole distribution – example of the Velebit mountain range, Dinarides, Croatia. In: Beck BF, Pettit AJ, Herring JG (eds) Proceedings of the seventh multidisciplinary conference on sinkholes and the engineering and environmental impacts on karst. AA Balkema, Rotterdam/Brookfield, pp 25–30

Faivre S, Reiffsteck Ph (2002) From doline distribution to tectonic movements – example of the Velebit mountain range, Croatia. Acta Carsologica 31(3):139–154

Faivre S, Bakran-Petricioli T, Horvatinčić N (2010) Relative sea-level change during the late Holocene on the Island of Vis (Croatia), Issa harbour archaeological site. Geodinamica Acta 23(5–6):209–223

Faivre S, Fouache E, Ghilardi M, Antonioli F, Furlani S, Kovačić V (2011a) Relative sea level change in Istria (Croatia) during the last millenia. Quat Int 232:132–143

Faivre S, Pahernik M, Maradin M (2011b) The gully of Potovošća on the Island of Krk: the effects of a short-term rainfall event. Geol Croat 64(1):67–80

Florido E, Auriemma R, Faivre S, Radić Rossi I, Antonioli F, Furlani S, Spada G (2011) Istrian and Dalmatian fishtanks as sea level markers. Quat Int 232:105–113

Fouache E, Faivre S, Dufaure J-J, Kovačić V, Tassaux F (2000) New observations on the evolution of the Croatian shoreline between Poreč and Zadar over the past 2000 years. Z Geomorph NF Suppl-Bd 122:33–46

Fouache E, Faivre S, Gluščević S, Kovačić V, Tassaux F, Dufaure JJ (2005) Evolution of the Croatian shore line between Poreč and Split over the past 2000 years. Archaeol Marit Mediterr 2:116–134

Herak M (1986) Geotektonski okvir zaravni u kršu. Acta Carsologica 14–15:11–15

Herak M (1991) Dinarides – mobilistic view of the genesis and structure. Acta geologica, Academia Scientiarum et Artium, Zagreb

Herak D, Herak M, Tomljenović B (2009) Seismicity and earthquake focal mechanisms in the North-Western Croatia. Tectonophysics 465:212–220

Janeković M, Juračić M, Sondi I (1995) Sedimentacijske osobitosti rijeke Mirne (Istra, Hrvatska). In: Proceedings of the first Croatian Geological Congress, vol 1, Opatija, 18–21 Oct 1995, pp 225–227

Juračić M, Benac Č, Pikelj K, Ilić S (2009) Comparison of the vulnerability of limestone (karst) and siliciclastic coasts (example from the Kvarner area, NE Adriatic, Croatia). Geomorphology 107(1–2):90–99

Lambeck K, Antonioli F, Purcell A, Silenzi S (2004) Sea-level change along the Italian coast for the past 10,000 yr. Quat Sci Rev 23:1567–1598

Majer D, Prelogović E (1993) Hydrogeology and seismotectonics. In: Klemenčić M (ed) A concise atlas of the Republic of Croatia the Republic of Bosnia and Hercegovina. The Miroslav Krleža Lexicographical Institute, Zagreb, 159 p

Mihljević D (1994) Analysis of spatial characteristics in distribution of sink-holes, as an geomorphological indicator of recent deformations of geological structures. Acta Geogr Croat 29:29–36

Mihljević D (1995) Relief reflection of structural reshaping during the recent tectonically active stage, in the north-western part of the Outer Dinarides mountain range. Acta Geogr Croat 30:5–16

Mihljević D (1996) Strukturno-geomorfološke značajke i morfotektonski model razvoja gorskog hrpta Učke (Structural-geomorphological characteristics and morphotectonic evolution model of the Mt. Učka ridge (Croatia)). Geografski Glasnik 58:33–50

Pahernik M (1998) Utjecaj klime i reljefa na intenzitet površinske korozije karbonata gorske skupine Velike Kapele (Climate and relief influence on the carbonate surface corrosion intensity in the Velika Kapela mountain group). Acta Geogr Croat 33:47–56

Pahernik M (2000) Prostorni raspored gustoća ponikava SZ dijela Velike Kapele – rezultati računalne analize susjedstva (Spatial distribution and density of dolinas on the NW part of Velika Kapela from GIS based buffer analysis). Geoadria 5:105–120

Pahernik M, Buzjak N, Faivre S, Bočić N (2010) Spatial disposition and morphological properties of bigger karst depression in Croatia. In: Mihevc A, Prelovšek M, Zupan Hajna N (eds) 18th International Karstological School "Classical Karst", Postojna, Slovenija, 14–18 June 2010. Abstracts. Karst Research Institute, Postojna, p 54

Pamić J (1986) Metamorfiti temeljnog gorja Panonskog bazena u Savko-dravskom međurječju na osnovi podataka naftnih bušotina, XI Kong.geo. Tara, Yugoslavia 2:259–272

Pamić J (1999) Kristalinska podloga južnih dijelova Panonskog bazna temeljena na površinskim i bušotinskim podacima. Nafta 50(9):291–310

Pamić J, Lanphere M, McKee K (1988) Radiometric ages of metamorphic and associated igneous rocks of the Slavonian Mountains in southern part of the Pannonian Basin. Acta Geol 18:13–39

Pamić J, Lanphere M, McKee K (1996) Hercynian I-type and S-type granitoids from the Slavonian Mountains (Southern Pannonian Basin, Croatia). Neues Jahrb Mineral Abh Stuttgart 171:155–186

Pamić J, Gušić I, Jelaska V (1998) Geodynamic evolution of the Central Dinarides. Tectonophysics 297:251–268

Pamić J, Gušić I, Jelaska V (2000) Basic geological features of the Dinarides and South Tisia. Vijesti hrvatskog geološkog društva 37(2):9–18

Pavičić S (1908) Pojava abrazije na istočnom izdanku Plješevice. Glasnik Hrvatskog naravoslovnog društva 20:103–113

Perica D (1998) Geomorfologija krša Velebita (Karst geomorphology of the Velebit Mountains). PhD dissertation, PMF, Geografski odsjek, Zagreb, 220 p

Perica D, Marjanac T, Mrak I (1999–2001) Vrste grižina i njihov nastanak na području Velebita. Acta Geogr Croat 34:31–58

Perica D, Lozić S, Mrak I (2002) Periglacijalni reljef na području Velebita (Periglacial relief in the Velebit area). Geoadria 7(1):5–29

Prelogović E, Cvijanović D, Aljinović B, Kranjec V, Skoko D, Blašković I, Zagorac Z (1982) Seismotectonic activity along the coastal area of Yugoslavia. Geol Vjesn 35:195–207

Prelogović E, Saftić B, Kuk V, Velić J, Dragaš M, Lučić D (1998) Tectonic activity in the Croatian part of the Pannonian basin. Tectonophysics 297:283–293

Prelogović E, Pribičević B, Ivković Ž, Dragičević I, Buljan R, Tomljenović B (2004) Recent structural fabric of the Dinarides and tectonically active zones important for petroleum-geological exploration. Nafta 4:155–161

Rajčić ST, Faivre S, Buzjak N (2010) Preoblikovanje žala na području Medića i Mimica od kraja šezdesetih godina 20. stoljeća do danas. Hrvatski geografski glasnik 72(2):27–48

Riđanović J (1983) Hydrogeografical characteristics of Croatia. Geografski Glasnik 45:33–42

Roglić J (1951) Unsko-koranska zaravan i Plitvička jezera – geomorfološka promatranja. Geografski glasnik 13:49–66

Roglić J (1957) Zaravni u vapnencima. Geografski glasnik 19:103–134

Roglić J (1988) Hydrogeography of SR Croatia. In: Encyclopedia of Yugoslavia, vol 5, 2nd edn. ugoslavenski leksikografski zavod "Miroslav Krleža", Beograd, pp 173–179

Royden L, Horváth F, Rumpler J (1983) Evolution of the Pannonian Basin: tectonics. Tectonics 2:63–90

Šegota T (1982) Razina mora i vertikalno gibanje dna Jadranskog mora od ris-virmskog interglacijala do danas. Geološki vjesnik 35:1–290

Statistical Yearbook of the Republic of Croatia (2010) Editor-in-Chief: Ljiljana Ostroški, Croatian Bureau of Statistics, Zagreb, p 40

Surić M, Juračić M, Horvatinčić N, Bronić IK (2005) Late Pleistocene-Holocene sea level rise and the pattern of coastal karst inundation:records from submerged speleothems along the Eastern Adriatic Coast (Croatia). Mar Geol 214:163–175

Tomljenović B (2000) Zagorje – Mid-Transdanubian Zone. PANCARDI Conference. Vijesti hrvatskog geološkog društva 37(2): 27–33

Tomljenović B, Csontos L (2001) Neogene-quaternary structures in the border zone between Alps, Dinarides and Pannonian Basin (Hrvatsko zagorje and Karlovac Basins, Croatia). Int J Earth Sci (Geologische Rundschau) 90:560–578

Tomljenović B, Pamić J (1998) Mixed Alpine and Dinaridic lithologies of Mt, Medvednica located along the Zagreb-Zemplen Line. PANCARDI conference, Europ. Sci. Union, Przegląd Geologiczny 119, Kraków

Velić I, Vlahović I, Tišljar J, Matičec D (2006) Introduction to the stratigraphy of the Karst (Outer) Dinarides. In: Menichetti M, Mencucci D (eds) Adria 2006 international geological congress on the Adriatic area. Field trip guide. Instituto di Scienze della Terra, Universitá di Urbino, Urbino, pp 15–17

Vlahović I, Tišljar J, Velić I, Matičec D (2002) The Karst Dinarides are composed of relics of a single mesozoic platform: facts and consequences. Geol Croat 55(2):171–183

Vlahović I, Tišljar J, Velić I, Matičec D (2005) Evolution of the Adriatic carbonate platform: palaeogeography, main events and depositional dynamics. Palaeogeogr Palaeoecol 220:333–360

Chapter 13
Recent Landform Evolution in Serbia

Radmila Pavlović, Jelena Ćalić, Predrag Djurović, Branislav Trivić, and Igor Jemcov

Abstract Serbia can be divided into two major geographical units: the Vojvodina Plain and the southern hills and mountains. In the plain fluvial and aeolian processes and landforms on loess are typical, while south of the Sava and Danube rivers, sheet and gully erosion, mass movements, fluvial, karstic, and fluvio-karstic landform evolution are most influential in shaping the present topography. Among the now inactive geomorphic processes tectonic movements, volcanic activity, coastal, glacial, and aeolian evolution have left behind traces still detectable in modern topography. Some examples of rapid topographic change resulting in the most spectacular landforms (landslides, rockfalls, natural bridges, and rock pinnacles) are presented and described in detail from various regions of Serbia.

Keywords Deluvial-proluvial processes • Landslides • Rockfall • Fluvial processes • Karst • Glacial features • Natural bridges • Rock pinnacles

---

R. Pavlović • B. Trivić • I. Jemcov
Faculty of Mining and Geology, University of Belgrade, Djušina 7,
11000 Belgrade, Serbia
e-mail: prada@rgf.bg.ac.rs; btrivic@rgf.bg.ac.rs; jemcov@gmail.com

J. Ćalić (✉)
Geographical Institute "Jovan Cvijić", Serbian Academy of Sciences and Arts,
Djure Jakšića 9, 11000 Belgrade, Serbia
e-mail: j.calic@gi.sanu.ac.rs

P. Djurović
Faculty of Geography, University of Belgrade, Studentski trg 3,
11000 Belgrade, Serbia
e-mail: geodjura@eunet.rs

13.1 Geographical Position

Serbia is a country of southeastern Europe of 88,361 km² area, situated within the geographical coordinates 41°53′ and 46°10′N, and 18°10′ and 23°00′E. Starting from the north, in clockwise direction, Serbia borders Hungary, Romania, Bulgaria, Macedonia, Albania, Montenegro, Bosnia and Herzegovina, and Croatia (Fig. 13.1).

Morphologically, Serbia is characterized by two major geographical units. North of the Sava and Danube Rivers, there is a vast *plain of Vojvodina*, which is a southern part of the Pannonian Plain. Within this unit, there are two mountains of horst structure: Fruška Gora on the southwest and Vršački Breg on the southeast.

South of the Sava and Danube Rivers, the relief is very dissected and belongs to the *hilly-mountainous relief* type. In north-south direction, there are remnants of former abrasion surfaces of the Miocene Pannonian Sea, positioned in step-like succession, and high mountains further to the south. Generally, the main morphological traits are the valleys of the Velika (Big) Morava and Južna (South) Morava Rivers, running in north to south oriented tectonic graben in central Serbia, and the

Fig. 13.1 Geographic position of Serbia, with locations of the landforms presented. 1, Jovac landslide; 2, Zavoj landslide; 3, Joc rockfall; 4, Vratna natural bridges; 5, Djavolja Varoš rock pinnacles

Table 13.1 Hypsometric zones in Serbia (Mladenović 1984)

Hypsometric zones (m)	Area (km²)	Share (%)
0–200	32,540	36.83
200–500	21,829	24.70
500–1,000	24,105	27.28
1,000–1,500	8,468	9.59
1,500–2,000	1,213	1.37
>2,000	206	0.23
Total	88,361	100

high mountains to the west and east of that graben. Next to the valleys of the Velika Morava and Južna Morava Rivers, the mountains and enclosed basins are parts of the tectonically fragmented Serbian–Macedonian Massif, which divides the Carpatho–Balkanides to the east from the Dinarides to the west.

The *Carpathians* extend from Romania to the eastern part of Serbia, where they have a general N–S strike. Further towards the south, they are connected to the Balkanides, which turn to SE and E direction towards Bulgaria. On the west, the Serbian-Macedonian Massif is gradually replaced by the *Vardar zone*, a morphological-tectonic unit rather specific by its geological and geomorphological evolution. In the southern part of Serbia, the Vardar zone, with its ranges and depressions, has an overall north-south, while northwards from central Serbia it turns to southeast–northwest, almost east–west, direction.

The southwestern part of Serbia belongs to the *Dinarides*, and is practically the southeastern margin of this tectonic unit. Lithologically, this part of the Dinarides differs from their main part, but tectonic and other characteristics remain unchanged. The Dinarides in Serbia have a mountainous character, and the river courses are affected by tectonic and neotectonic influences.

Generally, mountains rise from the north (Avala, south of Belgrade, 511 m) towards the south (Šar planina 2,640 m, Besna Kobila 1,922 m). The whole mountainous part of Serbia is characterized by mildly undulating depressions of elongated or elliptical shape with sediment remnants of Tertiary lakes. Their orientation is generally linked to younger tectonic structures.

A hypsometric analysis based on the 1:200,000 scale topographical map (Mladenović 1984 – Table 13.1) shows the dissection of relief in Serbia. Excluding the plain of Vojvodina, the overwhelming part of Serbia lies within the hypsometric belt of 200–1,000 m (over 82%), i.e., are hilly and hilly-mountainous areas.

13.2 Climate and Hydrology

Geomorphic processes and, consequently, recent relief characteristics, depend not only on lithological composition and neotectonic pattern, but also on climatic conditions. Mean annual air temperature in lowlands (up to 300 m above sea level) is about

11°C; in the 300–500 m zone about 10°C, and it further decreases with elevation. Three *climatic types* are present in Serbia (Ducić and Radovanović 2005):

(a) *Continental climate* in the Pannonian Basin and Peri-Pannonian region, especially in the valleys of the Velika Morava, Kolubara, and Drina Rivers, and on the east, in the region of Negotinska Krajina, which is the western margin of the Lower Danube Basin;
(b) *Moderate continental climate* in western, central and southeastern Serbia, with the elements of *mountainous* climate at higher elevations;
(c) *Modified Mediterranean climate* in southern Serbia, in the Metohija Basin and its surroundings.

The area of continental climate is characterized by hot summers and cold winters, with high temperature range. Average annual precipitation ranges from 550 mm (536 mm in Kikinda, in the Banat region) to 700 mm, locally up to 800 mm (Fruška Gora). The main precipitation maximum occurs in May or June. The dominant vegetation is in accordance with climatic characteristics – grasslands and beech, spruce, and oak woods.

In the area of moderate continental climate, seasons are well expressed, but with lower temperature ranges than in the continental climate. At higher elevations, mountainous climate prevails. Average annual precipitation in western Serbia ranges from 800 mm to 1,000 mm; in central Serbia, average values are about 700 mm (with the extremely dry area in the Prokuplje and Leskovac Basins, below 600 mm); while in eastern and southeastern Serbia, precipitation increases from 600 mm to more than 1,000 mm (on Stara Planina). The highest mountains host alpine and boreal vegetation, with predominant coniferous species.

The area of modified Mediterranean climate, in the Metohija Basin, is influenced by warm air masses that protrude from the Adriatic Sea through the valley of the Beli (White) Drim River. The winters are mild and wet, while the summers are warm and dry. Precipitation is variable, ranging from 600 mm at lower elevations to over 1,000 mm on the highest mountains (Prokletije), with a pronounced winter maximum.

Serbia has a well-developed drainage network of uneven density. Considering a longer period of recent morphological evolution, the amount of precipitation has been relatively uniform over the whole area. Consequently, drainage density directly depends on lithological composition, tectonic pattern, and neotectonic characteristics. It is much higher in the areas built of cemented clastic, metamorphic and igneous rocks than over carbonates. Besides, the drainage network is less developed in areas with thick Quaternary sequences (Vojvodina, Kosovo and Metohija), and those with young poorly cemented clastic rocks (the margins of the Pannonian and other Tertiary basins).

Apart from the Danube and the Sava, whose basins are in their major part located in neighboring countries, the greatest rivers of Serbia are Velika Morava, Južna Morava, Zapadna (West) Morava, Drina, Kolubara, Mlava, and Timok, which together drain more than 90% of Serbian territory south of the Sava and the Danube.

13.3 Geology and Geomorphological Units

Geological conditions, including rock quality (lithology), the spatial pattern of structural elements (tectonics) and age relationships (chronostratigraphy), are another control of geomorphic evolution. The type and intensity of process and the resulting landforms depend on lithology and tectonics. On the other hand, the basic morphostructures of recent topography are the result of geotectonic movements in the Earth's crust during its geological evolution. The major geotectonic units and their lithological composition are shown in Fig. 13.2. All rock masses on the map and in the following text are grouped by their erodibility, i.e., their impact on geomorphic processes.

The territory of Serbia consists of several geotectonic and thus geomorphological units: the Carpatho–Balkanids, Serbian–Macedonian Massif, Vardar zone, Dinarids, and Pannonian Basin (Fig. 13.2, top inset map; Dimitrijević 1997).

The *Carpatho–Balkanides (CB)* in eastern Serbia spread from the Danube and the Romanian border towards the south, to the valley of the Timok River under the name Carpathians, and further towards the southeast, to the Bulgarian border under the name Balkanides. The boundary between the Carpatho–Balkanides and the Serbian–Macedonian Massif is clearly defined and marked by the nappes zone. Moving from west to east, the Carpatho–Balkanides start with the *Suprageticum zone* (Dimitrijević 1997). This zone is built of green schists, Devonian flysch, Permian red sandstones, Mesozoic limestones, and Jurassic flysch. It has been thrust over the *Geticum zone*, a large nappe complex, building the major part of eastern Serbia. The Geticum zone is subdivided into several zones of partly different geological evolution, and, therefore, some authors consider them as separate nappes. The Geticum zone consists of Proterozoic–Cambrian schists, Ordovician and Silurian metamorphic rocks, Devonian flysch, Permian and Triassic red sandstones, limestones ranging in age from the Middle Triassic to the Upper Cretaceous, as well as Upper Cretaceous volcanic-sedimentary formation and different types of Neogene volcanic rocks. Limestones are widely distributed and the karstification process is advanced. Apart from karstification, the specific features of this zone are the landforms connected to Upper Cretaceous magmatism. Two klippen of this zone, Sip and Tekija, overlie the unit of the *Infrageticum*, which is composed of ultramafic rocks and gabbros with marginal schists and Lower Cretaceous flysch. The Danubicum is autochthonous in relation to the above mentioned units, and it is composed of schists, Carboniferous and Permian clastic rocks, thin strata of Jurassic carbonates, and Cretaceous flysch.

The *Serbian–Macedonian Massif (SMM)* occupies central Serbia, from Vršac on the north to the Macedonian border on the south. It is a geotectonically stable region built of Paleozoic to Tertiary schists and magmatites. Pre-Tertiary rocks are fractured by longitudinal and transversal ruptures into block structures. By their differential movements, a number of horsts (Vršački Breg, Juhor, etc.) and grabens (e.g., the Velika Morava valley) have been formed. The Serbian–Macedonian Massif consists of various types of schists of different ages and degrees of

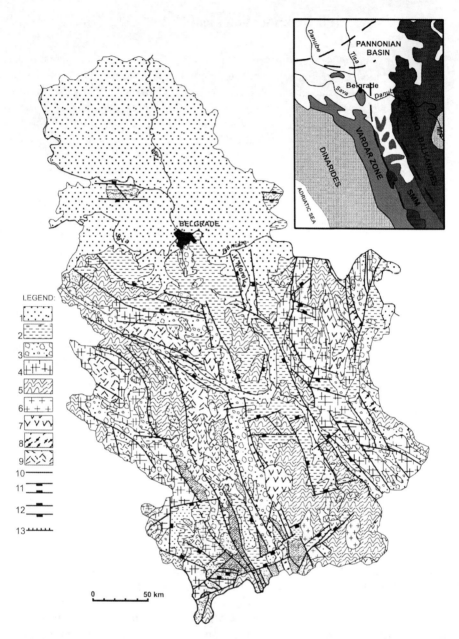

Fig. 13.2 Map of lithologically homogeneous units and regional ruptures (according to the Geological map of SFRJ 1:500,000, 1970). *Upper right corner*: draft map of geotectonic units (Dimitrijević 2002). 1, Unconsolidated Quaternary sediments of different genesis; 2, Poorly cemented Neogene clastic sediments; 3, Clastic sediments; 4, Carbonates; 5, Metamorphites; 6, Granites; 7, Volcanic rocks; 8, Basic rocks; 9, Ultramafites; 10, Fault; 11, Horst; 12, Graben; 13, Nappe

metamorphism. These rocks were intruded by granitoids. Lithology is simplified on the geological map (Fig. 13.2): metamorphites, granitoids, poorly cemented Neogene clastic sediments, and non-cemented Quaternary sediments of various origin are distinguished.

The *Vardar zone (VZ)* is the most complex belt of the Balkan Peninsula. It was a region of multiple opening and closure of an ocean basin (Dimitrijević 1997). It is located between the Serbian-Macedonian Massif on the east and the Dinarides on the southwest and west. The Vardar zone is composed of several blocks of diverse composition, origin, and geological evolution. From the border with Macedonia, up to the central part of Serbia, the Vardar zone stretches in north-northwestern–south-southeastern direction, and from central Serbia, it suddenly turns towards northwest and west. In the west–east direction, it can be divided to *External* (the Srem, Jadar, and Kopaonik blocks), *Central,* and *Inner subzones*.

The Vardar zone, a region of ancient oceanic sutures, is composed of peridotites and basalts (parts of an ancient oceanic crust), ophiolite mélange with ultramafites and limestone olistrostromes, flysch, and pelagic limestones. This zone is characterized by repeated and widespread Tertiary magmatism. Its landforms are still well preserved in the present-day topography.

The rocks are classified according to the degree of erodibility and type of erosion (Fig. 13.2). Two specific environments are ultramafites and ophiolite mélange (shown within the group of clastic rocks). Metamorphites are present in different parts of the Vardar Zone, and they are dominant on the boundary towards the Serbian–Macedonian Massif.

The western-southwestern parts of Serbia belong to the *Dinarides (D)*, and their boundary towards the Vardar zone is very clearly marked with high pressure and temperature metamorphism and mineral association of glaucophane-schist facies. Previously, the Vardar zone was thought to be part of the Dinarides but recently it is considered a separate unit (Dimitrijević 1997). The Dinarides are composed of several blocks of diverse geological history. Along the suture line with the Vardar zone, there is the *Drina–Ivanjica element*, to the southwest is the *Eastern Bosnia–Durmitor block* and in-between is the *Ophiolite belt*. The Drina–Ivanjica element consists of Paleozoic schists, overlain by Triassic and Cretaceous carbonates on the west, Cretaceous clastic rocks on the southeast towards the Vardar zone, while Tertiary volcanic rocks are present in its southern part. The Ophiolite belt is composed of large masses of Triassic limestones, ultramafites, shales, and cherts of the continental slope and blocks of oceanic crust embedded in the ophiolitic mélange all over the Mesozoic limestones. This belt of complex geological composition underwent a very complex geomorphic evolution. The Eastern Bosnia–Durmitor block occupies the farthest southwestern corner of Serbia. In the basement there are Paleozoic schists, Triassic clastic, volcanic and carbonate rocks, Cretaceous limestones, and flysch. Overthrusts are characteristic for this zone.

The *Pannonian Basin (PB)* is a depression between northern and southern branches of the Alpine orogene, and one of the basins of the Paratethys. Located between the Alps, Carpathians, and Dinarides, it was formed by subsidence of blocks composed of schists, granitoides, Mesozoic sediments, and ophiolites (on the boundary with the Vardar zone). The depression was occupied by the Miocene sea. The central subsidence started in the Miocene, followed with deposition of extensive sediment bodies, uplifting of some marginal blocks and forming of horst structures (Fruška Gora, Vršački Breg). The Pannonian Sea was protruding deep into the margin of the Pannonian Basin, covering the older formations of the Vardar zone, the Serbian–Macedonian Massif and the younger sediments of the Carpatho–Balkanides. The most recent geomorphic evolution of this region was controlled by geotectonic position and lithological composition of the Pannonian Basin and its marginal range.

The low hills and a plain to the east of the Carpatho–Balkanides in Serbia belong to the *Lower Danubian (Dacian) Basin*. Composed of poorly cemented Neogene clastics, it also includes some peripheral basins and grabens (ancient embayments), stretching to the foothills of the Carpatho–Balkanides (e.g., Timok graben).

13.4 History of Geomorphological Research in Serbia

Geomorphological research in Serbia and the Balkans began with *Jovan Cvijić*, who from the end of the nineteenth century until his death in 1927, published the results of his investigations on the karstic, fluvial, glacial, marine-limnic, aeolian, volcanogenic, and tectogenic relief of the Dinarides, Carpatho-Balkanides, and the adjacent areas. His detailed landform studies made the Balkan peninsula one of the best explored regions of the time. His most important works are his monographs on *karst* process, morphology, typology, hydrography, and underground karst evolution (Cvijić 1893a, 1895a, 1918, 1924–1926), based on extensive field experience in specific tectonic, lithological, and neotectonic conditions of the Carpatho–Balkanides, Dinarides and Vardar zone. Many of his papers deal with the karst relief of Serbia (Mt. Kučaj, Mt. Suva Planina, Lelić area, Pešter) and its caves (Cvijić 1891, 1893b, 1895b, c, 1912, 1914a). The first studies of *glacial* morphology on the Balkan peninsula are also connected to J. Cvijić, who studied glaciation phases and their morphological consequences in particular locations (e.g., in the Prokletije and Šar planina Mountains) (Cvijić 1899, 1900, 1903, 1913, 1914b, 1922).

The recent margin of the Pannonian Basin and its deep hinterland were interpreted by Cvijić as the remnants of the Pannonian Sea, its *abrasion* surfaces and cliffs, but also as the remnants of the lake which marked the final phase of the Pannonian regression (1909, 1910). Cvijić related the formation of abrasion surfaces on Mt. Miroč (eastern Serbia) and genesis of the Danube gorges (Đerdap, Iron Gates) to regression of the Pannonian and Dacian seas. The further subsidence of the Dacian Basin and entrenchment of the Danube have led to the formation of the Danube terraces in the region of Dunavski Ključ. *Fluvial* processes, and particularly the combined effects of endogenous forces and river action, were studied by Cvijić (1920, 1921, 1924–1926) as well.

The morphological evolution that followed the regression of the Pannonian Sea and its remnants on the territory of Serbia was studied by Cvijić's student

P.S. Jovanović (1922, 1951), who at first interpreted and supplemented Cvijić's work, but in later studies, developed a critical approach towards the marine-limnic abrasion surfaces on the Pannonian Basin margins. His approach was subsequently supported by M. Zeremski (1956), D. Petrović (1966), J. Roglić (1952) and others. Among the studies of recent fluvial relief, it is worth mentioning Jovanović (1955), who analyzed the impact of climate variations on fluvial processes, Ž. Jovičić (1960, 1966), Đ. Paunković (1953), R. Lazarević (1957, 1959), Č. Milić (1956, 1973), and the investigators of the fluvial morphology of Vojvodina, B. Bukurov (1948, 1950, 1953, 1954, 1982), B. Bulla (1937), and B. Milojević (1960). The last three authors analyzed the evolution of large rivers in Vojvodina and their drainage areas. Furthermore, their studies on *aeolian* relief and its evolution are also significant (Bulla 1938; Milojević 1948, 1949). Within the study of Carpathian evolution during the Pleistocene, Bulla (1938) mentioned the transport and redeposition of glacial material and dated the sand sheet Deliblatska Peščara in southeastern Vojvodina as post-glacial, i.e., that the sand was blown out from the Danube alluvium and deposited over the loess plains. Milojević (1949), however, links the formation of Deliblatska Peščara to the Pleistocene, when the winds winnowed the sand out of the deposits of the Banat rivers.

Among a number of papers about *loess* in Serbia as a geological substrate, conditions of its development and geological and geomorphological characteristics, the works of J. Marković-Marjanović (1949, 1953, 1954, 1965) must be stressed. They are illustrated with detailed profiles, and the presented data can be used for the interpretation of the evolution of loess regions in Vojvodina, in the areas south of the Danube and in the Velika Morava valley. The same author also studied the sand sheet of Deliblatska Peščara. From the correlation of Quaternary formations in the surroundings of Belgrade and in Vojvodina, V. Laskarev (1938, 1950, 1951) concluded that "marsh loess" was formed in a receding lake until its final disappearance, and subsequently subaerial loess was deposited in loess plateaus. Aeolian relief and loess in Serbia was also studied by D. Petrović (1976a, 1979), while D. Gorjanović–Kramberger (1921) wrote about the morphological and hydrogeological characteristics of loess in the Srem area. The role of Deliblatska Peščara as a groundwater collector and its potentials for water supply were examined by M. Košćal and Lj. Menković (1994). Recently, extensive studies of loess and paleosols in northern Serbia (Vojvodina), with the use of advanced methods, are carried out by S. Marković and co-workers (2006, 2008, 2009), providing valuable information on paleoclimatic variations and other paleoenvironmental aspects for the southern Pannonian Basin. Since the time of Cvijić, *karst* topography has attracted the attention of a large number of geographers, geologists, hydrologists, hydrogeologists, speleologists, and others. B. Milovanović (1964–1965) summarized karst evolution in the Dinarides, solved numerous problems and refuted previous erroneous views on the intensity and depth of karstification and the relation between the geological and geomorphological evolution of carbonate terrains. A series of papers on karst processes in general, on the specific features of some karst areas and their caves, were published by R. Lazarević (1978, 1996a, b), D. Petrović (1965), J. Petrović (1954, 1969, 1974), D. Gavrilović (1966, 1970, 1985), M. Zeremski (1967),

S. Milojević (1938), N. Krešić (1988), and others. The Speleological Atlas of Serbia, edited by P. Djurović (1999) comprises the data on 82 major caves in Serbia and chapters on various scientific aspects of speleology. The hydrogeological aspects of karst and its water supply or hydroenergetic potentials have been most thoroughly studied by P. Milanović (1979, 1981, 2000), B. Mijatović (1984, 1990), M. Komatina (1970–71), Z. Stevanović (1994, 2005), and I. Jemcov (2007).

The traces of Pleistocene glaciation are still preserved on the high mountains of Serbia. The effects of glaciation and both preserved and morphologically altered *glacial landforms* in the Šar planina, Suva Planina, Stara Planina, and Kopaonik Mountains were studied by R. Nikolić (1912), B. Milojević (1937), and D. Gavrilović (1976). Recent papers on glacial relief in Serbia (Menković 1985, 1988, 1994; Menković et al. 2004; Belij 1994, 1997) deal with the recent conditions of glacial processes and landforms, partially reconstruct the Pleistocene morphological evolution of the Šar planina, Koritnik, and Prokletije Mountains and study the effects of frost action (*periglacial* landform evolution). Others relate such features to the degradation of fossil glacial landforms.

The geology and topography of Serbia favors *slope processes,* mostly studied within geotechnical engineering research. Several geological and engineering-geological conferences focused on the investigations of the conditions of landslide development, methodology of research, mitigation measures, construction works in sliding areas, etc. The intensity of mechanical erosion by *fluvial action* and deluvial processes is studied by P. Manojlović and co-workers (Manojlović and Dragičević 2000; Manojlović et al. 2003), using calculations of sediment transport in river courses, in combinations with various parameters. The same author also studies *chemical erosion* processes (Manojlović 1992, 2002).

The *tectogenic relief* in Serbia was analyzed and well presented on the *Basic geological map of former Yugoslavia* (SFRJ), at a scale of 1:100,000 with memoirs. Tectonic structures with clear relief outlines, such as regional faults, horsts, grabens, depressions, and structural relief are all shown in detail. Volcanogenic relief was studied regionally, through particular forms or smaller areas related to certain phases of magmatic activity. Due to strong endogenous activity in the late Neogene, after the end of all magmatic phases in Serbia, Tertiary and older volcanic landforms were largely destroyed (partly due to intensive exogenous processes). The reconstruction of landforms is only locally possible, in the Vardar zone and the Timok volcanic area in eastern Serbia. The most important considerations on volcanogenic relief have been given, among others, by J. Cvijić (1924), M. Zeremski (1959), D. Petrović (1976b), M. Marković and Z. Pavlović (1967), R. Pavlović (1990), and M. Jovanović et al. (1973).

The project of *geomorphological mapping* of Serbia, which took place in the period 1978–2000 and was financed from the national budget, triggered a series of comprehensive studies of geomorphological characteristics, processes, temporal and spatial relations, quantitative parameters of present processes and forms. Before this project, the geomorphological characteristics had been cartographically presented on the *Overview geomorphological map of the Kingdom of Yugoslavia*

(Jovanović 1933) at the scale of 1:1,200,000, on the *Geomorphological map of Serbia* (Zeremski 1990) at 1:500,000 and on the geomorphological map of Yugoslavia (Group of authors 1992) at 1:500,000 – all were based on the analyses of topographical and geological maps, while in the project of 1978–2000 mapping was carried out sheet by sheet (field surveys at 1:50,000 and final map produced at 1:100,000). They remained unpublished, but have served as a basis for the 1:500,000 scale *Geomorphological map of Serbia* (Menković et al. 2003). It is expected that a larger scale version (1:300,000) will be published as well. The maps contain data on geological composition, the distribution of geomorphic processes, morphostructural and morphosculptural landforms, morphometric data, and landform evolution for Serbia. The mapping standards of some European countries were observed (Marković 1973).

13.5 Present-Day Geomorphic Processes and Relief Types

The present-day relief is a result of combined action of endogenous and exogenous processes during the neotectonic period that, on the present territory of Serbia, began in the early Pliocene. Since then, endogenous processes are manifested through moderate uplifts and subsidences, movements in different directions and of various intensity and had a direct impact on exogenous processes, their intensity of operation in space and time. The overview geomorphological map (Fig. 13.3) shows the most significant relief types and major landforms.

13.5.1 Weathering

Considering the geology and climate of Serbia, mechanical and chemical weathering processes of various intensity are presently active in most parts of the country – even in the areas covered with aeolian sediments of northern Serbia. Fragmented material produced by physical-chemical rock decomposition remains *in situ*; there are no changes in relief, erosion, transport or accumulation. The thickness of such material ranges from several centimeters to several tens of meters. *Weathering surfaces* are not distinguished as a relief type on the geomorphological map, but the significance of weathering is emphasized in the memoir.

In Serbia, *weathering mantle* is found on all flat and mildly inclined areas, as well as on slopes unaffected by rapid sheet wash. Thick weathering mantles occur in the Pannonian Plain of northern Serbia, on marine, lacustric, and old fluvial planation surfaces, as well as over loess plains. On steeper slopes south of the Sava and Danube, the weathering mantle is easily removed by gravity and runoff, and converted to deluvial, proluvial, and colluvial forms.

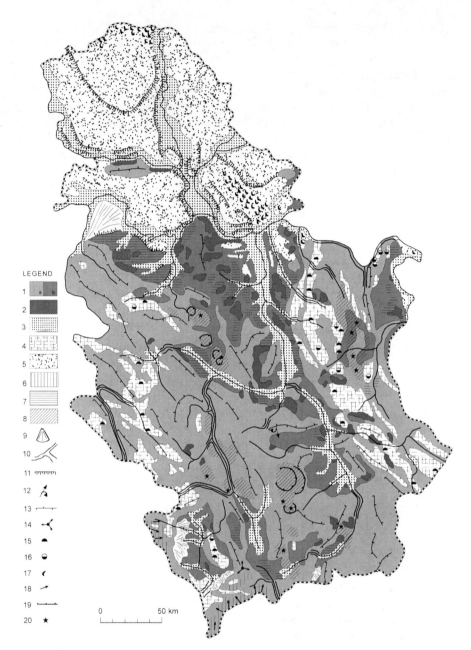

Fig. 13.3 Overview geomorphological map of Serbia (Modified after Menković et al. 2003). 1, Deluvial-proluvial relief: (*a*) deluvial, (*b*) proluvial; 2, Colluvial relief; 3, Fluvial relief, including marshes; 4, Karst relief; 5, Aeolian relief; 6, Glacial and periglacial relief; 7, Relief of marine and lacustrine origin; 8, Volcanogenic relief; 9, Alluvial fan; 10, Gorges and canyons; 11, Terrace riser; 12, River piracy; 13, Ridge; 14, Triple drainage divide; 15, Cave; 16, Vertical cave; 17, Fossil dune; 18, Direction of Pleistocene glacier movement; 19, Caldera remnants; 20, Volcanic neck

13.5.2 Deluvial-Proluvial Processes and Landforms

On non-soluble rocks temporary diffuse flows or sheet wash (deluvial processes) and temporary linear flows (proluvial processes) mostly act jointly in the same area, so their impacts are not easy to separate. Thus, both processes are often represented as a single process, especially in small-scale mapping. On the overview geomorphological map of Serbia (Fig. 13.3), two kinds of areas are distinguished within the deluvial-proluvial relief type: (a) areas with lower *sheet wash* intensity, and (b) intensive sheet wash and *gully erosion* on valley sides of hilly and mountainous areas (south of the Sava and Danube) (e.g., Manojlović and Dragićević 2000). The material is transported downhill and accumulated as deluvial-proluvial mantles or alluvial fans in areas of lower inclination. Climatic characteristics and geological composition explain the predominance of deluvial-proluvial process in recent geomorphic evolution. The intensity of the process is lower in the Mio-Pliocene basins of heterogeneous clastites (Fig. 13.3, 1a). Among all active geomorphic processes, intensive sheet wash and gully erosion cover the largest area.

Gullies, the predominant erosional features, are widely distributed over the entire area. Their lengths may even reach 1 km, while the depths range from several meters to several tens of meters (e.g., in the Kolubara River catchment – Dragićević 2007; Dragićević et al. 2008). Rare but extremely attractive erosional features are *rock pinnacles* at Đavolja Varoš, close to the town Kuršumlija in southern Serbia (see below). In the foothills of the areas prone to intensive sheet wash and gully erosion, there are vast *deluvial-proluvial mantles* and *alluvial fans* of various dimensions (from several meters to tens of kilometers).

13.5.3 Colluvial (Gravity) Processes and Landforms

Colluvial processes or mass movements play a considerable part in shaping the slopes of Serbia. A significant part of the territory is built of poorly cemented Tertiary clastic rocks very prone to weathering. With intensive precipitation, gentle slopes and moderate dips of heterogeneous clastite beds, weathering mantles of up to 10 m depth are a favorable setting for landslides. Intensive landsliding is enhanced by abundant sheet and rill erosion and the undercutting of slopes by temporary flows. *Landslides* mostly occur *in regolith*, while the landslides in bedrock are rather uncommon. In Serbia, the largest landslides (in area and depth) are formed *in Tertiary sediments*. Their depth can exceed 10 m. Some of the most significant landslides are situated in the zone of the direct impact of the Danube and Sava Rivers. Serbia is one of the countries with the highest density of landslides in Europe. In the general urban development plan of Belgrade, more than 2,000 landslides are registered in the urban area of 500 km^2. Sliding often affects weathered metamorphic rocks and pyroclastites.

The slopes with numerous active, temporarily stabilized or reclaimed landslides are often densely populated and bear significant infrastructure (highways, regional roads, railways). Apart from natural causes, human activity is also influential. During construction works, slope stability is endangered, and in certain cases, stabilized or reclaimed landslides are reactivated.

Besides landslides, other colluvial forms of Serbia are *talus cones* and *rockfalls* (rockslides), mainly on steep escarpments of carbonate rocks. Specific colluvial forms in northern Serbia occur on loess bluffs. In the literature, these forms are sometimes treated as earthfalls, and sometimes as landslides. In fact, this is *toppling* or sliding of weathered vertical loess bluffs, mostly to the impact of undercutting by large rivers, the Danube and the Tisa.

13.5.4 Fluvial Processes and Landforms

Permanent rivers often form marshes, swamps, oxbow lakes, and other similar features on valley floors and their sedimentation differs from channel accumulation. Therefore, fluvial processes and marsh processes are studied separately. On larger-scale maps they are distinguished, while on small-scale maps they appear as a single process. In the geomorphic evolution of Serbia, both fluvial and marsh processes have been of considerable importance. Permanent and intermittent streams are closely related to proluvial processes.

Fluvial processes are controlled by the amount of precipitation and geological composition. In the north, in Vojvodina, where thick sediment sequences dominate, the drainage network is sparse and permanent streams have extremely wide valley floors. The rivers are meandering and transitional towards marsh morphology. In central and southern Serbia, the river courses in insoluble rocks form valleys of normal type and a very *dense*, integrated *drainage network*. In the areas composed of soluble rocks (limestones and dolomites), surface drainage is disintegrated and, if present, runs through canyons and gorges. In such geological settings, the fluvial process is closely linked to karstification (*fluvio-karstic* landforms). Fluvial processes in Serbia also include the redeposition of till by permanent flows. In this way, *glacio-fluvial* terraces are formed in the foothills of the Šar planina and Prokletije Mountains (Belij 1994).

Fluvial and marsh processes in Serbia result in a variety of *landforms*: various types of valleys, trains of river terraces, entrenched meanders, river captures, vast alluvial plains, point bars, river islands, alluvial fans, marshes, swamps, oxbow lakes, etc. The overview geomorphological map (Fig. 13.3) only shows the most extensive and significant forms.

13.5.5 Karstic and Fluvio-Karstic Processes and Landforms

The Carpatho-Balkanides of eastern Serbia and the Dinarides of western Serbia host active karst areas. In the rocks with high $CaCO_3$ content, the karstification process is further intensified if the rocks are folded and fractured. The tectonic characteristics of the Carpatho-Balkanides and the Dinarides have enabled intensive and relatively deep-reaching karstification, producing a variety of surface and underground forms. Differential neotectonic movements have influenced karst evolution, depending on their direction, either by enhancing or by damping the process.

All karst landforms characteristic of the classical Dinaric karst morphology – karren, dolines, uvalas, and poljes – are present in Serbia, but with moderate dimensions and density. Among the various Serbian karst areas, the number, types and sizes of forms vary with microclimatic conditions, the proportion of insoluble components and local structural characteristics. Numerous *dolines* are intensively developed on extensive karst plateaus (e.g., Petrović 1965), while *uvalas* mostly have a polygenetic origin. There are few *karst poljes* (e.g., Odorovačko Polje in eastern Serbia, Peštersko Polje in southwestern Serbia), with fluvial and marsh processes in the sediment cover of their floors. On some *karst plateaus* it is possible to reconstruct the former surface drainage networks. As opposed to the classical Dinaric karst, with more or less uniform distribution of carbonates in extensive areas, karst in Serbia occurs in smaller, isolated portions, on limestone patches separated by insoluble rocks. Therefore, it abounds in features of karst and *fluviokarst* contact. Allogenic streams either flow through deep canyons and gorges, or sink in *ponors*, forming *blind*, dry, or hanging *valleys*. One of the peculiar characteristics of Serbian karst is a relatively large number of *natural bridges* (especially in eastern Serbia; see Sect. 7.3) and *intermittent springs* (both in western and eastern Serbia).

Most of the larger Serbian *caves* are found in contact karst, with strong throughputs of allogenic water and sediments (Djurović 1999). The longest explored caves are just below 10 km long (in eastern Serbia the Lazareva Pećina, on Mt. Kučaj, is 9,407 m long; while the Ušački cave system in western Serbia has a length of 6,185 m) (Krešić 1988). The vertical extension of the caves explored so far is up to 300 m (Rakin Ponor on Mt. Miroč is the deepest, −285 m). Several deep *siphonal springs* are found on the foothills of karst massifs (e.g., the spring Krupajsko Vrelo in eastern Serbia, dived to −123 m), an evidence of deep karstwater circulation.

13.6 Inactive Geomorphic Processes and Fossil Landforms

The present-day landscape in Serbia is not only a result of active exogenous processes, but an ever-changing combination of endogenous and exogenous forces. Landforms of formerly active processes are still recognizable and influence the operation of active processes. The oldest landforms in Serbia are related to endogenous processes: tectonic and igneous activities producing the morphostructural basis of the present-day relief.

The Dinaric and Carpatho-Balkanic mountain ranges and the Vardar zone were formed by *tectonic movements* (folding, uplifting, and subsidence) during the Alpine orogeny. With the Savian and Styrian orogenic movements in the late Oligocene and during the Miocene the phase of intense folding terminated in Serbia. In the early

Neogene vertical tectonic movements were activated to form *horst* (Fruška Gora and Jastrebac Mountains) and *graben* structures (e.g., graben valleys of the Velika Morava and Toplica Rivers). The youngest tectonic movements within the Pannonian Basin have formed the *smaller* Alibunar, Srem, and other *depressions*.

Volcanic activity in Serbia took place in the period from the late Cretaceous until the Pliocene with paroxysms in the late Cretaceous, the Miocene and early Pliocene. Apart from the large masses of pyroclastites and volcanic rocks, volcanic activity has left behind a number of *conical hills, volcanic necks* and *calderas*. Volcanic landforms in eastern Serbia have been considerably destroyed by exogenous processes, but those in the Vardar zone are well preserved – either entire forms or their remnants which enable reliable reconstruction (Petrović 1976b).

Marine-limnic coastal processes in Serbia ended in the late Pliocene, with final regression of the lake remnants of the Pannonian Sea on the north, the Dacian Sea on the east, and the Aegean Sea on the south. During the Pliocene, due to neotectonic uplifts in central Serbia, the seas were regressing, forming smaller lakes that finally disappeared in the late Pliocene and early Quaternary. *Abrasion surfaces* of marine or limnic origin have remained, positioned in step-like succession towards the central parts of the basins and still preserved in the relief, but very difficult to correlate because of neotectonic deformations in the Quaternary (Petrović 1966).

During the Pleistocene, *glacial processes*, i.e., combined ice, snow, and frost actions, were active in the high mountains of southern Serbia (at the elevations of 1,900–2,200 m). Snow concentration and Alpine-type glaciers started as cirques in the headwalls of watercourses above the snowline. The glaciers were moving along the former river valleys, eroding the bedrock in valley floors and sides and accumulated moraines below the snowline. On the Šar planina and Prokletije Mountains in southern Serbia, Pleistocene glacial processes produced erosional and aggradational forms (sharp peaks, arêtes, cirques, glacial valleys, terminal moraines, solifluction features) preserved to our time (Gavrilović 1976; Menković 1985; Belij 1997).

In Vojvodina and a narrow belt south of the Sava and the Danube, *aeolian processes* were predominant in the geomorphic evolution under the cold and dry Pleistocene climate (Petrović 1976a, 1979). Wind-blown dust and other particles were deposited at the bottom of the regressed Pannonian Sea, as thick loess sequences, subsequently dissected by rivers into *loess plateaus* (Gorjanović–Kramberger 1921). *River terraces* of the Danube and the Sava as well as lake *abrasion surfaces* along the margins of the Pannonian Basin, are partially *covered with loess*. Both sand sheets of northern Serbia, several kilometers across and considered to be of Holocene age, the Deliblatska Peščara and the Subotička Peščara, were still mobile some centuries ago, but today they are stabilized by human action. They are linked to southeastern (Deliblatska Peščara) and northwestern winds (Subotička Peščara), which transported and subsequently accumulated sands from the alluvial plains of large rivers. Aeolian sediments cover a considerable part of Serbia, but are exposed to heavy destruction by active fluvial and slope (deluvial, proluvial, and colluvial) processes.

13.7 Samples of Specific Landforms

In the last two centuries, endogenous activity (earthquakes) on the territory of Serbia caused death toll and material damage, but no relief change. Exogenous activity, however, considering the geological characteristics of Serbia and its present climate, favors the short-term development of new relief features. "Short-term" is to be considered in a relative sense: a rockfall happens within a few seconds, large landslide in a few days, badlands may form in a few months or decades – all just a blink in the course of geomorphic evolution. The most significant processes of *rapid topographic change* in Serbia are landsliding and proluvial processes combined with other active exogeneous (fluvial, deluvial, rockfall) processes.

13.7.1 Landslides

In Serbia most major landslides occur in poorly cemented Neogene clastic sediments and some in schists and other rocks. Construction works in Tertiary basins require particular attention and involve high additional investments in landslide control. Therefore, unstable slopes prone to landsliding are carefully studied.

The *Jovac landslide* is located in southern Serbia, between the towns of Niš and Vranje, on the right valley side of the Jovačka Reka river (Fig. 13.1, location 1). A catastrophic landslide started slowly on February 7, 1977, and intensified on February 15. In the first week of rapid movement the rate of movement was about 70 m day^{-1}, with occasional jumps of 200 m day^{-1}. The period of intensive movement lasted for more than a month. Under favorable geological conditions and slope inclination, after several months of heavy rains that completely saturated the ground, a rapid displacement of a large mass started on the top of the slope and *developed downhill* to the footslope (Petrović and Stanković 1981) – in opposite direction to the usual mechanism of regressive movement.

The landslide is situated in the area of Tertiary magmatism and affects clastic flysch-like sediments (of Upper Eocene age, E_3) that are only locally exposed. This formation is overlain by sedimentary-pyroclastic deposits (M_2) and pure pyroclastics, while the top of the slope is composed of dacites (Fig. 13.4). The sediments and pyroclastites are poorly cemented, with a thick weathered layer on the surface, subject to water infiltration. The underlying formation is considerably faulted, which increases the water capacity of the slope. In the period from November 1976 to January 1977, precipitation was 3.5 times higher than the annual average.

Water saturated rock, with increased load and weakened mechanical characteristics moved downhill. The movement extended over the entire slope and in 10 days the river was completely dammed (Fig. 13.4). In the first week of sliding, the upper parts of the sliding mass moved more than 500 m. The landslide is 3 km long and 1 km wide on average. The average depth is 50 m, total moved mass amounted to 150 million m^3 (300 million tons). The failure scarp was more than 50 m high, with inclinations between 30° and 70°. The landslide body is elongated, almost

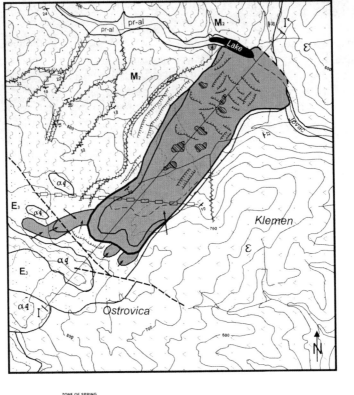

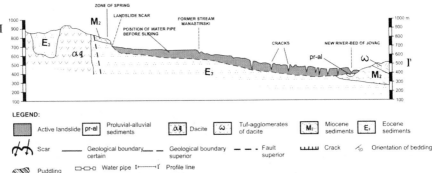

Fig. 13.4 Engineering-geological map and cross-section of the Jovac landslide (After Petrović and Stanković 1981)

rectangular, and is completely positioned *in degraded Middle Miocene sedimentary-volcanic formation*. The right edge was completely controlled by the geological boundary of this formation with pyroclastites. The Jovac landslide completely *changed* the morphology of *the slope* affected within just a month. In the soil a dense network of cracks, more than 2 m wide and more than 10 m deep, developed. Over 70 households, local roads, and agricultural facilities were destroyed, fortunately without a human toll.

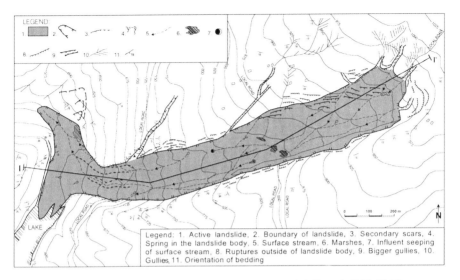

Fig. 13.5 Engineering-geological map of the Zavoj landslide (Gojgić et al. 1964–1965)

The *Zavoj landslide* is located on the right valley side of the Visočica River, on the slopes of the Stara Planina (Balkan) Mountains, close to the border with Bulgaria (Fig. 13.1, location 2). A large mass of non-cemented, degraded rocks was moving downhill to the narrow Visočica valley and dammed it with a natural rocky dam, *impounding a huge lake*. Intensive sliding started on February 23, 1963, and began to attenuate on March 16. The slope affected is composed of *Lower Triassic sandstones and shales*, well-bedded, slightly folded, with flysch-like characteristics and intensively fractured into blocks of various dimensions. Besides, the rocks are subject to strong physical-chemical weathering, with *regolith* thickness exceeding *10 m* (at the location Izvorski Do, where the sliding began). Uphill from the landslide failure, thick-bedded strongly karstified limestones overlie the Triassic clastites.

In mid-February 1963, fast snowmelt started to saturate the regolith and made it unstable. The fissure porosity of Triassic clastites enabled fast and easy penetration of water into depth. The fractured aquifer formed in karstic limestones uphill the failure front, amply fed the springs on the locality Izvorski Do, and thus hydraulically influenced the triggering of a huge landslide (Gojgić et al. 1964–1965). The swollen, saturated *kaolin clays*, formed by shale decomposition, favored sliding. In addition, human impact – uncontrolled deforestation and ill-designed irrigation systems – contributed to more intensive erosion, and at the same time, formation of the landslide.

Several years before the Zavoj landslide formation, there were minor sliding episodes in the upper part of the landslide, and some detachment of rock fragments in local gullies (Gojgić et al. 1964–1965). The first larger fissures and slope deformations at Izvorski Do (Fig. 13.5) occurred in mid-February, and in the night of February 23/24, 1963, a mudflow started to rush downslope, triggering a slide of immense dimensions. Exploiting slope and dip, the sliding mass first moved towards

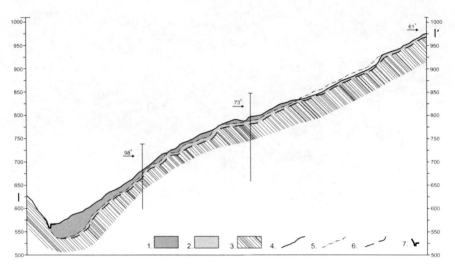

Fig. 13.6 Geological cross-section of the Zavoj landslide (Gojgić et al. 1964–1965). 1, Active landslide; 2, Zone of intensive rock weathering; 3, Stratified sandstones and shales; 4, Topographical cross-section after activation of landslide; 5, Topographical cross section before activation of landslide; 6 Line of supposed maximal weathering; 7, Tail-race

the southwest, and further on, in the lower part of the slope, towards the west at a rate of 20 m h^{-1}. It was quickly spreading and moving regressively towards the east and northeast. Finally, an enormous mass plunged into the Visočica River, whose energy was not sufficient to wash it away. The landslide continued to spread, with rockfalls in the landslide tongue and along the fault zone on the southern side. The excessive sliding finally attenuated on March 16, 1963. The tongue-shaped, elongated landslide was 1,500 m long and 120–200 m wide, with an area of 180,000 m^2 (Fig. 13.5). The top scarp was 10–12 m high and lateral failures 1–2 m on the northern side, and up to 30 m along the fault zone on the southern side. Sliding depth averaged 17–23 m, up to 35 m in maximum. The landslide toe created a natural dam 500 m long, 36 m high, and about 150 m wide at the crown. The dam consisted mainly of weathered clays and sandstone blocks of various dimensions. The impoundment of a lake of 30 million m^3 volume ended on March 14, 1963, when its artificial draining, overflow through an urgently constructed channel, began. The village of Zavoj was completely flooded, and a new threat became apparent, the natural dam breaching under lake water pressure.

The flood wave threatened several major settlements, communication lines, and agricultural areas downstream. An expert team managed to preserve the dam and to prevent the disaster. A tail-race channel in hard bedrock was built outside of the dam body. The channel was situated in the left valley side of the Visočica River, pressed by the landslide foot. The aim was to drain the lake and to relieve the pressure on the dam. A 600 m long *tunnel* was *constructed* in the left valley side, which finally drained the lake water into the Visočica downstream of the landslide. Subsequently, the landslide area was reclaimed, and, meeting the standards, a new dam, part of the hydroenergetic system "Zavoj," was built.

13.7.2 Joc Rockfall

During the night on December 12/13, 1974, a huge rockfall occurred on the right *valley* side of the *Danube* River, about 5 km downstream the town Golubac (Fig. 13.1, location 3) at the entrance to the Danube Gorges (Ðerdap). About 250,000 m³ of rock collapsed on the regional road Golubac–Kladovo and into the reservoir of the "Ðerdap Djerdap" hydroelectric system. The area is composed of massive and thick-bedded (1–8 m) *Cretaceous limestones*. The height of the slope in this part of the gorge exceeds 250 m, while slope angles vary from 30° to 70°. The present-day slope shape is controlled by lithological composition, the structural characteristics of limestones, neotectonic activity, and the action of exogenous forces (kinetic energy of the Danube, chemical erosion, gravity and human activities). The limestones are intensively fractured by faults and fissures, dividing the rock mass into blocks of up some meters diameter. Fault and fissure walls are covered with calcite or limonite crusts, often filled with clay. The limestones are deeply karstified, with conduits, caverns and small caves creating favorable conditions for groundwater circulation. The karst water table fluctuates considerably, from completely dry bedrock just above the Danube level to the water table almost reaching the surface (Sunarić 1976). The intensive *incision* of the Danube and the gorge formation since the end of the Neogene through the Quaternary (from the regression of the Pannonian Sea to our days), reflects neotectonic activity in the area. Steep, locally (sub)vertical valley sides, formed of fractured, karstified limestones with strong groundwater circulation and watertable fluctuations, are ideal settings for rockfall occurrences. Furthermore, the changes of groundwater hydraulic regime due to impoundment (reservoir filling and emptying) additionally contributed to rockfall hazard.

The Joc rockfall originated along several systems of ruptures, along an uneven slip plane with the average inclination of 70°. Most of the plane formed in limonite and calcite crusts or the remnants of clayey fracture infill. The first rockfall occurred on December 12, 1974 at 22:30 h, when 50–100 m³ of rock collapsed on the road. However, on December 13 at 2:31, about 250,000 m³ of rock collapsed in just a few seconds. The collapsed material covered the road and most of it reached the Danube, more precisely, the reservoir "Ðerdap 1." The rockfall mass consisted of blocks of various sizes (Fig. 13.7). As the material that covered the road had to be quickly removed, and the rest fell into the water, its exact volume was not determined. The dimensions can be estimated from the following data (Sunarić 1976.): the rockfall area was 18 ha, rockfall width in the riverbed 110 m, length 100–180 m, the displaced section of the road was about 180 m long, the road was blocked at the length of 50 m, while the failure scarp was about 150 m high (Fig. 13.7).

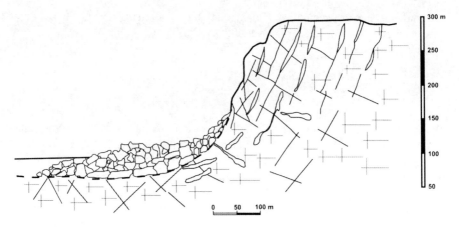

Fig. 13.7 Draft profile of the Joc rockfall (Sunarić 1976)

Fortunately, no human victims were involved but the material damage was considerable (interruption of traffic, regional road repair costs). However, there was no damage due to flood wave and no overflow above the dam. The flood wave had to follow a relatively long, winding path to reach the Đerdap dam (almost 100 km downstream), so its energy dropped without considerable effects to the dam. This event was a strong warning for natural hazards in the vicinity of the Đerdap reservoir.

13.7.3 Vratna Natural Bridges

In eastern Serbia, in the valley of the Vratna River, a right tributary of the Danube, three natural bridges (arches) are found (Fig. 13.1, location 4). The Vratna River is about 26 km long and has a drainage area of 185 km^2. The catchment is mostly built of Upper Palaeozoic crystalline schists, Jurassic and Cretaceous sandstones, shales and limestones, Neogene clastic sediments, thick Quaternary sediments of various origin and only about 10 km^2 are pure massive and thick-bedded *limestones* of Upper Jurassic age (Fig. 13.8). The limestones are karstified, with dry valleys, hanging valleys, and dolines on the surface. In this formation, the Vratna cut a 3.5 km long and about 150 m deep gorge, with steep, somewhere almost vertical walls exposing cave entrances of various dimensions (Petrović and Gavrilović 1969).

Particular landmarks of the *Vratna gorge* are its three natural bridges. In the upper gorge the Suva Prerast natural bridge (Fig. 13.8, location a) is 34 m long, 30 m high and 15 m wide. About 50 m upstream from the natural bridge, there is a ponor that captures the whole Vratna discharge in the dry season. In the downstream part of the Vratna gorge, there are two more natural bridges: Velika Prerast (Fig. 13.8, location b; Fig. 13.9) and Mala Prerast (Fig. 13.8, location c). Velika Prerast is 45 m

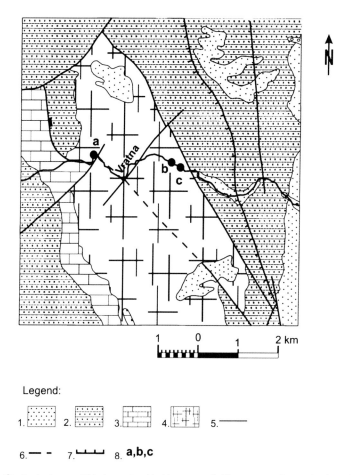

Fig. 13.8 Geological map of Vratna natural bridges area. 1, Neogene sandy-marly clays; 2, Lower Cretaceous clastic and carbonate rocks; 3, Middle Jurassic carbonates and sandstones; 4, Upper Jurassic limestones; 5, Geological boundary; 6, Fault; 7, Nappe; 8 Positions of natural bridges: a, Suva Prerast; b, Velika Prerast; c, Mala Prerast

long, 26 m high, and 23–33 m wide. The thickness of the limestone arch is about 30 m. The right wall hosts several cave entrances that all belong to the same cave system. The total length of horizontal, inclined, and vertical passages is 305 m with a total height difference of 61 m (Ćalić-Ljubojević 2000). Up to 26 m of relative height above the riverbed, crystalline schist pebbles have been found in the cave, which proves the underground flow of the Vratna at one stage of the evolution of the gorge. Mala Prerast is situated at the end of the Vratna gorge. In a downstream direction, after the contact of Upper Jurassic limestones and Lower Cretaceous sandstones, the Vratna valley becomes wide and spacious. The opening of the Mala Prerast is 34 m high and 33 m wide. The rocky arch shelters the riverbed at a length

Fig. 13.9 Natural bridge Velika Prerast on the Vratna River (Photo: P. Djurović)

of 15 m. The Vratna riverbed abounds in pots and smaller waterfalls (Petrović and Gavrilović 1969).

The Vratna gorge is a good example of a final stage of morphological evolution of *fluviokarstic caves* (Cvijić 1918). Presently, only two natural bridges (Velika Prerast and Mala Prerast) are the *remnants* of a former cave with an active river flow. Progressive collapse of the cave ceiling has destroyed the passage, leaving only two segments that prove its former existence.

13.7.4 Djavolja Varoš

Djavolja (Đavolja) Varoš is situated in southern Serbia, in the *Toplica River basin* on the western slopes of Radan Mountain, at 10 km distance from the regional road Kuršumlija–Priština, on the right valley side of the stream Žuti Potok, right tributary of the Velika Kosanica River. *Hydrothermally changed andesites* and pyroclastites and hornblende andesites dominate in the geological composition. Between two gullies that host temporary torrent streams (northern tributaries of the Žuti

Fig. 13.10 Djavolja Varoš rock pinnacles. Note the pinnacles on the right that lost their andesite cap and are likely to be destroyed soon (Photo: M. Milivojević)

Fig. 13.11 Top of the ridge of Djavolja Varoš (*thick line*), 1 m wide only three decades ago (Photo: M. Milivojević)

Potok), there is a ridge with more than 200 rock pinnacles formed in pyroclastic rocks (Fig. 13.10).

Rock pinnacles result from intensive proluvial processes: the formation of *gully networks*, narrowing intensively the inter-gully ridges and also transversally dissecting the terrain. The andesite *blocks* of some tens of centimeters dimensions *protect* the underlying *pyroclastic material* from rainsplash erosion and rock pinnacles with a maximum height of ca 15 m and base diameter of 2–6 m form. According to Lazarević (1995), the total sediment production in the two bordering gullies is 483.8 m^3 y^{-1} or 4987.6 m^3 km^2 y^{-1}. The same author claims that the main factor that directly triggered the formation of Djavolja Varoš was the expansion of *cattle breeding* and *deforestation* during the Middle Ages. A combination of favorable geological and topographical conditions (poorly cemented pyroclastites on highly inclined slopes) with the devastation of vegetation cover, has led to excessive proluvial erosion. Its intensity is evident even within a human lifetime, in several decades. Locals remember that about 30 years ago the top of the ridge, where the pinnacles are located, was wide enough for a safe walk (about 1 m), while today it is rather sharp (Fig. 13.11) and 1–2 m lower (Lazarević 1995). The site Djavolja Varoš is legally protected as a first category natural monument.

References

Belij S (1994) Savremeni periglacijalni procesi i oblici reljefa severozapadne Šar-planine. Šarplaninske župe Gora, Opolje i Sredska, Odlike prirodne sredine (Recent periglacial processes and relief forms of northwestern Šar Planina. Regions of Gora, Opolje and Sredska). Posebna izdanja Geografskog instituta "Jovan Cvijić" SANU, Beograd 41:113–144 (in Serbian)

Belij S (1997) Zaštita glacijalnog reljefa u Srbiji (Protection of glacial relief in Serbia). Zaštita prirode 48–49:59–70 (in Serbian)

Bukurov B (1948) Dolina Tise u Jugoslaviji (The Tisa Valley in Yugoslavia). Posebna izdanja Srpskog geografskog društva, Beograd 25:54 p (in Serbian)

Bukurov B (1950) Tri bačke doline: Krivaja, Jegrička i Mostonga (Three Bačka valleys: Krivaja, Jegrička and Mostonga). Glasnik Srpskog geografskog društva 30(2):77–86 (in Serbian)

Bukurov B (1953) Geomorfološke crte južne Bačke (Geomorphological characteristics of southern Bačka). Zbornik Geografskog instituta SAN, Beograd 26(4):1–63 (in Serbian)

Bukurov B (1954) Geomorfološki prikaz Vojvodine (Geomorphological overview of Vojvodina). Zbornik Matice srpske, Serija prirodnih nauka, Novi Sad 4:100–133 (in Serbian)

Bukurov B (1982) Sintetička razmatranja geomorfoloških problema na teritoriji Vojvodine (Synthetic investigation of geomorphological problems in Vojvodina). VANU, Novi Sad (in Serbian)

Bulla B (1937) Terassen und Niveaus am rechten Donauufer zwischen Adony und Mohács. Mathemat U Naturwiss Anzeiger d Ung Akademie d Wiss 55:193–224

Bulla B (1938) Der pleistozäne Löss im Karpathenbecken. Földtani Közlöny, Budapest 68(1–3):33–58

Ćalić-Ljubojević J (2000) Natural bridges on the Vratna River (Eastern Serbia), as the last remnants of a former cave. Acta Carsologica 29(2):241–248

Cvijić J (1891) The cave Prekonoška Pećna. Annales geologiques de la peninsule Balkanique 3:272–299 (in Serbian)

Cvijić J (1893a) Das Karstphänomen. Versuch einer morphologischen Monographie. Geographische Abhandlungen 5(3):1–114

Cvijić J (1893b) Geographical research in the area of Kučaj Mountain. Annales geologiques de la peninsule Balkanique 5:7–172 (in Serbian)

Cvijić J (1895a) Karst: geographical monograph. Royal publishing House, Belgrade, 176 p (in Serbian)

Cvijić J (1895b) Caves and underground hydrography in Eastern Serbia. Glas Serb Roy Acad Sci 46:1–101 (in Serbian)

Cvijić J (1895c) La grande grotte de Douboca, dans la Serbie Orientale. Bulletin de la Societe de Speleologie Paris 1:81–87

Cvijić J (1899) Glacial and morphological studies of the mountains of Bosnia, Herzegovina and Montenegro. Glas Serb Roy Acad Sci 57:1–196 (in Serbian)

Cvijić J (1900) L'epoque glaciaire dans la Peninsule des Balkans. Ann Geogr 9:359–372

Cvijić J (1903) Balkanian, Alpine and Carpathian glaciation. Glas Serb Roy Acad Sci 67:219–227 (in Serbian)

Cvijić J (1909) Lacustrine morphology of Šumadija. Glas Serb Roy Acad Sci 79:1–94 (in Serbian)

Cvijić J (1910) Life of the Dilluvial Eordeian lake. Glas Serb Roy Acad Sci 81:1–84 (in Serbian)

Cvijić J (1912) The karst of Lelić area. Bull Serb Geogr Soc 1:103–105 (in Serbian)

Cvijić J (1913) Ledeno doba u Prokletijama i okolnim planinama (Glacial epoch on the Prokletije and surrounding mountains). Glas Serb Roy Acad Sci 91:188–267 (in Serbian)

Cvijić J (1914a) Pešter na jugoistoku od Sjenice (The Pešter Cave, to the south-east of Sjenice). Bull Serb Geogr Soc 3–4:218–219 (in Serbian)

Cvijić J (1914b) Tragovi starih glečera u Srbiji (Traces of old glaciers in Serbia). Bulletin of the Serbian Geographical Society, Belgrade 3–4:211–212 (in Serbian)

Cvijić J (1918) Hydrographie souterraine et evolution morphologique du Karst. Recueil des Travaux de l'Institut de Geographie Alpine 6(4):1–56

Cvijić J (1920) Conform and inverse relief, polygenetic valleys, incised meanders. Bull Geogr Soc Belgrade 5:94–112 (in Serbian)

Cvijić J (1921) Platformes fluviales et pessauts d'erosion. Extrait des "Comptes Rendus des seances de l'Academie des Sciences", Paris 172:1–4

Cvijić J (1922) On nivation and glacial erosion. Bull Geogr Soc Belgrade 7–9:21–48 (in Serbian)

Cvijić J (1924–1926) Geomorphology I–II. National Printing House, Belgrade, 588+506 p (in Serbian)

Dimitrijević MD (1997) Geology of Yugoslavia. Special Publication/Geological Institute GEMINI, Beograd, 187 p

Dimitrijević MD (2002) Geološki atlas Srbije 1:2.000.000 (Geological Atlas of Serbia at 1:2,000,000). Ministarstvo za zaštitu prirodnih bogatstava i životne sredine Republike Srbije, Republički fond za geološka istraživanja, Edicija 1. Geološka karta, Beograd (in Serbian)

Djurović P (ed) (1999) Speleološki atlas Srbije (Speleological Atlas of Serbia). Posebna izdanja Geografskog instituta "Jovan Cvijić" SANU, Beograd 52:290 p

Dragičević S (2007) Dominantni erozivni procesi u slivu Kolubare (Dominant processes of erosion in the Kolubara River basin). Jantar grupa, Beograd, 246 p (in Serbian)

Dragičević S, Stepić M, Karić I (2008) Prirodni potencijali i degradirane povrsine opstine Obrenovac (Natural potentials and degraded areas of Obrenovac municipality). Jantar grupa, Beograd, 180 p (in Serbian)

Ducić V, Radovanović M (2005) Klima Srbije (Climate of Serbia). Zavod za udžbenike, Beograd, 212 p (in Serbian)

Gavrilović D (1966) Pećine na južnom odseku Beljanice (Caves in the southern escarpment of Mt. Beljanica). Zbornik radova Geografskog zavoda PMF, Beograd 13:51–65 (in Serbian)

Gavrilović D (1970) Relikti kupastog krasa u Karpatsko-balkanskim planinama Jugoslavije (Relicts of cone karst in the Carpatho-balkanides of Yugoslavia). Journal of the Geographical Institute "Jovan Cvijić" SANU, Beograd 23:117–126 (in Serbian)

Gavrilović D (1976) Glacijalni reljef Srbije (Glacial relief of Serbia). Glasnik Srpskog geografskog društva, Beograd 61(1):9–19 (in Serbian)

Gavrilović D, (1985) Elementi paleokrasa na teritoriji Jugoslavije (Paleokarst elements on the territory of Yugoslavia). Zbornik radova Instituta za geografiju PMF, Beograd 32:33–47 (in Serbian)

Gojgić D, Janić M, Luković S (1964–1965) Inženjerskogeološki uslovi stvaranja klizišta, prirodne brane i jezera Zavoj (Engineering-geological conditions for the formation of landslide, natural dam and Zavoj lake). Institut de recherches geologiques et geophysiques, Belgrade, Bulletin 4–5:147–173 (in Serbian)

Gorjanović–Kramberger D (1921) Morfološke i hidrogeološke prilike Srijemskog lesa (Morphological and hydrogeological characteristics of loess in the Srem area). Glasnik geografskog društva, Beograd 5:17–53 (in Croatian)

Group of authors (1992) Geomorphological Map of Yugoslavia at 1:500,000. Geografski institut "Jovan Cvijić", Odbor za geodinamiku SANU, Beograd

Jemcov I (2007) Water supply potential and optimal exploitation capacity of karst aquifer systems. Environ Geol 51(5):767–773

Jovanović PS (1922) Pribrežni jezerski reljef beogradske okoline (Lakeshore relief near Belgrade). Državna štamparija Beograd, 41 p (in Serbian)

Jovanović PS (1933) Pregledna geomorfološka karta Kraljevine Jugoslavije 1:1.200.000 (Overview geomorphological map of the Kingdom of Yugoslavia, at 1:1.200.000). Zbirke karata Geografskog društva, Beograd (in Serbian)

Jovanović PS (1951) Osvrt na Cvijićevo shvatanje o abrazionom karakteru reljefa po obodu Panonskog basena (Coup d'oeil sur la conception de Cvijić du caractere abrasif du relief de la bordure du bassin Pannonien). Zbornik radova GI SAN, Beograd 8(1):1–23 (in Serbian)

Jovanović PS (1955) Uticaj kolebanja pleistocene klime na proces rečne erozije (The impact of Pleistocene climate oscillations to the process of river erosion). Zbornik radova GI SAN, Beograd 46(10):19–65 (in Serbian)

Jovanović M, Karajičić Lj, Karamata S, Vukanović M (1973) Novi pogledi na razvoj vulkanizma u području Leckog andezitskog kompleksa (New approach to the volcanism development in the Lece andesite complex). Geološki anali Balkanskog poluostrva, Beograd 37(2):165–177 (in Serbian)

Jovičić Ž (1960) Reljef beogradskog Podunavlja i sliva Topčiderske reke (Relief in the drainage areas of Topčiderska Reka and of the Danube tributaries in Belgrade). Zbornik radova Geografskog instituta PMF, Beograd 7:27–55 (in Serbian)

Jovičić Ž (1966) Recentna erozija i akumulacioni procesi u Vranjskoj kotlini i Grdeličkoj klisuri (Recent erosion and accumulation processes in Vranje basin and Grdelica gorge). Posebna izdanja Narodnog muzeja, Vranje (in Serbian)

Komatina M (1970–1971) Uslovi razvoja karstnog procesa i rejonizacija karsta (The conditions of karst process development and karst regionalization). Vesnik Geozavoda, Beograd 10–11:13–35 (in Serbian)

Košćal M, Menković Lj (1994) Prirodno stanje Deliblatske peščare i izvorišta za vodosnabdevanje (Natural condition of Deliblatska Peščara and source areas for water supply). Deliblatski pesak, Zbornik radova, Pančevo 6(1) (in Serbian)

Krešić N (1988) Karst i pećine Jugoslavije (Karst and caves in Yugoslavia). Naučna knjiga, Beograd, 149 p (in Serbian)

Laskarev V (1938) Treće beleške o kvartarnim naslagama u okolini Beograda (Third notes on Quaternary sediments in the vicinity of Belgrade). Geološki anali balkanskog poluostrva Beograd 15:1–40 (in Serbian)

Laskarev V (1950) O ekvivalentima gornjeg sarmata u Srbiji (On the equivalents of the Upper Sarmatian in Serbia). Geološki anali Balkanskog poluostrva Beograd 18:1–16 (in Serbian)

Laskarev V (1951) O stratigrafiji kvartarnih naslaga Vojvodine (On the stratigraphy of Quaternary deposits in Vojvodina). Geološki anali Balkanskog poluostrva Beograd 19:1–18 (in Serbian)

Lazarević R (1957) Sliv Jezave, Ralje i Kanjske reke (Drainage areas of Jezava, Ralja and Kanjska Reka). Zbornik GI SAN, Beograd 57(13):95–163 (in Serbian)

Lazarević R (1959) Azanjska fosilna dolina (Azanja fossil valley). Posebna izdanja Srpskog geografskog društva, Beograd 36:64 p (in Serbian)

Lazarević R (1978) Zlotska pećina (Zlotska cave). Turistički savez opštine Bor, Bor (in Serbian)

Lazarević R (1995) Đavolja varoš. Erozija. Stručno-informativni bilten, Beograd 22:25–34 (in Serbian)

Lazarević R (1996a) Valjevski kras – pećine, jame, kraška hidrografija (Valjevo karst – caves and karst hydrography). Srpsko geografsko društvo, Beograd, 240 p (in Serbian)

Lazarević R (1996b) Kras Dubašnice, Gornjana i Majdanpeka. Srpsko geografsko društvo, Beograd, 302 p (in Serbian)

Manojlović P (1992) Hemijska erozija kao geomorfološki proces (Chemical erosion as a geomorphic process). Geografski fakultet PMF, Beograd, 112 p (in Serbian)

Manojlović P (2002) Intenzitet hemijske erozije u slivu Nišave (Intensity of chemical erosion in the drainage area of the Nišava River). Glasnik Srpskog geografskog društva 82(1):3–8 (in Serbian)

Manojlović P, Dragičević S (2000) Intenzitet mehaničke vodne erozije u slivu Veternice (Intensity of mechanical erosion in the drainage area of the Veternica River). Glasnik Srpskog geografskog društva 80(2):13–22 (in Serbian)

Manojlović P, Mustafić S, Dragičević S (2003) Pronos silta u slivu Jerme (Silt transport in the drainage area of the Jerma River). Glasnik Srpskog geografskog društva 83(2):3–10 (in Serbian)

Marković M (1973) Shvatanje o geomorfološkoj karti i predlog modela geomorfološke karte u nas (Perception of geomorphological map and a suggestion of a model of geomorphological map). Geološki anali Balkanskog poluostrva Beograd 38:219–236 (in Serbian)

Marković M, Pavlović Z (1967) Fotogeološka ispitivanja vulkanita u eruptivnoj zoni Kotlenik-Borač (Photogeological research of volcanites in the eruption zone Kotlenik-Borač). Geološki anali Balkanskog poluostrva Beograd 33:77–92 (in Serbian)

Marković SB, Oches E, Sümegi P, Jovanović M, Gaudenyi T (2006) An introduction to the Middle and Upper Pleistocene loess-paleosol sequence at Ruma brickyard, Vojvodina, Serbia. Quat Int 149:80–86

Marković SB, Bokhorst MP, Vandenberghe J, McCoy WD, Oches EA, Hambach U, Gaudenyi T, Jovanović M, Zöller L, Stevens T, Machalett B (2008) Late Pleistocene loess-paleosol sequences in the Vojvodina region, north Serbia. J Quat Sci 23(1):73–84

Marković SB, Hambach U, Catto N, Jovanović M, Buggle B, Machalett B, Zöller L, Glaser B, Frechen M (2009) Middle and Late Pleistocene sequences at Batajnica, Vojvodina, Serbia. Quat Int 198:255–266

Marković-Marjanović J (1949) Proučavanja kvartarnih naslaga južnog Banata i Požarevačkog Podunavlja (Studies of Quaternary sediments in southern Banat and Požarevac region). Glasnik SAN, Beograd 1(3) (in Serbian)

Marković-Marjanović J (1953) Prethodno saopštenje o geološkom sastavu i stratigrafiji lesnog platoa severne Bačke i Subotičke peščare (First notes on geological composition and stratigraphy of the northern Bačka loess plateau and of Subotička Peščara sand sheet). Glasnik SAN, Beograd 5(2):277–279 (in Serbian)

Marković-Marjanović J (1954) Severne padine Fruške gore od Slankamena do Vukovara, karakteristična oblast rasprostranjenja kvartarnih sedimenata. Glasnik SAN, Beograd 6(2):279–280 (in Serbian)

Marković-Marjanović J (1965) Osvrt na poznavanje lesnih problema Jugoslavije u doba Jovana Cvijića (Observation on the loess problems in Yugoslavia in the time of Jovan Cvijić). Glasnik Srpskog geografskog društva, Beograd 45(2):99–114 (in Serbian)

Menković Lj (1985) Glacijalna morfologija Koritnika (Glacial morphology of Koritnik). Priroda Kosova 6. Zavod za zaštitu prirode, Priština (in Serbian)

Menković Lj (1988) Reljef Šar-planine, geomorfološka studija (The relief of Šar Planina: a geomorphological study). Manuscript doctoral thesis. Geografski fakultet, Beograd (in Serbian)

Menković Lj (1994) Tragovi glacijacije u području Đeravice-Prokletije (Glaciation traces in the region of Djeravica peak, Mt.Prokletije) Geografski godišnjak, Kragujevac 30:139–146 (in Serbian)

Menković Lj, Košćal M, Mijatović M (2003) Geomorphological map of Serbia. 1:500.000. Geozavod-Gemini, Beograd

Menković Lj, Marković M, CČupković T, Pavlović R, Trivić B, Banjac N (2004) Glacial morphology of Serbia, with comments on the Pleistocene Glaciation of Montenegro, Macedonia and

Albania. In: Ehlers J, Gibbard PL (eds) Quaternary glaciations: extent and chronology. Elsevier, Amsterdam/San Diego, pp 379–384

Mijatović BF (1984) Hydrogeology of Dinaric Karst. International contributions to hydrogeology 4, Heise, pp 115–142

Mijatović BF (1990) Kras, hidrogeologija kraških vodonosnika (Karst: hydrogeology of karst aquifers). Geozavod, Beograd, 304 p (in Serbian)

Milanović P (1979) Hidrogeologija karsta i metode istraživanja (Hydrogeology of karst and research methods). HE Trebišnjica, Trebinje, 302 p (in Serbian)

Milanović P (1981) Karst hydrogeology. Water Resources Publication, Colorado, 434 p

Milanović P (2000) Geological engineering in karst. Zebra publishing, Belgrade, 347 p

Milić Č (1956) Sliv Peka (Drainage area of the Pek River). Posebna izdanja Geografskog instituta "Jovan Cvijić" SANU, Beograd 9, 125 p (in Serbian)

Milić Č (1973) Fruška Gora – geomorfološka proučavanja (Mt.Fruška Gora –geomorphological research). Matica srpska, odeljenje za prirodne nauke. Novi Sad, 74 p (in Serbian)

Milivojević S (1938) Pojavi i problemi krša. Proučavanje u Dinarskom kršu i kršu istočne Srbije (Features and problems of karst. Studies in the Dinaric karst and karst of eastern Serbia). Posebna izdanja Srpske Kraljevske Akademije, CXXIII, Beograd, 160 p (in Serbian)

Milojević BŽ (1948) Titelska lesna zaravan (Titel loess plateau). Glasnik Srpskog geografskog društva, Beograd 28(1) (in Serbian)

Milojević BŽ (1949) Lesne zaravni i peščare u Vojvodini (Loess plateaus and sand sheets in Vojvodina). Matica srpska, Novi Sad 2 (in Serbian)

Milojević BŽ (1960) Panonski Dunav na teritoriji Jugoslavije (The Pannonian Danube on the territory of Yugoslavia). Zbornik za prirodne nauke Matice srpske, Novi Sad 18:5–65 (in Serbian)

Milojević BŽ (1937) Visoke planine u našoj Kraljevini (High mountains in our kingdom), Beograd, 459 p (in Serbian)

Milovanović B (1964–1965) Epirogenska i orogenska dinamika u prostoru spoljnih Dinarida i problemi paleokarstifikacije i geološke evolucije holokarsta (La dynamique épirogénique et orogénique dans le domaine des Dinarides externes et les problèmes de la paléokarstification et de l'évolution de l'holokarst). Vesnik Geozavoda, Beograd 4–5:5–44 (in Serbian)

Mladenović T (1984) Visinska struktura zemljišta u SFR Jugoslaviji (Hypsometric structure of the SFR Yugoslavia). Zbornik radova VGI, Beograd 4:67–75 (in Serbian)

Nikolić PR (1912) Glacijacija Šar-planine i Koraba (Glaciation of Šar Planina and Korab). Glasnik Srpskog geografskog društva Beograd 2:72–79 (in Serbian)

Paunković Đ (1953) Reljef sliva Resave (Relief of the Resava catchment). Posebna izdanja Geografskog instituta SAN, Beograd 5:158 p (in Serbian)

Pavlović R (1990) Kompleksna analiza reljefa kao metod geološkog istraživanja (Complex analysis of relief, as a method of geological research). Doctoral thesis, Faculty of Mining and Geology, Belgrade (in Serbian)

Petrović J (1954) Kraška vrela u dolini Jerme (Karst springs in the valley of Jerma River). Glasnik Srpskog geografskog društva, Beograd 34(2):169–172 (in Serbian)

Petrović D (1965) Evolutivni tipovi kraških dolina na Kučaju (Evolutional types of karst dolines on Mt. Kučaj). Glasnik Srpskog geografskog društva 45(2):115–122 (in Serbian)

Petrović D (1966) Problem abrazionog reljefa Timočke krajine (Problem of abrasional relief at Timok). Zbornik radova Geografskog zavoda PMF, Beograd 13:37–49 (in Serbian)

Petrović J (1969) Pojava dubinske karstifikacije u južnom delu masiva Kopaonika (Deep karstification in the southern part of Kopaonik massif). Zbornik za prirodne nauke Matice srpske, Novi Sad 36 (in Serbian)

Petrović J (1974) Krš istočne Srbije (The karst of Eastern Serbia). Posebna izdanja Srpskog geografskog društva, Beograd 40:91 p (in Serbian)

Petrović D (1976a) Eolski reljef istočne Srbije (Aeolian relief in eastern Serbia). Zbornik radova Geografskog zavoda PMF, Beograd 23p (in Serbian)

Petrović D (1976b) Paleovulkanski reljef Srbije (Paleovolcanic relief of Serbia). Glasnik Srpskog geografskog društva Beograd 56(2):49–62 (in Serbian)

Petrović D (1979) Eolska morfologija Srbije (Aeolian morphology of Serbia). Zbornik radova Geografskog zavoda PMF, Beograd 26:41–57 (in Serbian)

Petrović D, Gavrilović D (1969) Reljef u slivu Vratne (Relief in the Vratna River catchment area). Zbornik radova Geografskog zavoda PMF, Beograd 16:7–27 (in Serbian)

Petrović V, Stanković S (1981) Veliko klizište u s. Jovac (A large landslide in the village of Jovac). Zbornik radova Prvog simpozijuma Istraživanje i sanacija klizišta, Bled (in Serbian)

Roglić J (1952) Problem neogenog abrazionog reljefa (The problem of Neogene abrasional relief). Kongres na geografite ot FNRJ. II, Skopje (in Serbian)

Stevanović Z (1994) Karst ground waters of Carpatho-Balkanids in Eastern Serbia. In: Stevanović Z, Flipović B (eds) Ground waters in carbonate rocks of the Carpathian–Balkan Mountain Range. Monograph. Special edition of CBGA, Ilston, Jersey, pp 203–237

Stevanović Z (2005) Jovan Cvijić's studies of eastern Serbian karst – a foundation of karst hydrogeology. In: Stevanović Z, Mijatović B (eds) Cvijić and karst/Cvijić et karst. Special edition of Board of Karst and speleology SANU, Beograd, pp 365–373

Sunarić D (1976) Analiza uzroka nastanka odrona Joc kod Golupca na Dunavu (Analysis of causes of the Joc rockfall, close to Golubac on the Danube). Proceedings of the Fourth Yugoslavian symposium of hydrogeology and engineering geology, Skoplje (in Serbian)

Zeremski M (1956) Fluvio-denudaciono ili abraziono poreklo Mačkatske površi (Fluvio-denudational or abrasional origin of Mačkat plateau). Zbornik radova Geografskog zavoda PMF, Beograd 4:87–105 (in Serbian)

Zeremski M (1959) Vulkanske kupe i mali denudacioni oblici na severozapadnoj podgorini Rogozne (Volcanic cones and small denudational forms on northwestern foothill of Mt. Rogozna). Zbornik radova GI PMF, Beograd 6:19–35 (in Serbian)

Zeremski M (1967) Tubića i Ušačka pećina (The caves Tubića and Ušačka pećina). Journal of the Geographical Institute "Jovan Cvijić", SANU, Beograd 21:245–265 (in Serbian)

Zeremski M (1990) Geomorfološka (morfostrukturna) karta Srbije 1:500.000 (Geomorphological [morphostructural] map of Serbia at 1:500,000). Geografski institut "Jovan Cvijić" i Odbor za geodinamiku SANU, Beograd

Chapter 14
Recent Landform Evolution in Bulgaria

Krasimir Stoyanov and Emil Gachev

Abstract Morphostructures exert a strong control on the landforms of Bulgaria as in the mountain landscapes planation surfaces prevail. In mountains valley incision is considerable, deep and narrow gorges and cross-section asymmetry of valleys is common. In major valleys two flood and seven upper river terraces have been identified. Regionally, active karst processes are significant in shaping the landscape, especially in the Predbalkan and Rhodope Mountains. Most (59.7%) of the Black Sea coastline is of abrasional type with cliffs, while 28% of its total length is accumulational with beaches, spits, limans, lagoons, and sand dunes. As the equilibrium line altitude (ELA) of glaciers in Bulgaria was the highest in the Carpatho-Balkan area, glacial landforms only developed in the Rila and Pirin Mountains, above 2,400 m, but periglacial features are more common. The southernmost occurrence of loess in Europe is in the Bulgarian Danube Plain. The most severe human interventions into the natural environment are related to mining activities. Improper tillage techniques in agriculture may lead to higher erosion hazard than predictable climate change.

Keywords Planation surfaces • Karst landforms • Valley network • River terraces • Coastal landforms • Glacial and periglacial processes • Loess • Human impact • Geomorphological hazards

K. Stoyanov • E. Gachev (✉)
Southwestern University "Neofit Rilski", Faculty of Mathematics and Natural Sciences, Department of Geography, Ecology and Environmental Protection, 66 Ivan Mihailov str., 2700, Blagoevgrad, Bulgaria

Institute of Geography, Bulgarian Academy of Sciences, Sofia, Bulgaria
e-mail: krasi_sto@yahoo.com; e_gachev@yahoo.co.uk

14.1 Geomorphological Units

Bulgarian territory presents an extraordinary diversity of landforms produced by various geomorphic systems, in both horizontal and vertical dimensions (Fig. 14.1). Landform diversity reflects morphostructure development of Bulgaria's territory in a complicated geological setting on the Balkan peninsula between the Eurasian and African lithospheric plates, the long-term geodynamical evolution as well as the environmental history.

The main *morphostructural features* of the territory differ not only in the configuration of macroforms but also in morphometric parameters (Fig. 14.1). The decisive factors in the present differentiation of morphostructures are geology and tectonics (crustal movements and the related morphogenesis). In general, morphostructures in Bulgaria show a subparallel arrangement and a shift from north to south, while particular features are often aligned in west-northwest to east-southeast direction. Average elevations decrease in the same direction. The only exception is the Danube (Moesian) plain, which has its highest parts to the east.

Studies of morphostructures in Bulgaria date back to the beginning of the twentieth century, when J. Cvijić (1903) identified four main structures: the Rhodope massif, the Transitional area, the Balkan folded system and the Bulgarian plate. Later St. Bonchev (1929) identified only three large units: the Bulgarian plate, the Balkanides, and the Macedonian–Rhodope massif. In 1946 E. Bonchev suggested that the name Moesian plate should replace the Bulgarian plate. He also outlined the

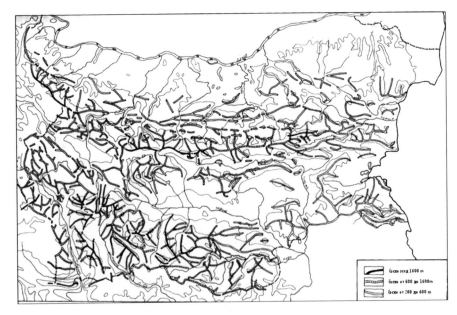

Fig. 14.1 Main orographic features in Bulgaria. Hypsometric lines at 200, 600, and 1,600 m above sea level (After Galabov 1982)

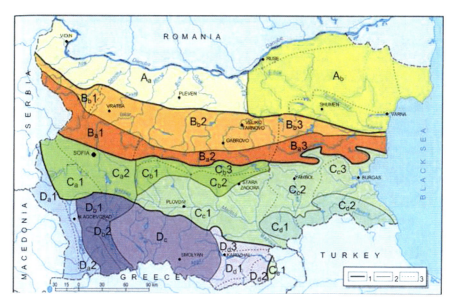

Fig. 14.2 Geomorphological units in Bulgaria (After Alexiev 2002b). *1*, boundary of zone; *2*, subzone; *3*, region. A: Zone of the Danube plain, subzones: a, Lom Depression; b, Ludogorie–Dobrudja Hills; B: Zone of the Balkan mountain system, subzones: a, Stara planina (regions: 1, western; 2, central; 3, eastern); b, Predbalkan (regions: 1, western; 2, central; 3, eastern); C: Transitional mountain and depression zone, subzones: a, Western (regions: 1, southwestern; 2, northeastern); b, Sredna gora–Zadbalkan (regions: 1, Sashtinska Sredna gora; 2, Sarnena gora; 3, Zadbalkan basins); c, Upper Thracia–Middle Tundzha (regions: 1, Upper Thracia Lowland; 2, Tundzha Valley; 3, coastal plain of Burgas); d, Sakar–Strandja (regions: 1, Sakar; 2, Strandja); e, Eastern Thracia; D: Zone of the Macedonia–Rhodope Massif: a, Dardanes subzone (regions: 1, Osogovo–Belasica mountain group; 2, Struma Valley); b, Rila–Pirin subzone (regions: 1, Rila and Pirin Mountains; 2, Mesta Valley); c, Western Rhodopes subzone; d, Eastern Rhodopes subzone (regions: 1, Middle Arda; 2, Snezhnik–Maglenik; 3, Chukata–Gorata)

following subunits within the Balkanides: Predbalkan, Stara planina, and Srednogorie. He also defined the Southern Carpathians, the Kraishtides, and the Rhodope area. Galabov (1956, 1966) was the first to identify geomorphic units on the basis of morphostructures. Also known are tectonic regionalizations by Muratov (1949), Yordanov (1956), Yovchev (1960), E. Bonchev (1966), Jaranov (1960), and Alexiev (2002b). Other morphostructure-based classifications are those by Gerasimov (1957), Mishev and Vaptsarov (1984, 1997).

Geomorphological units comprise landforms with similar morphographic and morphogenetic characteristics. The morphostructure approach, which is accepted by most authors for geomorphic regionalization, defines the following hierarchy of geomorphological units, as used by Alexiev (2002b) (Fig. 14.2):

1. *Geomorphological zones:* distinct units developed over essentially different geological structures and related large morphostructural features (epiplatform, block-fault etc.);

2. *Geomorphological subzones:* defined on the basis of intraregional topographic diversity caused by differences in structure, neotectonic evolution and related morphogenesis;
3. *Geomorphological regions:* defined on the basis of mesoscale topographic variation, determined by the Neogene-Quaternary evolution of lower-rank morphostructures and by specific exogenic processes.

Four large morphographic zones are distinguished on the territory of Bulgaria. The largest of them is the *Danubian hilly plain*, situated to the north. It has been formed on the background of the Moesian platform. This combination of lowlands, hills and plateaus has a relatively moderate relief. The Danubian plain can be divided into two sections:

1. In the *western* part (west of the Yantra River) the main topographic features are flat watershed ridges between the valleys of the Danube tributaries. The western part of the subzone is dissected by valleys with asymmetrical cross-sections (gentle left and steep right slopes), while in the eastern part the wide valleys of the meandering Osam and Yantra rivers are formed. A specific element of the landscape here is the 14 basalt hills of Neogene age along the Svishtov–Suhindol line. Elevations are low (below 130 m above sea level), and so are the rates of relief fragmentation (dissection), both horizontal and vertical (0.5–1.0 km km^{-2} and 15–100 m km^{-2}, resp.);
2. In the *eastern* part elevation is higher and topography more diverse. Its highest central section coincides with the top of the North-Bulgarian flat dome. The main features of the relief are hills and plateaus, reaching up to 500 m elevation (Shumen Plateau). The tributary valleys to the Danube are deeply incised (70–100 m) canyons and gorges. Due to the well-developed karstic features, the rivers have intermittent discharge. The horizontal and vertical fragmentation of relief reaches high values (1.5–2.0 km km^{-2}; 150–200 m km^{-2}, respectively) with mean altitude of 204 m above sea level.

Several riverside lowlands are formed along the right banks of the Bulgarian Danube. They are part of Danube floodplain with elevation from 35 to 15 m.

The *Balkan mountain system* (*range*) stretches between the Timok River Valley (to the west) and the Black Sea (to the east) across the whole Bulgarian territory. It consists of a series of parallel ridges of west to east strike, separated by long valleys, valley branches and structural depressions. Two subparallel macrofeatures can be outlined within the Balkan mountain range: the Predbalkan ridges to the north and the Main Balkan range (Stara planina) to the south.

The main ridge, 530 km long and 15–45 km wide, is situated excentrically along the southern margin of the mountain system. In its major part it hosts the principal watershed of the Balkan Peninsula between the catchments of the Mediterranean and Black Seas. In general, the altitude of the Stara planina drops towards the east, with two main "bulges" that exceed 2,000 m elevation: the massifs of Midzor–Kom (2,168 m, in the western part) and Vezhen–Botev–Triglav (2,376 m, in the center). These highest sections correspond to the main domes of anticlinoria, or represent a morpological expression of high tectonic style nappes: the Stara planina granite nappe and the Stara planina face strip (Stefanov 2002).

The *Stara planina* can be divided into three sectors: the Western, Central, and Eastern. Situated between the passes of Belogradchik (to the west) and Botevgrad (965 m, to the east), The Western Stara planina is the widest with an arcuate, 170 km long main ridge. The northern slope rises steeply above the subparallel upper valleys of the Lom and Prevalska Ogosta Rivers, and above the structural depression of Botunja. In its Bulgarian part the long and gentler southern slope descends towards the depression of Sofia in steps of three karstic geomorphological surfaces (Galabov 1982). The horizontal dissection (valley density) is ranging from 1–1.5 km km^{-2} (on dividing ridges) to 2–2.5 km km^{-2} (on the northern slope). At some locations along the N slope relative relief reaches 500 m km^{-2}. The central Stara planina is the narrowest but with the greatest and most marked orographic expression (Fig. 14.3). Situated between the passes Botevgradski (to the west) and Vratnik (1,107 m, to the east), this chain of four mountain massifs, separated by relatively high cols stretches in a length of 205 km with west to east strike. At most places the ridge has flat planated summits in strong contrast with the very steep fault slopes. The mountain range has an asymmetrical cross section: the southern slopes are steeper at their lower segments (below 1,200–1,400 m), while the northern are steepest at their upper segments. Relative relief amounts to 600 m km^{-2}. The eastern sector of Stara planina, situated between Vratnik Pass and the Black Sea (Cape Emine) is 155 km long. Here the mountains fork into two parallel branches, separated by the valleys of the Luda Kamchija and Hadjijska Rivers, which form the synclinal depression of the Luda Kamchia. Highest elevations are found to the west at Slivenska planina (1,181 m), while to the east summits are not higher than 500–700 m.

Fig. 14.3 The main ridge of Central Stara planina (Kupena peak, 2,169 m)

The *Predbalkan* (*Fore-Balkan*) area occupies the northern strip of the Balkan mountain system and consists of hills and low mountain ridges, most often of marked structural and substructural character (monoclines, synclines, anticlines). Several parallel west to east ridges are separated by depressions of structural and erosional origin (Galabov 1982). The Predbalkan is also divided into western, central, and eastern sections. The Western Predbalkan, west of the Vit River, comprises numerous monocline ridges defined by the fault pattern and generally lowering eastward (Vrachanska planina, 1,482 m). The relative relief along these ridges is in the range of 200–250 m km^{-2}, while in structural depressions and valley extensions drops to 50 m km^{-2}. The central Predbalkan (between the Vit and Stara Rivers) has a better orographic expression (Vasilyov vrah, 1,490 m). Here several subparallel rows of ridges descend towards the north. In general, they are divided into two zones: an external zone of structural plateaus and solitary hills to the north (often categorized as the southern periphery of the Danubian plain – Stefanov 2002), and an internal zone of low and middle mountains to the south. Crossing the rows of ridges, transverse river valleys have cut deep and narrow gorges, while their tributaries follow straight and wide anticlinal and synclinal valleys. The Eastern Predbalkan has a typical low hilly relief. Two large structural and denudational depressions are found here with floors at 200–300 m. Valley density ranges from 1.5 to 2.5 km km^{-2} and relative relief is 100–300 m km^{-2}.

The *transitional mountain and depression basin zone* lies between the Balkan mountain system to the north and the Rhodope massif to the south. This is a zone of high vertical and horizontal differentiation of large and diverse landforms: elongated mountain ranges (Sredna gora), solitary dome massifs (Vitosha), isolated sharp or flattened ridges, separated by a multitude of hollow depressions and valleys of various size and shape. The Western subzone, bordered to the east by the Topolnica River, is characterized by small-scale block faulting and sharply controversial vertical tectonic movements. The area has a jigsaw pattern of marked mountainous topography (Vitosha, 2,290 m; Koniavska planina, 1,487 m) and several elliptic elongated basins. The mountain ridges have a general northwest to southeast strike. Drainage density and relative relief show great variations (from 0.5 to 2.5 km km^{-2} and from 25 to 400 km km^{-2}, resp.). The Sredna gora–Zadbalkan subzone has better expressed linear morphographic features: the chain of the Sredna gora with two separate massifs (Sashtinska Sredna gora, 1,604 m, and Sarnena gora, 1,236 m). The southern slope descends step-like towards the Upper Thracian Lowland, while the northern towards the chain of Zadbalkan (Back-Balkan) basins, the most marked structure in Bulgaria, stretching from the Serbian border west of Sofia to the Black Sea. This linear corridor is fragmented by several transverse mountain ridges connecting the ranges of the Sredna gora and the Stara planina. The altitude of basin floors decreases towards the east. Relative relief in this subzone also shows great variations, similarly to the previous zone. The Upper Thracia–Middle Tundzha subzone includes the lowlands and hills along the Middle Maritsa and Tundzha Rivers, along with the coastal lowland of Burgas. Several solitary hills and low ridges rise above the flat lowland surfaces to altitudes of 416–600 m. The subzone of the Sakar and Strandja mountains, heavily dissected low mountains, occupies the southeastern corner of the country. Its main topographic units are Sakar mountain (856 m), and the northern and lower part of Strandzha mountain (710 m). Relative relief is low (up to 200 m km^{-2}), while drainage density is high (2–2.5 km km^{-2}).

The *Macedonia–Rhodope zone* is part of the vast Macedonia–Thrace mountain massif, which is shared by several Balkan countries. The present territory of Bulgaria includes parts of the Dardanian massif, represented by the Osogovo-Belasica mountain group and the Rila–Rhodope massif (the Rila, Pirin, and Rhodopes Mountains). Here the greatest topographic contrasts are observed in Bulgaria. Submeridional fault systems divide the massif into several subzones. The northern margin of the zone is sharp, defined by steep and straight slopes along the North-Rhodope Fault. The Osogovo–Belasica mountain group includes all the mountains along the Bulgarian–Macedonian border: Osogovo (2,252 m), Vlahina (1,924 m), Maleshevska (1,803 m), Ograzhden (1,644 m on Bulgarian territory), and Belasica (2,029 m). Mountain slopes are formed along fault-lines and, although having several planation surfaces, they are very steep, especially in the Belasica, where relative relief exceeds 500 m km^{-2} on the northern side. East of the mountain group lies the submeridional graben valley of the Struma River, with a series of valley embayments and basins, divided by transverse ridges crossed by gorges.

The subzone of Rila and Pirin includes both highest peaks in Bulgaria: Musala (2,925 m) and Vihren (2,914 m). With deeply incised river valleys relative relief often exceeds 600 m km^{-2}. During the Quaternary the high mountain regions above 2,150–2,300 m elevation were subject to mountain glaciation and glacial tongues on the northern slopes reached down to 1,100–1,300 m. Further to the east, the valley of the Mesta River is structurally similar to that of the Struma, but lies at considerably higher elevation with classical intramontane basins (the Razlog at 900 m and the Gotse Delchev Basin at 550 m).

The next subzone comprises the Western Rhodopes, a vast, highly uplifted mountain massif (2,191 m) with complex internal orographic divisions. River valleys are deeply incised, mostly along the fault network. Mountaintops are flat, forming the most extensive summit surfaces in the country (1,450–1,650 m above sea level). These properties are reflected in the low dissection of flat ridges (1–1.5 km km^{-2}) and the sharp increase of relative relief at deep valleys (up to 500 m km^{-2}).

The Eastern Rhodopes subzone comprises low mountains and hills along the Middle Arda River. Highest elevations here (up to 1,463 m) are reached on the ridges along the border with Greece. The ridges are gradually lowering towards the north. The widespread loose volcanic deposits allow minute surface dissection (drainage density: 2–3 km km^{-2}), while relative relief varies between 50–100 and 300 m km^{-2}.

14.2 History of Geomorphological Research

Five main stages can be outlined in the history of research into the geomorphology of Bulgaria. The *first stage* (before the Liberation in 1878) is represented by separate and scattered descriptions of Bulgarian lands by Western travelers and scientists who visited the Ottoman empire during the nineteenth century. The first studies were mostly performed by geologists: A. Boue, O. Viquesnel, F. von Hochstetter, F. Toula, H. and K. Scorpil, and J. Cvijić. An important tool for exploration of landforms was the Russian topographic map of Bulgarian lands at the scale of 1: 126,000, prepared during the Liberation war (1877–1878).

During *the second stage (1878–1944)* geomorphology in Bulgaria was affirmed and developed into a modern science. It was strongly influenced by leading geomorphologists (A. Penck, K. Ostreich, J. Gellert, H. Louis) with the participation of scientists from the Balkan countries (like J. Cvijić). Bulgarian geomorphology was strengthened through the contributions of Zh. Radev, G. Gunchev, Zh. Galabov, D. Jaranov, and others and supported by the foundation of the Bulgarian Geographical Society and several institutions specialized in complex research of Bulgarian nature and economy.

During *the third stage (1944–1989)* numerous specialized studies in geology and geomorphology were carried out. Most of them were initiated by the communist government for the needs of central planning. In the beginning of the 1950s two principal geographical research centers took shape: the Institute of Geography of the Bulgarian Academy of Sciences, and the Faculty of Geology and Geography at the Sofia University "St. Kliment Ohridski." Pure and applied research in geomorphology was focused on neotectonic movements, morphostructures and the different types of morphosculpture, planation surfaces, karst, periglacial, coastal, and aeolian processes. Among the main achievements of this period are the studies of underwater topography of Bulgarian Black Sea shelf, the geomorphological surveys and regionalizations made among others by Zh. Galabov, I. Vaptsarov, K. Mishev, Vl. Popov, A. Dinev, M. Georgiev, H. Spiridonov, D. Kanev, D. Parlichev, Il. Ivanov, M. Glovnia, T. Krastev, and A. Velchev.

The *fourth stage* started in the *early 1990s* under the hard conditions of economical transition and continuous crisis. Although limited by lack of funding, research was maintained by the passion and enthusiasm of researchers. In this period Bulgarian geomorphology started to leave the Russian influence behind and to open up to global science. Remarkable contributions from this period are by A. Velchev, R. Kenderova, G. Alexiev, G. Baltakov, Tz. Tzankov, Y. Evlogiev, and others. Hopefully, Bulgarian geomorphology has already entered its *fifth stage*, marked by the end of the economic transition and the EU accession in *2007*. In this stage national knowledge of landforms should be incorporated into regional and global structures, thus exchanging scientific experience and contributing to a better and more complete understanding of the environment of Europe, including the Carpatho-Balkan region.

14.3 Recent Geomorphic Processes

14.3.1 Surface Planation

Remains of several *planation surfaces*, diverse in age and origin, have been recognized in the complex topography of Bulgaria and investigated since the first decades of the last century. The pioneer studies were done by both German and Bulgarian geomorphologists: in 1924 Ostreich described four planation surfaces in the Western Stara planina, Wilhelmy increased their number and tried to estimate the age of their formation, Louis identified four planation surfaces and two lower deposition levels in the Western Rhodopes, Rila, and Pirin Mountains. Regional studies on the problem during the period 1925–1933 were done also by Gellert, A. Penck and Zh. Radev. As a result, four main planantion surfaces, aged from the Older Miocene, Younger Miocene, Older Pliocene, and Younger Pliocene, were assumed to exist. This initial theory was later upgraded and improved by Galabov and his supporters. In the two revisions of his studies, made for the monograph "Geography of Bulgaria" (Galabov 1966, 1982), he proposed the idea of the polygenetic origin of some of these surfaces. Vaptsarov (1975) described examples of tilted footslope surfaces of "glassy type" near Kyustendil and elsewhere, being of upper Villafranchian to Pleistocene age. Some new insights concerning denudation surfaces are found in the works of Mishev (1992, 1993), Vaptsarov (1997), Mishev and Vaptsarov (1984), Alexiev (2002a), and Alexiev and Spiridonov (2002). According to Alexiev (2002a, b, c, 2009) there is a single basic (initial) planation surface, which has suffered complicated transformations of uplift, deformation, fragmentation and defossilization during the Late Alpine orogeny and by further block faulting and burial under the conditions of extension in the following stages. The same author distinguishes relics of the following planation surfaces:

1. *Initial planation surface of Mesozoic–Eocene age.* Formed over the consolidated crust of the Hercynian crystalline basement. On the Moesian plate it is buried under Mesozoic and Cainozoic sediments of great thickness, while in the Balkan mountain system and the Macedonian-Thracian massif this surface has suffered complicated transformations of uplift and repeated planation (Alexiev 2002a). In the central zone of the Rila dome its fragments are preserved at elevations of 2,400–2,600 m, and towards the periphery of the structure it is gradually dropping to 1,800–2,000 m, to the south to about 1,700 m, while to the north to about 1,400 m. In the western Rhodopes the rhyolite cover of the Oligocene Bratsigovo–Dospat eruption built up a plateau that had fossilized the flat surface of the crystalline basement.
2. *Basic Late Eocene–Neogene planation surface.* It has the best morphological expression in the Danube plain, Predbalkan, Eastern Stara planina, and Strandja–Sakar zone. The correlative molass sediments from the renewed planation of the initial surface, widespread throughout the country, are assumed to be of the same age.
3. *Plio-Pleistocene levels on hills and foothills.* They are predominantly formed on unconsolidated sediments and closely linked to the evolution of valley networks and hillslopes. Two genetic types are distinguished: (a) an erosional-depositional type, related to active faults and the formation of active stepped alluvial-proluvial foothills (Alexiev 2002a), and (b) an erosional-denudational type, developed over heterogenic rocks, but most of all on the unconsolidated sediments of foothills, mostly in northern Bulgaria, below the loess cover, as well as in the valleys of the Struma and Mesta Rivers, in the Kraishte–Sredna gora area, the Eastern Rhodopes and other locations to the south. Corresponding to these levels are some old pre-Pleistocene river terraces 150–200 m above present river channels.

According to Tsankov et al. (2000), Tzankov (2002), the end of Late Alpine tectonic collisions and compressions was marked by the formation of large multiple *nappes* in central and southern Bulgaria, arranged in a staircase pattern descending towards the north to the southern flank of the Moesian plate. This led to the formation of a *landscape* of hardly dissected plains, hills and low mountains, spotted by preserved plateaus and solitary remnant hills, preserved almost unchanged until the early Neogene. Then the processes of folding and nappe formation ceased and a stable autochtonous tectonic setting came into existence with intensive exogenic processes. In the beginning of the late Miocene the regional tectonics switched to extension and the microcontinental fragment of the eastern Balkan peninsula was faulted and *dismembered* into blocks. At the end of the early Pleistocene extensional tensions acquired a subhorizontal direction and the blocks started to differentiate under the influence of lystric tectonics (Tzankov et al. 1998, 2000). Circular structures were formed (Tzankov 1995), Tzankov et al. (2005) and new mountain and plain morphostructures appeared. The present mountain morphostructures result from processes in extensive Neogene–early Pleistocene plains active in physical environments similar to the present-day savannas (Tzankov and Nikolov 2000; Tzankov 2002; Tzankov et al. 2005).

14.3.2 Fluvial Erosion and Landforms

The *valley network*, formed by river erosion, is the dominant morphosculptural element in the present Bulgarian landscape. Due to several million years of morphogenesis there is a great diversity of river valleys. In general, valleys in mountains are deeply incised, with steep slopes, and bedrock riverbeds are usually narrow. In plains and lowlands valleys are wide, slopes are gentler and wide channels are covered by thick alluvial deposits. River valley morphology and evolution, its spatial and temporal characteristics have been controlled by the levels of the sea in the Precarpathian depression, the Black Sea, and the Aegean Sea (Alexiev 2002a, b, c).

Drainage pattern in northern Bulgaria is concordant with the tilt of the initial surface. Most of the valleys are developed in *meridional* or *submeridional* direction. They start from the high areas of the Balkan mountain system, cross the folded structures of the Predbalkan and the horizontally bedded strata of the Moesian plain, and finally reach the Danube. Within the Danube plain rivers are incised 180–200 m deep, which gives the plain a hilly-ridge appearance. The particular depths and shapes of valleys depend on the resistivity of bedrock to erosion. Usually erosional spurs alternate with gorge-like passages. A *cross-section asymmetry* with a gentle left slope and a steeper right slope is specific for the western Danube plain best expressed in the middle and lower sections of river valleys. The drainage pattern of the eastern Danube plain is concordant with the North-Bulgarian dome (uplifted in the Neogene-Quaternary) and has a well expressed *radial pattern*: rivers start from the center of the dome in the vicinity of Popovsko–Samuilovski visochini Highland and flow towards the periphery. River incision intensifies in the same direction, reaching 180–200 m along the lower sections. There most of the valleys are incised in carbonate rocks and obtain a canyon character, with rocky cliff slopes (especially on the external section of bends) and flat floodplains. Incised meanders are especially typical for the Rusensli Lom River and its tributaries.

The drainage network in the Balkan mountain system reflects the elongated nature of the mountains and has a *feather-like pattern*. River valleys in the Stara planina are deeply incised with very steep slopes. In the Predbalkan slopes are gentler and valley bottoms wider. Here drainage pattern is modified by fold structures and turns more rectangular. The most extraordinary features are the gaps of the Iskar and Luda Kamchiya Rivers, which both cut entirely across the main Balkan ridge in deep and long *antecedent gorges*.

In southern Bulgaria the drainage network is predominantly controlled by the main orographic units. In the transitional zone, within the extent of the dome structures (Vitosha, Plana, Sakar, etc.) the upper sections of river valleys have radial arrangement, great longitudinal tilts and unbalanced valley profiles. Entering the basins, valleys drastically change both in longitudinal profile and type of cross-section. The ridges and hills around and in basins are crossed by *epigenetic* or *antecedent gorges*. The Maritsa River has a wide alluvial valley and appears as a main convergent axis of the valley network in the central and eastern part of the transitional zone.

Drainage in the Rila–Rhodope massif is closely related to the block morphostructures and their evolution during the Neogene–Quaternary. The large rivers, the Struma, Mesta, and Arda, have great variations in their *longitudinal profiles*, caused by fault tectonics and bedrock properties. Within the gorges anomalous sections with high gradients are found, while within extended and flat areas gradients rapidly drop. The gradient is highest for the rivers in the Rila and Pirin Mountains: in the northwestern Rila up to 280 m km^{-1} (Ivanov 1954), while in the foothill zone it decreases to 40–100 m km^{-1}. The type of bedrock is of great importance for valley morphology. In marble massifs *narrow gorges* (Trigradsko, Buynovsko) form and their entire floors are occupied by the river channel. In areas built of weakly consolidated Paleogene sediments and volcanic sediments river valleys are much wider, while on hard volcanic rocks rivers also form gorges. The most representative features of the latter type are found in the catchment of the Arda River. Young fault structures determine the existence of *straight valleys* (of the Shirokolashka, Vucha, and Dospat Rivers). Valley morphology in the middle and high parts of the Rila and Pirin was strongly affected by the extensive valley glaciation during the Quaternary, resulting in the development of typical *U-shaped valleys* (Beli Iskar, Demjanica, Bunderitsa, Rilska, etc.).

In unconsolidated sediments along footslopes sheet wash, rill and gully erosion has formed landforms, some of which are valuable as tourist attractions, such as the *earth pyramids* near Melnik (at the southwestern footslopes of the Pirin Mountains) and at Stob (the western footslopes of the Rila – Fig. 14.4). The combined action of erosion, weathering, and deflation shaped the cliffs of Belogradchishki skali (Western Predbalkan) in red Triassic conglomerates, the *"rock forest"* of Dikilitash (Rhodope Mountains), and numerous rock formations in the Eastern Rhodopes, carved in Paleogene volcanic sediments.

Fig. 14.4 The rock pyramids of Stob (Rila Mountains)

River terraces are the best indicators of the intensity of tectonic movements and the Quaternary evolution of the landscape. On the territory of Bulgaria traces of two flood terraces and 6–8 upper (erosional or erosional-depositional) terraces are found (Alexiev 2002a, b, c). The latter terraces have the following relative altitudes above the present riverbeds: T_7: 105–110 m; T_6: 80–85 m; T_5: 55–65 m; T_4: 35–45 m; T_3: 25–30 m; T_2: 15–20 m; T_1: 7–12 m; T_{0B}: 3–4 m (high flood terrace); T_{0A}: 1–2 m (low flood terrace). The oldest terraces (T_7, T_6) are considered to date from the Pleistocene, the lower (T_5, T_4) from the Eopleistocene, and the younger ones formed in the Middle and Upper Pleistocene. The two flood terraces are of Holocene age. Differential tectonic movements along with climate changes have caused regional differences in river terrace series. Traces of a maximum of five upper terraces are discovered along the Bulgarian bank of the Danube (in the area around Novo selo, Vidin district), while between Ruse and Silistra there are three of them. Some of the older terraces are covered by loess or have been destroyed by landslides and rockfalls. Flood terraces are best developed in the Danube lowlands. A series of two flood and seven upper river terraces is found in the Danube plain and also in the Balkan mountain system, where the altitude of the older terraces decreases from the slopes of Stara planina towards the Predbalkan. Within the Transitional zone a series of two flood and six upper terraces has been formed. High terraces are not found in the Zadbalkan basins, while in the areas of Pazardjik–Plovdiv and Yambol–Elhovo only the lowest

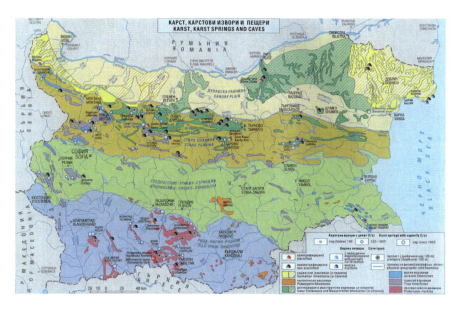

Fig. 14.5 Distribution of carbonate rocks, karst springs and caves in Bulgaria (By Popov 1997)

terraces exist. Holocene subsidence led to the appearance of swamps and marshlands in some river valley sections. In the Rila–Rhodope massif an other assemblage of two flood and seven upper terraces is found. High terraces occur at higher relative altitudes within gorge sections (up to 200 m). Flood terraces are well developed along the main valleys.

The traditionally accepted terrace scheme described above still involves a great deal of *uncertainties*. Assumptions are often exclusively based on relative position (above the present level of the river bed) instead of absolute dating. Especially uncertain are the cases when erosion terraces are identified without corresponding river deposits. The determination of the relative altitudes can be incorrect when the initial terrace surface had been tilted, or when it is not known which part of the terrace is preserved (the riser or the tread). The correlations between terrace series of the main river and those of its tributaries (each having a specific local base level of erosion) or between two or more tributary valleys are also problematic (Stoyanov 2004).

14.3.3 Karst Processes and Landforms

Karstic areas in Bulgaria occupy nearly 23% of the total area (Fig. 14.5). Karst is developed in two main types of carbonate rocks: sedimentary (limestones and rarely dolomites) and metamorphic (marbles). The observed features and landforms show considerable regional differences. In each region their particular characteristics reflect to some extent the nature of the corresponding geomorphic unit.

Karst in the Danube plain is developed in Jurassic and mainly of Cretaceous as well as Tertiary (Sarmatian) limestones. Most areas are specific with small tilts of the carbonate strata and heavy jointing. Small precipitation amounts and the shallow river incision do not favor deep karstification, with the exception of the eastern parts (within the North-Bulgarian dome), the Dobrudja and Ludogorsko plateaus, where karstified rocks are found to depths of 4000 m.

The karst in the western and central section of the Danube plain is open, forming flat *karren fields* of barren rock and scarce vegetation. Buried deep karst complexes, fossilized by a cover of Plio-Pleistocene clays and loess deposits are typical for the Dobrudja and Ludogorsko plateaus. Here a well-developed system of canyons and dry valleys can be recognized under young deposits. Karstified rocks are exposed only on canyon walls and along the coastline where *abrasion combines with karstification*, thus forming a complex of coastal caves, rock arches and steep cliffs (Fig. 14.6). The most prominent karst landforms are situated in the canyon system of the Rusenski Lom River, where limestone walls reach heights up to 70 m. Some of the several large *cave systems* here have 2–4 stories that mark consequent stages of karst incision and correspond to river terraces. One of them is Orlova chuka (13 km), near the village of Pepelina, which is the second longest in Bulgaria.

Karst in the Balkan mountain system is found most of all in the Predbalkan area with massive folded sequences of Mesozoic limestones and dolomites and rarely

Fig. 14.6 The Black Sea coast at Tjulenovo: a combination of abrasion and karst

along the main ridge of the Stara planina. In the Balkan mountain system climate is more humid, and the various structures with a sufficient expression in elevation allow deep karstwater circulation (Popov 1997). Some deep *caves* are found here: Raychova dupka (−372 m) near the village of Cherni Osam, Belyar (−246 m) near Vratsa, Lednika (242 m) near Kotel, and 17 more abyss caves deeper than 100 m. Other prominent cave systems are the Saeva dupka, Bacho Kiro, Ledenika, and the Magurata, where prehistorical paintings are found. All these caves are open for tourists. Other impressive caves are found at Devetashko plato (the system of Devetashka cave), the area of Emen (near Veliko Turnovo), Morovitsa (near Teteven), and Parnicite (Bezhanovsko plato). Some of the most impressive *rock arches* in Bulgaria are found in the Predbalkan area: Prohodna near Karlukovo village with a height of 45 m (the highest cave ceiling), Bozhia most close to Vratsa, the stone bridges of Devetashka cave (Pleven area), and others.

Surface karst landforms include the 400 m high *cliffs* of Vratsata (south of Vratsa), Ritlite (the Ribs) in the Iskar gorge. Canyons with *sinter cascades* are situated in the typically plain-hilly areas of the central Danube plain.

The Stara planina is specific with smaller distribution of karst due to its lithology (more limited extension of suitable rocks), particularly in the western Stara planina, however, several typical karst regions are present, such as the Lakatnik area with Temnata dupka cave (in the Iskar gorge, total length: 9 km Gachev and Gachev (2005)), Zimevishko plato, Ponor planina, the area of Zabrge (southern slopes of the Stara planina along the border with Serbia). The Balkan mountain system (Predbalkan and Stara planina) hosts 53% of the mapped caves in Bulgaria.

Karst features in the Transitional mountain-basin zone are developed mainly in Triassic limestones in the Kraishte area as well as in small and scattered areas in the hills and low mountains along the border with Turkey. Here is the longest cave in Bulgaria, Duhlata (18 km), situated on the southern slope of Vitosha Mountain (one of the most beautiful in Bulgaria, but closed for the public).

In contrast to the mentioned three geomorphic units, karst in the Rila–Rhodope massif is developed in metamorphic rocks, pre-Mesozoic *marbles* of the Pirin, Slavyanka, and Western Rhodope Mountains, and only small features in the Triassic and Paleogene limestones of the Vlahina, Osogovo, and Eastern Rhodopes. In general, due to the cooler climate and the dense forest cover classical surface karst landforms in the Rhodopes are rare, but subsurface forms are well developed. The three large karst areas in the West Rhodopes, Dobrostan, Trigrad, and Velingrad-Shiroka poljana, serve as vast underground water reservoirs due to the larger precipitation amounts and lower evaporation. Outside these main areas karst landforms are found in numerous smaller spots: near Smolyan, in Dabrash, west of Chepelare, and elsewhere. The Rhodope Mountains host some of the most prominent caves in Bulgaria. Popular tourist sites are the 8.2 km long Yagodinska cave, Djavolskoto garlo (Devil's throat), where the highest Bulgarian *subterranean waterfall* is found, Uhlovica, Snezhanka. Among the most interesting karstic phenomena are Chudnite mostove (the Bridges of Wonder), two *rock arches* on the eastern slope of Chernatica mountain (Fig. 14.7), and the gorges Trigradsko and Buynovsko, cut deeply in the marbles of Trigradsko Plateau.

Fig. 14.7 Karst features in the Rhodope massif: Chudnite mostove rock arch (*left*), Uhlovica cave (*right*)

In the Pirin Mountains built of marble, and especially around Vihren peak, where the mountain ridge rises above 2,800 m, karst is strongly influenced by high mountain conditions and a combination of karstic, periglacial, and glacial landscapes evolved. The marble strata, tilted to the northeast, are ca 2,000 m thick. Although barren rock surfaces prevail at the highest elevations, surface karst forms such as karren and gorges are less marked, because periglacial processes prevail over karst corrosion. Relict glacial cirques on the northern slope of the mountain represent a combination of *cirques and dolines*. Cirque bottoms are dotted by numerous caves, most of which are vertical *shafts*. There is a chance to find here the deepest cave system in Bulgaria (snowmelt and rainwater from the cirques drain 2,000 m below in the Razlog Basin), the area, however, is still not well studied. Research until now has pointed out that Propast (Abyss) No 9 in Banski suhodol cirque is the deepest (−226 m).

Karst processes are significant agents in shaping the Bulgarian landscape, especially in the Predbalkan area and Rhodope mountains. They are valuable attractions for tourism.

14.3.4 Coastal Processes and Landforms

The Bulgarian Black Sea coast is a strip of various width (5–10 km or 40–50 km) extended between Cape Sivriburun (to the north) and the mouth of Rezovska river (to the south). The total length of the shoreline is 378 km. The coast has been sculpted over zones of diverse bedrock and structure, typical of the main morphostructure units of Eastern Bulgaria, continued underwater towards the Blak Sea. The fold zones of the Stara planina and Srednogorie structures sharply turn towards the southeast (Krastev 2002).

According to Popov and Mishev (1974) the predominant part of the Bulgarian sea coast (59.7%) has a character of an abrasion cliff, while 28% of the total length is estimated to be accumulational type of coast, and 12.3% to the landslide-abrasion type. Sand strips (beaches) cover an area of 16 million m². For structure and geomorphology the coastal zone is divided into four subzones (from north to south): the Moesian, Balkan, Burgas, and Medni rid–Strandja subzones.

The shore of the Moesian plate (the Danube plain) is of neutral type. The morphology of the shore is determined by the horizontal strata of limestones, marls and clays of predominantly Neogene age. The exact configuration of the shoreline is also influenced by the fault system of the block-fault geological structure. Several sections with specific morphology can be identified along the Moesian coast: between the Capes Sivriburun and Shabla the coast is relatively straight, shaped in loess deposits, gradually rising to 10–15 m towards the south. From Cape Shabla (the easternmost point of Bulgaria) the shoreline follows a southwestern direction, controlled by the seismically active Kaliakra dislocation. The coast height rises to 60–65 m at Cape Kaliakra. It is steep, almost vertical, and *abrasion* has carved numerous caves and notches (see Fig. 14.6). Here occurred two big *landslides* (Krastev 2002). Beyond Cape Kaliakra the coast increases its height to 110 m. The almost vertical cliff is due to the rigid middle and upper Sarmatian limestones. After the town of Kavarna the rim of the Dobrudja Plateau abruptly turns away from the shore. Here landslide processes are very active. To the southeast is the wide valley of the Batova River with a large sandy beach in its mouth. To the south the shoreline follows the margin of the Frangensko Plateau, where some of the most extensive and most active landslides in Bulgaria occur.

At the Bay of Varna the Varna–Beloslav lake district deeply penetrate the continent. South of the bay the coast rises again to a height of 125 m along the rim of Avren Plateau, and then follows a low accumulation section at the mouth of the Kamchija River with the largest, 11-km long sandy beach in Bulgaria. According to Popov and Mishev (1974), along the Moesian coast a full complex of *marine terraces* are present. Relics of the terraces Chaudinska (85–90 m), Staroevksinska (55–60 m), Evskino–Uzunlarska (30–40 m), Starokarangatska (20–25 m), and Mladokarngatska (8–12 m) are found along different sections of the coast. The Novochernomorska terrace (4–6 m) is most often found in complex with Karangat terraces (see below).

In the Stara planina area the coast crosses the anticlines and synclines of the Balkan mountain system at almost right angles (classical *transverse type* of coast). Abrasion coast predominates especially along anticlinal structures, which are offset into the sea and culminate at Cape Emine, the easternmost point of the Stara planina. Sandy beaches developed near river mouths and landslides are found at several spots. Here the full complex of marine terraces is present, including two high Chaudinska terraces (Popov and Mishev 1974; Krastev 2002).

The next section of the coast forms the eastern margin of the extensive lowland of Burgas. The shoreline is highly dissected by bays (the Bay of Burgas is the westernmost point of the Black sea; Bay of Pomorie). The low coast is of accumulation type with *limans* and *lagoons*. In the northern part, on the 5.5 km long beach of Slanchev

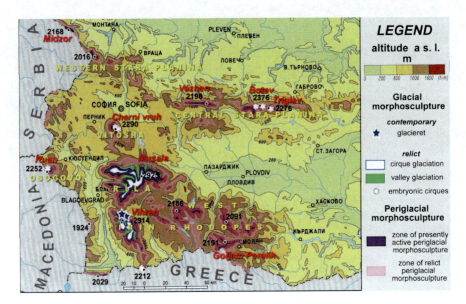

Fig. 14.8 Distribution of present and relict glacial and periglacial processes and landforms

bryag and on the beach of Nessebar *sand dunes* are formed with height up to 11 m. Remarkable landforms are the *double tombolo* and the sand bar of the lagoon of Pomorie, the peninsula of Nessebar and the big liman lakes of Atanasovsko, Burgasko, and Mandra. Older marine terraces are only recognized in the periphery of the lowland, while the Upper Pleistocene terraces are located in the vicinity of the shore.

The coast between Medni rid and Strandja has the most dissected by a series of small rocky peninsulas ending in steep and hardly accessible *headlands* with small arcuate bays, fine sand beaches and small-size lagoons between them. The mouths of the Ropotamo, Dyavolska, Veleka, Rezovska, and other rivers are typical limans, filled with marine and fluvial deposits. Here high terraces are poorly preserved, while the lower Old Karangat terrace is widely observed.

14.3.5 Glacial and Periglacial Processes and Landforms

Glacial features in Bulgaria are almost entirely relicts of the Pleistocene glaciations, when air temperatures were 5–7°C lower than at present and the snowline in mountains descended to 2,100–2,300 m elevation. As the equilibrium line altitude (ELA) of glaciers in Bulgaria was highest in relation to all the other countries in the Carpatho–Balkan area, classical glacial landform complexes were developed only in the highest mountains Rila and Pirin, which rise above 2,400 m (Fig. 14.8). Some embryonic forms are found in other high mountains such as the Stara planina, Vitosha, and Osogovska planina.

Most researchers of glacial landforms (Glovnia 1958, 1962a, b; Popov 1962; Choleev 1982; Velchev 1999) share the opinion that traces from two glaciations (Riss and Würm) are evident in the Rila and Pirin. The Riss glaciation is only assumed from some existing exaration features such as high cirque and valley shoulders, but no glacial deposits have been preserved. On the contrary, Würm glaciation left behind whole complexes of well preserved, relatively "fresh" glacial landforms, several sets of moraines that mark the main phases of the last glaciation, the largest extension of valley glaciers during the Last Glacial Maximum (LGM – 23–18 ka BP) and the following retreat stages in the end of the Pleistocene. The LGM age of the lowest *terminal moraines* in the valleys of the Rila mountains was confirmed by ^{10}Be dating (Kuhlemann et al. 2008). The positions and characteristics of relict glacial landforms in the Bulgarian mountains are determined by the direction of prevailing winds and moisture supply during glaciations as well as by the local patterns of lithology and tectonics. During the LGM valley glaciers reached down to 1,100–1,300 m along the northern slopes of the Rila and Pirin and on the western slopes of the Rila. The longest was the Beli Iskar valley glacier in the Rila (ca. 22 km). The widest and deepest *glacial cirques* are developed in northern and northeastern aspects due to shaded conditions and the leeward accumulation of snow by southwestern winds that prevailed during the LGM. Here a most complete set of glacial landforms is present.

Glacial landforms appear in two different bedrock settings: on silicate (granitoids) and on carbonate rocks (marbles). Silicate rocks build all the Rila mountains and most parts of Northern Pirin. In favorable aspects several large complex stepped cirques ending in classical *U-shaped valleys* are seen. Numerous *tarns* are scattered on cirque bottoms between 2,050 and 2,710 m elevation – their total number is about 270 (140 in the Rila and 110 in the Pirin). The largest tarns have surfaces of 21.2 ha (Smradlivo lake, Rila) and 11.5 ha (Popovo lake, Pirin), and the deepest is 38 m deep (Okoto, Rila). Some glacial depressions are fully or partly filled by fossilized *rock glaciers*. From the northern side high peaks often have a typical karling (horn) shape with rock faces up to 250 m high, while their southern sides are most often steadier. However, there are also some exceptions: the most impressive *vertical wall* in the Rila Mountains on the southern slope of Maljovica massif (Fig. 14.9). Inaccessible *arêtes* are formed along mountain ridges where two adjacent cirques contact.

Several sets of *moraines* are deposited in cirques and valley bottoms that mark consequent stages of glacier retreat after the LGM. Some of them are identified as traces of the Oldest Dryas (about 15 ka BP) and the Younger Dryas (10–11 ka BP) stadials. The uppermost moraines in the highest cirques are considered to be of Holocene age (Kuhlemann et al. 2008) – related to the so called "8k" cooling event, or even to the Little Ice Age. The underwater crescent-shaped ridge found on the bottom of Ledeno ezero (2,709 m), near Musala peak, is such a case (Gachev et al. 2008). In addition, typical glacial features such as roches moutonnées, confluence steps, valley trimlines, and others are also abundant.

On the marble surfaces (in the highest part of the Northern Pirin Mountains, Vihren, 2,914 m) karstified bedrock did not favor the development of long valley glaciers as ice flow was hindered by the drainage of subglacial meltwaters down to

Fig. 14.9 The south wall of Zlia zab peak

Fig. 14.10 Varieties of glacial landscape: silicate bedrock (granitoids) – Musala, Rila (*left*), carbonate bedrock (marble) – Vihren and Kutelo peaks, Pirin (*right*)

karst caverns. Consequently, in northern and northeastern exposures several short but overdeepened cirques formed instead. The highest vertical rockwalls in Bulgaria, such as the 420 m high northeastern wall of Vihren peak are found here (Fig. 14.10). Vihren area is the only place in Bulgaria where glacierets exist. According to Grunewald et al. (2006) a *glacieret* or a *microglacier*, is a permanently existing firn (névé) body, with the presence of stratified annual layers and with sufficient dimen-

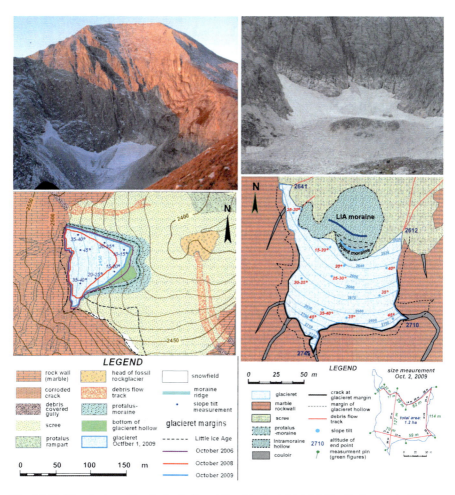

Fig. 14.11 The glacierets Snezhnika (*left*) and Banski suhodol (*right*) (By Gachev et al. 2009)

sions (area at least 1,000 m^2 and several meters of firn thickness). Although not having the typical characteristics of mountain glaciers, glacierets are in fact embryonic forms of recent mountain glaciation. Research (Popov 1962, 1964; Grunewald et al. 2006, 2008, 2009; Grunewald and Scheitchauer 2008; Gachev et al. 2009) has shown that they have been stable and persistent at least during the last few centuries, that they are typical with a slow downward movement of firn producing modern deposition forms (moraines) at their margins. Two glacierets are found and studied: the Banski suhodol glacieret (firn area was 1.2 ha in autumn 2009) and the Snezhnika glacieret (firn area fluctuating from year to year between 0.82 and 0.30 ha), the latter having already a 50-year tradition of research (Fig. 14.11).

Glacier meltdown at the end of the last glaciation produced *fluvioglacial landforms* in river valleys and at footslopes. As glaciers in Bulgaria did not reach beyond mountain margins even at their maximum extent, the most extensive deposits of fluvioglacial materials are in the form of *large fans* at valley outlets (e.g., Bansko in the Pirin and Dupnitsa in the Rila). Fluvioglacial deposits near the village of Ploski (southwestern Pirin) were TL dated at 111 ka BP (Riss–Würm interglacial, OIS3) (Velchev 1997). Similar deposits were also described from the northern foot of Vitosha and the massif of Midzor (Velchev et al. 2000), but their origin and age are still under debate.

Glacial features from the Pleistocene in Vitosha, Osogovo, and the Stara planina are rare and embryonic. Recently it has been accepted that during the coldest phase of the Würm an ice cap existed on the summit plateau of the 2,290 m high Vitosha mountain (ELA during LGM estimated at ca. 2,100 m) but it was thin and no glacial tongues descended from it. However, several clearly expressed embryonic cirques (nivation hollows) can be recognized: Velchev and Todorov (1994) reported about the existence of small cirques and U-shaped valleys on the Bulgarian side of the Osogovo Mountains (2,252 m). In the Stara planina small but well shaped cirques are found on the eastern slope of the highest point, Botev peak.

At present *periglacial processes* (in the broad sense of this the term) are active in the mountains above 1,900–2,000 m (Glovnia 1959, 1962, 1968), where freeze-thaw alterations in the active layer (20–60 cm) are frequent and annual precipitation is sufficient (at least 900–1,000 mm, mostly snow). In the Rila and Pirin Mountains intensive periglacial processes (cryofracture) are promoted by the sharp topographic contrasts of the glacial relief. Periglacial landforms and processes are both recent and relict, combined in the high mountain belt, while at lower altitudes (1,200–1,300 m) they are entirely Pleistocene relicts. Frost action on coarse-grained crystalline rocks induced the formation of thick layer of regolith on tops of mountain ridges, on high uplifted flat summit surfaces and especially over high mountain cols. Large *screes* cover cirque bottoms in the Rila and Pirin and usually form linear series of cones. Some of them represent relict rock glaciers that once had cores of dead ice (Fig. 14.11). However, none of them has been found recently active. Vast scree surfaces of in situ weathered or weakly transported debris, *felsenmeer*, are spread on southern slopes above 2,000 m elevation but in some cases are also observed below this altitude. Among the most typical forms are the so-called *"stone rivers,"* debris-covered strips of various length and width that follow river and stream valleys. Water currents are often hidden under the debris blocks of several meters in diameter. In this case prolonged frost weathering is supported by the continuous outwash of weathered particles by stream waters. Stone rivers (Fig. 14.12) are mostly but not entirely formed, by periglacial processes. They are found at various elevations, usually between 1,200 and 1,900 m. Stone rivers have different size, shape, and roundness of stone blocks in dependence of rock type and microclimate. Especially interesting are these features in Vitosha, where they are widespread and typical. Here huge round blocks are formed from the sienite and montsonite of coarse fabric mainly through exfoliation. On silicate rocks of the Rila and Pirin Mountains stone rivers of more angular and smaller blocks are found above the mid-mountain forest belt.

Fig. 14.12 The stone river Zlatni mostove in the Vladajska River valley (Vitosha)

The group of cryonival forms includes *nivation hollows*, embryonic cirques, ridgetop hollows, *protalus ramparts*. *Couloirs* on steep slopes serve as tracks for avalanches and debris flows, usually short and only exceptionally initiate hazardous events. The most dangerous locations in terms of avalanche risk are the slopes of Todorka peak near Bansko and the Kominite couloir in Vitosha, because these areas receive many visitors in winter.

Active periglacial high mountain areas show a variety of *gelisolifluctional features*: stone strips, grass hummocks, terraces, creeping boulders. Until now the existence of permafrost has not been proved anywhere in the Bulgarian mountains, but more studies are needed in this matter.

14.3.6 Aeolian Processes and Landforms

Although not playing a primary role in geomorphic evolution in Bulgaria, aeolian processes modify landforms and contribute to landform diversity. The clearest evidence of aeolian activity is the formation of loess from aeolian deposition and of dunes in the sandy beaches along the Black Sea shore.

Loess and loess-like formations cover about 19,000 km^2 of Northern Bulgaria, thus representing the southernmost occurrence of loess in Europe (Mitev and Iliev 2002).

Loess in Bulgaria appears as a continuous strip of variable width along the Danube River and the state border with Romania from the Timok River to the Black Sea coast (Fig. 14.13). Remarkable regional differences are observed from west to east, as well as from north to south. Small spots of loess-like formations are found also at several locations south of the Balkan range. Everywhere the Bulgarian–Romanian border serves as the northern limit of loess distribution, while the southern limit can be traced along the southern margin of the Danube Plain. To the west the loess strip is only 5–10 km wide but broadens to 100 km in Eastern Bulgaria. Loess thickness is greatest near the Danube River (80–120 m), and decreases towards the south, strictly controlled at meso and microscale on subsurface paleotopography. Absolute maximum thickness is at Orjahovo (145 m), while average thickness is about 50 m (Minkov 1968). Four types of loess follow from north to south: sandy loess, typical loess, clayey loess and loessy clays.

The most typical loess landforms are:

1. The *loess plateaus* in the western Danube Plain with a thick loess cover and expressed slope asymmetry (steep to 15–20° western and gentle 3–5° eastern slopes), predefined by the patterns of river erosion.
2. *Loess humps* – extensive ground rises of increased loess accumulation, situated in the vicinity of the Danube River. They appear as northern spurs of karst plateaus extending towards the river, usually rising 20–30 m above the plateau surface and descending towards the Danube, often terminating in steep bluffs with active landslides.
3. *Loess gredas*, i.e., small, elongated, parallel, submeridional ridges developed through the reworking of loess sands along the valley of the Danube.
4. *Loess pots* are widespread hollows, several hundred meters long and 1–6 m deep. Most of them are associated with *piping*, while in Dobrudja they are suggested to reflect the underlying fossilized karst features.
5. *Funnels* are small, ravine-like features, usually about 200–300 m long, 30–70 m wide and 15–40 m deep (Mitev and Iliev 2002). Spread along plateau margins, next to the valley slopes, they are produced by the erosion of temporary watercourses on the unconsolidated loess surfaces.

Fig. 14.13 Distribution of loess in Bulgaria (Mitev and Iliev 2002). Loess types: 1, sandy loess; 2, typical loess; 3, clayey loess; 4, loessy clays. Loess morphosculpture: 5, large loess funnels; 6, loess potholes; 7, isolines of average loess thickness

6. *Bluffs* develop as a result of the typical vertical jointing of loess. Some of them are associated with large landslides. The largest is the 4 km long and 80 m high Kozloduy bluff (Minkov 1968).

Wind action in the present is associated with the formation of *sand dunes* in coastal areas, favored by frequent winds blowing from the open sea and by the widespread occurrence of fine sand along the Black Sea coast. Beaches up to several kilometers long and 50–100 m wide develop.

Wind action also redistributes snow cover in mountains during the winter and builds temporary features of snow: gullies, dunes, aprons, snow patches. Thus by transferring snow from one side of a mountain crest to the other, wind action strongly influences avalanche activity.

14.4 Direct Human Impact on Landscape

At present human activity is no doubt the most active geomorphic agent. On the territory of Bulgaria it has a history several millennia, acting in various ways and at various rates. The oldest evidence for anthropogenic changes of topography date back to ca 3,000 years BP, related to fortification, construction of roads, burial tumuli, slope terracing and river channel corrections (Fig. 14.14). Such modifications, however, did not have a major impact on the environment until the nineteenth century, when industrial development started.

14.4.1 Impact of Mining

Although spatially limited, it is the most severe intervention into the natural environment, often resulting in total destruction of the initial landscape.

Excavation landforms result from the opencast extraction of mineral resources. Although *quarries* are often small in size, they are quite large in number – about 800 all over the country. Districts with the highest number of quarries are those of Sofia (88), Lovech (58), Silistra (54), and Sliven (51). The total area directly destroyed by quarries is estimated to about 60,000 ha and another 60,000 ha are indirectly disturbed by servicing quarries and storing quarry waste (Vlaskov 2002). The total number pits for extraction of inert materials along flood river terraces is 380 for the year 1998, with a total area of 11,400 ha, mainly located along the Danube and the lower sections of its tributaries and around Sofia and Burgas. Many such abandoned pits to the east of Sofia are turned into artificial lakes, some of them used for recreation. The largest excavation forms are *opencast mine pits*. Most of them are related to the extraction of coal and kaoline. The greatest pit near Kremikovci (20 km east of Sofia, iron ore mining, Fig. 14.15) is more than 1 km in diameter, and the deepest "Asarel" pit reaches a depth of 700 m. In terms of surface area the most affected is

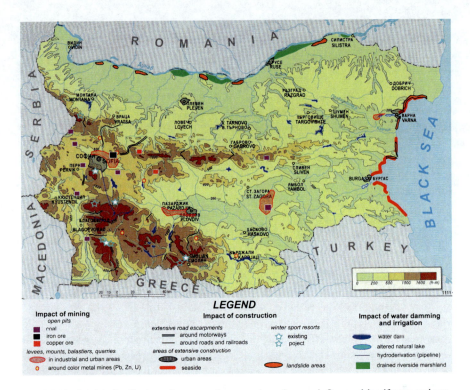

Fig. 14.14 Spatial distribution of some anthropogenic or human influenced landforms and processes categorized by the type of human impact

the eastern part of Maritsa valley, where open pits occupy 178,000 ha (Vlaskov 2002). Ponds or marshes often form at their bottoms.

Supporting landforms appear when waste material from construction or mining is piled. *Levees* are elongated artificial elevations of various height. *Platform mounds* (*tabans*) are the most extensive of anthropogenic landforms, usually with steep slopes and a flat top. There are ca 3000 *waste disposal sites* in Bulgaria, of which only 560 are legal (Vlaskov 2002). According to the Ministry of Environment and Water the total area covered by various types of waste is ca 130,000 ha.

14.4.2 Impact of Infrastructure and Housing Construction

Linear transport infrastructure feature does not affect large areas, but forms dense networks and their great total length results in millions of square meters of altered landscape. The construction of motorways, especially in hilly and mountainous areas, when escarpments, cuttings, tunnels, bridges and barrages are made, involve major alterations in the landscape. Most affected areas are along *motorways*. The

Fig. 14.15 The excavation pit of Kremikovtsi iron mine: a futurist landscape

motorway network in Bulgaria is still of an insufficient length and density, so at national scale these effects are negligible, and the same also valid for the density of road network in mountains. Major transformations of the natural environment are especially observed on passes, gorges, deep valleys and other naturally narrowed sections. *Forest roads* in mountains serve as channels for gully erosion (especially in soft bedrock).

Tourism infrastructure has a considerable effect on the mountain environments. Modern winter sport centers in Vitosha, Rila and Pirin have a particularly disastrous impact. The worst is the case with Bansko resort, where between 2001 and 2005 several kilometers of new ski tracks were established in the Pirin National Park, an UNESCO Natural Heritage site. Tourist trails also induce rill and gully erosion, particularly in thick unconsolidated regolith on the flat main ridges of the Rila, Stara planina and Vitosha Mountains. The northern slopes of Vitosha, the area mostly preferred for hiking, are considerably affected.

By the intensity of human impact, *housing construction* follows mining. In 2008 Bulgaria had a total of 5,304 settlements (of which 255 towns). Major transformations are observed in urban areas. In mountains settlements are concentrated in valleys that suffer severe human impact, while mountain slopes and ridges are much less affected. Intensive housing construction on the Black Sea shore exerts a particularly strong impact, in addition to the pollution and the negative effects on aesthetics, at many sites such works trigger extensive landslides. The most typical sad example is the southern part of Zlatni pjasaci (Golden Sands) resort. Excessive constructions along the Danube, around Bansko, Razlog, and Pamporovo result in pollution, increase of erosion, landslide, and rockfall risks.

14.4.3 Impact of River Damming and Agriculture

Other major direct impact on natural landscapes is associated with river damming and irrigation. River regulation dates back to ancient times, but it only started to play a significant role in the twentieth century. Most considerable changes were made along the main rivers. In the 1990s the overall length of channelized rivers was about 2,836 km. Building of dams started in the 1930s, when numerous but relatively small reservoirs were created. The largest dams of *cascade systems* date from the early stage of the communist period (from the 1950s to the 1970s). The biggest one is the Iskar dam (between Samokov and Sofia), built in 1956. Cascades of reservoirs were built on the largest rivers and summits of the Rhodope Mountains to collect rainwater and snowmelt (see Table 14.1). The highest dams are those of Vucha (145 m), Krichim (104 m) and Kardzali (103 m). Additional damming has increased the surface of some natural lakes. Building of an artificial channel has turned the freshwater Beloslavsko Lake into the Bay of Varna (Black Sea). Dam construction is combined with creation of long *canals for water transfer* from one river system to another (the longest being the Rila–Belmeken that collects water from all the Rila Mountains, the main water pipeline of Sofia from the Rila Mountains and the Dospat–Vucha in the Western Rhodopes).

In the first decades of the twentieth century almost all floodplain *swamps* and marshes along the Danube were *drained* and the floodplains began to be cultivated. To achieve this many kilometers of *dykes* were built along riverbanks. Similar measures were also performed along other main rivers. Many kilometers of artificial canals were built in lowland agricultural areas for irrigation and drainage. They change natural topography and serve as paths for the transport and accumulation of sediments.

Although at a small rate, *agriculture* has changed the natural landscape over extensive areas, especially in hilly and low mountain regions. The most widespread forms are vineyard terraces. Rills and gullies are formed in steep slope sections,

Table 14.1 Largest river dams in Bulgaria (Source: Geography of Bulgaria 2002)

Name	River	Total volume – 10^6 m^3	Year of opening
Iskar	Iskar	673	1956
Kardzali	Arda	538	1964
Ogosta	Ogosta	505	1985
Studen Kladenets	Arda	489	1957
Dospat	Dospat	446	1967
Zhrebchevo	Tundzha	400	1970

especially as a result of improper tillage. Numerous traces of old roads and stone walls modify local topography throughout the country.

14.4.4 Impacts of Changes in Forest Cover

Until the first decades of the twentieth century mass *deforestation* around villages for the sake of gaining agricultural land, particularly for sheep and goat keeping had resulted in extensive erosion as well as in the formation of large proluvial and colluvial fans at footslopes. With the adoption of the Law of Forests in the 1930s most of these degraded lands were planted with conifers (predominantly with *Pinus silvestris,* subordinately with *Pinus nigra, Robinia pseudoacacia*). Although sometimes widely criticized, this had led to a drastic drop of erosion and ground stabilization. At present many traces of old rills and gullies can still be found in these "subrural" forests, but almost none of them have been currently active. Overgrazing had also led to the formation of "grazing terraces" in high mountain slopes above the timberline, as well as "parallel paths" in forests. After 1989 mountain grazing activity has been reduced not only in the Rila, Pirin and central Stara planina, where national parks were established, but also in other parts of the country due to the collapse of agriculture. As a result natural landscapes started to recover and many former pasture lands are at present covered with bushes and young forests. Recently, with the gradual recovery of stock breeding and the increased rate of forest clearance (mostly illegal) human influence started to intensify again.

14.5 Perspectives on Landscape Evolution

At present the main issues in geomorphic evolution are the following: (1) the regional aspects of global environmental change and future climate; (2) human impact in the light of the new economic conditions in Bulgaria as a member of the European Union, and (3) endogenic processes as triggers of landform evolution.

14.5.1 Landscape Evolution and Climate Change

Instrumental climatic observations in Bulgaria indicate that mean annual air temperatures (MAAT) remained relatively stable when comparing periods 1931–1960 and 1961–1990, but have a moderate 0.1–0.4°C rise for 1979–2008 compared to 1961–1990 (Velev 2010). On a decade scale the periods 1900–1920 and 1940–1980 can be considered relatively cool, while the periods 1920–1940 and 1990 to present relatively warm. A slight long-term tendency towards *warming* is more obvious *in high mountains*. In the period 1933–2007 MAAT at Musala peak (2,925 m) rose from −3.1°C to −2.3°C when comparing 10-year moving averages, and 0.5°C of this rise occurred over the last decade (Nojarov 2008a). Similar is the situation at Cherni vruh (2,290 m) and Botev peak (2,376 m) (Nojarov 2008b, 2009). At the same time, lowland areas suffered quite limited or no warming at all. The most evident climate change in Bulgaria for the last three decades concerns *winter temperatures*, especially monthly averages for January – almost all stations registered a considerable *rise*, +0.6°C for South-Bulgaria and +1.8°C for North-Bulgaria. This can be called "*maritimization*" of climate, as at the same time summer temperatures in fact do not change and MAAT ranges lowered (Nikolova 2002). Some drop is observed in autumn by 0.5–1.4°C and spring shows controversial tendencies (Nikolova 2002).

Changes in precipitation are much more complicated. For the longer part of the period of instrumental observation (the standard period 1961–1990 compared to previous periods) no serious changes in *precipitation* amounts and regime had been observed. Decreasing trends appeared in the past 30 years, as –1% to –15% *drop* of annual sums is registered for the period 1971–2000 compared to 1961–1990 (Velev 2002). The trends are most expressed *in Southwest-Bulgaria*, while there is almost no change in the northern and northeastern parts of the country. In the past two decades the greatest decrease is measured in high mountain areas (according to some authors exceeding 25%, but here precipitation data is quite uncertain and evidence of measurement errors have also been revealed).

Although future changes are difficult to predict and evaluate, the following statements can be made at this stage:

1. As stated in the Fourth Assessment Report of IPCC (2007) one of the predictable effects of global warming is the increasing frequency of *extreme climatic events* (intensive rainfalls, thunderstorms and droughts). In Bulgaria such phenomena occurred quite often in the past several years. A two-fold increase in the frequency of *thunderstorms*, heavy rains and snowfalls were observed from 1993 to 1998 (Nikolova 2001). Intensive rainfalls increase probability of floods and debris flows, which are known as main triggers for geomorphic evolution: river channel modifications, increased slope or riverbank erosion and deposition, landsliding, and rockfalls. Prolonged *droughts* make the soils more susceptible to *aeolian* blowout and *erosion* during floods.
2. Temperature rise in high mountains *intensifies periglacial processes* (frost weathering, landslides, rockfalls, debris flows, solifluction, and creep), because of the increase of freeze-thaw cycles in the active layer (Gachev et al. 2008).
3. The expected shrinkage of snow cover will lead to *water shortage* through lower river discharge. Snowmelt will occur earlier and snow accumulation later causing a disequilibrium in high mountain natural systems and increasing amounts of rainfall in certain months (such a tendency is already observed for September), thus accelerating erosion and mass movement processes.

In all cases climate predictions and the consequences of future climate change on landscape and landform evolution are still quite uncertain, particularly for Bulgaria.

14.5.2 Landscape Evolution and Human Impact

Deleterious economic practices in land management are still quite common in Bulgaria and more dangerous for the natural environment than climate change. They include the *construction* of hotels and private buildings in vulnerable mountains (including protected areas), terrains with high landslide risk (seashore, river banks), construction of numerous small hydropower plants, deforestation in hilly and mountain areas. Although predictions for a decrease of snow amounts and shortening of the ski season are already confirmed, state services and private companies are still engaged in new projects (e.g., for ski resorts in areas where snow supply is more than uncertain). Developments in the existing resorts of Borovets, Pamporovo, and especially Bansko, lead to high pressure on the landscape, devastating natural forests and grasslands. The tourist bed capacity is too high. The same applies to the seashore where relatively intact natural sites have been completely destroyed by the construction of large hotels and recreation facilities.

At the same time, most of the Bulgarian territory is constantly *depopulating* and the pressure on the landscape is relieved. At many sites manmade features (buildings, roads, factories, river barrages) are completely abandoned and natural processes (weathering, erosion, corrosion, deflation, and deposition) slowly destroy artificial structures. *Abandoned* arable *lands* turn into bushland, rill erosion ceases and old roads narrow down to paths or become completely unusable. However, these tendencies are expected to occur just on the short term. As Bulgaria is in a process of integration into the densely populated European Union with a lack of space, especially in western countries, it is expected that the empty spaces to the east will be filled and human pressure on the landscape will intensify on the long term.

14.5.3 Impact of Endogenic Processes

Slow epeirogenic uplift and sinking, earthquakes and faulting have ever been influencing geomorphic evolution. Epeirogenic movements slowly but continuously maintain mass movements. Main zones of *uplift* are the mountains of Southwest-Bulgaria and partly the North-Bulgarian dome. Deformations of slope lystric prisms develop along rising mountain slopes (Alexiev 2002a, b, c; Tzankov 1995)", which make slope morphology more complicated and generate favorable conditions for landslides and rockfall development. The main subsidence areas are most of the Black Sea coast (especially around Burgas and Varna), and basins in West-Bulgaria (around Sofia, Ihtiman, Kazanlak, and Radomir), where these processes induce deposition and rise of groundwater level (involving soil salinization).

Bulgaria is situated in a *seismically active* zone, confirmed by several strong earthquakes (Sofia: 1818, 1858; Shabla: 1901; Krupnik: 1904; Gorna Orjahovica: 1913; Popovitsa: 1928, Chirpan: 1928; Svishtov and Velingrad: 1977; Strazhitsa:

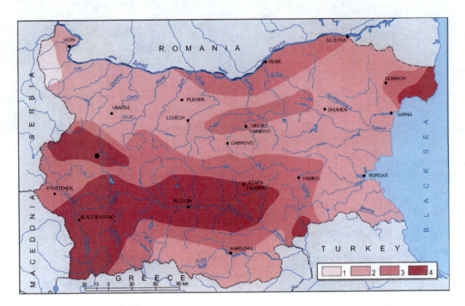

Fig. 14.16 Map of seismic risk and expected seismic impacts for a period of 1,000 years (After Bonchev et al. 1982). Seismicity: 1, low; 2, moderate; 3 high; 4, very high

1986). Seismic events originate from sources outside (deep hypocenters are mainly in Greece, Turkey, and Romania) but also within Bulgaria (primarily shallow hypocenters at 15–20 km depth). Earthquakes trigger vertical and horizontal movements along faults, first of all in mountain footslopes, where landslides and rockfalls develop. Among the frequently observed consequences are the changes in underground water circulation and in location of springs. Figure 14.16 shows the seismic risk zones outlined by Bonchev et al. (1982).

Conclusions of the impact of natural endogenic processes in the future can only be made on the basis of past observations. No sudden changes will occur in near future.

14.6 Geomorphological Hazards

Landslides are well spread throughout the country. According to Alexiev (2002c) the number of registered landslides in Bulgaria is 920, covering a total area of 20,000 ha but at present their number is considered to have risen to about 1,500. Along the high Danube bluffs and along the steep right banks of the tributary valleys catastrophic landslides occur, favored by the vertical jointing of loess, the exhumation on steep slopes of Pliocene clays (slip planes), the strong seasonal variation of precipitation and groundwater level, earthquakes and active piping. In the Predbalkan area landslide formation is supported by river incision and tilted clay and marl strata. Seaside landslides are associated with high cliffs that transversely cut sediment bedding and fault structures. The length of the shore affected by landslides is 52 km and the area is ca 5,400 ha (Alexiev 2002c). As it has been already mentioned, here human activities also contribute to the hazard. In the rest of Bulgaria landslides are developed in deeply incised valleys, along fault lines that mark mountain margins. *Collapses* are most abundant in the loess cover of North Bulgaria and in the central and western Danube Plain. *Block movements* are related to active normal faulting on the piedmont surfaces.

Soil erosion and deposition are the mostly widespread exogenic processes in the country. According to Alexiev (2002a) the rate of erosion is below 150 $t km^{-2} y^{-1}$ in most river catchments. At present the highest rates of erosion (1,000–2,000 $t km^{-2}$) are recorded in the Eastern Rhodopes and along the Struma River, explained by high slope inclination, slope deforestation, easily erodible sediments, precipitation regime (abundant winter rains). River floods with debris flows present a particular hazard.

Intensive *deflation* is another hazard for agriculture over vast (loess) areas (1.5 million ha in total) free of vegetation cover and improperly managed. Ninety percentage of these surfaces are agricultural areas with human-induced deflation (Alexiev and Vlaskov 1997).

The damaging effect of *wave erosion* is especially strong during extreme weather, sea storms with strong northeastern winds that generate high sea waves. According to Alexiev (2002c) destructive wave erosion affects 72% of the length of the Bulgarian coast.

Avalanches are considered to be the greatest natural hazard in high mountains. Favorable conditions for them are steep slopes between 25° and 45° (Mardirosyan 2007), suitable structure of snow cover, sufficient amount of snow, appropriate synoptic conditions. In Bulgaria avalanches are most common in the highest parts of the Rila Stoyanov (2009), Pirin, Central Stara planina, and Slavyanka Mountains.

References

Alexiev G (2002a) Erosional morphosculpture. In: Geography of Bulgaria. BAS, Sofia, pp 52–57 (in Bulgarian)
Alexiev G (2002b) Geomorphological regionalization. In: Geography of Bulgaria. BAS, Sofia, pp 104–105 (in Bulgarian)
Alexiev G (2002c) Unfavourable and dangerous endogenic processes. In: Geography of Bulgaria. BAS, Sofia, pp 70–81 (in Bulgarian)
Alexiev G (2009) Morphotectonics of the Balkan Peninsula. Autoreferat 68. (in Bulgarian)
Alexiev G, Spiridonov H (2002) Denudation surfaces. In: Yordanova M, Donchev D (eds) Geography of Bulgaria. Bulgarian Academy of Sciences, Sofia, pp 44–51 (in Bulgarian)
Alexiev G, Vlaskov V (1997) Modern morphogenetic processis. Geomorphological risk. In: Geography of Bulgaria. BAS, Sofia, pp 92–103 (in Bulgarian)
Bonchev St. (1929) The buildup of the mountains in Bulgaria. Nature and Science, XXIX, book 9–10 (in Bulgarian)
Bonchev E (1966) Revue generale de la structure geologique de la Bulgarie. Bulletin Inst Geol Bulgarian Academy of Sciences 15:5–21

Bonchev E, Bune V, Christoskov L, Karagyuleva U, Kostadinov V, Reisner G, Rizhkova S, Shebalin N, Sholpo V, Sorekova D (1982) A method for compilation of seismic zoning prognostic maps for the territory of Bulgaria. Geologia Balcanica 12(2):12–24

Choleev I (1982) About some glacial landforms in the valley of Pirinska Bistrica. Ann Univ Sofia Fac Geol Geogr 74(2):31–46

Cvijić J (1903) Die Tektonik der Balkanhalbinsel. Comptes Rendus de IX Congres geol Int. Vienna

Gachev E, Gachev V (2005) Lakatnik karst massif – a land of secrets hidden under ground. International conference "Protected karst areas – state, problems and perspectives. Nature park "Shumensko plato", Shumen (in Bulgarian)

Gachev E, Gikov A, Gacheva I, Nojarov P, Popov M (2008) Morphology of the bottom of Ledeno ezero (Rila mountain) and its relation to Quaternary evolution of relief. Probl Geogr 2008(3–4): 97–104 (in Bulgarian)

Gachev E, Gikov A, Zlatinova C, Blagoev B (2009) Present state of Bulgarian glacierets. Landf Anal 11:19–27

Galabov Zh (1946) A short physical geographic characteristic of Bulgaria. Fundamentals of the geology of Bulgaria. Annual journal of the Direction of Geology and Mining reserach, part A. No. 4.(in Bulgarian)

Galabov Zh (1966) A short complex characteristic of the separate natural geographic regions. Geography of Bulgaria, vol. 1. Physical Geography, Bulgarian Academy of Sciences, Sofia

Galabov Zh (1982) Headlines of relief. In: Geography of Bulgaria. BAS, Sofia, pp 13–33 (in Bulgarian)

Glovnia M (1958) Geomorphological research in the southwestern Rila Mountains. Ann Univ Sofia Fac Geol Geogr 51(3):7 50 (in Bulgarian)

Glovnia M (1959) About periglacial relief in Bulgaria. Announc Bulgarian Geogr Soc 2:5–15 (in Bulgarian)

Glovnia M (1968) Glacial and periglacial relief in the southern central Rila Mountains. Ann Univ Sofia Fac Geol Geogr 61(2):5–44 (in Bulgarian)

Glovnia M (1962a) Glacial and periglacial relief in eastern Rila Mountains. Ann Univ Sofia Fac Geol Geogr 55(2):7–62 (in Bulgarian)

Glovnia M (1962b) A contribution to the studies of periglacial morphosculpture in Rila mountain. Announc Bulgarian Geogr Soc 3:47–55 (in Bulgarian)

Grunewald K, Scheithauer J (2008) Klima und Landschaftsgeschichte Südosteuropas. Rekonstruktion anhand von Geoarchiven im Piringebirge (Bulgarien). Rhombos Verlag, Berlin, 180 p

Grunewald K, Weber C, Scheithauer J, Haubold E (2006) Mikrogletscher als Klimaindikatoren und Umweltarchive in südosteuropäischen Hochgebirgen? – Untersuchung des Vichrengletschers im Piringebirge (Bulgarien). Zschr Gletscherkunde und Glazialgeologie 39:99–114

Grunewald K, Scheithauer J, Gikov A (2008) Microglaciers in the Pirin Mountains. Probl Geogr 2008(1–2):159–174, Sofia (in Bulgarian)

Grunewald K, Scheithauer J, Monget J-M, Brown D (2009) Characterisation of contemporary local climate change in the mountains of southwestern Bulgaria. Clim. Change 95 (3-4):535–549

IPCC (2007) Fourth assessment report "climate change 2007". In: Parry M et al (eds) International Panel on Climate Change (IPCC). Cambridge University Press, Cambridge/New York

Ivanov, I (1954). Geomorphological research in the western part of Northwestern Rila mountains. Announcements of the Institute of Geography, Bulgarian Academy of Sciences. pp 7–84 (in Bulgarian)

Jaranov D (1960) Tectonics of Bulgaria. Tehnika, Sofia. (in Bulgarian)

Krastev T (2002) Morphosculpture of the Bulgarian Black Sea Coast. In: Geography of Bulgaria. BAS, Sofia, pp 57–60

Kuhlemann J, Gachev E, Gikov A, Nedkov S (2008) Glacial extent in the Rila Mountains (Bulgaria) as part of an environmental reconstruction of the Mediterranean during the Last Glacial Maximum (LGM). Probl Geogr 2008(3–4):87–96

Mardirosyan G (2007) Natural calamities and ecological catastrophes. Acad. Publishing House "Prof. M. Drinov", Sofia (in Bulgarian)

Minkov M (1968) Loess in Northern Bulgaria. A complex study. Bulgarian Academy of Sciences, Sofia, 202 p (in Bulgarian)

Mishev K (1992) History of research of planation surfaces in Bulgaria. Probl Geogr 1992(2):3–19 (in Bulgarian)

Mishev K (1993) Planation denudation surfaces in the mountains of the Balkan Peninsula. Probl Geogr 1:3–19 (in Bulgarian)

Mishev K, Vaptsarov I (1984) The Vilafrankian stage in Bulgaria: paleogeomorphologic and tectonic peoblems. Announcements of Bulg Geogr Soc, book XXII (XXXII) (in Bulgarian)

Mitev J, Iliev M (2002) Loess morphosculpture. In: Geography of Bulgaria. BAS, Sofia, pp 125–134 (in Bulgarian)

Muratov M (1949) Tektonika SSSR (Tectonics of the USSR). vol. II. AN SSSR, Moscow. (in Russian)

Nikolova M (2001) Natural hazards in Bulgaria. Probl Geogr 2001(1–2):70–71 (in Bulgarian)

Nikolova M (2002) Air temperatures. In: Geography of Bulgaria. BAS, Sofia, pp 146–148 (in Bulgarian)

Nojarov P (2008a) Air temperature variability and change at peak Musala for the period 1933 – 2007. Geography 21(2):14–19, Sofia (in Bulgarian)

Nojarov P (2008b) Air temperature variations at Cherni vruh top for the period 1936–2007. Probl Geogr 2008(3–4):141–148

Nojarov P (2009) Air temperature changes at Botev top for the period 1941–2007. Probl Geogr 2009(1):28–35

Popov V (1962) Morphology of the cirque Golemia Kazan, Pirin Mountains. In: Announcements of the Institute of Geography. BAS, Sofia, 6: 86–99 (in Bulgarian)

Popov V (1964) Observation over the perennial snow patch in the cirque Golemia Kazan, Pirin mountains. In: Announcements of the Institute of Geography. BAS, Sofia, 8: 198–207 (in Bulgarian)

Popov V (1997) Karst morphosculpture. In: Geography of Bulgaria. BAS, Sofia, pp 78–82 (in Bulgarian)

Popov V, Mishev K (1974) Geomorphology of Bulgarian Black sea coast and shelf. Bulgarian Academy of Sciences, Sofia, p 267. (in Bulgarian)

Stefanov P (2002) Morphographic features. In: Geography of Bulgaria. BAS, Sofia, pp 29–43 (in Bulgarian)

Stoyanov Kr (2004) About the river terraces correlation of Sredna Struma river and its tributaries. Scientific conference with international participation 5:340–344, Stara Zagora (in Bulgarian)

Stoyanov Kr (2009) Evaluation of the avalanche danger in Northwest Rila Mountains. In: Sustainable development of the mountains of Southeastern Europe. Springer, Heidelberg/Berlin, pp 95–101

Tzankov Tz (1995) Generations of circle structures in the areas of Sofia and Kyustendil (West Bulgaria). Scientific Conference 'Mineral source analysis of Kyustendil region – state and perspectives'. Kyustendil, 26–27 May 1995 (in Bulgarian)

Tzankov Tz (2002) Preconditions for the Neogene-Qaternary Geodynamic evolution of Bugaria. In: International scientific conference in Memory of Prof. Dimitar Jaranov 1: 134–141

Tzankov Tz, Nikolov G (2000) On the Late Oligocene-Middle Neogene Fundamental Change in the Regional Stress Field in the Northeastern Part of the Balkan Peninsula. In: Geological Conference, Sofia. Session 1. Book of abstracts, pp 167–168

Tzankov Tz, Katskov N, Kuleva S (1993) Circular structures in the Fore-Balkan and Stara Planina Mountains between the Rivers Barzija and Yantra. Comptes Rendues, Bulg Acad Sci 46(11): 73–76

Tzankov Tz et al (1998) Explanatory Note to the Neotectonic (Quaternery) map of Bulgaria, scale 1:500000. Publishing House Grafica, Sofia, 12 p

Tzankov Tz, Popov N, Nikolov G (2000) On the Neogene tectonic evolution of Bulgaria. Comptes Rendues, Bulg Acad Sci 53(3):63–66

Tzankov Tz, Spassov N, Stoyanov Kr (2005) Neogene-Quaternary Geodynamics and Paleogeography of Middle Struma River Valley Area. University Publishing House "N. Rilski", Blagoevgrad. 135 p. (in Bulgarian)

Vaptsarov I (1972) Staircased footslope surfaces of the 'glassy' type in the hollow of Kyustendil. Problems of Geography, Sofia 1972/2 (in Bulgarian)

Vaptsarov Iv (1975) Staircased slopefoot surfaces of glassy type in the hollow of Kyustendil. Problems of Geography, Bulgarian Academy of Sciences, vol. 2.

Vaptsarov I (1997) Planation surfaces. In: Geography of Bulgaria. BAS, Sofia, p 85 (in Bulgarian)

Velchev A (1997) The Pleistocene glaciations in the Bulgarian mountains. Ann Univ Sofia Fac Geol Geogr 87(2):53–65 (in Bulgarian with English summary)

Velchev A (1999) Glacial and cryogenic relief in some parts of the Musala watershed in the Rila mountains. Ann Univ Sofia Fac Geol Geogr 89(2):45–53 (in Bulgarian)

Velchev A, Todorov N (1994) Subalpine landscapes in Osogovo mountains. Ann Univ Sofia Fac Geol Geogr 84(2):53–65 (in Bulgarian)

Velchev A, Konteva M, Penin R, Todorov N (2000) Landscape characteristics of Midzor (Chiprovska planina). Ann Univ Sofia Fac Geol Geogr 91(2):97–104 (in Bulgarian)

Velev St (2002) Contemporary variations of air temperature and precipitation. In: Geography of Bulgaria. BAS, Sofia, pp 157–160 (in Bulgarian)

Velev St (2010) Climate of Bulgaria. Heron Press, Sofia

Vlaskov V (2002) Anthropogenic morphosculpture. In: Geography of Bulgaria. BAS, Sofia, pp 67–69 (in Bulgarian)

Yordanov T (1956) Tectonics of Bulgaria. Geology of Bulgaria. N.I. Sofia (in Bulgarian)

Yovchev Y (1960) Mineral deposits of PR Bulgaria. Tehnika. Sofia (in Bulgarian)

Chapter 15
Recent Landform Evolution in Macedonia

Dragan Kolčakovski and Ivica Milevski

Abstract In the Republic of Macedonia, where mountains are predominant in topography, morphostructures control hillslope processes, also promoted by weathered rocks prone to erosion and steep slopes (39.5% of the area steeper than 15°). Fluvial erosion is of equal importance since rivers are short but of torrential character and flow in composite valleys. Some of the deepest canyons (1,000 m deep) and the deepest underwater cave (190 m deep) of Europe are found in Macedonia. The highest polje is at 2,050 m elevation. Lake shore erosion and deposition can be studied on the largest lakes of the Balkan Peninsula and glacial and periglacial features in the highest mountains above 2,000 m. The most typical direct and indirect human interventions in the landscape are accelerated erosion and deposition, opencast mining, road building, canal, dam and reservoir constructions on rivers. Together with the influences of changing climate, human impact will be decisive in future landform evolution.

Keywords Morphostructures • Hillslope processes • Fluvial erosion • Karst • Glacial and periglacial processes • Lakes • Human impact

15.1 Geotectonic Setting

Ivica Milevski and Dragan Kolčakovski

Tectonic movements of various direction and intensity have been active on the relatively small territory of the Republic of Macedonia (25,713 km^2) through geological history, culminating in the Alpine orogeny. In his first tectonic regionalization

---

D. Kolčakovski • I. Milevski (✉)
Institute of Geography, University "Ss. Cyril and Methodius", Skopje, Gazi Baba bb., 1000, Skopje, Macedonia
e-mail: kolcak@iunona.pmf.ukim.edu.mk; ivica@iunona.pmf.ukim.edu.mk

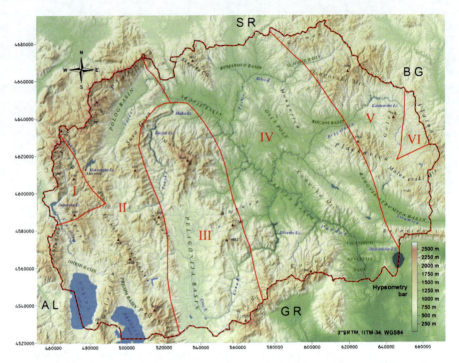

Fig. 15.1 The topography of Macedonia on a Digital Elevation Model (based on 3"SRTM; Jarvis et al. 2006), showing the major tectonic units (After Arsovski 1997). I, Chukali–Krasta zone; II, Western Macedonian zone; III, Pelagonian anticlinorum; IV, Vardar zone; V, Serbian–Macedonian massif; VI, Kraishte zone

Kossmat (1924) identified four tectonic zones. This zonation was recently refined by Arsovski (1997), in whose system the western and central part of Macedonia tectonically belongs to the Dinarides (Helenides), while the easternmost (Kraishte) zone is part of the Carpatho-Balkanic unit. The Serbian-Macedonian Massif lies between them. The tectonic units following the northwest-southeast strike of the Dinarides, from east to west are the Vardar Zone, the Pelagonian horst anticlinorium, the West-Macedonian zone, and the Tsukali-Krasta zone (Fig. 15.1).

The highly unstable *Vardar zone*, mostly along the Vardar River, is a lineament of first order, 60–80 km wide and several hundred kilometers long (from Belgrade on north to Thessaloniki on south). It is very distinctive from other tectonic units, including fragments of Precambrian crust, a Paleozoic volcanic-sedimentary complex, and Mesozoic structural complexes. Intensive volcanic activity ended in the early Quaternary (Arsovski 1997).

The ca 40-km wide *Pelagonian Massif* (or horst anticlinorium) extends nearly 150 km in north-northwest–south-southeast direction. As a relict of the Precambrian crust, it is mostly composed by metamorphic crystalline rocks, gneiss, mica-schists, and marbles. Deep regional faults clearly separate it from the neighboring units.

The *West-Macedonian zone* is recognized by Arsovski (1960) as a distinct tectonic unit, reaching from the Šara Mountains (Šar Planina) on the north to the Pelister Mountains on the south. It is generally composed of a Palaeozoic metamorphic complex, with volcanic and sedimentary formations in the lower parts, and carbonaceous formation on top.

Finally, the *Tsukali-Krasta Zone* is a small area on the border with Albania. This is a complex tectonic zone of Cretaceous sediments, overlain by Eocene conglomerates (Arsovski 1997).

The *Serbian-Macedonian Massif* represent a fragment of the Rhodopian Massif between the Dinarides and the Carpatho-Balkanides, both parts of the Alpine geosynclinal belt. Dimitrijevic (1974) claims that in the Precambrian and Paleozoic this tectonic unit was entirely separated from the Dinarides and Rhodopes, while others assume a continuous connection. The Serbian-Macedonian Massif is characterized by the domination of very old Precambrian and Riphean-Cambrian lithological complexes: gneisses, mica-schists, amphibolites, green-schists, and others. During the Mesozoic and the Paleogene this massif was temporary under transgression, while in the Oligocene and Miocene it was faulted and longitudinal depressions formed. These processes were accompanied by intensive volcanic activity with several volcanic centers.

The *Kraishte zone*, on the border with Bulgaria, is of limited extension (about 800 km^2) in Macedonia near Delčevo (Pianets Basin). With its geotectonic characteristics and different development of green schists formations with presence of meta-gabbros, meta-sandstones, and large Hercynian granitoid masses, this zone clearly belongs to the Carpatho-Balkanides. The deep mid-Alpine Pianets graben was filled with Eocene marine and Pliocene-Quaternary lake sediments. Graben formation was accompanied by major volcanic activity.

The complex geotectonic evolution of this (inner) part of Balkan Peninsula is a product of numerous *orogenic cycles*, which involved transgressions and regressions and accompanying volcanic activity. The Pelagonian and Serbian-Macedonian Massifs were created by powerful granite and granodiorite intrusions as well as regional and contact metamorphism (Kolčakovski 2004b) during the Precambrian Baikal orogeny. The separation of these tectonic units from basic Rhodope mass, and from one another by the Vardar Zone took place during the Caledonian orogeny in the early Paleozoic. The Hercynian orogeny especially powerfully affected the western parts of Macedonia, where the sediment complex is folded and metamorphosed.

The Cainozoic tectonic evolution of Macedonia consists of two periods of *extension*, separated by two episodes of crustal *shortening* (Burchfiel et al. 2004). The Paleogene extension and related basin development was diachronous and began in the late Eocene time and continued into the early Oligocene. The Paleogene basins show a general northwest-southeast strike and were formed mainly within the Vardar zone and the Serbian-Macedonian tectonic units. They are filled with Eocene–Oligocene molass sediments in thicknesses of 3,500–4,000 m (Dumurdzanov et al. 2004). Southwest-vergent folding and thrusting occurred during late Eocene and latest Oligocene–early Miocene times. Following shortening deformation, orogenic relief was greatly reduced by the end of the early Miocene. The *planated surfaces* were disrupted during the second period of extension, the predominant mode of tectonic deformation from the early Miocene to present, and are presently preserved at different elevations up to 2,000 m above sea level. Volcanic activity in eastern Macedonia occurred during the first period of extension, the two periods of crustal shortening, and continued into the second period of extension in the late Cainozoic during deposition in the Neogene extensional basins. The second period of extension, from the middle Miocene to the present, is marked by deposition in the deepest basins and extended to the whole central Balkan Peninsula (Zagorchev 1995; Burchfiel et al. 2004). Late Cainozoic extension and associated basin formation is characterized by differential vertical tectonic movements along normal, oblique and strike-slip faults, producing numerous horsts and grabens, rapidly filled with lakes. Most of these lakes (except the Ohrid, Prespa, and Dojran Lakes) desiccated or drained by the early Pleistocene (Kolčakovski 2004b). The Quaternary period saw stages of tectonic uplift of various intensity. Mountain uplift was faster than that of the neighboring depressions and this directly influenced fluvial processes and promoted valley incision. Extensional movements are still rapid, indicated by both seismic events and GPS measurements (Gospodinov et al. 2003).

15.2 Geomorphological Units

Dragan Kolčakovski

As a result of intense local and regional geotectonic movements in the past, the landscape in the Republic of Macedonia (25,713 km^2) has a typical *checkerboard topography* of mountains and basins (Fig. 15.1). Hills dominate over 78.8% of the country, while mountains only cover 47.7% (12254.5 km^2) of the total area (Markoski 1995, 2004). Among the about 40 mountain peaks, 13 rise above 2,000 m elevation (Kolčakovski 2004b), the highest, Korab, reaching to 2,753 m. Thus, the mean altitude of Macedonia is 831 m and mean slope angle is 13.5° (Milevski 2007b). There are also 16 basins, filled with lacustrine deposits (remnants from Neogene lacustrine phase), and surrounded and closed by mountains from almost all sides (Fig. 15.2). As a result of steep slopes and the human impact on the landscape, hillslopes are rapidly eroding with increased footslope accumulation. Basins are usually connected with composite valleys, deeply incised as gorges and canyons along mountain sections and broad cross-sections along basin sections. Because carbonate rocks constitute 10% of Macedonia, karst landforms (dolinas, poljes, pits, caves, etc.) also occur, mostly in the west. In high mountains (above 2,000 m) there are remnants of glacial landforms (cirques, glacial troughs) as well as fossil and recent periglacial phenomena.

Fig. 15.2 Neotectonic regions of the Republic of Macedonia (After Arsovski 1997). *Morphostructures of uplift (blocks)*: 1, Šara; 2, Korab; 3, Bistra; 4 Stogovo; 5, Jablanica; 6, Galičica; 7, Ilinska; 8, Pelister; 9, Šemnica; 10, Ljuben; 11, Pesjak; 12, Suva Gora; 13, Žeden; 14, Jakupica; 15, Babuna; 16, Selečka; 17, Skopska Crna Gora; 18, Raduša; 19, Kadina; 20, Bregalnica; 21, Klepa; 22, Mariovo; 23 Plauš; 24, Kožuf; 25, Furka; 26, Kozjak; 27, Osogovo; 28, Ruen; 29, Golak; 30, Maleševo; 31, Plačkovica; 32, Belasica. *Morphostructures of subsidence (basins)*: I, Polog; II, Ohrid; III, Debar; II–III, Drim graben; IV, Prespa; V, Kičevo; VI, Belica; VII, Pelagonija; VIII, Poreče; IX, Skopje; X, Kumanovo; XI, Ovče Pole; XII, Tikveš; XIII, Lakavica; XIV, Valandovo; XV, Slavište; XVI, Kočani; XVII, Pehčevo; XVIII, Strumica; XIX, Dojran. *Paleovolcanic areas*: a, Zletovo; b, Vitačevo; c, Šopur; d, Venec; e, Pehčevo

As uplifting morphostructures *mountains* are the predominant elements of the landscape in Macedonia (Fig. 15.2). The highest mountain massifs are mostly found in the western and central parts; there are only two from 13 mountains above 2,000 m in eastern Macedonia: Osogovo (2,252 m) and Belasica (2,029 m). According to altitude, mountains in Macedonia are divided into high mountains (2,000–2,753 m) with subgroup of five very high mountains (2,500–2,753 m), medium-high mountains (1,000–2,000 m), and low mountains (below 1,000 m). Lowest altitude limit for a mountain in Macedonia is 700 m of absolute elevation and 500 m of relative altitude. Because of the geotectonic setting, mountains in western and central parts (in the West-Macedonian zone, the Pelagonian massif and the Vardar Zone) have a general northwest-southeast strike. In contrast, the ranges in eastern Macedonia are east-west aligned (because of the predominant north-south extensional regime). Mountains in the *western and central parts* are generally composed of marbles (Jakupica, Suva Gora), limestones (Bistra, Jablanica, Galičica, Šara), granites (Pelister) or other very resistant rocks. For that reasons, these mountains have usually *narrow, sharp ridges* and peaks and deeply incised valleys. Mountains in *eastern Macedonia* are dominantly composed of more erodible crystalline rocks (schists) and, consequently, show a more *subdued relief*, rounded ridges and peaks, and less deeply incised valleys. However, both groups of mountains were shaped during the Neogene–Pleistocene (Lilienberg 1966).

Depressions or *basins* are also basic structural units in the landscape of Macedonia. Tectonic grabens generally formed by subsidence during the Aegean crustal extension (from the middle Miocene to the Pliocene and the Quaternary). Initially basins were closed and gradually collected freshwater, forming *lakes* (Petkovski 1998). But later, with Pleistocene glacial and interglacial erosion, disintegration and subsidence of the Aegean crust and the formation of the Vardar River system, many of these lakes were connected, filled and desiccated (except the Ohrid, Prespa, and Dojran Lakes). As a consequence of the lacustrine phase, there are lacustrine deposits (sands, sandstones, clays) on the bottoms of basins up to 2,000 m thickness, which indicate stages of their development, intensity and directions of tectonic movements, climatic changes, etc. (Dumurdzanov et al. 2004). The largest basin, Pelagonija extends over 3,682 km^2, Tikveš Basin is 2,518 km^2, while the smallest are Slavište Basin (768 km^2) and Prespa Basin (559 km^2) (Markoski 1995). Altitude of the bottoms is also variable, ranging from 44 m for the Gevgelija Basin, through the Skopje Basin (220 m), Kumanovo Basin (340 m), Pelagonija Basin (600 m), Berovo Basin (700 m) to 840 m for the Prespa Basin. Similar to mountains, depressions have variable strikes: in the West-Macedonian zone and in the Pelagonian massif northwest-southeast, while in the Vardar zone and especially in the Serbo-Macedonian massif east-west directed, because of meridional extension (Kolčakovski 2004b).

Palaeovolcanic landscapes are more or less preserved remnants of Tertiary (rarely Pleistocene) volcanism. The forms include *volcanic cones, calderas, necks* and *plateaus*, highly eroded by post-volcanic fluvial erosion. Largest and morphologically best expressed is the Kratovo-Zletovo palaeovolcanic area in the northeast. From about 20 volcanic cones in this area, Plavica (1,297 m) and especially Lesnovo (1,167 m) with its top calderas are particularly well preserved (Milevski 2005a). These two volcanic centers, together with Uvo-Bukovec cones, Zdravči Kamen, Živalevo, and other volcanic necks, belong to the older, Oligocene, volcanic phases (Serafimovski 1993). Younger (Miocene-Pliocene) centers are located in the south part of the palaeovolcanic area, from Zletovo to Spancevo village (Boev and Yanev 2001), the Preslap (1,117 m), and Rajčani (867 m) cones with some remnants of (double) calderas. Here violent volcanic activity produced thick tuff and breccia deposits. Volcanic activity began in the late Eocene or early Oligocene, and with some pauses lasted up to the early Pliocene (Stojanoviḱ 1986), while it was *moving from the northeast to the southwest* (Boev and Yanev 2001), changing in intensity (violent eruptions followed by production of pyroclastic material as well as dormant and effusive phases). On the western edge of Kratovo-Zletovo volcanic area near Kumanovo, a unique phenomenon of an eroded Miocene basaltic plateau persist fragmented by erosion into eight hills of volcanic cone shape. Another palaeovolcanic landscape is south of Kavadarci stretching to the Kožuf Mountain on the border with Greece. It consists of several Pliocene volcanic cones and the extensive volcanic upland of Vitačevo. Volcanic features of polygenetic origin have been modified and destroyed by intensive erosion (Milevski 2005a).

15.3 History of Research of Recent Geomorphic Processes

Ivica Milevski and Dragan Kolčakovski

Geomorphological research in the Republic of Macedonia can be divided into *three* distinct *periods*. Before the twentieth century, only some basic descriptions were characteristic. In the beginning of the twentieth century until World War I, Jovan Cvijić set up a real geomorphological research basis for Macedonia. Between World Wars (1918–1941), several detailed geomorphological investigations were performed in the Poreče, Jakupica, Skopje Basin, and other areas (Jovanović 1928, 1931; Radovanović 1931). After World War II, these investigations intensified and extended to recent geomorphological processes too.

In the middle of August 1898, the Serbian geographer Jovan Cvijić, well known all over Europe, started academic excursions through the territory of the present-day Macedonia and neighboring areas leading to an extensive study entitled "*Basics of Geography and Geology of Macedonia and Old Serbia,*" published in three volumes on 1372 pages (Cvijić 1906). In this very significant monograph, 320 pages are devoted to the geomorphological features of Macedonia. Despite some inconsistencies, this capital work has enormous scientific significance and is widely used by following generations of researchers.

Between the two world wars, some very valuable regional geomorphological studies were completed on the glacial landscape of Jakupica Mountain (Jovanović 1928), the fossil coastal and karst landscapes in the Poreče Basin (Jovanović 1927, 1928), the Skopje Basin (Jovanović 1931), the denudation of Selečka Mountain (Radovanović 1928), the karst landscape of Kožuf Mountain (Radovanović 1931) etc. Unlike the studies of Cvijić, which are mostly general in approach, the latter are more detailed and usually elaborate on one type of landscape predominant in the area.

After World War II, there are several generations of researchers who work extensively on recent geomorphic processes in Macedonia, starting with studies in the Debarska Basin (Gaševski 1953) and on the gypsum landscape in the valley of Radika (Stojadinoviќ 1958), the periglacial landscape of Pelister (Stojadinoviќ 1962, 1970), the coastal landscape of Lake Ohrid (Stojadinoviќ 1968). However, the greatest contribution was made by Dušan Manakoviќ, who worked on a wide range of issues: from karst, fluvial and coastal to glacial and periglacial landscapes. It is important to point out that he also treats the problem of recent relief evolution and anthropogenic influence of the landscape. From the late 1970s to the 1990s, in addition to further research on karst processes in Macedonia (Andonovski 1977; Kolčakovski 1988, 1992; Andonovski and Milevski 1999; Kolčakovski et al. 2007), significant attention is devoted to soil erosion due to human impact (Andonovski 1982). From the 1990s there are some investigations of the karst landscape in the Skopje Basin (Kolčakovski 1992), the periglacial landscape on the Jablanica, Stogovo, Galičica and Pelister (Baba) Mountains (Kolčakovski 1994, 1999). In recent years, geomorphological research in Macedonia has also moved toward processes of accelerated erosion (Milevski 2001, 2005b), denudation (Kolčakovski 2005) and overall human impact on the landscape with significant environmental implications (Dragičević and Milevski 2010).

15.4 Recent Geomorphic Processes

Dragan Kolčakovski and Ivica Milevski

15.4.1 Hillslope Processes

With large areas of erodible rocks, steep slopes, semi-arid climate and weak vegetation, hillslope processes are very intensive in Macedonia. Thus, most hillslopes are composed of crystalline rocks (gneiss, mica-schists, and other schists), sandstones, lacustrine and river deposits – all highly erodible. Mean topographic slope of the country is also very high, 13.5°, with 39.5% of the area steeper than 15°. In addition to relative relief, southern slopes are predominant (Milevski 2007b). Climate is semi-arid with 500–700 mm precipitation, but with frequent summer storms and heavy rains. In combination with anthropogenic influence and a sparse vegetation, this intensifies bedrock weathering and produces erosional landforms, the most famous being "Markovi Kuli" on Zlatovrv-Babuna Mountain (Kolčakovski 2005), included in the UNESCO World Natural Heritage list, as well as on the mountains of Selečka (Radovanović 1928), Jakupica (Kolčakovski 1987), Mangovica (Milevski and Miloševski 2008) and Osogovo (Milevski 2006). The most significant hillslope processes are *rain splash, sheet wash, and rill erosion*. At numerous sites *rill, gully erosion*, and even *badlands* are formed. There are frequent mass movements represented by *landslides, slumps*, and *soil creep*, especially on the rims of basins, where Pliocene lacustrine sands and sandstones are superimposed over inclined clay layers. Large landslides are present in the Tikveš, Berovo and Delčevo Basins, in the Radika Valley. In more compact weathered rocks (igneous, limestones, marbles), usually in the western mountainous part of the country, *rockfalls, rockslides, debris flows* and other phenomena occur (Stogovo, Šara Mountain, Korab, Jakupica). As a consequence of hillslope erosion, variable deposition landforms were created on the sides or bottoms: *talus cones, scree slopes, alluvial fans, colluvial,* and *debris deposits*. Although most hillslope processes are natural in origin, human impact greatly increase their intensity and makes them even predominant over large areas. According to the erosion map by Djordjevic et al. (1993), the average material removal by erosion in Macedonia is 690 m^3 km^{-2} year^{-1}, reaching even more than 2,000 m^3 km^{-2} year^{-1} in some regions, which means a very high erosion rate. Most of this material is moved by processes of hillslope erosion.

15.4.2 Fluvial Erosion and Landforms

Fluvial erosion is the prevailing geomorphic process in Macedonia. The drainage network is relatively dense (1.3 km km^{-2}), but most of the rivers are rather short. The largest is the Vardar River with 301 km length in Macedonia and 388 km to the mouth in Aegean Sea. Rivers have low and very variable discharges (mostly below 10 m^3 s^{-1}) and are usually torrential in character. The reason for that lies in the semi-arid climate (mean precipitation: 680 mm year^{-1}), reduced forest cover and intense denudation. However, after the draining and drying of most lakes that existed in depressions during the Neogene (Dumurdzanov et al. 2004), the newly established (post-Pliocene) river network created a typical fluvial landscape with valleys as the most remarkable features. Because of the mosaical structural landscape, there are two major valley types: *valleys in basins* are broad, shallow and usually with well-preserved river terraces, while in mountains, they are deeper, narrow, with typical V-shaped profile, frequently with convex sides (Manakoviḱ et al. 1998). The convexity of valley sides indicates intensive Quaternary incision (at 1–2 mm year^{-1} rate) caused by base level lowering due to differential tectonic uplift as well as Pleistocene climate changes (Milevski 2007a). That is especially obvious in incised valleys through mountains and hills, connecting basins. Among the *deeply incised canyons*, that of Radika River (in the western part of Macedonia) is 1,000 m deep, while the also deep canyon of the Treska River (a right tributary of the Vardar River) incised into Precambrian marbles. There are another 15 gorges and canyons on major rivers. In these sections, riverbeds are mostly narrow, rocky with steps, waterfalls and cascades. It is estimated that there are ca 150 typical waterfalls; best known are the tectonic waterfalls of the Smolarska (39.5 m high) and the Kolešinska River (15 m) on Belasica Mountain, while the highest is Projfel waterfall (136 m) on Korab Mountain. As river slopes rapidly reduce at entering basins, rivers build large *alluvial fans* (as at the footslopes of Šara and Belasica Mountains) or alluvial plains. Because of the frequent alternation of "normal" shallow *valley* sections with gorges and canyons, most of the valleys in Macedonia, especially of larger rivers are *composite* in character. Thus, the Vardar River passes through four gorges and five basins. The recent evolution of the fluvial landscape mostly depends on tectonic movements, climate change and human impact. In line with geophysical measurements showing differential tectonic uplift with higher rates for mountains than for grabens (Jančevski 1987; Arsovski 1997), rivers tend to incise more intensively in mountainous areas. Further dropping precipitation and runoff, however, will probably reduce fluvial erosion and disturb concavities in

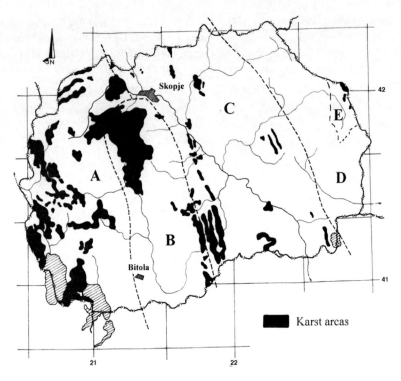

Fig. 15.3 Map of karst areas in the Republic of Macedonia. A, Western Macedonian zone; B, Pelagonian zone; C, Vardar zone; D, Serbian–Macedonian massif; E, Kraishte zone

longitudinal profiles. On the other hand, anthropogenic influence intensifies soil erosion, alters river energy and transport capacity, promoting increased deposition, especially in basins.

15.4.3   Karst Processes and Landforms

In Macedonia there are ca 2,724 km^2 (10.6%) karst landscapes in unequal distribution (Kolčakovski and Boskovska 2007) (Fig. 15.3). Most of the karstic areas are found in the western and central parts, i.e. in the West-Macedonian zone (Žeden, Bistra, Galičica, Jablanica Mountains) of Triassic limestones and in the Pelagonian Massif (Jakupica Mountain, Poreče Basin, Mariovo Basin) of Precambrian marbles (Fig. 15.4). In the Vardar zone there are only few karst areas (Taorska and Demir Kapija gorges of the Vardar River, the Veles gorge and the nearby hills, the Bislim gorge on the Pcinja River), while in the Serbian-Macedonian massif and the Kraishte zone (in the eastern part of the country) karst is insignificant with only several small patches of few square kilometers.

Fig. 15.4 Karst landscape on Jakupica Mountain

As karst areas are localized, these occurrences are referred to as "karst oases" (Manakoviḱ 1980a). Surface karst landforms in Macedonia are represented by a range of features from small *karren* through medium-sized *sinkholes* to karst *poljes*. Most of the 28 karst poljes are small (0.5–9.5 km^2) and located in mountains above 1,000 m elevation, except Brce (at 790 m; area: 1.2 km^2) on Kožuf Mountain and Cersko Pole (at 950 m; 9.5 km^2) on Babasač Mountain (Andonovski 1989). With its bottom at 2,050 m, Šilegarnik is highest karst polje on the Balkan Peninsula (Kolčakovski 1988). Underground karst landforms are represented by the ca 200 *caves* and pits described and surveyed to date but their total number is estimated at 400–500 (Kolčakovski 1989). According to the latest data, the longest cave is Slatinska Spring with 4.0 km length (Kolčakovski 2006), and the deepest pit is Slovačka Jama on Jakupica mountain, 540 m deep. Curiously, recent research (in 2009) showed that the 190 m deep Koritište (Vrelo) cave in Matka canyon (with entrance below the water surface of Matka reservoir) is one of the deepest underwater cave in Europe. It is obvious that recent karstification is slower than before because of the decreased amount of precipitation and increased temperatures. Thus, because of climate change and the dropping karst water table, most speleothems became dry.

15.4.4 Lake Shore Processes and Landforms

In the Republic of Macedonia both fossil and recent lake shores are found. *Fossil* (Neogene) *lake shores* and erosional and depositional *terraces* encircle almost all basins. The terraces are especially prominent around the Skopska Basin (Jovanović 1931), in the Berovska and Delčevska Basins (Manakoviќ 1980b), in the Tikveš Basin (Manakoviќ 1968) and elsewhere. However, most of them are heavily dissected and eroded by subsequent post-Pliocene fluvial erosion processes. Large deposition terraces, as those in the Berovo (Maleš) Basin, have been destroyed by anthropogenic erosion over the past millennia (Milevski et al. 2008). Recent shore processes are related to the *existing lakes*, particularly to the three largest: the Ohrid (348 km^2), Prespa (275 km^2) and Dojran Lakes (48 km^2). The shoreline of Ohrid Lake is 86 km long, generally steep and rocky (along 63.1 km length). In the remarkable cliffs numerous shallow caves formed (Stojadinoviќ 1968). Only the southern and especially the northern shores show gently sloping sand and gravel beaches. Several short rivers form small *deltas* (Čerava, Rače, Koselska Rivers). For a higher water influx into the Ohrid Lake, recently the Sateska River is redirected from its natural channel and inflow in the Crni Drim River directly into the lake. Because of high amount of deposits, the offshore belt near the river mouth gradually extends and the lake-shore is shifted. The shore of the Prespa Lake is very similar to that of the Ohrid Lake, with steep cliffs on the east and west, and gentle slopes with sandy-gravel beaches on the northern and southern sides. The northeastern shore zone (Ezerani) is shallow and overgrown with dense vegetation (*phytogenic shore*). Because of high wave energy, the shore of the small island Golem Grad in the southwestern part of the Prespa Lake is entirely encircled by *cliffs* up to 10 m high. The Dojran Lake is somewhat different, with lower shores and no distinct cliffs. Most of the shore is flatter, with sand and gravel beaches. *Lake shore retreat* is very fast due to natural processes and recently to human impact (enormous amounts of water used for irrigation, especially in 1988). Since the water level in all of the three lakes is slowly dropping under natural and anthropogenic influence, recent lake-shore landforms are exposed. Except of the mentioned tectonic lakes, contemporary shore processes are recorded in larger *reservoirs* (Tikveš, Debar and Kalimanci Lakes), generally built in the 1960s. Thus, on most of them, micro-cliffs are formed on steep sides, while in rivers mouths deltas were built (e.g., at the Kalimanci Lake).

15.4.5 Periglacial Processes and Landforms

In the mountains of Macedonia above 2,000 m elevation (380.3 km^2 area), previous research recorded numerous fossil and recent periglacial landforms (Belij and Kolčakovski 1997). Among fossil landforms impressive *rockflows* (Stojadinoviќ 1962) on Pelister Mountain (2,601 m), a remarkable *rock glacier* on Jakupica

(Manakoviḱ 1962), one of the largest on the Balkan Peninsula and smaller rock glaciers on Jablanica Mountain (Kolčakovski 1999) can be mentioned. On the northern slopes of most mountains, *nivation cirques* formed, where snow can persist in small patches up to the end of July. However, the mentioned landforms are generally created during the latest glacial phases in the Pleistocene (Kolčakovski 2004a), although precise measurements and observations will show if they are still active. Also several *recent periglacial landforms* were identified: *nivation niches and hollows*, *rockflows*, "*plowing blocks*," *periglacial terraces* (on Galičica Mountain, 2,288 m), *solifluction lobes*, *polygonal soils* (on Jablanica Mountain, 2,256 m) (Kolčakovski 1994, 1999). With regard to the recent activity of some landforms like rockflows and plowing blocks, precise comparison of photographs of these landforms on Osogovo Mountain (2,252 m) made in 1997 and 2004 showed that they move very slowly, ca 20 cm within this period. On the other hand, certain morphological traces (furrows, depressions, wrinkled land), particularly related to some blocks next to the rockflows, show that they were sliding (moving) in much larger amounts in the recent past (Milevski 2008), possibly during colder and more humid periods, such as the 1950s and 1960s, or even more in the Little Ice Age in the sixteenth and seventeenth centuries, when mean temperatures in Europe were about 2°C lower than today.

15.4.6 Glacial Landforms

Under the cold Würm climate, in almost all mountains in Macedonia higher than 2,100 m, mountain glaciers existed (Fig. 15.5). Most of them were *cirque glaciers* (on Pelister, Jablanica, Stogovo Mountains) or occupied short and small *U-shaped valleys* downslope (on Šara, Jakupica, Korab Mountains) as the snowline was located high (at about 2,000 m) and precipitation was relatively low (Kolčakovski 2004a). From glacial times numerous landforms have survived (Fig. 15.6): ca 30 cirques, generally between 2,000 and 2,400 m elevation, some of which are filled with lakes; U-shaped valleys with lengths of up to 2.5 km, most characteristic of Šara Mountain (2,747 m). Depositional landforms are mostly represented by *moraines* of different types: terminal, lateral, recessional, etc. Terminal moraines are significant for the determination of maximum advance of glaciers as well as for the reconstruction of Pleistocene snowline. According to Kolčakovski (2004a), who used the methods of Hoffer and Messerli, mean snowline on the mountains in Macedonia is estimated at about 1,950–2,050 m. However, the remnants of the glacial landscape are gradually destroyed mainly through processes of weathering and fluvial erosion due to Holocene climate warming. With further climate change, periglacial and glacial landforms will be destroyed even more efficiently by fluvial processes. Even now, some recent landslide occurrences that partly result from direct or indirect human impact on high mountain landscapes, rapidly mask and obliterate original landforms (Milevski 2008). Glacial and periglacial landforms are also threatened by intensive farming, overgrazing, road building and tourism activities.

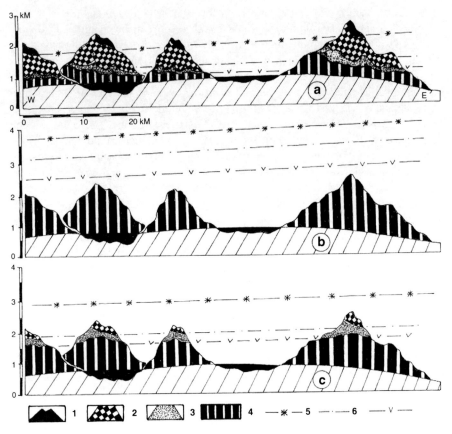

Fig. 15.5 Vertical zoning of natural geocomplexes on the mountains in southwestern Macedonia (Jablanica on the west, Stogovo, Galičica and Pelister on the east): (**a**) During the last glacial maximum – 19,000 BP; (**b**) Holocene climatic optimum – 6,500 BP; (**c**) Recent situation after Kolčakovski (2001). 1, Nivation-glacial geocomplex; 2, Periglacial geocomplex; 3, Transitional periglacial-fluvial geocomplex; 5, Snowline; 6, Treeline

15.5 Direct Human Impact on the Landscape

Ivica Milevski

In Macedonia the most typical modes of direct and indirect human impact on the landscape are: accelerated erosion and deposition, opencast mining, road building (particularly roadcuts), canal construction, river dam and reservoir construction, tailings, dump sites, influences on fluvial processes and other activities.

Fig. 15.6 Glacial landscape on Galičica Mountain with cirques (*above*) and on Šara Mountain with U-shaped valley (*below*)

15.5.1 Accelerated Erosion and Deposition

In Macedonia large areas are under accelerated erosion caused by intensive human impact on suitable environments over centuries. Such areas are steep, southern slopes of depressions and valleys, generally below 1,000 m elevation, which were most appropriate for early settlement as well. Because of accelerated erosion, the landscape is very often dissected into rills, gullies, *badlands*, *earth pyramids* etc. (Milevski 2004; Fig. 15.7). According to Djordjevic et al. (1993) some 36% of total country area is under severe and excessive erosion, the mean erosion rate is 16.9 million m^3 $year^{-1}$ (0.7 mm $year^{-1}$), and net *sediment delivery* is 7.5 million m^3 $year^{-1}$. In some landscapes extreme soil erosion has a devastating impact: on the catchments of the Upper

Fig. 15.7 Excessive erosion (badlands, gullies) and deposition in the Bregalnica catchment

Bregalnica (Milevski et al. 2008), of the Lower Crna (Trendafilov 1996) and in the Pčinja catchment (Manakoviḱ 1969; Milevski et al. 2007) and many others. On the other hand, lower parts of these catchments and valleys suffer severe *deposition* of eroded material with notable impact on fluvial processes. The lower sections of the Vardar, Pčinja, Bregalnica and Crna Rivers are bordered by extensive alluvial plains with fresh deposits, where lateral erosion, meandering and even channel accretion prevail (Fig. 15.7). As a result of excessive deposition, the Kamenička and Radanjska beds (left tributaries of the Bregalnica) are now more than 10–15 m higher (Milevski 2009). Large tracts of agricultural land were abandoned with significant negative socioeconomic impacts on the rural environment. However, in the past 30–40 years there is an opposite trend in anthropogenic influence on soil erosion processes. Namely, because of decrease in rural areas, land dereliction, aging population as well as numerous measures of soil conservation (afforestation, dam construction, bioamelioration, canalization), there is a notable *drop in the rate of erosion* and sediment yield (which is still high) obvious from the analyses of gullies, badlands, alluvial fans almost fossilized and from the comparisons of older erosion maps and more recent surveys. Human activities have also modified the intensity of mass movements, quite often unconsciously. For instance, numerous *landslides* were *activated* recently in Macedonia with road and canal construction on susceptible terrain, or by building major structures on sloping terrain. A typical example is the landslide Ramina in the city of Veles, where sliding appeared in 1999 as a consequence of housing construction on unstable steep slopes and another is the large landslide near Bituše and Velebrdo villages in the Radika valley (Manakoviḱ 1974). Because of a deep roadcut in tuffs on Stracin pass (on the road Kumanovo–Kriva Palanka), a shallow landslide occurred recently, affecting an international road. Landslides quickly change the local landscape through planation at their upper section and through deposition at their tongue. One of the largest recent landslides in Macedonia on the hill Gradot near Kavadarci (with volume of 15 million m^3), even impound the valley of Bunarska River, causing ponding and significant landscape change (Manakoviḱ 1960). The overall impact of accelerated erosion in Macedonia is *surface planation*.

15.5.2 Impact of Mining and Waste Disposal

In Macedonia there are several huge opencast mines, where the landscape is significantly changed. In the Suvodol coal basin, east of Bitola city, since 1982 more than 140 millions m^3 of coal have been excavated from an area of 10 km^2. A large depression was created with almost 3 km in diameter and 50–100 m in depth, while in its vicinity, high *spoil heaps* rise. Here large mass movements occur triggered by slope undercutting, and slides of several millions m^3 volume cause significant landscape change, like the huge landslide (26.4 millions m^3) in 1995 (Manasiev et al. 2002). Similar examples are in Oslomej coal mine near Kičevo, the Sivec marble quarry near Prilep, the Bučim mine near Radoviš, the Ržanovo mine south of Kavadartci. Numerous *limestone quarries* are spread throughout the country, representing artificial depressions and elevations and also modify natural karst processes (Fig. 15.8). Extraction of gypsum on the sides of Krčin Mountain (southern extension of Korab), not only change karst processes but destroyed a unique gypsum cave called Alčia. In Macedonia there are some 10 major *tailing* sites where waste from ore mining is disposed – usually on valley floors as is the case of the Sasa lead and zinc mine near the town of Makedonska Kamenica. Here the tailing site in the Kamenička valley is 1.5 km long, 200 m wide and about 40–60 m high (Fig. 15.9). Recently, because of the collapse of an underground river tunnel, huge quantities of waste are washed downstream leading to an ecological disaster. Further tailing sites with local landscape change are found at the Toranica mine in the Kriva River valley and at the Zletovo mine with a tailing site on the Probištipska River. Some abandoned tailing sites, the Tajmište mine near Kičevo or the Bučim mine near Radoviš, have already been reclaimed, covered with soil and vegetation, and virtually cannot be distinguished from landforms of natural origin. All tailings will be reclaimed in the near future. Besides tailings, *waste landfills* also involve considerable landscape changes. In developed countries, regional waste landfills are permanently covered with soil and vegetation, and perfectly assimilated to their environment. Recently the reclamation of landfills began in Macedonia too. For instance, the large Drisla landfill near Skopje has been partly reclaimed. For landscapes under such intense human impact the term "technological landscapes" is used (Dragičević and Milevski 2010).

Fig. 15.8 Impact of stone quarrying and ore mining. Limestone quarry south of Skopje

Fig. 15.9 Tailings in the valley of the Kamenička River

15.5.3 Impact of Dam and Reservoirs Construction

Artificial lakes or *reservoirs* in Macedonia have a significant impact on the landscape in many ways. There are six reservoirs with storage capacities above 100 millions m^3, and areas larger than 4 km^2. Reservoirs also influence hillslope, fluvial, lake-shore and even karst processes. Their shorelines are often eroded, while raised base levels modify the rate of upstream fluvial erosion. For example, significant water-level regression on the Debar reservoir in 1993 activated a major landslide on the unstable eastern sandy shore near Pralenik village (Andonovski and Vasileski 1996). This reservoir largely influenced fluvial process along the Crni Drim (impounded into a lake) and its right-hand tributary, the Radika. On the other hand, the Mavrovo reservoir on Bistra Mountain, constructed on limestones, immediately affect karst processes in the area in such way that some holes and caverns appear in the cracks on the bottom and cause water losses. A similar effect on limestone processes is shown on the Matka and the newly built Kozjak reservoirs near Skopje (Fig. 15.10). In some reservoirs excessive sedimentation is also a problem.

Thus, the Bregalnica River and its tributaries, the Kamenička, Lukovica and Ribnica Rivers, yielded ca 17 millions m^3 of deposit in the Kalimanci reservoir (120 millions m^3) since its construction in 1969 (Fig. 15.11). The loss of useful storage capacity amounts to ca 14% (Blinkov 1998; Milevski et al. 2008). A similar situation occurs in the Tikveš reservoir, where about 30 millions m^3 of sediment accumulated from 1968 to 1991, or 6.5% of the total storage capacity of 475 millions m^3 (Trendafilov 1996). With such sedimentation, after some 200–300 years these reservoirs will ultimately turn into extensive alluvial plains. The tributaries to these reservoirs transport large amount of deposits and have already built minor *deltas* in their mouths.

15.5.4 Impact of Road Construction (Roadcuts)

Because of the checkerboard structural morphology of Macedonia, road construction is frequently very difficult. Across hilly and mountainous terrains numerous *roadcuts* have to be established, increasing natural slope inclination and leading to slope instability. As huge areas are composed of erodible rocks, landslides or intensive rill and even gully erosion occur on steep slopes during or after construction works. For example, along the Kumanovo–Stracin–Kriva Palanka main road (73.5 km) (Fig. 15.12), and the Stracin–Probištip regional road (30 km) across terrain of volcanic tuffs and breccias, many rills, gullies, earth pyramids, landslides, and even rockfalls affect roadside slopes. At footslopes the eroded material accumulates as scree, debris, or minor colluvial fans. Similar occurrences are recorded along the highway Tabanovce–Gevgelija (174.2 km), with number of smaller landslides, deep rills and gullies. On the other hand, the material removed during road construction is often deposited downslope in *artificial screes, talus*

Fig. 15.10 The dam of the Kozjak reservoir

Fig. 15.11 Shore process and delta formation in the Kalimanci reservoir

15 Recent Landform Evolution in Macedonia

Fig. 15.12 Road incision in tuffs on Kumanovo–Stracin road

cones, and *fans*. This is well expressed on the local road to the newly constructed Matka 2 (Sv. Petka) dam in the spectacular Treska canyon (Fig. 15.13). Such features are also present down the road Zletovo–Kneževo dam (20 km) near the town of Probištip, with large masses of blocks and boulders on the hillslopes up to the valley floor of the Zletovska River. Here road construction also triggers major mass movements of scree slopes above the road. Along many inadequate aligned macadam and dirt roads, severe rill formation followed by gully erosion occur and lead ultimately to slope instability and failure. Considerable impact on the landscape is also manifested along the railway corridor under construction between Kumanovo and Devebair (76 km), where deep cuts into igneous clastic rocks and schists are common.

15.5.5 Other Processes and Forms of Human Impact

Natural fluvial processes are also altered by human impact in Macedonia. The intensive *sedimentation* of basins and lower river sections, valley floors rise and floodplains extend through intensified meandering. As a consequence, floods occur more frequently, and involve more severe damage. For flood control, some riverbed

Fig. 15.13 Debris transported downslope of the road to Sv. Petka dam

sections were channelized, embankments, concrete walls were raised (The Vardar through Skopje Basin and the Veles and the Bregalnica through Delčevo, the Dragor River through Bitola, the Kumanovska River through Kumanovo, the Strumica, Luda Mara and other rivers). Also, there are hundreds of river *sand and gravel extraction* sites, partly legal and partly illegal, along floodplains with deep alluvia. Dredging produces small depressions and hollows, most of which are then ponded, and natural channels change their directions (as in the case of rivers Pčinja, Vardar, Bregalnica and Crna).

As a result of huge accumulation of fine sediment on the Vardar alluvial plain downstream of the Demir Kapija gorge, even aeolian landforms and blown sand features (near Gevgelija) occur, a rarity in Macedonia. However, they are the consequence of indirect human impact, accelerated erosion in the Vardar catchment and the subsequent deposition.

In recent times significant human impact is evident on shore processes on the Dojran and Prespa Lakes. Thus, exaggerated use of the waters of the Dojran Lake (43 km^2; 10 m deep) for irrigation purposes in the drought year 1988, reduced water level by about 3 m and caused *shore retreat* by more than 200 m (Stojanoviḱ 1995). A new shore with a total area of 5–6 km^2 formed, and the lake itself was almost drained. But with artificial water supply from nearby springs, water level is gradually increasing again. A similar situation is characteristic for the Prespa Lake

(275 km², 54 m deep), which has lost significant amount of water, and water level has dropped by about 10 m since 1963. An even larger shore area resulted than that at the Dojran Lake.

Recently, some *ground subsidence* is recorded in major cities built on clastic rocks (mostly Pliocene lacustrine sediments), such as in Skopje, due to the weight of structures and the consolidation of sediments after construction (Petkovski 2002).

15.6 Perspectives on Landscape Evolution

Ivica Milevski

15.6.1 Predictable Climate Change

There are several scenarios for climate change on the territory of the Republic of Macedonia. To describe the relationship between large-scale climatic variability across southeastern Europe and in Macedonia, the results of four Global Circulation Models (GCMs) were used together with the NCEP/NCAR reanalysis data. As the simulations of future climate with GCMs are based on a limited number of emission scenarios, usually SRES A2 and B2, the regional climate change projection, developed for the first time, was additionally scaled to other marker SRES emission scenarios (A1T, A1b, A1Fl, B1) using the pattern scaling method (Mitchell 2003). Particularly summer precipitations and temperature changes seem to be dramatic. According to the scenario based on direct GCM output (Bergant 2006), a mean *temperature increase* of 1.9°C is expected for 2050 and 3.8°C for 2100, while precipitation will decrease for about −5% in 2050 to −13% in 2100 with significant seasonal variations in temperature and precipitation. The highest temperature increase for the entire country and precipitation decrease is foreseen for the summer. Thus, temperature will increase by 2.5°C until 2050 and by 5.4°C until 2100 compared to 1990, and the highest rise is expected for maximum daily temperatures (3.0°C by 2050 and 6.2°C by 2100 compared to 1990). On the other hand, a *drastic drop in summer precipitation* is predicted (−17% in 2050 and −37% in 2100). In autumn, mean daily temperatures will increase by 1.7°C until 2050 and by 4.2°C until 2100 in respect to 1990, while mean precipitation will drop by −4% until 2050 and by −13% until 2100. Winter mean daily temperature increase may reach 1.7°C by 2050 and 3.0°C by 2100, with no major change in winter precipitation. In spring, the mean daily temperature is forecast to increase by 1.5°C until 2050 and by 3.2°C until 2100 compared to 1990. Precipitation will decrease by −6% until 2050 and by −13% until 2100. As this scenario shows, there will be some regional differences too with higher temperature increase in northwestern and southern region (ca 8.5°C), while a major drop in precipitation will occur in the central and southern regions (−26 to −27% by 2100, compared to 1990) (Bergant 2006).

15.6.2 Direct Impact of Climate Change

The exact consequences of climate change on landform evolution are hardly predictable because they depend on a range of variables. Higher temperatures and increased temperature range induce more intensive rock weathering. Although at the first glimpse, lower precipitation would imply lower erosion and transport rate, in reality the situation is much more complicated. Seasonal variations in precipitations (mostly in rainfall) would reduce vegetation cover and density, and more severe storm events would cause flash floods with intense sheetwash, rill erosion, mass movements and other hillslope processes. *Extreme hydrological events*, floods and droughts, becoming more frequent with climate change (Donevska 2006), increased erosion would affect upper catchments and deposition lower catchments. Thus, the overall erosion and deposition rate will be greater, with considerable impact on the landscape in the form of desertification. That is in line with the scenarios by Zhang and Nearing (2005), Favis-Mortlock and Savabi (1996) and others. Torrential rivers and their catchments would be even more torrential with flashy discharge by the end of the twenty-first century. After Donevska (2006), it can be concluded that generally there is a drop in annual average discharges for all river basins in the Republic of Macedonia for the past half a century. The same trend is found for annual minimum and maximum discharges for the whole territory of Macedonia and will affect the nature and rate of fluvial processes (Ministry of Environment and Physical Planning 2008). *Lateral accretion* would be more pronounced than vertical accretion and this would result in the planation of the relief. Recent water levels in major natural lakes will continue to decrease with implication on shore evolution, particularly in the case of the Prespa and Dojran Lakes (Fig. 15.14), and some (post-glacial) mountain lakes and river channels may even dry out (Fig. 15.15). With estimated climate change, the rate of *karst processes* will also reduce; some stream (riverine) caves (a trend already observed) will get dry and some ice pits in high mountains may lose their ice fill. A lower precipitation would reduce the rate of karst denudation, the formation and growth of speleothems (Shopov et al. 2006). The recent lower limit of periglacial zone, at present is at 1,600–1,800 m elevation, will be significantly higher in the future. Hillslope and fluvial processes would take over the role of periglacial geomorphic action.

15.6.3 Indirect Impact of Climatic Change

In addition to the above direct influences, the indirect impact of climate change on landscape evolution will be exerted through changes of vegetation cover, soils, biota and human activities. In recent times, the number of *forest fires* as well as the size of the area burned have significantly increased. Thus, in 2000 there were 1,187 forest fire events with 37,920 ha of burned area, while in the very hot year 2007 ca 40,000 ha forest burned – much more than before 1990 (usually less than 100 fires and 2,000 ha of burned forest). It is expected that Mediterranean and sub-Mediterranean elements of the pseudo-maquis will broaden their range towards the central and northern regions,

Fig. 15.14 Recent retreat of the Prespa Lake and the formation of a new shore, mostly as a result of lower precipitation

Fig. 15.15 The dry channel of the Radanjska River: a result of erosion and climate change

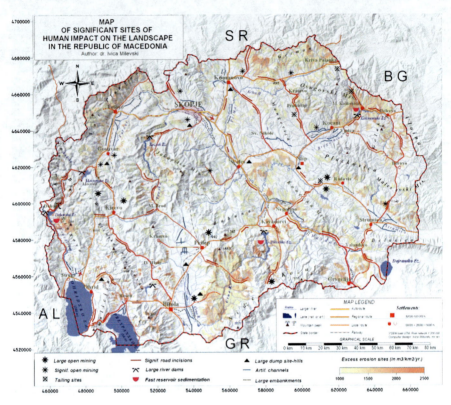

Fig. 15.16 Map of significant sites of human impact on the landscape in the Republic of Macedonia (after Dragičević and Milevski 2010)

i.e. climate change will cause an *extension of zonal vegetation* (Adžievska et al. 2008). With the reduction of forest area and overall vegetation quality and density, runoff and overland flow would increase raising the rate of erosion. Further, in lack of proper vegetation protection, soils will become more prone to erosion. As an adaptation measure to higher air temperatures and decreased precipitation, the *migration of* certain *tree species* to higher elevations (above 2,000 m) and latitudes is evident over the past decade. Periglacial processes and landforms as well as fossil glacial remnants may gradually also modify. As water resources shrink with lower precipitation, *more water* will be *extracted* from groundwater, river channels and lakes, promoting *ground subsidence*, altering fluvial and lake-shore processes. Because of a greater water demand for irrigation and decreased discharge as a result of lower precipitation, some *rivers* (even larger channels) may *partly dry out* in summer (the Bregalnica River downstream of Štip; the Pčinja, Strumica and Crna Rivers). One important issue is that because of increased summer heat in urban settlements located in basins, more and more development (houses, farms, villas) takes place in mountains and cause the visual deterioration of the natural landscape.

References

Andonovski T (1977) Underground karst landforms in the Radika Valley. Ann Fac Geogr Skopje 23:97–120 (in Macedonian)
Andonovski T (1982) Erosion areas in SR Macedonia. In: Proceedings from the 11th Congress of FRY Geographers, Budva, 1982 (in Macedonian)
Andonovski T (1989) Karst poljes in SR Macedonia. Geogr Rev Skopje 27:3–13 (in Macedonian)
Andonovski T, Milevski I (1999) Underground karst landforms in Bislim Gorge. Geogr Rev Skopje 34:5–21 (in Macedonian)
Andonovski T, Vasileski D (1996) Sliding (slump) of earth near village Pralenik in Debarska Župa. Geogr Rev Skopje 31:11–21 (in Macedonian)
Arsovski M (1960) Some characteristics of the tectonic features of the central part of Pelagonian horst anticlinorium and its relation to the Vardar Zone. Papers of Geologic Survey of SRM, Skopje 7:76–94 (in Macedonian)
Arsovski M (1997) Tectonics of Macedonia. Faculty of Geology and Mining, Stip. 306 p (in Macedonian)
Adžievska et al. (2008) Second national communication on climate change, Ministry of Environment and Physical Planning, Skopje, 118 pp
Belij S, Kolčakovski D (1997) Methodology of complex research of periglacial zone on the high mountains in the Balkan Peninsula. In: Scientific meeting "Perspectives and directions in the development of Geographic science". Brezovica, FR Yugoslavia, 1997, 7 p
Bergant K (2006) Climate change scenarios for Macedonia – review of methodology and results. University of Nova Gorica, Centre for Atmospheric Research, Nova Gorica, Slovenia, pp 1–50
Blinkov I (1998) Influence of the rains on the intensity of soil erosion in the Bregalnica watershed up to the profile "Kalimanci Dam". Manuscript doctoral thesis. Faculty of Forestry, Skopje, 127 p (in Macedonian)
Boev B, Yanev Y (2001) Tertiary magmatism within the Republic of Macedonia: a review. Acta Volcanologica 13(1–2):57–71
Burchfiel BC, Dumurdzanov N, Serafimovski T, Nakov R (2004) The Southern Balkan Cenozoic Extensional Region and its relation to extension in the Aegean Realm. Geol Soc Am Abs Prog 36(5):52
Cvijić J (1906) Basics of geography and geology of Macedonia and old Serbia. Book I. SKA, Belgrade (in Serbian)
Djordjevic M, Trendafilov A, Jelic D, Georgievski S, Popovski A (1993) Erosion map of the Republic of Macedonia. Memoir. Water Development Institute, Skopje, 89 p (in Macedonian)
Donevska K (2006) Vulnerability assessment and adaptation for water resources sector. In: Report on second communication on climate and climate changes and adaptation in the Republic of Macedonia. Ministry of Environment and Physical Planning, Skopje, 128 p
Dragičević S, Milevski I (2010) Human impact on the landscape – examples from Serbia and Macedonia. In: Zlatic M (ed) Global change – challenges for soil management, vol 41, Advances in geoecology. Catena Verlag, Reiskirchen, pp 298–309
Dumurdzanov N, Serafimovski T, Burchfiel C (2004) Evolution of the Neogene-Pleistocene basins of Macedonia. Geological Society of America, Boulder, CO. pp 1–20
Favis-Mortlock DT, Savabi MR (1996) Shifts in rates and spatial distribution of soil erosion and deposition under climate change. In: Anderson MG, Brooks SM (eds) Advances in hillslope processes, vol 1. Wiley, New York, pp 529–560
Gaševski M (1953) Debar Basin – geomorphological research. Annal SGS Belgrade 33(1):31–44 (in Serbian)
Gospodinov S, Zdravchev I, Alksandrov B, Peneva E, Georgiev I, Tzenkov Z, Dimitrov D, Pashova L (2003) Multidisciplinary investigation of the recent movements between basic tectonic structures on the Southwest part of Bulgaria. Report from the Project funded by UACEG, Sofia, pp 1–9

Jančevski J (1987) Classification of fault structures according to genesis, age and morphology, with review to their seismicity on the territory of Macedonia. Ph.D. thesis (manuscript), Sts. Cyril and Methodius University, Skopje, 247 p (in Macedonian)

Jarvis A, Reuter HI, Nelson A, Guevara E (2006) Hole-filled SRTM for the globe, Version 3, available from the CGIAR-CSI SRTM 90m Database: http://srtm.csi.cgiar.org

Jovanović P (1927) Coastal and fluvial elements in Poreče Basin. Annals SGS 13:169–194 (in Serbian)

Jovanović P (1928) Karst features in Poreče. Ann SGS Belgrade 4:1–46 (in Serbian)

Jovanović P (1931) Relief of Skopje Basin. Annals SGS Skopje 11:62–116 (in Serbian)

Kolčakovski D (1987) Denudation landforms in the basin of the Kadina River. Geogr Rev Skopje 25:229–238 (in Macedonian)

Kolčakovski D (1988) High-mountain karst of the Karadzica Mountain. Geogr Rev Skopje 26:95–113 (in Macedonian)

Kolčakovski D (1989) Historical review of speleological research on the territory of SR Macedonia with bibliographic prospective. Geogr Rev Skopje 27:133–144 (on Macedonian)

Kolčakovski D (1992) Karst relief in Skopje Basin – a geomorphological study. Special edition of FNSM, Skopje, 47 p (in Macedonian)

Kolčakovski D (1994) High-mountain areas of the mountains Jablanitsa, Galichitsa and Pelister. Ecol Environ Prot Skopje 1(1–2):43–51 (in Macedonian)

Kolčakovski D (1999) Glacial and periglacial relief on the mountain Jablanica. Annu Inst Geogr FNSM Skopje 33–34:15–38 (in Macedonian)

Kolčakovski D (2001) Chronological zoning of periglacial geocomplex in the southwest part of the Republic of Macedonia and its comparison with neighbour and distant areas. Annal Inst Geogr FNSM Skopje 35–36:61–83 (in Macedonian)

Kolčakovski D (2004) Snowline – theoretical views and its reconstruction during Würm climatic minimum on the mountains of southwestern Republic of Macedonia. In: Proceedings of the second congress of ecologists of Macedonia, Skopje, 2004, pp 23–28 (in Macedonian)

Kolčakovski D (2004b) Geotectonic characteristics of the relief in the Republic of Macedonia. Bull Phys Geogr Skopje 1:7–23 (in Macedonian)

Kolčakovski D (2005) Contribution to a research of micro-relief forms of the granodiorite rocks on the locality "Markovi Kuli". Bull Phys Geogr Skopje 2:5–23 (in Macedonian)

Kolčakovski D (2006) Geomorphology. University text-book, Skopje, 488 p (in Macedonian)

Kolčakovski D, Boskovska G (2007) Extension and age of carbonate rocks in Republic of Macedonia. Bull Phys Geogr Skopje 3–4:77–80 (in Macedonian)

Kolčakovski D, Petreska B, Temovski M (2007) New informations about the underground karst forms in the central parts of the Republic of Macedonia. Bull Phys Geogr Skopje 3–4:55–66 (in Macedonian)

Kossmat F (1924) Geologie der zentralen Balkanhalbinsel. Die Kriegsschauplätze 1914–1919 geologisch dargestellt, Berlin, 198 p

Lilienberg DA (1966) General tendency of recent tectonic movements in Macedonia. Ref. sest. SGD, Ohrid 245–270 (in Russian)

Manakoviќ D (1960) Landslide of Gradot hill. Annu SAS Belgrade 17:119–128 (in Serbian)

Manakoviќ D (1962) Nivation processes and landforms on the mountain Jakupica. Ann FNSM Skopje 13:47–57 (in Macedonian)

Manakoviќ D (1968) Middle-Vardar Lake. In: Proceedings from VIII congress of geographers from SFRY, Skopje, 1968, pp 155–164 (in Macedonian)

Manakoviќ D (1969) Soil erosion in the downstream part of the Pčinja catchment. Geogr Rev Skopje 7:43–54 (in Macedonian)

Manakoviќ D (1974) Landslide in the village Bituša. Annu FNSM Skopje 20:37–46 (in Macedonian)

Manakoviќ D (1980) Oasis type of karst hydrography in Macedonia. In: Proceedings of VII Yugoslavian speleological congress, Titograd, 1980, pp 293–309 (in Macedonian)

Manakoviќ D (1980) Geomorphology of Maleš and Pianec. Maleš and Pianec II. Natural and socio-geographical characteristics of Maleš and Pianec, MASA, Skopje, pp 47–69 (in Macedonian)

Manakoviḱ D, Andonovski T, Stojanovic M, Stojmilov A (1998) Geomorphologic map of the Republic of Macedonia. Memoir. Geographical Review, Skopje 32–33:37–70 (in Macedonian)

Manasiev J, Jovanovski M, Gapkovski M, Novacevski T, Petreski Lj (2002) Landslide on the NE part of the coal mine "Suvodol": phenomenology of the event and experiences. In: Proceedings of 1st symposium of the Macedonian Association for Geotechnics, Ohrid, 2002, pp 68–77 (in Macedonian)

Markoski B (1995) Hypsometry of the space and population in the Republic of Macedonia – a cartographic method. Makedonska Riznica, Skopje, 316 p (in Macedonian)

Markoski B (2004) Cartographic definition and differentiation of the mountain areas in the Republic of Macedonia. Bull Phys Geogr Skopje, pp 25–34 (in Macedonian)

Milevski I (2001) Some aspects of morphology and genesis of gullies in the Pčinja catchment. Geogr Rev Skopje 36:197–207 (in Macedonian)

Milevski I (2004) Soil erosion in the Želevica catchment. Bull Phys Geogr Skopje 1:59–75 (in Macedonian)

Milevski I (2005a) Basic features of palaeovolcanic relief in the western part of Osogovo massif. Geogr Rev Skopje 40:47–67 (in Macedonian)

Milevski I (2005b) Characteristics of recent erosion in Kumanovo Basin. Bull Phys Geogr Skopje 2:25–45 (in Macedonian)

Milevski I (2006) Geomorphology of the Osogovo Mountain Massif. Doctoral dissertation, manuscript, Institute of Geography, FNSM, Skopje

Milevski I (2007a) Quantitative-geomorphologic characteristics of longitudinal profiles of Osogovo massif. Bull Phys Geogr Skopje 3–4:31–48 (in Macedonian)

Milevski I (2007b) Morphometric elements of terrain morphology in the Republic of Macedonia and their influence on soil erosion. In: Proceedings from international conference "Erosion 2007", Belgrade, 2007, 10 p

Milevski I (2008) Fossil glacial landforms and periglacial phenomena on the Osogovo Mountain massif. Annal Inst Geogr Skopje 37:25–49

Milevski I (2009) Excess erosion and deposition in the catchments of Kamenička and Radanjska River, Republic of Macedonia. Annu Serbian Geogr Soc Belgrade 89(4):109–120

Milevski I, Miloševski V (2008) Denudation landforms in the Mavrovica Catchment. Bull Phys Geogr Skopje 5:87–100 (in Macedonian)

Milevski I, Dragičević S, Kostadinov S (2007) Digital elevation model and satellite images in assessment of soil erosion potential in the Pčinja catchment. Annal SGS Belgrade 87(2):1–20

Milevski I, Blinkov I, Trendafilov A (2008) Soil Erosion processes and modelling in the Upper Bregalnica Catchment. In: Proceedings of the 24th conference of the Danubian Countries on the hydrological forecasting and hydrological bases of water management, Bled, Slovenia. 2008. http://ksh.fgg.uni-lj.si/bled2008/cd_2008/index.htm

Ministry of Environment and Physical Planning of the Republic of Macedonia (2008) Second National Communication on Climate Change. Project strategy led by Adzievska M. Skopje, 118 pp

Mitchell TD (2003) Pattern scaling. An examination of the accuracy of the technique for describing future climates. Climatic Change 60(3):217–242

Petkovski R (1998) Connection between neotectonic movements and lake stadiums in Macedonia. In: Proceedings from 1st congress of ecologists of Macedonia, Ohrid, 1998, pp 855–867 (in Macedonian)

Petkovski R (2002) Possible damages of the construction objects because of unequal soil settlements. In: Proceedings of 1st symposium of the Macedonian Association for Geotechnics, Ohrid, pp 427–435 (in Macedonian)

Radovanović VS (1928) Small denudation landforms in gneiss. Annal SND Skopje 4(1):53–121 (in Serbian)

Radovanović VS (1931) Holokarst Huma bellow Kožuf. Annal SND Skopje 9(3):108–159 (in Serbian)

Serafimovski T (1993) Structural metalogenic features of the Lece Chalkidiki zone: types of mineral deposits and distribution. Faculty of Mining and Geology, Stip, Macedonia. Special Issue 2, 328 p (in Macedonian)

Shopov Y, Stoykova D, Tsankov LT, Marinova E, Sauro U, Borsato A, Cucchi F, Forti P, Piccini L, Ford DC, Yonge CS (2006) Past annual variations of the Karst denudation rates. In: Duran JJ, Andreo B, Carrasco F (eds) Publicaciones del Instituto Geologico y minero de Espana. Serie: Hidrogeologia y aguas subterraneas. Madrid, vol. 18, pp 487–494

Stojadinoviḱ C (1958) Geomorphologic observation of the gypsum relief in the Radika Valley. Annals FNSM Skopje 11:73–118 (in Macedonian)

Stojadinoviḱ C (1962) Stone rivers and screes on Pelister. Geogr Rev Skopje 1:45–51 (in Macedonian)

Stojadinoviḱ C (1968) Genesis of littoral relief on the basin of Ohrid Lake, and its tectonic and coastal elements. Annu FNSM Skopje 16(4):103–127 (in Macedonian)

Stojadinoviḱ C (1970) Geologic-geomorphic evolution of the relief of Pelister. In: Proceedings of the symposium for Molika, Skopje, 1970, pp 39–48 (in Macedonian)

Stojanoviḱ M (1986) Areas of fossil volcanoes in the territory of SR Macedonia. Ann FNSM Skopje 30:148–166 (in Macedonian)

Stojanoviḱ M (1995) Dojran Lake: origin, evolution and extinction. Geogr Rev Skopje 30:81–90 (in Macedonian)

Trendafilov A (1996) Erosion in the Crna River catchment and sedimentation of the Tikveš reservoir. Manuscript doctoral dissertation, Faculty of Forestry, Skopje, 256 p (in Macedonian)

Zagorchev I (1995) Pre-Paleogene Alpine tectonics in Southwestern Bulgaria. Geologica Balcanica Sofia 25(5–6):91–112

Zhang XC, Nearing MA (2005) Impact of climate change on soil erosion, runoff, and wheat productivity in central Oklahoma. Catena 61(2–3):185–195

Part III
Conclusions

Chapter 16
Conclusions

Dénes Lóczy, Miloš Stankoviansky, and Adam Kotarba

This monograph could not have been completed without a long-term collaboration of geomorphologists in the countries situated within the Carpatho–Balkan–Dinaric region. The first and the most important step towards this collaboration was the founding of the *Carpatho–Balkan Geomorphological Commission (CBGC)* in 1963. The common objective of geomorphologists of member countries was the study of the mountain systems of the Carpathians and Balkanides (Stara Planina) and the adjacent depressions. The second milestone was the acceptance of the *Carpatho–Balkan–Dinaric Regional Working Group (CBDRWG)* in the International Association of Geomorphologists (IAG) in 2005. At the preparation of its proposal, on the basis of an initiative from some successor countries of the former Yugoslavia, the competence of the CBGC was broadened to cover the Dinarides too. Today 12 countries adhere to it, namely Austria, Bulgaria, Croatia, Czechia, Hungary, Macedonia, Poland, Romania, Slovakia, Slovenia, Serbia, and Ukraine. Apart from Austria all mentioned countries participated in the preparation of the present monograph.

D. Lóczy (✉)
Institute of Environmental Sciences, University of Pécs, Ifjúság útja 6,
H-7624 Pécs, Hungary
e-mail: loczyd@gamma.ttk.pte.hu

M. Stankoviansky
Department of Physical Geography and Geoecology, Faculty of Natural Sciences,
Comenius University in Bratislava, Mlynská dolina, 842 15 Bratislava 4, Slovakia
e-mail: milos.Stankoviansky@test.fns.uniba.sk

A. Kotarba
Department of Geomorphology and Hydrology of Mountains and Uplands,
Institute of Geography and Spatial Organisation, Polish Academy of Sciences,
ul. Św. Jana 22, 31-018 Kraków, Poland
e-mail: kotarba@zg.pan.krakow.pl

The monograph relies on several *publications* as predecessors. Some years after the foundation of the CBGC a two-volume monograph entitled *Geomorphological Problems of the Carpathians* was issued. The first volume was issued by Veda Publishers of Bratislava in 1965 and was dedicated to landform evolution in the Tertiary period. Its regional coverage extended over the territory of the Western Carpathians and Stara Planina, the contributors represented Czechoslovakia, Poland, Hungary, the Soviet Union (where the present Ukrainian Carpathians used to belong) and Bulgaria. The second volume was published as the No. 10 issue of the journal Geographia Polonica in Kraków in 1966 and dealt with landform evolution in the Quaternary. The same countries contributed, with the exception that Bulgaria was substituted by Romania. Both books assessed the state of research in the mentioned topics and summarized the results acquired until that time. Their authors found that neotectonic (Miocene and Pliocene) movements and selective erosion depending on climatic changes are the most important factors in the landform evolution of the Carpatho–Balkanic system.

As far as the topics investigated are concerned, *geomorphic research* had different *focuses* in early times: mostly planation surfaces, river terraces and young tectonic movements, karst, glacial, glacifluvial, periglacial, and aeolian landforms, river network changes and the expression of structural and rock properties (passive geological structure) in relief stood in the forefront. Investigations were closely connected to developments in *geomorphological mapping*. Mapping projects finally led to establishing geomorphological divisions in each country presented on national maps of geomorphological units, often included in the national atlases published at that time.

In the course of the 1960s, under the influence of the *Soviet* geomorphological school, the *morphostructural approach* became prevalent and strove to assess the course of relief evolution, taking into account the consequences of active geological structure or, in other words, the influence of tectonic movements on shaping landforms. This approach gradually acquired a predominant position in the geomorphology of most of the Carpatho–Balkan–Dinaric countries and manifested itself not only in mapping geomorphological divisions but also in the preparation of independent morhostructural maps of (group of) countries (e.g., the morphostructural map of the Western Carpathians). The most ambitious international mapping project was the preparation of the *Geomorphological Map of the Danubian Countries* starting from 1978, included in the *Atlas of Danubian Countries*, issued in Vienna, representing both morphostructural relief types and individual landforms. The map indicates the approximate geological age of the different landforms and this information, combined with the symbol of the legend representing the origin of the individual landforms, shows the morphodynamic character of the geomorphological unit. Thus, the map could also function as a starting point for studies on active geomorphic processes.

However, in some countries of the region investigations have traditionally focused on endogenic processes or, even more often, exogenic geomorphic processes are viewed from an overwhelmingly *geological perspective*: the impact of neotectonic movements on landform evolution, mass movements or fluvial erosion processes controlled by rock quality. An alternative approach places more emphasis on *climatic circumstances*. Through the incorporation of the climatic environment a more complex evaluation of the physical background to landform evolution becomes possible.

Consequently, another important research direction of geomorphic investigation of participating countries that played a decisive part in the preparation of this monograph and was employed as an orientation principle is the *morphodynamic approach*. Though the beginnings of the study of some partial exogenic geomorphic processes date back to the past, their systematic research mostly began in the post-World War II period. Out of the Carpatho–Balkan–Dinaric countries such investigations became most popular in Poland. The Polish Carpathians was most thoroughly explored from this viewpoint and the findings of Polish geomorphologists served as a model for morphodynamic investigations in other countries. It is not a mere coincidence that at that time the CBGC working group for the unification of techniques of study of *present-day geomorphic processes* was chaired by a representative of Poland, T. Gerlach. The study of geomorphic processes was also intensively pursued in Romania and in some other countries like the former Czechoslovakia, Hungary and, to a considerable extent, also the Ukrainian (Carpathian) part of the former Soviet Union. This research direction was not really practiced in Bulgaria and the former Yugoslavia. In short, in the past and also today highly variable importance has been attached to the study of present-day geomorphic processes in the individual countries of the Carpatho–Balkan–Dinaric region. In some of them the morphodynamic approach have acquired predominance, while in the rest other directions in geomorphology have become more pronounced. Such developments can be explained by different traditions cherished during the historical evolution of geomorphic research and markedly different geomorphological schools.

This is the reason why the treatment of geomorphological research in the national chapters *cannot be homogenized*. However, it is nothing new for a publication of this type. We can soothe ourselves by citing some ideas from the Foreword to the first volume of the above mentioned monograph *Geomorphological Problems of the Carpathians* from 1965: "*Even though geomorphological studies in the individual Carpathian countries have had a long tradition and have been intensified within the last years, we still feel the need of a synthetic work on the subject. Research work went on rather isolately in the past, with various thematic aims, so that the results, even though they are considerable as to quantity, are rather heterogeneous and hard to use in a comparative study.*" Our predecessors also emphasized that "*it is not a work that would embrace the results of a coordinated international geomorphological study of the Carpathians, nor does this work pretend to give a complex synthesizing picture of the Tertiary genesis of the Carpathian relief. The aim of the publication is to give a general outline of the present-day situation in the knowledge of at least the fundamental features and development of the Carpathian relief, especially of the Western Carpathians, as this outline appears in the individual countries of the Carpathian region.*" The editors of the mentioned volume also claim that "*they want to express the wish that this publication will be a successful first step on the road to future international cooperation of Carpathian geomorphologists.*"

All that was said to describe the contents of the publication dedicated to the Tertiary relief evolution of the Western Carpathians and Stara Planina is even more valid for the present volume on the recent landform evolution of the Carpatho–Balkan–Dinaric area. The heterogeneity of national chapters, the disparities between them and the resulting *low* level of *comparability* reflect the *diversity of* the *traditions* of geomorphic research and the variety of approaches employed in the study of this rather extensive area. On the other hand, it can be regarded a success that so many countries could be involved in the project, in fact, much more than in the early years of the CBGC.

Naturally, the research of processes varied in character and intensity for the individual countries. In the countries which can be considered most advanced from this aspect a well-developed network of *research stations* was founded to serve the monitoring of current processes in selected catchments, in test areas of hillslopes or plot scale (e.g., Szymbark, Hala Gąsienicowa, Homerka, Łazy in Poland; Pătârlagele, Piatra Neamț, Perieni–Bârlad in Romania; Csákvár, Aggtelek and Bodrogkeresztúr in Hungary, and others). The outcome of monitoring efforts is the estimation of the rate of recently active processes as described in most of the national chapters. For this purpose, in addition to monitoring itself, various field and laboratory measurements using a range of techniques and experiments proved to be instrumental. The spatial distribution of processes, their temporal course and behavior have also been analyzed in increasingly quantitative approaches. Processes were also evaluated as *geomorphological (natural) hazards*. Attention was also paid to documentation of geomorphic effect and environmental impact of particular extreme events. An important components of the study of processes was the assessment of the cumulative geomorphic effect of a series of consecutive (often poorly documented) events within the study period (decades, centuries) manifested in reducing surface convexities and increasing concavities associated with changes in the sediment budget.

The *processes* studied in the individual chapters can be referred into three groups on the basis how much space is devoted to them. In most chapters (in eight countries) gravitational (mass movements) and runoff processes (water erosion) are of outstanding *significance*, the second group represents fluvial and aeolian processes, treated in detail in six national chapters, while the third, periglacial (cryogene and nival), karstic (pseudokarst), and littoral processes, appear in three chapters. Biogenic processes are only assessed in one of the national chapters. Important sections of Part II focus on direct human interventions into the relief, a topic that is not missing from any of the national chapters. The history of geomorphological investigations in each country has strongly influenced decisions on the selection of topics and on the structure of the chapters. Countries with lesser tradition in morphodynamic research prefer to focus on landforms rather than on processes and not only on recent evolution but also long-term relief development in the geological past.

The *changes* of the *political and economic* order after World War II had a serious indirect impact on geomorphic processes. The general trend of transformation was similar in all countries of the Carpatho–Balkan–Dinaric territory: the nationalization of industry and agriculture. The new land use regulations had a crucial influence on the physical environment, unknown in Western Europe. As privately owned

16 Conclusions

arable plots were incorporated into collective or state farms, *accelerated soil erosion* ensued and sometimes reached catastrophic dimensions. The intensive erosional processes were analyzed by the geomorphologists of the CBD countries. The only exception is the Polish Carpathians, which have never been affected by the collectivization of agriculture and the resulting large-scale farming, but traditional small-plot cultivation survived to our days. Slovak, Czech and Hungarian scientists documented that intensive soil degradation was caused by human impact. This aspect of human intervention on recent landform evolution is clearly reflected in the national chapters of the monograph.

Landslides are the most common and spectacular forms of mass movement in the mountains and exert the greatest impact on relief in areas of active *young tectonics*. High seismicity and neotectonic activity in the flysch mountains is responsible for the most widespread landslides, first of all, in the Eastern Carpathian Curvature area of Romania and partly in Macedonia. In contrast, in tectonically stable areas of flysch mountains (i.e., the Polish and Ukrainian Carpathians, partly in Slovakia), the main factors triggering substantial mass movement types are heavy rainfall, high relief and suitable lithology.

In several countries geomorphic processes are studied in mutual interaction. Landform evolution depends on the interaction of various geomorphic agents like river undercutting and the landslides induced by this process; deep weathering and its corollary: deep-seated landslides on slopes where conditions are favorable. The significance of *inherited landforms* on present-day geomorphic processes is studied in several countries focusing on the remnants of glacial and periglacial landforms in the high mountains, like the Polish and Romanian Carpathians, the Bulgarian and Macedonian mountains and the Slovenian Alps.

The most intensive and abrupt *human intervention* into the physical environment is represented by river regulations, the constructions of dams and reservoirs as well as industrial plants. Some regions, rich in mineral deposits are extremely strongly affected by landform transformation and chemical pollution caused by extraction industries. This process is observed in each country but seems to have the greatest dimensions in Romania.

The share of the assessment of investigation of recent geomorphic processes in total geomorphic research of individual countries is also markedly reflected in the determination of the *length of the period interpreted as recent*, which is not uniform for all CBD countries. The shortest period is defined in the *Polish* chapter but it essentially means the time interval when high-intensity, extreme geomorphic events took place with catastrophic consequences for human society. Geomorphic and hydroclimatic events were mentioned in written sources from the last millennium, but relatively precise descriptions deriving from travelers and natural scientists and the conclusions were not published before the beginning of the nineteenth century. This explains why the 200 years time span had been decided on by the authors of the Polish chapter. The Polish geomorphologists have ample findings even for such a short period. The *Slovak* chapter assesses the effect of processes in the Carpathians within the last eight centuries, in lowland areas exclusively within the last two millennia. In *Hungary* recent geomorphic processes only refer to the past two or three centuries, which is considered to be a convenient time period for their modeling (particularly in the case of accelerated water and wind erosion). The concept also includes present-day relief evolution and, thus, allows us to reveal future perspectives and points to the practical significance of actual geomorphological studies. In the southern countries where the morphostructural school has had especially strong tradition (like Croatia and Bulgaria) neotectonic movements are emphasized as the main drivers of landform evolution. Consequently, the time span investigated as recent is considerably longer.

In general, most of the contributing countries concentrate on the period of human interference. As to the last centuries, more national chapters refer to the influence of the Little Ice Age on the significant intensification of precipitation-induced processes and some of them on the influence of the above mentioned large-scale land use changes associated with collectivization in agriculture on the acceleration of water erosion.

Finally, some remarks on the *motivation* for writing a monograph on recent geomorphic processes. Several impulses have contributed to this decision. The first was purely emotional, the current generation wanted to honor with the book the predecessors who have laid the foundations of geomorphology in the Carpatho–Balkan–Dinaric area. On the other hand, they want to prove their abilities and not only in the fields studied by the teachers but also in topics of recent geomorphic processes which in their time was not quite common – and in some countries has not become widely investigated to our days either. Placing the CBDRWG under the "protecting wings" of the IAG, an organization that promoted our endeavor, we wanted to show that we had been accepted into this high-ranking international geomorphological body not without merits.

The editors leave it to the readers of this book to decide how successful they have been to fulfil their objectives. Perhaps the greatest achievement of the volume is the tremendous potential for mutual learning, enriching experience and influencing studies in the field of recent landform evolution, starting with the unification of terminology and finishing with the creation of international teams in thematic research. Each country has a chance to borrow methodological tools tested in other countries and to use the acquired experience of its partners to assess the same phenomena on their own territory. This particularly refers to neighboring countries, as the spatial distribution of operation of geomorphic processes is not limited by administrative boundaries. However, it can also concern countries that are farther away from each other, but show similar natural conditions and history of anthropogenic transformation.

It is easier to provide a comprehensive overview of research activities in the individual CBD countries than to summarize the regional variations of the individual groups of geomorphic processes. However, if there is a discrepancy between both attitudes (i.e., a marked gap between performed research and the topics with remarkable research potential) can be perceived as indicative of future perspectives and, possibly, the most promising fields of academic collaboration in the study of recent landform evolution.

16 Conclusions

Thus, in general, this book has a great chance to encourage geomorphologists of participating countries to look for ways to gradual *harmonization* of research of recent geomorphic processes and not only within individual evaluated megaunits, but perspectively in the whole Carpatho–Balkan–Dinaric area. Echoing again the statement of editors of the first volume of the monograph *Geomorphological Problems of the Carpathians* from 1965, we would like to express the wish that this publication will represent the next successful step towards an improved international cooperation between geomorphologists active in the landscapes of the Carpatho–Balkan–Dinaric region.

Index

A

Abandoned land, 233, 407, 428
Abrasion, 333, 339, 346, 352, 353, 360, 390, 393
Accelerated erosion, 104, 160, 163, 420, 426–429, 434
Accidents, 127, 132, 276
Adria Microplate, 16, 208
Aeolian deposition, 49, 266, 399
Aeolian processes, 49, 70–72, 82, 86, 151, 164, 166–167, 224–227, 323, 360, 384, 399–401, 448
Afforestation, 43, 149, 153, 339, 428
Agricultural land, 41, 43, 86, 111–113, 129, 149, 163, 165, 200, 217, 218, 250, 428
ALCAPA Mega-Unit, 15
Alluvial centers, 191
Alluvial fans, 32, 33, 75, 78, 79, 148, 158, 210–212, 224–226, 234, 356–358, 420, 421, 428
Alluviation, 78, 79, 214, 216
Alpine belt, 58, 72, 152, 153, 186, 331
Alpine foreland, 210
Alpine karst, 292, 293
Alpine orogeny, 4, 359, 385, 413
Amelioration, 129
Antecedent gorges, 387
Anthropogenic geomorphic processes, 110
Applied research, 207, 224, 384
Apuseni Mountains, 15, 250–252, 255, 256, 263–266, 275–277
Arêtes, 360, 395
Asymmetry, 333, 386, 400
Avalanches, 58, 59, 83, 84, 123, 152–154, 184, 186, 187, 201, 202, 269, 399, 401, 409
Azonal processes, 150

B

Badlands, 161, 299, 361, 420, 428
Balkanides, 4, 6, 11, 347, 349, 352, 358, 359, 378, 379, 415
Balkan Zone, 5, 11
Banat Mountains, 251, 263–265
Bank erosion, 158, 160, 181, 182, 185, 193, 197, 200, 406
Basement, 5–10, 13–16, 143, 145, 208, 251, 252, 263, 315, 316, 321, 336, 351, 385
Beaver habitats, 80
Bedload, 33, 35, 57, 76, 78, 79, 83, 110, 126, 148, 157–160, 213, 273
Bimodality, 255, 272–274
Biogenic landforms, 50, 80
Biogenic processes, 80, 448
Blind valleys, 229, 265, 292, 293, 296, 304, 328
Block movements, 165, 270, 409
Blockslides, 191
Block streams, 193
Blown-sand regions, 224, 225, 234
Bosnian Unit, 5, 12–13
Budva zone, 12
Burrowing, 130

C

Calderas, 221, 223, 356, 360, 418
Canyons, 194, 199, 228, 292, 294, 332, 335, 356, 358, 380, 387, 390, 391, 416, 421, 423, 433
Carbonate karst, 201, 262
Carbonate platform, 12, 13, 263, 315–317, 332
Carpathian foredeep, 4–5, 108

Carpatho–Balkanides, 347, 349, 352, 358, 359, 415
Cascade systems, 404
Cauldrons, 70
Caves, 70, 73, 75, 107, 118, 129, 131, 165, 228, 229, 232, 254, 265, 290, 292, 294–297, 303, 304, 326, 328–332, 339, 352–354, 356, 359, 365–368, 389–393 416, 423, 424, 429, 436
Cave sediments, 297
Central Moravian Carpathians, 105–107, 110, 113, 127
Channel alterations, 216
Channelized streams, 78
Channel length, 32, 116, 212
Channel pattern, 212, 214
Channel width, 213
Check dams, 78
Checkerboard topography, 416
Chemical denudation, 56, 73, 164, 188
Cirque glaciers, 302, 425
Cliffs, 129, 152, 156, 160, 233, 263, 299, 300, 333, 352, 387, 390, 391, 393, 409, 424
Climate change, 78, 149, 154, 167–168, 224, 291, 294, 298, 388, 405–407, 421, 423, 425, 435–438
Climatic regionalization, 22
Collapse of cave ceilings, 165, 368
Collectivization, 105, 117, 125, 127, 130, 149, 161–165, 167, 449, 450
Complex rehabilitation, 233
Composite valleys, 335, 416
Compound landslides, 75, 119
Concavity index, 269
Conical hills, 292, 294, 304, 360
Contour tillage, 69
Core mountains belt, 143
CORINE, 40
Correlative sediments, 162, 336
Corrosion, 164, 228, 229, 294–297, 326, 331, 335, 337, 339, 392, 407
Couloirs, 399
Creep, 56, 58, 59, 72, 84, 122, 152, 155, 156, 207, 210, 222, 225, 269, 271, 299, 406, 420
Crevice-type caves, 73, 75, 118
Crişana Hills, 250, 252
Cross-section, 62, 75, 175, 197, 212–214, 216, 228, 255, 274, 362, 364, 380, 381, 386, 387, 416
Crustal extension, 208, 418
Crustal shortening, 415, 416

Cryonival morphoclimatic system, 152
Cryoplanation zone, 266
Crystalline-Mesozoic Zone (CMZ), 5, 9

D

Dams, 35, 50, 76–78, 83, 85–87, 109, 157–160, 186, 200, 212, 250, 256, 274, 276, 363–366, 404, 426, 428, 431, 432, 434, 449
Danube Delta, 33–35, 42
Danubian hilly plain, 380
Debris avalanches, 58, 59, 83, 84, 123, 152, 184, 186, 202, 399
Debris floods, 190
Debris flows, 190–191
Deep-seated creep, 155
Deep-seated rockslides, 59
Deflation, 57, 71, 72, 86, 167, 207, 225, 226, 232, 339, 387, 407, 409
Deflation niches, 71
Deforestation, 82, 86, 104, 112, 118, 125–127, 149, 188, 191, 196, 197, 199, 201, 226, 256, 363, 370, 405, 407, 409
Deltas, 16, 33–36, 42, 76–78, 86, 145, 157, 335, 424, 431, 432
Deluvial-debris centers, 191
Denudation rate, 186, 257, 297, 330
Derasion, 207, 324
Dinaric karst, 292, 293, 296, 304, 326, 328, 359
Dinarides, 4, 12, 315–319, 321, 324–326, 328, 332, 334, 337–339, 347, 351–353, 358, 359, 414, 415
Dinarid Ophiolite Belt, 13
Dirt roads, 162, 433
Dolines, 73, 227–229, 292–297, 304, 326, 328–331, 339, 359, 366, 392
Double tombolo, 394
Downslope plowing, 69, 85
Drainage, 31–36, 54, 76, 80, 82, 129, 132, 133, 157, 161, 165, 194, 196, 199, 200, 208, 210, 211, 215, 216, 254–257, 265, 269, 272–275, 277, 294, 333, 334, 348, 353, 356, 358, 359, 366, 382, 383, 386, 387, 395, 404, 421
Drainage density, 194, 334, 348, 382, 383
Drainage pattern, 82, 196, 386, 387
Drava, 32, 33, 35, 41, 211, 298, 319, 322, 335, 338, 339
Dredging, 197, 214, 217, 434
Drina–Ivanjica Unit, 13
Dyjsko-svratecký úval Graben, 108, 110, 112, 113, 116, 117, 130, 132

E

Earth pyramids, 387, 431
Earthquakes, 72, 250, 252, 257, 292, 299, 304, 319, 322, 361, 407, 408
East Bosnian–Durmitor Unit (EBD), 5, 13
Eastern Carpathians, 6, 8–9, 20–25, 35, 36, 49, 52, 53, 77, 86, 143, 144, 154, 178, 181, 250, 255–257, 263, 265, 269, 272, 273, 449
Éboulis ordonée, 266
Effluent caves, 229
Embankments, 78, 116, 133, 159, 167, 188, 255, 269, 434
Ephemeral gully erosion, 49, 85, 162
Epigenetic valleys, 198, 228, 229, 291
Erosional surfaces, 207, 337
Erosional-tectonic valleys, 198
Erosion rates, 67, 68, 129, 188, 219, 253, 261, 262, 421, 427
Excavation landforms, 401
External Molasse Zone, 4–5
Extreme climatic events, 406
Extreme rainfalls, 120, 123, 147, 155, 160–162

F

Feather-like pattern, 387
Felsenmeers, 193, 202, 398
Field pattern, 68
Field roads, 67, 69, 82, 84, 86
Fishponds, 112
Flash floods, 55, 109, 110, 112, 126, 128, 147, 148, 188, 202, 217, 298, 436
Flexural downbending, 4
Flood dynamics, 76
Flood frequency, 125
Flood hazard, 199–201
Floodplains, 33, 50, 54, 75, 76, 78–80, 85, 86, 104, 105, 108, 109, 111–113, 115–117, 125–127, 133, 158–161, 193, 196, 198, 208, 210, 215–217, 234, 276, 325, 335, 380, 387, 404, 433, 434
Floods, 55, 57, 61, 63, 75, 76, 80, 83, 84, 86, 104, 109–112, 116, 117, 125–128, 131, 133, 147, 148, 155, 157–162, 165, 168, 181, 185–188, 190, 191, 193, 194, 199–202, 211, 213, 214, 216–218, 290, 291, 298, 300, 303, 364, 366, 388, 389, 401, 406, 409, 433, 436

Flows, 32, 33, 35–37, 48, 49, 58–60, 63, 65, 67, 70, 76–79, 83, 84, 87, 109, 118, 123, 124, 126, 127, 133, 152–159, 163, 168, 181, 183, 184, 190, 191, 194, 197, 199–202, 211, 212, 215, 217, 228, 229, 233, 234, 253, 257, 298, 299, 303, 305, 328, 332, 333, 357–359, 367, 368, 386, 395, 399, 406, 409, 420, 438
Flowstones, 165, 294
Fluvial processes, 49, 50, 54, 82, 109, 147, 150–152, 157, 159, 210, 217, 269, 274, 275, 297, 323, 335, 358, 416, 425–427, 433, 436
Fluvio-glacial fans, 303
Fluviokarst, 264, 265, 296, 323, 326, 335, 359, 368
Flysch zone, 5, 6, 8, 257
Forest, 32, 39–43, 57–59, 66, 72, 73, 77, 81, 83, 110, 111, 113–117, 125, 129, 148, 149, 151, 153–167, 186–188, 194, 196, 197, 202, 210, 218, 231, 256, 261, 265, 302, 328, 338, 387, 391, 398, 403, 405, 407, 421, 436, 438
Forest clearance, 149, 161, 187, 197, 231, 405
Frost creep, 56, 152, 269, 271
Frost weathering, 72, 83, 152, 254, 267, 303, 398, 406
Funnels, 70, 400

G

Gelisolifluction, 207, 337, 399
Gemer Belt, 8, 143, 145
Geomorphological cycle, 293
Geomorphological maps, 121, 178, 179, 207, 252, 256, 288, 314, 326, 327, 339, 354–358, 446
Getic nappes, 5, 10, 11
Getic Piedmont, 250, 251, 257
Geticum, 349
Glaciations, 152, 206, 226, 291, 300–303, 336, 337, 352, 354, 383, 387, 394, 395, 397, 398
Glacieret, 396, 397
Glaciers, 152, 254, 265–267, 269, 271, 291, 301–303, 326, 356, 360, 394–398, 424, 425
Glacis, 207, 211, 335, 336
Gravel beaches, 333, 424
Gravel extraction, 79, 82, 158, 196, 217, 434
Grazing, 81, 104, 111, 113, 125, 129, 153, 161, 187, 199, 226, 405, 425
Great colonization, 149, 150, 160

Great Hungarian Plain, 210, 211, 216
Gully erosion, 49, 82, 85, 86, 110, 127, 129, 161–163, 181, 182, 185, 186, 201, 210, 218–220, 232, 252, 257, 262, 357, 387, 403, 420, 431, 433
Gully formation, 109, 155, 161, 162, 165, 218

H
Headlands, 394
Headward erosion, 63
High Karst Unit, 5, 12
High magnitude/low frequency events, 297
Honeycomb weathering, 267
Horst and graben, 251
Housing construction, 402–404, 429
Hron, 32, 35, 147–149, 155, 157, 158
Human action, 123, 129, 153, 157, 210, 261, 360
Human impact, 40, 50, 63, 71, 80–84, 131, 150, 158, 186–188, 212, 214, 217, 231–234, 274, 333, 363, 401–405, 407, 416, 420, 421, 424–435, 449
Human interventions, 76, 148–150, 159, 160, 162, 212, 214, 274, 448, 449
Hydrological regime, 148, 157

I
Incision, 63, 65, 69, 78, 84–86, 118, 126, 162, 194, 197, 212–214, 229, 250, 256, 274, 275, 365, 386, 390, 408, 416, 421, 433
Infrageticum, 349
Inner Carpathian depressions, 108
Inner Carpathians, 145, 211
Inner Somogy, 224–226
Insular relief, 325
Intermittent springs, 359
Inundation, 85, 151, 327
Isolated karst, 292
Isolated massifs, 263–265

J
Jadar Unit, 5, 13–14

K
Karren, 73, 202, 229, 265, 326, 359, 390, 392, 423
Karst
 karst gorges, 229
 karst lakes, 164
Kettles, 202

Kiskunság, 211, 224–226
Kontas, 303
Kopanice, 113, 149, 161
Kraishte zone, 414, 415, 422
Kučaj-Sredna Gora Zone, 5, 11

L
Lagoons, 34, 393, 394
Lake impoundment, 364
Lakes, 16, 33, 42, 152, 164, 206, 217, 219, 220, 222–224, 228, 231–233, 252, 266–268, 290, 299, 301, 347, 352, 353, 358, 360, 363, 364, 393–395, 401, 404, 415, 416, 418, 419, 421, 424, 425, 431, 434–438
Land cover, 39–44, 149, 150, 166, 218, 219, 299
Landform typology, 288
Landslide classification, 224
Landslides, 48, 54, 59–65, 70, 75, 80, 82, 84–86, 104, 107, 109, 118–123, 129, 151, 154–158, 165, 168, 179, 181–186, 190–193, 197, 200, 202, 220–222, 224, 231, 252, 253, 255–260, 291, 292, 299, 304, 305, 335, 337, 346, 354, 357, 358, 361–365, 388, 393, 400, 401, 403, 404, 406–409, 420, 425, 428, 429, 431, 449
Land use, 39–44, 67, 69, 76, 78, 82, 85, 86, 104, 110–117, 127, 129, 149, 157, 161–163, 165, 196, 257, 295, 448, 450
Large woody debris, 50, 80
Lateral accretion, 127, 436
Lateral spreading, 260
Levees, 58, 70, 76, 78, 79, 83, 157, 217, 402
Limans, 393, 394
Limestone bars, 263–265
Little Plain, 210
Loess, 32, 41, 53, 115, 117, 127, 129, 146, 158, 164–166, 192, 198, 206, 207, 210, 218, 220, 222, 223, 230, 234, 266, 314, 325, 332, 335–339, 353, 355, 358, 360, 385, 388, 390, 393, 399–401, 408, 409
Loess bluffs, 32, 220, 358
Loess gredas, 400
Loess humps, 400
Loess plateaus, 325, 335–337, 339, 353, 360, 400
Loess pots, 400
Logging, 65, 76, 81, 82, 84, 129
Longitudinal valleys, 183, 198, 199
Lower Danube, 22, 32–35, 214, 348
Lower Danubian Basin, 352
Lower Danubian Unit, 10

M

Macedonia-Rhodope zone, 383
Magura Belt, 6, 7
Marine terraces, 393, 394
Marsh, 353, 358, 359
Material transport, 67, 68, 75, 82, 217
Meandering, 35, 79, 86, 157–159, 188, 207, 211, 212, 214, 215, 326, 358, 380, 428, 433
Mediterranean mountain system, 3, 4
Mehedinţi Plateau, 251, 264, 265
Meliatic Unit, 8
Metal smelting, 81
Microclimate, 331, 332, 398
Microcontinent, 14, 386
Microglacier, 396
Middle Danube, 32, 34, 42
Mid-Hungarian Zone, 15
Mining, 81, 104, 108, 130–132, 149, 150, 201, 220, 221, 231–233, 255, 256, 258, 275–277, 401–403, 426, 429–430
Moldavian Plateau, 250, 252, 260–262
Moraines, 79, 152, 181, 201, 302, 360, 395, 397, 425
Morava, 11, 32–35, 106–108, 111–113, 115, 117, 125, 126, 127, 131, 133, 147, 148, 157, 158, 346–349, 353, 360
Moravian Gate, 133
Moravian-Slovakian Carpathians, 105–107
Morphodynamic units, 150, 151
Morphosculpture, 179, 384, 400
Morphostructures, 146, 193, 330, 337, 349, 378–380, 384, 386, 387, 392, 417
Muddy floods, 161, 162, 168
Mudflows, 54, 59, 84, 111, 179, 183, 184, 186, 191, 200, 202, 257, 258, 363
Mudstream centers, 191
Myjava Hills, 144, 149, 156, 161, 162, 165

N

Nappes, 4–12, 15, 50, 51, 108, 143, 145, 165, 349, 350, 367, 381, 386
Nappe systems, 7, 8, 12
Natural bridges, 346, 359, 366–368
Needle ice, 57, 72, 255, 267
Neotectonic movements, 104, 186, 214, 335, 359, 384, 446, 450
Neotethys, 4, 6, 8, 9, 12–15
Nival niches, 73, 83
Nivation, 49, 73, 152, 255, 269, 271, 398, 399, 425, 426
Nivation hollows, 269, 398, 399
North-Hungarian Mountains, 209, 211, 221
Nyírség, 224–226

O

Opencast mining, 131, 426
Orogenic cycle, 415
Outer Carpathian Depressions, 106, 108, 130–133, 134
Outer (Flysch) Carpathians, 50, 51, 53, 105–108

P

Paleo-channels, 80
Paleo-drainage, 215
Paleovolcanic landscape, 417
Pan-European land cover monitoring (PELCOM), 40, 42
Pannonian Basin, 4, 14–16, 22, 27, 28, 143–146, 288, 315, 316, 319–323, 325, 328, 329, 335, 336, 338, 348, 349, 351–353, 360
Pannonian Lake, 16, 252
Pastures, 41, 43, 70, 113, 153, 161–164, 188, 210, 241, 405
Patterned ground, 73, 266, 267
Peat bogs, 50, 79, 80
Pediments, 106, 107, 207, 211, 335–338
Pelagonian Massif, 414, 417, 418, 422
Penninic Ocean, 4, 6
Pericarpathian region, 249–277
Periglacial processes, 49, 72–73, 206, 207, 266–269, 291, 292, 300, 303, 332, 392, 394–399, 406, 424–425, 438
Permanent gullies, 128, 130, 161, 162
Phytogenic shore, 424
Pieniny Klippen Belt, 5, 7, 8, 50, 51, 145
Piping, 49, 64, 69–70, 84, 110, 129–130, 152, 165, 228, 327, 337, 339, 400, 408
Piping fans, 70
Pits, 70, 117, 133, 165, 229, 232, 233, 277, 330, 401–403, 416, 423, 436
Planation surfaces, 106, 146, 337, 355, 383–385, 446
Plateaus, 33, 165, 210, 223, 228, 250–252, 260–265, 288, 290, 292, 293, 297, 301–304, 324, 325, 328, 329, 335–339, 353, 359, 360, 380, 382, 385, 386, 390, 391, 393, 398, 400, 418
Plate tectonic models, 208
Platform mounds (tabans), 402
Plot boundaries, 82, 85
Podhale Basin, 50, 51
Poljes, 227, 290, 292–294, 299, 303, 326, 328, 329, 339, 359, 416, 423
Polygenetic relief, 337–338
Polygonal soils, 425

Ponor contact karst, 296, 297
Ponors, 164, 227–229, 296, 297, 326, 328, 359, 366, 391
Post-rift stage, 208
Precipitation threshold, 48, 49, 60
Predbalkan, 379, 380, 382, 385–388, 390–392, 408
Pre-Karst Unit, 12
Permafrost, 23, 72, 224, 254, 266, 399
Preparatory factors, 118
Protalus ramparts, 73, 399
Pseudokarst, 73–75, 107, 110, 129, 150, 164–166, 335, 339, 448

Q

Quarries, 81, 132, 232, 233, 401, 429
Quaternary research, 207, 314

R

Radial pattern, 386
Rainfall duration, 147
Rainfall intensity, 67, 119, 147, 187
Ravines, 70, 188–190, 192, 193, 218, 400
Reactivation of landslides, 109, 154, 260
Relative relief, 52, 126, 164, 209, 218, 224, 256, 328, 337, 381–383, 420
Research stations, 48, 49, 188, 191, 254, 261, 448
Reservoirs, 35, 42, 77, 78, 82, 86, 109, 117, 132, 133, 157, 160, 193, 217, 232, 255, 256, 258, 269, 274, 365, 366, 391, 404, 424, 426, 431, 432, 449
Rias, 335
Rill erosion, 49, 64–69, 82, 84, 128, 161–163, 218, 357, 407, 420, 436
Rimstone bars, 229
River channels, 32, 48, 54, 63, 67, 68, 75–80, 82–87, 104, 116, 148, 157, 159, 186, 211–215, 255, 274, 276, 385, 387, 401, 406, 436, 438
River engineering, 49, 78, 81, 85, 86
River systems, 196, 199, 404, 418
Roadcuts, 121, 221, 426, 429, 431–433
Rock arches, 390, 391
Rockfall centers, 191
Rockfalls, 55, 58, 59, 64, 72, 83, 109, 118, 123, 152, 156, 191, 193, 194, 253, 257, 298, 299, 301, 304, 305, 335, 337, 346, 358, 361, 364–366, 388, 404, 406–408, 420, 431
Rock forest, 387

Rock glaciers, 254, 266, 267, 269, 271, 395, 398, 424, 425
Rock pinnacles, 346, 357, 369, 370
Rockslides, 54, 59, 63, 64, 335, 337, 358, 420
Rockwall retreat, 55, 72, 299
Rockwalls, 55, 56, 72, 83, 299, 396
Rocky slopes, 52, 55, 58, 83, 152, 156
Rotation, 15, 67, 69, 125, 208, 297, 319, 330
Rotational landslides, 59, 118, 121
Roughness, 288
Runoff processes, 150, 152, 160–164, 448

S

Salt extraction, 256
Salt karst, 185, 201
Sand covers, 166, 222, 234, 339
Sand dunes, 113, 166, 167, 210, 226, 234, 394, 401
Sand mobilization, 226
Sava, 14, 32, 33, 35, 41, 298, 301, 302, 319, 322, 323, 334, 335, 338, 346, 348, 355, 357, 360
Scree slopes, 266, 269, 420, 433
Sedimentation, 14, 126, 127, 157, 158, 208, 214, 217, 261, 262, 273, 296, 297, 300, 301, 303, 321, 332, 338, 358, 431, 433
Sediment delivery, 83, 157, 186, 194–196, 218, 254, 262, 427
Sediment waves, 214
Sediment yield, 298, 428
Seismic activity, 132, 210
Seismically active zones, 317, 407
Seminival belt, 152
Serbian-Macedonian Massif, 11, 347, 349, 351, 352, 414, 415, 422
Serbian-Macedonian-Rhodope Zone, 5, 11
Severin Nappe, 5, 9, 10
Shafts, 70, 118, 131, 229, 232, 265, 328, 329, 392
Sheet erosion, 49, 53, 54, 65, 67, 69, 72, 84–86, 127, 188–190, 201, 202
Shoreline evolution, 332–333
Shore retreat, 424, 434
Silesian-Krosno Belt, 5, 143
Sinkholes, 232, 265, 423
Sinter cascades, 391
Siphonal springs, 359
Sliding centers, 191
Slumps, 222, 257, 259, 420
Snow-slide centers, 191
Soil erosion, 67, 69, 82, 85, 104, 110, 111, 128–129, 151, 191, 207, 208, 217–220, 253, 260–262, 299, 305, 409, 420, 422, 427, 428, 449

Index 459

Solifluction, 49, 53, 56, 65, 72, 73, 207, 255, 266, 267, 299, 335, 337, 360, 399, 406, 425
Southern Carpathians, 6, 8–11, 20–25, 27, 33, 34, 251, 263–265, 379
South Moravian Carpathians, 105, 106, 129
Splash, 49, 67, 207, 219, 370, 420
Spoil heaps, 131, 133, 232, 233, 429
Stara planina, 11, 21, 23, 36, 41, 348, 354, 363, 379–382, 385, 387, 388, 391–394, 398, 403, 405, 409, 446, 448
Step-like slope profile, 64, 163
Stone rivers, 398, 399
Strike-slip movements, 15
Structural landslides, 59, 62
Subalpine belt, 58, 72, 152, 153, 186
Subcarpathians, 9, 250–252, 255, 257–259
Submarine relief, 326
Submerged notches, 333
Subsidence, 16, 108, 131, 132, 151, 165, 208, 210, 214, 232, 272, 319, 333, 335, 337, 351, 352, 355, 359, 389, 407, 417, 418, 435, 438
Subterranean waterfall, 391
Suburbanization, 117
Sulfate karst, 201
Supragetic nappes, 10
Suprageticum, 349
Suspended load, 33, 35, 78, 79, 82, 83, 86, 112, 158–160, 217, 274, 275, 298
Swamps, 16, 42, 217, 358, 389, 404
Synrift stage, 208

T
Tailing, 250, 256, 275, 276, 426, 429, 430
Talus cones, 152–154, 156, 266, 269, 358, 420
Talus slopes, 55, 57, 59, 73
Talus-type caves, 73, 75
Tatra–Fatra Belt, 7
Tatra Mountains, 21, 23, 28, 48–50, 54, 58, 59, 70–73, 80, 143, 146, 152–154, 164
Tectonic phases, 296, 322
Terminal moraines, 302, 360, 395, 425
Terraced slopes, 68, 180
Terraces, 64, 68, 69, 72, 75, 80, 82, 86, 87, 104, 107, 111, 163, 180, 185, 189, 192, 193, 198–199, 206, 207, 210, 222, 234, 267, 269, 273, 291, 299, 300, 302, 303, 325, 327, 335, 339, 352, 356, 358, 360, 385, 388–390, 393, 394, 399, 401, 404, 405, 421, 424, 425, 446

Terracettes, 71, 73, 81
Test areas, 219, 253, 448
Threshold limits, 290
Thufurs, 73, 152
Tillage, 67, 69, 128, 129, 150, 151, 162–164, 166, 219, 226, 405
Timberline, 23, 54, 59, 70–72, 123, 152, 153, 187, 405
Tisza Mega-unit, 14, 15
Tisza (Tisa, Tysa), 14, 32, 33, 34, 35, 41, 42, 185, 188–190, 194, 195, 199–201, 211–218, 225, 250, 276, 319–321, 358
Toppling, 118, 358
Tourism, 81, 232, 392, 403, 425
Transdanubian Hills, 209, 210
Transdanubian Mountains, 209–211, 227, 232
Translational landslides, 118, 119
Transversal valleys, 198, 199
Transverse coast, 393
Transylvanian nappes, 5, 9
Travertine mounds, 229
Triggering factors, 118, 123, 253, 304
Troughs, 9, 12, 152, 167, 192, 202, 232, 317, 416
Tsukali-Krasta Zone, 414, 415
Tunnels, 64, 70, 130, 231, 267, 364, 402, 429
Turnaic Unit, 8

U
Underground mining, 131
Undermining, 231
Unroofed caves, 292, 297
Upper Danube, 32, 34, 213
Upper Danubian Unit, 10
Upper Moravian Graben, 108, 133
Urbanized areas, 39
U-shaped valleys, 387, 395, 398, 425, 427
Uvalas, 228, 292, 293, 328, 329, 359

V
Valley floors, 85–86, 360
Vardar zone, 5, 9, 13, 14, 347, 349, 351, 352, 354, 359, 360, 414, 415, 417, 418, 422
Vegetation cover, 63, 65, 67, 69, 76, 81, 152, 166, 186, 191, 199, 210, 226, 261, 331, 370, 409, 436
Vepor Belt, 8, 143
Vertical climate zonation, 23–26
Vertical mobility, 274
Volcanic necks, 356, 360, 418

W

Walachian colonization, 104, 111, 112, 149, 153
Waste disposal sites, 402
Waste landfill, 429
Water-and-stone flows, 190
Water management, 35, 133, 212, 216
Water transfer, 404
Wave erosion, 160, 333, 409
Weathering mantle, 152, 164, 355, 357
Weathering-type caves, 75
Western Beskids, 77, 106, 107, 144
Western Carpathians, 6–8, 21–23, 27, 35, 36, 52–55, 77, 104, 132, 143–146, 152, 153, 446–448
West-Macedonian zone, 414, 415, 417, 418, 422
Wind erosion, 110, 130, 152, 153, 166, 167, 225–227, 450
Woodlands, 41–43, 81, 162, 163, 192

Z

Zonal processes, 150